VOLUME ONE HUNDRED AND TWENTY THREE

CURRENT TOPICS IN DEVELOPMENTAL BIOLOGY

Protein Kinases in Development and Disease

CURRENT TOPICS IN DEVELOPMENTAL BIOLOGY

"A meeting-ground for critical review and discussion of developmental processes"
A.A. Moscona and Alberto Monroy (Volume 1, 1966)

SERIES EDITOR
Paul M. Wassarman
Department of Developmental and Regenerative Biology
Icahn School of Medicine at Mount Sinai
New York, NY, USA

CURRENT ADVISORY BOARD

Blanche Capel
Wolfgang Driever
Denis Duboule
Anne Ephrussi

Susan Mango
Philippe Soriano
Cliff Tabin
Magdalena Zernicka-Goetz

FOUNDING EDITORS
A.A. Moscona and Alberto Monroy

FOUNDING ADVISORY BOARD

Vincent G. Allfrey
Jean Brachet
Seymour S. Cohen
Bernard D. Davis
James D. Ebert
Mac V. Edds, Jr.

Dame Honor B. Fell
John C. Kendrew
S. Spiegelman
Hewson W. Swift
E.N. Willmer
Etienne Wolff

VOLUME ONE HUNDRED AND TWENTY THREE

CURRENT TOPICS IN DEVELOPMENTAL BIOLOGY

Protein Kinases in Development and Disease

Edited by

ANDREAS JENNY

*Developmental and Molecular Biology,
Albert Einstein College of Medicine,
New York, NY, United States*

ACADEMIC PRESS
An imprint of Elsevier
elsevier.com

Academic Press is an imprint of Elsevier
50 Hampshire Street, 5th Floor, Cambridge, MA 02139, United States
525 B Street, Suite 1800, San Diego, CA 92101-4495, United States
The Boulevard, Langford Lane, Kidlington, Oxford OX5 1GB, United Kingdom
125 London Wall, London EC2Y 5AS, United Kingdom

First edition 2017

Copyright © 2017 Elsevier Inc. All rights reserved.

No part of this publication may be reproduced or transmitted in any form or by any means, electronic or mechanical, including photocopying, recording, or any information storage and retrieval system, without permission in writing from the publisher. Details on how to seek permission, further information about the Publisher's permissions policies and our arrangements with organizations such as the Copyright Clearance Center and the Copyright Licensing Agency, can be found at our website: www.elsevier.com/permissions.

This book and the individual contributions contained in it are protected under copyright by the Publisher (other than as may be noted herein).

Notices

Knowledge and best practice in this field are constantly changing. As new research and experience broaden our understanding, changes in research methods, professional practices, or medical treatment may become necessary.

Practitioners and researchers must always rely on their own experience and knowledge in evaluating and using any information, methods, compounds, or experiments described herein. In using such information or methods they should be mindful of their own safety and the safety of others, including parties for whom they have a professional responsibility.

To the fullest extent of the law, neither the Publisher nor the authors, contributors, or editors, assume any liability for any injury and/or damage to persons or property as a matter of products liability, negligence or otherwise, or from any use or operation of any methods, products, instructions, or ideas contained in the material herein.

ISBN: 978-0-12-801513-1
ISSN: 0070-2153

For information on all Academic Press publications
visit our website at https://www.elsevier.com/books-and-journals

  Working together to grow libraries in developing countries

www.elsevier.com • www.bookaid.org

Publisher: Zoe Kruze
Acquisition Editor: Zoe Kruze
Editorial Project Manager: Shellie Bryant
Production Project Manager: Vignesh Tamil
Cover Designer: Greg Harris

Typeset by SPi Global, India

Transferred to Digital Printing in 2017

CONTENTS

Contributors xi
Preface xiii

1. WNK Kinases in Development and Disease 1
Aylin R. Rodan and Andreas Jenny

1. Introduction 2
2. The WNK-SPAK/OSR1 Kinase Cascade: Roles in Physiology and Disease 2
3. Emerging Functions of the WNK Signaling Axis in Development 20
4. Functions of WNKs in Cancer 33
5. Conclusions 34
Acknowledgments 35
References 35

2. SGK1: The Dark Side of PI3K Signaling 49
Antonio Di Cristofano

1. Introduction: The Glucocorticoid-Regulated Kinase Family 50
2. SGK1: Expression and Stability Control 50
3. SGK1 Is Activated in a PI3K-Dependent Manner 51
4. AKT and SGK1: Target Overlap and Selectivity 52
5. Roadblocks to Defining Specific SGK1 Functions 56
6. SGK1 in Development and Differentiation 57
7. SGK1 and Cancer 61
8. Concluding Remarks 64
Acknowledgments 64
References 65

3. Homeodomain-Interacting Protein Kinases: Diverse and Complex Roles in Development and Disease 73
Jessica A. Blaquiere and Esther M. Verheyen

1. Introduction 74
2. Requirements for Hipk Proteins During Development 78
3. Hipk Roles in Regulation of Diverse Signaling Pathways 82
4. Regulation of Hipk Activity 88
5. Hipks in Disease 91
6. Conclusions 93

Acknowledgments	94
References	94

4. ROR-Family Receptor Tyrosine Kinases   105
Sigmar Stricker, Verena Rauschenberger, and Alexandra Schambony

1. Domain Architecture and Expression Patterns	106
2. ROR Function as a WNT Receptor and Its Role in WNT Signal Transduction	110
3. Human Inheritable Syndromes Caused by *ROR2* Mutation	114
4. The Role of ROR-Family RTKs in Embryonic Development	119
References	133

5. Regulation of *Drosophila* Development by the Golgi Kinase Four-Jointed   143
Yoko Keira, Moe Wada, and Hiroyuki O. Ishikawa

1. Introduction	144
2. Genetic and Molecular Identification of *Drosophila* Four-Jointed	145
3. The Fat/Dachsous/Four-Jointed Pathway	149
4. Biochemical Characterization of Fj	156
5. Modulation of Fat-Ds Binding by Fj	158
6. Fj Polarizes Fat Activity	164
7. Vertebrate Four-Jointed, Fat, Dachsous, and Other Kinases in the Secretory Pathway	165
8. Conclusion	171
Acknowledgments	171
References	171

6. The Hippo Pathway: A Master Regulatory Network Important in Development and Dysregulated in Disease   181
Cathie M. Pfleger

1. Introduction	182
2. The Highly Conserved Core Kinase Cassette	183
3. The Hippo Pathway Restricts Mass Accumulation and Proliferation	186
4. The Hippo Pathway Regulates Apoptosis	187
5. The Hippo Pathway Restricts Organ Size and Maintains Organ Homeostasis	187
6. Molecular Mechanisms and Additional Roles of the Hippo Pathway: Posttranslational Targets and Transcriptional Outputs	188

7. Upstream Signals and Regulators and Mechanisms of Pathway Homeostasis	197
8. Dysregulation of the Hippo Pathway in Disease	207
9. Conclusions and Open Questions	210
Acknowledgments	212
References	212

7. Regulation of Embryonic and Postnatal Development by the CSF-1 Receptor — 229

Violeta Chitu and E. Richard Stanley

1. Introduction	230
2. The CSF-1R in Mononuclear Phagocyte Development	233
3. CSF-1-Regulated Macrophages in Tissue Morphogenesis and Organismal Growth	239
4. Regulation of Osteoclasts and Bone Development by CSF-1	246
5. Direct Regulation of Nonhematopoietic Cells by the CSF-1R	256
6. Conclusions	258
Acknowledgments	260
References	260

8. Glycogen Synthase Kinase 3: A Kinase for All Pathways? — 277

Prital Patel and James R. Woodgett

1. The Early Years	278
2. GSK-3 Isoforms, Orthologues, and Expression	278
3. GSK-3 Structure	280
4. Regulation	282
5. Signaling Infidelity	284
6. So How Is Specificity Achieved?	292
7. Lessons From GSK-3 KO Mice	292
8. Therapeutic Perspectives	294
9. Conclusions and Perspectives	297
Acknowledgments	297
References	297

9. CK1 in Developmental Signaling: Hedgehog and Wnt — 303

Jin Jiang

1. Introduction	304
2. CK1 in the Regulation of Hh Pathway	308
3. CK1 in Wnt Signaling	317

4. Regulation of CK1 in Hh and Wnt Signaling	320
5. Conclusion	322
Acknowledgments	323
References	323

10. Ligand Receptor-Mediated Regulation of Growth in Plants — 331
Miyoshi Haruta and Michael R. Sussman

1. Introduction	332
2. Novel and Unique Plant Signaling Pathways	333
3. Plasma Membrane as the Site of Peptide Ligand Sensing by Receptors	341
4. FERONIA and Its Gene Family in Plant Growth and Development	342
5. Organ Size and Growth Are Determined by Stimulation and Inhibition by Signaling Pathways in Plants and Animals	345
6. RALF Family and Function in Plant Growth and Development	347
7. RALF-Like Peptides Are Not Only Produced by Plants But Also by Pathogenic Microbes	348
8. Other Endogenous Hormone-Like Peptides and Their Receptor Pairs for Plant Growth, Development, and Physiology	349
9. Conclusion and Future Perspectives	353
References	354

11. Regulation of Cell Polarity by PAR-1/MARK Kinase — 365
Youjun Wu and Erik E. Griffin

1. Introduction	366
2. Structure and Regulation of PAR-1/MARK Kinases	367
3. Regulation of Cell Polarity by the PAR Proteins	371
4. Asymmetric Division of the *C. elegans* Zygote	372
5. Establishment of the Anterior/Posterior Axis During *Drosophila* Oogenesis	379
6. MARK Kinases and Neurogenesis	385
7. PAR-1 and Disease	386
8. Concluding Remarks	388
References	389

12. Receptor Tyrosine Kinases and Phosphatases in Neuronal Wiring: Insights From *Drosophila* — 399
Carlos Oliva and Bassem A. Hassan

1. Introduction	400
2. Model Circuits Used to Study the Genetic Control of Neuronal Wiring in *Drosophila*	401

3.	Receptor Tyrosine Kinase	405
4.	Receptor Protein Tyrosine Phosphatases	414
5.	Future Directions	424
6.	Conclusions	425
Acknowledgments		425
References		425

13. VEGF Receptor Tyrosine Kinases: Key Regulators of Vascular Function 433

Alberto Álvarez-Aznar, Lars Muhl, and Konstantin Gaengel

1.	Introduction	434
2.	Structure and Function of VEGFRs	437
3.	An Evolutionary Perspective on VEGFR Function	450
4.	Neuropilins: Coreceptors Modulating VEGFR Signaling	453
5.	VEGFR Ligands	455
6.	Perspective	463
Acknowledgments		464
References		464

CONTRIBUTORS

Alberto Álvarez-Aznar
Rudbeck Laboratory, Uppsala University, Uppsala, Sweden

Jessica A. Blaquiere
Department of Molecular Biology and Biochemistry, Centre for Cell Biology, Development and Disease, Simon Fraser University, Burnaby, BC, Canada

Violeta Chitu
Albert Einstein College of Medicine, Bronx, NY, United States

Antonio Di Cristofano
Albert Einstein College of Medicine, Bronx, NY, United States

Konstantin Gaengel
Rudbeck Laboratory, Uppsala University, Uppsala, Sweden

Erik E. Griffin
Dartmouth College, Hanover, NH, United States

Miyoshi Haruta
University of Wisconsin-Madison, Madison, WI, United States

Bassem A. Hassan
Sorbonne Universités, UPMC Univ Paris 06, Inserm, CNRS, AP-HP, Institut du Cerveau et la Moelle (ICM)—Hôpital Pitié-Salpêtrière, Boulevard de l'Hôpital, Paris, France

Hiroyuki O. Ishikawa
Graduate School of Science, Chiba University, Chiba, Japan

Andreas Jenny
Albert Einstein College of Medicine, New York, NY, United States

Jin Jiang
University of Texas Southwestern Medical Center at Dallas, Dallas, TX, United States

Yoko Keira
Graduate School of Science, Chiba University, Chiba, Japan

Lars Muhl
Karolinska Institutet, Stockholm, Sweden

Carlos Oliva
Biomedical Neuroscience Institute, Faculty of Medicine, Universidad of Chile, Santiago, Chile

Prital Patel
Lunenfeld-Tanenbaum Research Institute, Sinai Health System & University of Toronto, Toronto, ON, Canada

Cathie M. Pfleger
The Icahn School of Medicine at Mount Sinai; The Graduate School of Biomedical Sciences, The Icahn School of Medicine at Mount Sinai, New York, NY, United States

Verena Rauschenberger
Developmental Biology, Friedrich-Alexander University Erlangen-Nuremberg, Erlangen, Germany

Aylin R. Rodan
UT Southwestern, Dallas, TX, United States

Alexandra Schambony
Developmental Biology, Friedrich-Alexander University Erlangen-Nuremberg, Erlangen, Germany

E. Richard Stanley
Albert Einstein College of Medicine, Bronx, NY, United States

Sigmar Stricker
Institute for Chemistry and Biochemistry, Freie Universität Berlin, Berlin, Germany

Michael R. Sussman
University of Wisconsin-Madison, Madison, WI, United States

Esther M. Verheyen
Department of Molecular Biology and Biochemistry, Centre for Cell Biology, Development and Disease, Simon Fraser University, Burnaby, BC, Canada

Moe Wada
Graduate School of Science, Chiba University, Chiba, Japan

James R. Woodgett
Lunenfeld-Tanenbaum Research Institute, Sinai Health System & University of Toronto, Toronto, ON, Canada

Youjun Wu
Dartmouth College, Hanover, NH, United States

PREFACE

As the ancient Greek word κινειν (kinein; to move) suggests, kinases are true movers (or blockers) in a cell and are broadly grouped into protein, lipid, and carbohydrate kinases (plus a few others such as nucleoside-phosphate kinases). The roughly 518 human protein kinases comprise seven major subfamilies and represent roughly 2% of the genome (Manning, Plowman, Hunter, & Sudarsanam, 2002; Manning, Whyte, Martinez, Hunter, & Sudarsanam, 2002; Taylor & Kornev, 2011; Ubersax & Ferrell, 2007).

Posttranslational phosphorylation likely is the most widespread way of regulating protein function. Phosphorylation state affects every basic process in a cell including transcription, translation, cell division, inter- and intracellular communication, differentiation, metabolism, and so on. Not surprisingly, kinases and their counterparts, phosphatases, are crucial for normal development of multicellular organisms and aberrant kinase function or regulation can cause diseases.

The first kinase to be discovered in the 1950s by Fischer and Krebs was Phosphorylase kinase, which converts Phosphorylase B to the more active Phosphorylase A that mediates degradation of glycogen. This discovery paved the way forward for this previously unappreciated mode of regulation (Krebs, 1998; Krebs, Graves, & Fischer, 1959). Over the years, work in many labs has contributed to the identification, and biochemical, structural, and physiological characterization of a variety of kinases (reviewed in Taylor & Kornev, 2011; Ubersax & Ferrell, 2007). All kinases are characterized by the presence of a kinase domain, the activity of which is tightly regulated by intra- and intermolecular interactions. Kinase domains span about 250 amino acids and consist of a smaller N-terminal lobe composed of mostly β-sheets and a larger α-helical C-lobe (Knighton et al., 1991; reviewed in Taylor & Kornev, 2011; Ubersax & Ferrell, 2007). Sandwiched between these lobes is the hydrophobic ATP-binding site with the γ-phosphate oriented toward the substrate that binds in the cleft. Auto- or transactivation of a protein kinase generally occurs via phosphorylation of an activation segment within the C-lobe. Phosphorylation orders and moves the loop structure to allow access of the substrate to the binding cleft (Adams, 2003; Taylor & Kornev, 2011). This off/on switch type of regulatory mechanism thus allows for tight control of kinase activity and offers the opportunity for intricate regulation of cellular signaling networks.

The 13 chapters of this issue of *Current Topics in Developmental Biology* highlight the roles of some familiar and some less well-known kinases in development and disease.

This volume begins with WNK kinases that are characterized by an atypical placement of a critical lysine residue in the catalytic domain (Rodan and Jenny; Chapter 1) and have recently been shown to have developmental functions in addition to their role in ion transport regulation in the kidney. Activation of PI3 (Phosphoinositide 3)-kinase is central to many physiological and pathological processes including growth control, motility, and differentiation. Although Akt (aka Protein kinase B) has long been thought to be the key mediator of PI3K effects, Di Cristofano in Chapter 2 highlights a more recently discovered PI3K effector, serum, and glucocorticoid-regulated kinase 1 (SGK1), and emphasizes both roles shared with Akt and effects that are mediated exclusively by SGK1.

Blaquiere and Verheyen shed light on the diverse and sometimes conflicting roles of Homeodomain-interacting protein kinases (Hipk; Chapter 3) and discuss involvement of these kinases in the regulation of a variety of signaling pathways. Chapters 4 and 7 by Stricker et al. and Chitu and Stanley, respectively, discuss regulatory roles of the tyrosine kinases Ror (Receptor tyrosine kinase-like orphan receptor) and CSF1-R (Colony-stimulating factor-1 receptor) during embryonic development in vertebrates, the former affecting gastrulation and the latter fulfilling macrophage-dependent and -independent functions.

In Chapter 5, Keira *et al.* summarize current knowledge of Four-jointed, an intriguing kinase originally identified in *Drosophila* that acts in the Golgi lumen where it phosphorylates extracellular receptors involved in growth and epithelial planar polarity. In addition to Four-jointed, the Hippo/Salvador/Warts kinase module also affects cell and tissue growth. Recent advances toward the mechanistic basis of the evolutionarily conserved Hippo signaling pathway and functions of Hippo to prevent aberrant cell growth are illustrated by Pfleger in Chapter 6.

GSK3s (Glycogen synthase kinases) are two largely redundant kinases originally identified as regulators of glycogen metabolism that intersect with most signaling pathways in multicellular organisms. Given that most roads apparently converge upon these kinases, Patel and Woodgett (Chapter 8) discuss the puzzling matter of how two kinases that are—unusually—chiefly regulated by their inhibition can lead to pathway-specific output and functional specificity. They also outline possible utility and risks associated with application of GSK inhibitors for disease treatment. In Chapter 9, Jiang explains that Casein kinases 1 (CK1s) not only serve as priming kinases for GSK3 during Wnt signaling but are also critical for Hedgehog signaling during development.

Once rooted, plants spend their entire life at the same location and therefore rely on unique mechanisms to adapt to their environment, for example, by adjusting growth rate. In Chapter 10, Haruta and Sussman discuss plant hormones, their receptors, and functions with a particular emphasis on FERONIA tyrosine kinase that may play a role in the transduction of a mechanosensory signal during growth.

Establishment of cell and organismal polarity is highly reliant on the function of Par1 (Partitioning defective 1) in *C. elegans*, *Drosophila*, and vertebrates, as becomes evident from the contribution by Wu and Griffin (Chapter 11). Continuing with polarity, Oliva and Hassan (Chapter 12) review the functions of tyrosine kinases and phosphatases, some of which have lost catalytic activity, in neuronal wiring. The issue closes with a review by Álvarez-Aznar et al. in Chapter 13 of the functions of Vascular Endothelial Growth Factor (VEGF) receptors during the development of the mammalian vascular system.

Collectively, this series of reviews aims to provide an overview of the remarkable recent advances in our understanding of protein kinase (and phosphatase) functions during development. A mission of this collection of articles written by experts in their fields is to demonstrate the persisting utility and merit of traditional model organisms in the "omics" era.

I am indebted and grateful to all of the authors for their hard work and dedication that allowed compilation of this set of very interesting and high-quality reviews. I also would like to thank the reviewers for critically and quickly reading the manuscripts. I would like to take the opportunity to thank my mentors, collaborators, and past and present lab members, all of whom continue to be important for the research in my lab. Last, but not least, I am grateful to Paul Wassarman for giving me the opportunity to assemble this volume for *Current Topics in Developmental Biology* and to Shellie Bryant and the production team for their assistance.

ANDREAS JENNY
Developmental and Molecular Biology,
Albert Einstein College of Medicine,
New York, NY, United States

REFERENCES

Adams, J. A. (2003). Activation loop phosphorylation and catalysis in protein kinases: Is there functional evidence for the autoinhibitor model? *Biochemistry*, *42*, 601–607.

Knighton, D. R., Zheng, J. H., Ten Eyck, L. F., Ashford, V. A., Xuong, N. H., Taylor, S. S., & Sowadski, J. M. (1991). Crystal structure of the catalytic subunit of cyclic adenosine monophosphate-dependent protein kinase. *Science, 253*, 407–414.

Krebs, E. G. (1998). An accidental biochemist. *Annual Review of Biochemistry, 67*, xii–xxxii.

Krebs, E. G., Graves, D. J., & Fischer, E. H. (1959). Factors affecting the activity of muscle phosphorylase b kinase. *The Journal of Biological Chemistry, 234*, 2867–2873.

Manning, G., Plowman, G. D., Hunter, T., & Sudarsanam, S. (2002). Evolution of protein kinase signaling from yeast to man. *Trends in Biochemical Sciences, 27*, 514–520.

Manning, G., Whyte, D. B., Martinez, R., Hunter, T., & Sudarsanam, S. (2002). The protein kinase complement of the human genome. *Science, 298*, 1912–1934.

Taylor, S. S., & Kornev, A. P. (2011). Protein kinases: Evolution of dynamic regulatory proteins. *Trends in Biochemical Sciences, 36*, 65–77.

Ubersax, J. A., & Ferrell, J. E.; Jr. (2007). Mechanisms of specificity in protein phosphorylation. *Nature Reviews. Molecular Cell Biology, 8*, 530–541.

CHAPTER ONE

WNK Kinases in Development and Disease

Aylin R. Rodan[*,1,2], **Andreas Jenny**[†,1]
*UT Southwestern, Dallas, TX, United States
†Albert Einstein College of Medicine, New York, NY, United States
[1]Corresponding authors: e-mail address: aylin.rodan@hsc.utah.edu; andreas.jenny@einstein.yu.edu

Contents

1. Introduction	2
2. The WNK-SPAK/OSR1 Kinase Cascade: Roles in Physiology and Disease	2
2.1 Overview of the WNK-SPAK/OSR1 Kinase Cascade	2
2.2 WNK-SPAK/OSR1 Signaling in Invertebrates	7
2.3 WNK-SPAK/OSR1 Signaling in Osmoregulation	14
2.4 The Role of Mouse Protein-25 in WNK-SPAK/OSR1 Signaling	18
3. Emerging Functions of the WNK Signaling Axis in Development	20
3.1 Mammalian WNKs	20
3.2 Zebrafish Wnk1a/b Have Roles in Angiogenesis and Neural Development	23
3.3 Insights from the *Drosophila* Wnk-Frayed Axis	26
4. Functions of WNKs in Cancer	33
5. Conclusions	34
Acknowledgments	35
References	35

Abstract

WNK (With-No-Lysine (K)) kinases are serine–threonine kinases characterized by an atypical placement of a catalytic lysine within the kinase domain. Mutations in human WNK1 or WNK4 cause an autosomal dominant syndrome of hypertension and hyperkalemia, reflecting the fact that WNK kinases are critical regulators of renal ion transport processes. Here, the role of WNKs in the regulation of ion transport processes in vertebrate and invertebrate renal function, cellular and organismal osmoregulation, and cell migration and cerebral edema will be reviewed, along with emerging literature demonstrating roles for WNKs in cardiovascular and neural development, Wnt signaling, and cancer. Conserved roles for these kinases across phyla are emphasized.

---

[2] Current address: University of Utah, Salt Lake City, UT, United States.

1. INTRODUCTION

The With-No-Lysine (K) (WNK) kinases are a family of serine/threonine kinases, first identified in 2000 by Cobb and colleagues in a screen for novel mitogen-activated protein kinase (MAPK) kinases (Xu et al., 2000). The characteristic feature of the "With-No-Lysine" kinases is the absence of the catalytic lysine found in subdomain II in most other kinases; instead, this lysine is found in subdomain I (Fig. 1A and B) (Xu et al., 2000). Another unique feature of WNKs is their regulation by chloride, which directly binds to the kinase active site and inhibits autophosphorylation and kinase activation (Fig. 1C) (Bazua-Valenti et al., 2015; Piala et al., 2014; Terker et al., 2016). WNKs are evolutionarily ancient: in the initial description, homologs from the nematode *Caenorhabditis elegans*, the plants *Oryza* and *Arabidopsis*, and the fungus *Phycomyces* were noted (Xu et al., 2000). Here, aspects of the roles of WNKs in physiology, development, and disease, with an emphasis on recent discoveries in model organisms, will be reviewed.

2. THE WNK-SPAK/OSR1 KINASE CASCADE: ROLES IN PHYSIOLOGY AND DISEASE

2.1 Overview of the WNK-SPAK/OSR1 Kinase Cascade

WNK homologs are found throughout the animal kingdom. Mammalian genomes encode four WNK paralogs, some of which are duplicated in the zebrafish *Danio rerio* (discussed in Section 3.2). There is a single WNK homolog in the genomes of the invertebrates *Drosophila melanogaster* and *C. elegans*. The WNK kinase domain is highly conserved (Figs. 1A and 2A). In contrast, the C-terminus, which is of varying length, has lower sequence homology. Common features include predicted coiled-coil domains, PXXP motifs, and an RFX(V/I) motif required for binding to the downstream kinases Ste20/SPS1-related proline/alanine-rich kinase (SPAK, also known as PASK) and oxidative stress responsive-1 (OSR1) (Fig. 2A and B; reviewed in McCormick & Ellison, 2011). SPAK and OSR1 are closely related Sterile 20 (Ste20)-related kinases that arose from gene duplication (Delpire & Gagnon, 2008) and have highly conserved orthologs in *D. melanogaster* and *C. elegans* (Fig. 2B).

In 2001, Lifton and colleagues published their finding that two of the four human WNK paralogs, *WNK1* and *WNK4*, are mutated in a syndrome

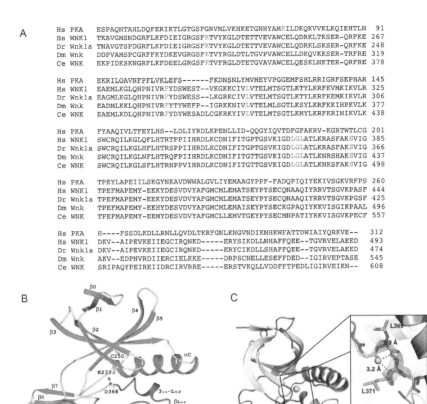

Fig. 1 Unique features of WNK kinases. (A) Alignment of the kinase domains of human WNK1 (Hs WNK1), zebrafish Wnk1a (Dr Wnk1a), fruit fly (Dm Wnk), and worm (Ce WNK) WNKs with human protein kinase A (Hs PKA). Atypical placement of the subdomain II lysine in subdomain I is indicated by *magenta*. Chloride-binding residues are indicated in *red*. The site of autophosphorylation, required for kinase activation, is indicated in *blue*. (B) Crystal structure of the kinase domain of rat WNK1. The atypically placed catalytic lysine, Lys233 in rat WNK1, is shown in comparison to Cys250, the usual lysine position. (C) Kinase domain of WNK1, showing Cl$^-$ binding (*green ball*). Enlargement shows hydrogen-bonding distances to Leu369 and Leu371. Note that Leu369 is a substitution of Phe in the "DFG" motif that is characteristic of most protein kinases in subdomain VII, including PKA (see (A)). *Panel (B) From Min, X., Lee, B. H., Cobb, M. H., & Goldsmith, E. J. (2004). Crystal structure of the kinase domain of WNK1, a kinase that causes a hereditary form of hypertension. Structure, 12, 1303–1311 and panel (C) modified from Piala, A. T., Moon, T. M., Akella, R., He, H., Cobb, M. H., & Goldsmith, E. J. (2014). Chloride sensing by WNK1 involves inhibition of autophosphorylation. Science Signaling, 7, ra41.*

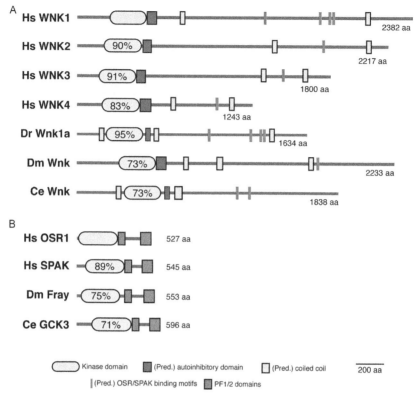

Fig. 2 Schematic representation of selected WNK (A) and OSR1 (B) kinase family members. Sequence identities of kinase domains to human WNK1 and OSR1, respectively, are given in percent based on BLASTP scores (Johnson et al., 2008). Note that other protein splice-isoforms also exist (reviewed in McCormick & Ellison, 2011) and that some of the indicated motifs are predictions (Pred.) and have not been functionally verified. *PF1/2*, PASK/Fray homology domains 1/2; *Hs*, Homo sapiens; *Dr*, Danio rerio (zebrafish); *Dm*, Drosophila melanogaster; *Ce*, Caenorhabditis elegans.

variously known as pseudohypoaldosteronism type II, familial hyperkalemic hypertension (PHAII/FHHt), or Gordon's syndrome. The *WNK* mutations are transmitted in an autosomal dominant fashion and result in high blood pressure and high serum potassium concentrations in affected individuals (Wilson et al., 2001). This phenotype suggested that WNKs may play a role in renal physiology, and this has been substantiated in extensive subsequent research, as recently reviewed (Dbouk et al., 2014; Hadchouel, Ellison, & Gamba, 2016). Consistent with this, mutations in the E3 ubiquitin ligase complex components, Kelch-like 3 and Cullin 3, were also found to cause PHAII/FHHt

(Boyden et al., 2012; Louis-Dit-Picard et al., 2012), likely due to their role in WNK degradation (reviewed in Ferdaus & McCormick, 2016).

The best-understood function of WNKs is their ability to phosphorylate SPAK and OSR1 (Anselmo et al., 2006; Moriguchi et al., 2005; Vitari, Deak, Morrice, & Alessi, 2005). Phosphorylation of the SPAK/OSR1 T-loop threonine, T243 (SPAK) or T185 (OSR1), is required for SPAK/OSR1 activation, while the function of phosphorylation on a C-terminal serine, Ser373 (SPAK) or Ser 325 (OSR1), in the PF1 domain (PASK and *Fray*; Fig. 2B), is less clear (Gagnon & Delpire, 2010; Gagnon, England, & Delpire, 2006; Moriguchi et al., 2005; Vitari et al., 2005). Activated SPAK and OSR1 phosphorylate members of the SLC12 family of cation-chloride cotransporters (CCCs). These include the three related sodium-coupled chloride cotransporters, NCC (sodium chloride cotransporter), and the sodium-potassium-2-chloride cotransporters NKCC1 and NKCC2 (Anselmo et al., 2006; Dowd & Forbush, 2003; Gagnon & Delpire, 2010; Gagnon et al., 2006; Moriguchi et al., 2005; Richardson et al., 2008, 2011), as well as the potassium-coupled chloride cotransporters, KCC1–4 (potassium chloride cotransporters) (de Los Heros et al., 2014; Melo et al., 2013). Phosphorylation of NCC, NKCC1, and NKCC2 results in transporter activation, whereas phosphorylation of KCCs results in transporter inactivation (Fig. 3A). Regulation of these transporters by WNKs is important for cell volume control, transepithelial ion transport, and the regulation of intracellular chloride concentration (Kahle et al., 2006). In neurons, for example, intracellular chloride concentration determines whether activation of ligand-gated chloride channels, such as the $GABA_A$ or glycine receptors, results in a hyperpolarizing or depolarizing effect. When intracellular Cl^- concentration is low, GABA or glycine binding to their receptors results in Cl^- influx, resulting in hyperpolarization. Conversely, when intracellular Cl^- concentration is high, GABA or glycine binding to their receptors results in Cl^- efflux and neuronal depolarization. In cell volume control, activation of NKCCs results in inward ion flux, due to the low intracellular sodium concentration generated by the activity of the Na^+/K^+-ATPase, which pumps three Na^+ ions out of the cell in exchange for two K^+ ions in. Similarly, the high intracellular K^+ concentration generated by the Na^+/K^+-ATPase generates an outward driving force for K^+ and Cl^- through KCCs, and inhibition of KCCs by WNK-SPAK/OSR1 signaling decreases this outward ion flux. WNK-SPAK/OSR1 signaling also plays important roles in the regulation of transepithelial ion transport through SLC12 transporters.

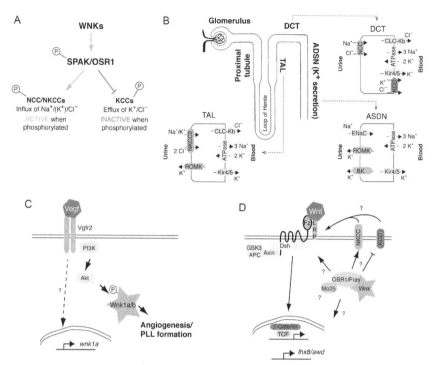

Fig. 3 WNK pathway. (A) WNKs phosphorylate the two related Ste20 kinases, SPAK and OSR1, on a T-loop threonine in the active site, which is required for SPAK/OSR1 activation, and on a serine in the PF1 domain in the C-terminus of the protein. SPAK and OSR1 phosphorylate conserved serines and threonines in the sodium-coupled SLC12 chloride cotransporters, NCC, NKCC1, and NKCC2 in mammals, increasing transport activity. Phosphorylation of the potassium-coupled SLC12 chloride cotransporters, KCCs 1–4 in mammals, results in transporter inactivation. (B) Schematic of the nephron showing sites of WNK action. WNK-SPAK/OSR1 signaling positively regulates NKCC2 in the thick ascending limb (TAL) of the loop of Henle, and NCC in the distal convoluted tubule (DCT), promoting sodium chloride reabsorption. Decreased sodium delivery to the K^+-secretory principal cell of the aldosterone-sensitive distal nephron (ASDN), where K^+ secretion depends on a lumen-negative charge generated by Na^+ reabsorption through the epithelial Na channel (ENaC), likely contributes to the hyperkalemia observed in patients with PHAII. WNKs also regulate ENaC and the K^+-secretory channels renal outer medullary potassium channel (ROMK) and big potassium channel (BK, also known as maxi-K) (Carrisoza-Gaytan, Carattino, Kleyman, & Satlin, 2016; Welling, 2013). Also pictured are the Na^+/K^+-ATPase, which generates the driving force for sodium reabsorption in the TAL, DCT and the principal cell of the ASDN; Clc-Kb, a chloride channel allowing basolateral exit of Cl^-, and the heterodimeric Kir4.1/5.1 potassium channel, which is important for recycling K^+ entering through the Na^+/K^+-ATPase and setting the basolateral membrane potential. KCC4 could also play a role in basolateral KCl exit (see text). (C) Summary of Wnk1 function in zebrafish. During angiogenesis, Wnk1 is regulated by Vegfr signaling via Akt phosphorylation and is also a transcriptional target of Vegf signaling. Vgfr2 and PI3K are encoded by the *flk1* and *pi3kc2α* genes, respectively. (D) *Drosophila* Wnk regulates Wnt signaling and the expression of *Awh/Lhx8* via Fray/OSR1.

In PHAII/FHHt, overexpression of WNK4 or WNK1, either due to gain-of-function alleles of those genes or loss-of-function of Kelch-like 3 or Cullin 3, results in increased phosphorylation of NCC in the kidney (Ferdaus & McCormick, 2016; Hadchouel et al., 2016; Huang & Cheng, 2015). Phosphorylation of NCC by the WNK-SPAK/OSR1 kinase cascade results in increased NCC activity (Richardson et al., 2008). Overactivation of NCC results in increased NaCl reabsorption, causing hypertension. Concomitant decreased sodium delivery to the downstream aldosterone-sensitive distal nephron, where potassium secretion is dependent on sodium delivery, results in hyperkalemia (Fig. 3B); direct effects of WNKs on potassium channels may also contribute. Furthermore, activation of WNK-SPAK/OSR1 signaling in the vasculature results in vasoconstriction through effects on NKCC1 (Bergaya et al., 2011; Susa et al., 2012; Yang et al., 2010; Zeniya et al., 2013), which may contribute to the hypertensive phenotype, particularly in individuals with WNK1 mutations. Additional effects of WNK signaling on other transport processes in the nephron, such as regulation of the epithelial sodium channel, the chloride/bicarbonate exchanger pendrin, and paracellular chloride reabsorption through claudins, may also contribute to the PHAII/FHHt phenotype, but this is less well established (reviewed in Hadchouel et al., 2016).

Polymorphisms in serine–threonine kinase 39 (*STK39*), which encodes the human SPAK ortholog, have been associated with essential hypertension (Xi et al., 2013). One of these, rs375477, was shown to increase *STK39* mRNA and SPAK protein expression when introduced into human embryonic kidney cells using CRISPR technology (Mandai, Mori, Sohara, Rai, & Uchida, 2015), again suggesting a connection between increased WNK-SPAK/OSR1 signaling and elevated blood pressure.

2.2 WNK-SPAK/OSR1 Signaling in Invertebrates
2.2.1 *WNK-SPAK/OSR1 Signaling Regulates NKCC in* Drosophila *Renal Tubule Function*

D. melanogaster and *C. elegans* each has a single *wnk* ortholog, called *wnk* in *Drosophila* and *wnk-1* in *C. elegans,* and a single SPAK/OSR1 ortholog, called *frayed (fray, CG7693)* in *Drosophila* and *gck-3* in *C. elegans* (Fig. 2). Using bacterially expressed, purified components, Sato et al. and Serysheva et al. demonstrated that *Drosophila* Wnk phosphorylates Fray in vitro (Sato & Shibuya, 2013; Serysheva et al., 2013). Similarly, Fray phosphorylates the N-terminus of Ncc69 (Wu, Schellinger, Huang, & Rodan, 2014), a fly NKCC (Leiserson, Forbush, & Keshishian, 2011; Sun, Tian, Turner, &

Ten Hagen, 2010). As discussed in further detail in Section 3.3, examination of developmental phenotypes placed *fray* downstream of *wnk* (Sato & Shibuya, 2013; Serysheva et al., 2013). In addition, loss-of-function mutations in both *fray* and *Ncc69* result in similar axon bulging phenotypes in the *Drosophila* larval nervous system, suggesting that they may act in the same pathway.

The Malpighian (renal) tubule is part of the iono- and osmoregulatory system of the fly. Unlike the mammalian nephron, the Malpighian tubule is aglomerular and blind ended. Urine generation therefore occurs through the isosmotic secretion of KCl-rich fluid across the main segment of the tubule, from the hemolymph to the tubule lumen. Transepithelial cation flux occurs through the principal cell of the main segment, whereas chloride flux occurs through the neighboring stellate cells (Fig. 4A) (Cabrero et al., 2014; Linton & O'Donnell, 1999; O'Donnell et al., 1996; Rheault & O'Donnell, 2001).

It was proposed that WNK-SPAK/OSR1 signaling regulates transepithelial ion flux through NKCC2 and NCC in the mammalian kidney, although there are few studies directly demonstrating this, due to the technical difficulty of directly assaying transepithelial ion flux in the mammalian nephron (Cheng, Truong, Baum, & Huang, 2012; Cheng, Yoon, Baum, & Huang, 2015). The fly thus affords the opportunity to study the molecular physiology of WNK-SPAK/OSR1 signaling in a genetically manipulable transporting epithelium. Indeed, it was demonstrated that the fly NKCC, Ncc69, is required in the cation-conducting principal cell for normal transepithelial fluid and potassium secretion in the fly renal tubule, where it functions as a secretory NKCC (as compared to the absorptive NKCCs in the mammalian kidney) (Rodan et al., 2012). Tubule Ncc69 is regulated by the WNK-SPAK/OSR1 pathway (Fig. 4B). Knocking down either *wnk* or *fray* in the tubule principal cell decreases transepithelial potassium flux, similar to the *Ncc69* null phenotype. As in developmental processes, *fray* operates downstream of *wnk*. Mutation of the predicted Wnk phosphorylation site in Fray, Thr206, to a phospho-mimicking Asp results in constitutive kinase activity toward Ncc69 in vitro and restores normal transepithelial potassium flux to *wnk* knockdown tubules. Importantly, *wnk* or *fray* knockdown do not reduce potassium flux in *Ncc69* null tubules, indicating that the NKCC transporter is the target of Wnk and Fray regulation (Wu et al., 2014). Thus, WNK and Fray regulate transepithelial ion flux through the regulation of the fly renal tubule NKCC.

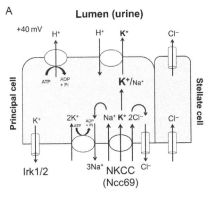

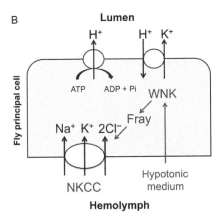

Fig. 4 Wnk function in the Malpighian (renal) tubule in *Drosophila*. (A) The Malpighian tubule main segment secretes a KCl-rich fluid into the lumen (urine). Transepithelial cation transport in the *Drosophila* Malpighian tubule occurs through the principal cell, whereas transepithelial chloride transport occurs through the neighboring stellate cells. The apical vacuolar H^+-ATPase drives fluid secretion (Dow et al., 1994) and generates a lumen-positive transepithelial potential difference (O'Donnell, Dow, Huesmann, Tublitz, & Maddrell, 1996). This drives exchange of protons for cations (K^+ or Na^+, primarily K^+ in *Drosophila* renal tubules). Chloride secretion is also driven by the lumen-positive charge. The fly NKCC, Ncc69, is required for normal transepithelial K^+ flux. Na^+ entering through the NKCC is recycled by the basolateral Na^+/K^+-ATPase (Rodan, Baum, & Huang, 2012). Cl^- may also be recycled through Cl^- channels or through basolateral Cl^-/HCO_3^- exchangers (not shown) (Romero et al., 2000; Sciortino, Shrode, Fletcher, Harte, & Romero, 2001). The inwardly rectifying potassium channels Irk1 and Irk2 are also required for normal transepithelial K^+ flux (Wu, Baum, Huang, & Rodan, 2015). (B) The WNK-SPAK/OSR1 (=Fray in flies) pathway regulates transepithelial K^+ flux in the *Drosophila* renal tubule principal cell. Hypotonic bathing medium stimulates transepithelial K^+ flux in a WNK/Fray/NKCC-dependent manner.

The WNK–Fray signaling pathway has also been shown to regulate transport processes in the *Drosophila* prepupal salivary gland. Farkaš et al. made the surprising observation that the salivary gland at this developmental stage secretes a calcium oxalate-rich fluid, which may form part of the secretory "glue" that allows puparia to fix themselves to a substrate during metamorphosis (Farkaš et al., 2016). Based on prior work from the Romero laboratory showing that the Slc26a5/6 transporter, Prestin, functions as a chloride/oxalate exchanger and is involved in calcium oxalate secretion by the *Drosophila* Malpighian tubule (Hirata, Cabrero, et al., 2012; Hirata, Czapar, et al., 2012; Landry et al., 2016), the investigators tested the hypothesis that Prestin was involved in salivary gland calcium oxalate excretion. Indeed, *prestin* knockdown in the salivary gland decreased calcium oxalate excretion (Farkaš et al., 2016). The Romero laboratory had also previously demonstrated that Prestin is positively regulated by Fray (Hirata, Czapar, et al., 2012), and Farkaš et al. demonstrated that knocking down either *wnk* or *fray* in the salivary gland decreased calcium oxalate excretion (Farkaš et al., 2016). The salivary gland secretory process, like that of the Malpighian tubule (Dow et al., 1994), is dependent on the vacuolar H^+-ATPase (Farkaš et al., 2016). Thus, salivary gland transport has elements that are conserved with the renal tubule, including the requirement for the H^+-ATPase and regulation by WNK-SPAK/OSR1 signaling.

2.2.2 WNK-SPAK/OSR1 Signaling in C. elegans and Chloride Channel Regulation

Elegant studies in *C. elegans* have uncovered roles for WNK-SPAK/OSR1 signaling in multiple physiological processes. As is the case with the mammalian and *Drosophila* proteins, *C. elegans* Wnk-1 phosphorylates the worm SPAK/OSR1 ortholog, GCK-3, in vitro (Hisamoto et al., 2008). *C. elegans* with mutations in *wnk-1* or *gck-3* prematurely terminate the excretory canal, which forms part of the nematode renal system (Hisamoto et al., 2008; Kupinski, Muller-Reichert, & Eckmann, 2010). The defect in *wnk-1* mutant worms is rescued by reexpression of wild-type *wnk-1*, but not by *wnk-1* that is kinase dead, or that carries a mutation in the RFXV motifs that are required for GCK-3 binding. GCK-3 carrying a phosphomimicking mutation in the Wnk target T-loop threonine, T280E, which has increased kinase activity in vitro, also rescues the *wnk-1* mutant phenotype, again indicating that *gck-3* lies downstream of *wnk-1*. In the *gck-3* mutants, expression of wild-type *gck-3*, but not kinase dead or T280A mutant *gck-3*, rescues the excretory canal phenotype. In contrast, *gck-3* with a mutation in the

Wnk-1-phosphorylated serine (S419A) in the PF1-domain is able to rescue, suggesting that phosphorylation of this residue by Wnk-1 is not necessary in this in vivo context (Hisamoto et al., 2008).

Interestingly, extension of the *C. elegans* excretory canal during development is modulated by osmolarity: placing worms on a hyperosmolar medium (e.g., 500 mM NaCl), and then returning them to an isotonic medium (50 mM NaCl), promotes the fusion of vesicles with the apical surface and canal extension. These processes fail to occur in *gck-3* mutant worms. This suggests that during development, excretory canal extension depends on the ability to sense extracellular osmolarity, and *gck-3* mutant worms are unable to either sense or respond to changes in osmolarity, resulting in stalled excretory canal extension (Kolotuev, Hyenne, Schwab, Rodriguez, & Labouesse, 2013).

Because Strange and colleagues had previously identified the ClC chloride channel, CLH-3, as a target of GCK-3 in worm oocytes (Denton, Nehrke, Yin, Morrison, & Strange, 2005), Hisamoto et al. examined whether the shortened excretory canal phenotype in *gck-3* mutant worms was due to dysregulation of CLH-3. Indeed, a *clh-3* mutation partially suppressed the mutant phenotype of *gck-3* mutant worms, indicating that *clh-3* is downstream of *gck-3* and is negatively regulated by the SPAK/OSR1 ortholog (Hisamoto et al., 2008). Similar suppression was also observed for decreased fertility observed in *gck-3* mutant worms, but not for early larval lethality or additional developmental phenotypes resulting from loss of *gck-3* function, indicating that *gck-3* likely has additional targets (Hisamoto et al., 2008; Kupinski et al., 2010). NKCC1 does not appear to be such a target, at least for excretory canal extension, as *nkcc-1* mutant worms have normal excretory canal morphology (Hisamoto et al., 2008). Whether WNK-SPAK/OSR1 signaling in the excretory canal regulates NKCC-1 in nondevelopmental contexts, for example, in excretory canal function, has not been determined.

CLH-3b is a splice variant of the *clh-3* gene in *C. elegans*. It is expressed in worm oocytes and is activated by serine/threonine dephosphorylation during oocyte meiotic maturation, or in response to cell swelling. GCK-3 negatively regulates CLH-3b by phosphorylating the channel on Ser 742 and Ser 747 and inducing conformational changes that decrease channel activity (Denton et al., 2005; Falin, Morrison, Ham, & Strange, 2009; Miyazaki & Strange, 2012; Miyazaki et al., 2012; Yamada, Bhate, & Strange, 2013). Interestingly, GCK-3 activity toward CLH-3b appears to be Wnk-independent and rather is downstream of the *C. elegans* ERK (extracellular signal regulated) MAPK, MPK-1 (Falin, Miyazaki, & Strange, 2011).

SPAK/OSR1 regulation of chloride channels may also have relevance to the mammalian kidney, where chloride channels play important roles in renal physiology, such as the transepithelial reabsorption of sodium chloride in the thick ascending limb (TAL) of the loop of Henle and the distal convoluted tubule (DCT; see schematic in Fig. 3B) (Zaika, Tomilin, Mamenko, Bhalla, & Pochynyuk, 2016). *C. elegans* CLH-3b is a member of the CLC-1/2/Ka/Kb chloride channel family. In the mammalian kidney, CLC-Ka and CLC-Kb are expressed in the loop of Henle and distal nephron, and mice lacking the CLC-Ka ortholog (CLC-K1 in mice) have nephrogenic diabetes insipidus. In humans, simultaneous mutations in the genes encoding CLC-Ka and CLC-Kb, or mutations in CLC-Kb alone, cause Bartter's syndrome, a salt-losing tubulopathy characterized by hypokalemic metabolic alkalosis and secondary hyperaldosteronism. Bartter's syndrome can be also be caused by mutations in the gene encoding Barrtin, which is a CLC-K channel regulatory subunit (reviewed in Andrini et al., 2015; Zaika et al., 2016).

Recent work in mice has suggested that Clc-Kb plays a role in potassium sensing by the DCT. Dietary potassium intake has natriuretic effects (Barker, 1932; Keith & Binger, 1935; Krishna, Miller, & Kapoor, 1989; Womersley & Darragh, 1955), likely contributing to the antihypertensive effect of a high-potassium diet (Aburto et al., 2013; Mente et al., 2014). Potassium infusion or ingestion results in decreased sodium reabsorption in the proximal tubule and the TAL (Battilana et al., 1978; Brandis, Keyes, & Windhager, 1972; Cheng et al., 2012; Higashihara & Kokko, 1985; Stokes, 1982), which promotes increased distal delivery of sodium and, therefore, potassium secretion (Fig. 3B). More recently, the effect of potassium on the NCC, which reabsorbs sodium chloride in the DCT, has been examined. Like the proximal tubule and the TAL, the DCT lies upstream of the potassium-secretory portion of the nephron. Therefore, changes in NaCl reabsorption by NCC influence potassium secretion by affecting sodium delivery to the potassium-secretory segment, where potassium is secreted in exchange for sodium. As described earlier, NCC phosphorylation by SPAK/OSR1 results in increased NCC transport activity. High dietary potassium results in decreased expression and reduced phosphorylation of NCC, which is predicted to decrease NCC activity (Castaneda-Bueno et al., 2014; Rengarajan et al., 2014; Sorensen et al., 2013; van der Lubbe et al., 2013; Wade et al., 2011), while low dietary potassium increases NCC expression and phosphorylation (Castaneda-Bueno et al., 2014; Frindt, Houde, & Palmer, 2011; Terker et al., 2015;

Vallon, Schroth, Lang, Kuhl, & Uchida, 2009; Wade et al., 2015). Consistent with the role of WNK-SPAK/OSR1 signaling in NCC phosphorylation, a high-potassium diet alters the subcellular distribution of phosphorylated (activated) SPAK in the DCT (van der Lubbe et al., 2013), and a low-potassium diet increases WNK4 levels (Terker et al., 2015), SPAK abundance and phosphorylation (Castaneda-Bueno et al., 2014; Terker et al., 2015; Wade et al., 2015), and the apical abundance of OSR1 in the DCT (Wade et al., 2015). The effect of dietary potassium on NCC phosphorylation is blunted in SPAK knockout or SPAK/OSR1 knockout mice (Terker et al., 2015; Wade et al., 2015) and abolished in a mouse with SPAK knockout and inducible renal OSR1 knockout (Ferdaus et al., 2016). These data indicate that a low-potassium diet activates WNK-SPAK/OSR1 signaling in the DCT, increasing NCC phosphorylation and activity. While this decreases potassium secretion by decreasing sodium delivery to the potassium-secretory portion of the nephron (Fig. 3B), renal salt reabsorption is increased and can result in increased blood pressure, particularly in individuals consuming the high-salt/low-potassium diet typical of the modern diet (Cogswell et al., 2012; Mente et al., 2014).

How is low dietary potassium sensed by the DCT? Ellison and colleagues have proposed a model in which a low-potassium diet hyperpolarizes the basolateral membrane of DCT epithelial cells by increasing the driving force for potassium efflux from DCT cells through the basolateral inwardly rectifying potassium channel, Kir4.1/5.1 (Fig. 3B). This in turn is expected to increase chloride efflux through CLC-Kb, lowering intracellular chloride and activating WNK (Terker et al., 2015). WNK4, which is the predominant regulator of NCC in the DCT, is particularly sensitive to changes in chloride (Terker et al., 2016). This model is supported by experiments in cultured HEK cells expressing wild type or mutant variants of Kir4.1 and CLC-Kb, as well as by mathematical modeling (Terker et al., 2015). The functional coupling of Kir4.1 with CLC-Kb is also supported by studies in Kir4.1 knockout mice, in which the basolateral chloride conductance of the DCT was strongly diminished. Interestingly, SPAK and NCC expression was very low in these mice (Zhang et al., 2014).

An as-yet unexplored topic is whether SPAK/OSR1 could be regulating CLC-Kb, since, as mentioned earlier, GCK-3 negatively regulates CLH-3b in *C. elegans*. If this were the case, activation of SPAK/OSR1 under low-potassium/low intracellular chloride conditions could inhibit CLC-Kb, putting a brake on further chloride efflux from the DCT. Alternatively, if SPAK/OSR1 acts downstream of MAPK signaling rather than WNK, as is the case for GCK-3/CLH-3b, this would afford additional opportunities for regulation of CLC-Kb independent of WNK. Another unknown is whether KCC4, which has been localized to the basolateral membrane of DCT in the rabbit kidney (Fig. 3B) (Velazquez & Silva, 2003), plays a role in chloride efflux in low dietary potassium conditions. Since WNK-SPAK/OSR1 signaling negatively regulates KCCs, this could serve as another negative feedback mechanism to avoid ongoing activation of the WNK pathway. Finally, WNK-SPAK/OSR1 signaling also modulates sodium chloride reabsorption through NKCC2 in the TAL (Fig. 3B) (Cheng et al., 2012, 2015; Rafiqi et al., 2010). Whether the pathway also regulates CLC-Kb in this segment is unknown. Mammalian CLC-2 appears to be negatively regulated by SPAK and OSR1, based on decreased chloride conductance, as measured by two-electrode voltage clamp in *Xenopus* oocytes coexpressing CLC-2 with SPAK or OSR1 (Warsi et al., 2014). CLC-Ka and CLC-Kb have predicted SPAK/OSR1-binding RFXI motifs, but their regulation by SPAK/OSR1 has not been studied.

2.3 WNK-SPAK/OSR1 Signaling in Osmoregulation

Activation of the WNK-SPAK/OSR1 pathway has been observed in cells under both hypertonic and hypotonic conditions (Chen, Yazicioglu, & Cobb, 2004; Dowd & Forbush, 2003; Lenertz et al., 2005; Moriguchi et al., 2005; Naito et al., 2011; Richardson et al., 2008; Zagorska et al., 2007). Hypertonicity also results in redistribution of WNK1 and WNK4 in cells (Sengupta et al., 2012; Shaharabany et al., 2008; Zagorska et al., 2007), although the functional significance of this is unknown. In hypotonic conditions, intracellular chloride initially falls due to the dilutional effect of water moving into cells. Subsequently, during the process of regulatory volume decrease, intracellular chloride falls further as K^+ and Cl^- efflux from the cell, followed by osmotically obliged water, to allow the cell volume to return toward normal (Hoffmann, Lambert, & Pedersen, 2009). Piala et al. demonstrated, surprisingly, that chloride binds directly to the kinase domain of WNK1 and stabilizes the kinase in an inactive conformation that prevents autophosphorylation and kinase activation (Fig. 1C) (Piala et al., 2014). The fall in intracellular chloride that occurs during hypotonicity and the subsequent regulatory volume decrease response likely explains at least part of the mechanism for WNK activation in hypotonic conditions (Bazua-Valenti et al., 2015; Ponce-Coria et al., 2008); the physiological significance of WNK activation under these conditions is under investigation. The mechanism by which hypertonicity, causing cell shrinkage, activates the WNK-SPAK/OSR1 pathway is unknown. However, activation of the WNK-SPAK/OSR1-NKCC1 pathway after a hypertonic challenge stimulates ion influx into cells, allowing recovery of cell volume (Cruz-Rangel, Gamba, Ramos-Mandujano, & Pasantes-Morales, 2012; Roy et al., 2015).

Fluid secretion from the main segment of the *Drosophila* renal tubule decreases in hypertonic conditions and increases in hypotonic conditions (Blumenthal, 2005). Consistent with this, transepithelial potassium flux in the main segment also decreases in hypertonic conditions and increases in hypotonic conditions (Wu et al., 2014). The decrease in hypertonic conditions occurs in *fray* knockdown tubules, suggesting that this effect is independent of WNK-SPAK/OSR1 signaling in principal cells (Wu et al., 2014). Blumenthal demonstrated that the hypertonic effect on fluid secretion is due to decreased tubule sensitivity to the diuretic effects of tyramine (Blumenthal, 2005). In contrast, the hypotonic stimulation of transepithelial potassium flux is abolished in tubules in which *wnk* or *fray* is knocked down in the principal cells, or in tubules carrying a null mutation in the NKCC, *Ncc69* (Wu et al., 2014), indicating that hypotonicity stimulates transepithelial potassium flux in a WNK-SPAK/OSR1-NKCC-dependent manner (Fig. 4B). Because urine generation occurs through the transepithelial secretion of a KCl-rich fluid in the main segment of the aglomerular fly renal tubule, the hypotonic stimulation of urine generation in the main segment may allow for more efficient excretion of a water load following ingestion of a hypotonic meal, if ions are reabsorbed in subsequent segments that the urine passes through (tubule lower segment and hindgut) to allow generation of a hypotonic excreta (Larsen et al., 2014).

Roles for WNK-SPAK/OSR1 signaling in osmoregulation have also been described in *C. elegans*. Worms in which *wnk-1* or *gck-3* are knocked down have impaired survival during hypertonic stress; interestingly, survival on sorbitol is less impaired than survival on an iso-osmolar concentration of sodium chloride. Wild-type worms shrink and then recover volume after exposure to hypertonic stress, whereas recovery was impaired in *wnk-1* and *gck-3* knockdown worms. The survival and volume regulatory defects of *gck-3* knockdown worms were rescued by preventing knockdown in the intestine or hypodermis, the worm epidermis, suggesting that the skin or gut is critical for the response to ionic stress (Choe & Strange, 2007).

Presumably, the WNK-SPAK/OSR1 pathway is regulating ion channels or transporters in these organs to mediate the response to hypertonic stress, but the identity of these channels/transporters has not been determined.

A subsequent study from the Strange laboratory demonstrated that in worms exposed to hypertonic sodium chloride stress, protein translation is inhibited by GCN1/2 (general control nonderepressible) kinase complex-mediated phosphorylation of the eukaryotic translation initiation factor eIF2α. Through unknown mechanisms, decreased protein translation activates WNK1-GCK3 signaling, which then results in increased expression of the glycerol synthesis enzyme glycerol-3-phosphate dehydrogenase-1, which allows the accumulation of the organic osmolyte glycerol (Lee & Strange, 2012). Interestingly, in mouse inner medullary collecting duct cells, hyperosmolar urea stress results in a similar increase in eIF2α phosphorylation by GCN2, which is protective for cell survival under the osmotic stress faced by cells in the renal medulla (Cai & Brooks, 2011). Whether WNK is activated in this circumstance has not been examined.

In mammals, several studies have connected cellular osmoregulation, WNK-SPAK/OSR1 signaling, and disease. ASK3 (apoptosis signal-regulating kinase 3) is a mammalian MAPK kinase that is activated under hypotonic conditions and repressed under hypertonic conditions. ASK3 is a negative regulator of WNK-SPAK/OSR1 signaling, and therefore is expected to increase WNK-SPAK/OSR1 activation under hypertonic conditions. Consistent with its role as a negative regulator of WNK-SPAK/OSR1 signaling, ASK3 knockouts have increased SPAK/OSR1 phosphorylation and hypertension (Naguro et al., 2012). In addition, an ASK3 phosphorylation site on WNK4, Ser 575, has been identified, although the functional consequence of this phosphorylation event has not been described (Maruyama et al., 2016).

Two studies have linked activation of WNK-SPAK/OSR1 signaling and cell volume regulation to the migration of glioma cells. Gliomas are locally invasive glial cell tumors, and the malignant glial cells undergo dynamic cell volume changes during migration. Glioma cells express WNKs 1, 3, and 4, SPAK and OSR1, and NKCC1. Inhibiting NKCC1 with bumetanide, or knocking down WNK1 or WNK3, inhibits cell volume recovery after a hypertonic challenge (Haas et al., 2011; Zhu et al., 2014). Bumetanide or WNK3 knockdown decreased glioma cell migration in a Transwell assay (Haas et al., 2011). The chemotherapeutic agent temozolomide, which is used to treat gliomas, stimulated migration and serum-induced microchemotaxis by activating the WNK1-OSR1-NKCC1 pathway in some glioma cell lines, an effect which could diminish temozolomide's antineoplastic properties (Zhu et al., 2014). Bumetanide treatment, or knockdown of WNK1 or OSR1, abolished this effect, suggesting that inhibition of the WNK1-OSR1-NKCC1 pathway could be beneficial as adjunctive treatment with temozolomide for some patients with glioma (Zhu et al., 2014).

A role for the WNK1-SPAK/OSR1-NKCC1 pathway has also been demonstrated in T cell migration. Knocking down WNK1, SPAK, OSR1, or NKCC1, or treatment with bumetanide, decreased T cell migration in multiple assays. Furthermore, *Wnk1*-deficient T cells have decreased homing and migration in lymph nodes in vivo in mice (Kochl et al., 2016). Whether this is due to alterations in cell volume was not examined, but is a possible explanation for the altered migration. WNK1 is also a negative regulator of T cell adhesion, but this effect was independent of SPAK/OSR1 or NKCC1 (Kochl et al., 2016), suggesting multiple roles for WNK1 in T cell biology and immune system function.

In stroke, activation of WNK3-SPAK/OSR1 signaling is deleterious. Cerebral edema accompanies severe strokes and is associated with mortality rate of up to 80%, leading to increased interest in treating this complication (Bardutzky & Schwab, 2007). An early component of cerebral edema is cytotoxic edema (cell swelling) (Stokum, Gerzanich, & Simard, 2016). After middle cerebral artery occlusion in mice, the WNK3-SPAK/OSR1-NKCC1 pathway is activated in neurons and glial oligodendrocytes by unknown mechanisms, potentially increasing cerebral edema. Indeed, mice in which WNK3 or SPAK are knocked out have decreased edema, as well as decreased infarct size and axonal demyelination. Importantly, functional neurological outcomes after stroke are also improved, suggesting that the WNK3-SPAK/OSR1-NKCC1 pathway could be a therapeutic target in stroke (Begum et al., 2015; Zhao et al., 2016). The currently available NKCC1 inhibitor, bumetanide, has low blood–brain barrier permeability, and a poor side effect profile due to inhibition of renal NKCC2 (Donovan, Schellekens, Boylan, Cryan, & Griffin, 2016; Pressler et al., 2015). Attempts to develop compounds that distinguish between NKCC1 and NKCC2 have been complicated by the structural similarity of the transporters (Lykke et al., 2016). Despite the presence of WNK3 in the kidney, WNK3 knockout has minimal effect on renal function, probably because of compensation by WNK4 and WNK1 (Mederle et al., 2013; Oi et al., 2012). Thus, WNK3 may be an attractive target for inhibition in the setting of stroke.

2.4 The Role of Mouse Protein-25 in WNK-SPAK/OSR1 Signaling

Mouse protein-25 (Mo25, also called calcium binding protein 39 or Cab39) is a scaffold protein that binds to the pseudokinase STE20-related adaptor (STRAD) and liver kinase B1 (LKB1) to activate LKB1 (Boudeau et al., 2003; Zeqiraj et al., 2009). SPAK/OSR1/Fray are additional members of the STE20 kinase family (Delpire & Gagnon, 2008) and a 2008 study in *Drosophila* revealed that *Mo25* and *fray* work together in the process of asymmetric cell division (reviewed in greater detail in Section 3.3.3) (Yamamoto, Izumi, & Matsuzaki, 2008). A subsequent study demonstrated that Mo25α increases the in vitro kinase activity of SPAK and OSR1 by 70- to 90-fold (Filippi et al., 2011). The related Mo25β also stimulated SPAK and OSR1 in vitro, though to a somewhat lesser degree. These experiments utilized SPAK and OSR1 mutants in which the T-loop threonine targeted by WNKs was mutated to a phospho-mimicking glutamic acid. Mo25 did not stimulate the activity of wild-type OSR1, unless WNK1 was coincubated in the reaction. These experiments suggested that Mo25 and WNKs synergistically increase SPAK/OSR1 kinase activity. In cultured human embryonic kidney cells, NKCC1 activity was decreased in both baseline and stimulated (hypotonic low-chloride) conditions when Mo25α was knocked down (Filippi et al., 2011).

The crystal structure of dimerized OSR1 demonstrated domain swapping of the activation loop in OSR1 (Lee, Cobb, & Goldsmith, 2009). Domain swapping allows exchange of identical structural elements between monomers within a protein dimer, without disrupting chemical interactions present in monomeric forms. Lee et al. proposed that OSR1 domain swapping may allow for *trans*-autophosphorylation. Based on structural and mutational analysis of MST4, another STE20 kinase that complexes with Mo25, Shi and coworkers proposed that Mo25 may facilitate the *trans*-autophosphorylation of MST4 dimers in order to fully activate MST4 (Shi et al., 2013). Indeed, the crystal structure of a SPAK mutant in which the WNK target T-loop threonine is mutated to a phospho-mimicking aspartic acid (T243D) demonstrated a partially active conformation, supporting the hypothesis that Mo25 binding allows for full activation of SPAK/OSR1 kinases after partial activation by WNK phosphorylation (Taylor et al., 2015). Further support for the hypothesis that Mo25 facilitates domain swapping in SPAK/OSR1 dimers was provided by an elegant series of experiments by Delpire and colleagues. They examined NKCC1 activation in *Xenopus* oocytes injected with cRNAs for NKCC1, Mo25, and wild-type or mutated SPAK monomers or concatemerized dimers. In the presence of Mo25, a wild-type Thr in the swap domain of SPAK could substitute for a mutated Thr (to Ala) in the SPAK in the other half of the concatemerized dimer, allowing for NKCC1 activation. This did not occur if wild-type SPAK and the Thr-to-Ala mutant SPAK were introduced as separate monomers, indicating that prior dimerization (experimentally recapitulated by concatemerization) is required to observe the Mo25 effect. The authors proposed that WNK phosphorylation of SPAK allows it to assume a domain swapping-competent conformation, which is further facilitated by Mo25 (Ponce-Coria, Gagnon, & Delpire, 2012), consistent with the results of the structural studies of SPAK T243D described earlier (Taylor et al., 2015). An additional study from the Delpire group, examining mouse and sea urchin OSR1 with or without Mo25, adds additional insights into OSR1 activation mechanisms (Gagnon, Rios, & Delpire, 2011).

The Delpire group also observed that WNK4 contains a domain that resembles the PF2 WNK binding domain of SPAK and OSR1. They therefore wondered whether WNK4 could bind to NKCCs directly, independently of SPAK/OSR1, and phosphorylate and activate the transporters. Indeed, while WNK4 alone did not stimulate NKCC1 or NKCC2 activity when cRNAs were coinjected into oocytes, the combination of WNK4 and Mo25 was able to stimulate both NKCC1 and NKCC2. *Xenopus* oocytes express an endogenous OSR1 (Pacheco-Alvarez et al., 2012), but the WNK4/Mo25 stimulation of NKCC1 was not inhibited by coinjection of kinase-dead SPAK nor was it abolished by mutating the WNK4 RFXV motif required for SPAK/OSR1 binding, suggesting independence from SPAK/OSR1 activity. Similarly, Mo25 mutants lacking the ability to bind to SPAK/OSR1 were still able to stimulate NKCC1 activity when coexpressed with WNK4. However, a WNK4 mutant lacking NKCC1 binding was not able to stimulate NKCC1 activity, even in the presence of Mo25. Together, these results suggest that WNK4 could directly activate NKCC1 independently of SPAK/OSR1, in the presence of Mo25 (Ponce-Coria et al., 2014). The role of Mo25 in transepithelial ion transport has not been elucidated, but Mo25 is expressed in both the TAL and the DCT (Grimm et al., 2012), suggesting that it could play a modulatory role in regulation of NKCC2 and NCC by WNK-SPAK/OSR1 signaling in the mammalian nephron.

3. EMERGING FUNCTIONS OF THE WNK SIGNALING AXIS IN DEVELOPMENT

3.1 Mammalian WNKs

As discussed in the previous sections, a substantial amount of knowledge about the function of the WNK-SPAK/OSR1 kinase axis in the regulation of ion transport has been discovered. Potential additional roles of WNKs important for the development of vertebrates and invertebrates have started to emerge only recently, and surprisingly little is known about embryonic functions of WNKs.

Human WNK1 is widely expressed in most tissues, including in the embryonic heart, skin, spleen, and the small intestine (Verissimo & Jordan, 2001). WNK2 is expressed in the fetal brain and heart, WNK3 in fetal brain, while WNK4 appears more restricted to the embryonic liver and skin (Verissimo & Jordan, 2001). While the phenotype of *Wnk2* knockout mice is unknown, *Wnk3* mutant mice are homozygous viable and show no gross abnormalities (Mederle et al., 2013; Oi et al., 2012). Similarly, mice lacking WNK4 are born at Mendelian ratios and show no overt developmental or behavioral defects (Castaneda-Bueno et al., 2012; Takahashi et al., 2014). WNK3 and WNK4 are thus either not required for early development or their functions may be redundant with other WNK family members.

In contrast, a gene trap allele of *Wnk1* in mice is embryonic lethal prior to day E13 with heterozygotes showing no developmental phenotype (but a reduced blood pressure as adults) (Zambrowicz et al., 2003). A more detailed time-course analysis by Xie et al. showed that homozygous *Wnk1* mutant embryos start to show growth retardation at E9.5 and are all abnormal by E10.5, displaying pericardial edema and hemorrhage (Xie et al., 2009). Importantly, the lack of detectable blood flow suggested cardiovascular developmental defects. Indeed, while the four cardiac chambers and the dorsal aortae and cardinal veins form, the heart chambers are hypoplastic and show significantly reduced trabeculation and thinner outer myocardial walls (Fig. 5A and B) (Xie et al., 2009). The dorsal aortae and cardinal veins are smaller or collapsed, the latter likely caused by secondary blood circulation defects. Furthermore, *Wnk1* mutant vessels of the yolk sac do not properly remodel and embryonic arteries and veins show defective angiogenesis including coexpression of the arterial and venous markers Neuropilin-1

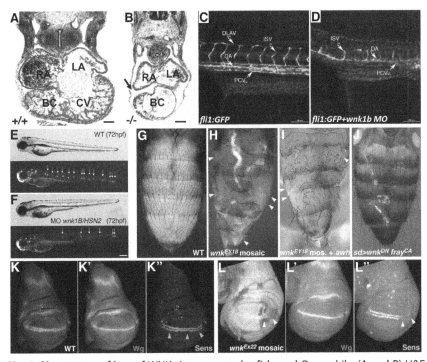

Fig. 5 Phenotypes of loss of WNKs in mouse, zebrafish, and *Drosophila*. (A and B) H&E staining of transverse sections of an E10.5 wild type (A) and *Wnk1* mutant embryo (B). Compared to WT (A), *Wnk1* mutant embryos show reduced ventricular trabeculation (*yellow arrows*) and dilatation of pericardial sac (*black arrows*). RA/LA, right/left atrium; BC, bulbus cordis; CV, common ventricle. (C and D) Lateral views of the trunk of uninjected zebrafish control embryos (C) and *wnk1b* morphants (D) at 33 hpf (hours postfertilization). Growth of the intersegmental vessels (ISVs) and formation of the dorsal longitudinal anastomotic vessel (DLAV) are inhibited in *wnk1* morphants. Vessels formed by the vasculogenesis process, including the dorsal aorta (DA) and the posterior cardinal vein (PCV), are unaffected. (E and F) Compared to WT zebrafish (E), embryos specifically lacking the *Wnk1/HSN2* isoform at 72 hpf (F) show posterior lateral line defects (neuromasts stained with vital dye 4-di-2-ASP are indicated with *yellow arrows* in lower panels). (G–L) *Drosophila wnk* phenotypes. (G and J) Compared to the abdomen of a WT fly covered with cuticle and bristles (G), homozygous *wnk* mutant tissue (identified by the absence of pigment due to concomitant lack of the *yellow* gene; (H)) is unable to form cuticle and bristles. (I) Reexpression of Awh in clones mutant for *wnk* largely restores cuticle formation and partially suppresses bristle defects. (J) Coexpression of constitutively active Fray restores the cuticle and bristles on abdomina expressing dominant-negative Wnk, which lack abdominal cuticle and bristles (not shown). *Yellow arrowheads* indicate mutant tissue in (H) and (I). (K and L) Loss of *wnk* leads to a reduction in expression of the Wnt target gene Sens in 3rd instar wing imaginal discs. (K) WT wing discs express Wg in a line along the dorsoventral boundary

(Continued)

and EphB4, which are usually expressed in a mutually exclusive manner. WNK1 is thus either involved in venous vs arterial fate specification or maintenance of those fates (Xie et al., 2009).

Even though *Wnk1* is expressed in all layers of the developing heart, endothelial-specific knockout of *Wnk1* using a conditional allele recapitulates all phenotypes of the global knockout, suggesting that the observed phenotypes are due to an endothelial-specific requirement of WNK1 (Xie et al., 2009). Consistent with this, the heart and vascular phenotypes of the global *Wnk1* knockout were rescued by a *Wnk1* transgene specifically expressed in endothelial cells, but not by reexpressing *Wnk1* in somatic embryonic cells only. However, these animals are smaller at birth and die perinatally for uncharacterized reasons, suggesting additional roles for WNK1 beyond the cardiovascular system.

WNK1 functions either via OSR1/SPAK or through kinase-independent mechanisms, such as the activation of SGK kinase or by modulating GPCR signaling (An et al., 2011; Xu, Stippec, Chu, et al., 2005; Xu, Stippec, Lazrak, Huang, & Cobb, 2005). It was thus important to determine the mechanism by which WNK1 regulates cardiovascular development. Intriguingly, homozygous *Osr1* mutant mice in which the catalytic domain is truncated show indistinguishable phenotypes to the *Wnk1* mutants (Xie, Yoon, Yang, Lin, & Huang, 2013). Moreover, expression of a

---

Fig. 5—Cont'd (*red*; single channel shown in K′) where it induces the expression of its target gene Sens in abutting cells (*green*; *green arrowheads* in single channel image K″). (L) Homozygous *wnk* mutant cells marked by the absence of GFP (*green*) in mosaic discs cell autonomously express reduced levels of Sens (*blue*; single channel in L″; *yellow arrowheads* indicate mutant areas). Note that there is no effect on Wg expression (*red*; single channel in L′). Scale bars are 100 μm in (A–E). *Panels (A) and (B) Modified from Xie, J., Wu, T., Xu, K., Huang, I. K., Cleaver, O., & Huang, C. L. (2009). Endothelial-specific expression of WNK1 kinase is essential for angiogenesis and heart development in mice. The American Journal of Pathology, 175, 1315–1327. Panels (C) and (D) After Lai, J. G., Tsai, S. M., Tu, H. C., Chen, W. C., Kou, F. J., Lu, J. W., et al. (2014). Zebrafish WNK lysine deficient protein kinase 1 (wnk1) affects angiogenesis associated with VEGF signaling. PloS One, 9, e106129. Panels (E) and (F) After Bercier, V. (2013). WNK1/HSN2 isoform and the regulation of KCC2 activity. Rare Diseases, 1, e26537. Panels (G–J) Modified from Sato, A., and Shibuya, H. (2013). WNK signaling is involved in neural development via Lhx8/Awh expression. PloS One, 8, e55301. Panels (K) and (L) After Serysheva, E., Berhane, H., Grumolato, L., Demir, K., Balmer, S., Bodak, M., et al. (2013). Wnk kinases are positive regulators of canonical Wnt/beta-catenin signalling. EMBO Reports, 14, 718–725.*

constitutively active form of OSR1 in endothelial cells is sufficient to suppress the heart and angiogenesis defects of global *Wnk1* mutant embryos, showing that WNK1 acts via OSR1 during mouse embryonic development (Xie et al., 2013). The mechanistic cause of the heart and angiogenesis defects downstream of OSR1 remains to be determined. In particular, SLC12 CCC knockouts do not show similar phenotypes, although redundant roles cannot be excluded (reviewed in Arroyo, Kahle, & Gamba, 2013; Delpire & Mount, 2002; Gamba, 2005).

3.2 Zebrafish Wnk1a/b Have Roles in Angiogenesis and Neural Development

The zebrafish genome encodes two paralogs each of *wnk1* and *wnk4*, and a single *wnk2* gene (Howe et al., 2013), with only *wnk1a* and *wnk1b* expression being detectable during early embryogenesis (prior to 48 h postfertilization) (Lai et al., 2014). Knockdown of either *wnk1a* or *wnk1b* causes significant defects in angiogenesis of head and trunk blood vessels. In particular, intersegmental vessels (ISVs) that sprout and elongate dorsally from the dorsal aorta and the posterior cardinal vein (PCV) fail to form or do not properly extend (Fig. 5C and D) (Lai et al., 2014). This phenotype can be significantly rescued by reexpression of *wnk1*, suggesting that the morpholino effect is specific (Kok et al., 2015). In contrast to the angiogenesis defect, vasculogenesis, the de novo formation of blood vessels, is normal as judged by normal expression of the vasculogenesis marker *etv2* (Sumanas & Lin, 2006) and the presence of the dorsal aorta or the caudal and posterior cardinal veins (Fig. 5D) (Lai et al., 2014). Intriguingly, the knockdown phenotype of *wnk1a* or *wnk1b* is similar to the knockdown of *flk1*, the gene encoding Vegfr2 (vascular endothelial growth factor receptor 2). Vegfr2 mediates most of the angiogenic effects of Vegf via the activation of phosphoinositide-dependent protein kinase PI3K and Akt/protein kinase B (PKB), and knockdown of *pi3kc2α* causes similar angiogenesis defects as reduction of *flk1* or *wnk1* (Lai et al., 2014). Human WNK1 has been shown to be an Akt substrate (Vitari et al., 2004), and zebrafish Wnk1 contains a putative Akt phosphorylation site, suggesting that Wnk1 could be downstream of Akt in the VEGF signaling pathway (Fig. 3C). Interestingly, the vascular phenotype of *flk1* knockdown is partially rescued by injection of mRNA encoding wild-type Wnk1a, but not by kinase-dead Wnk1a or Wnk1a with a mutation in the putative Akt site. This is consistent with a role for Wnk1 kinase activity downstream of Vegfr2 during angiogenesis in zebrafish (Fig. 3C). In addition, VEGF signaling also appears to play a role in transcriptional regulation of *wnk1*, as *wnk1* mRNA is downregulated upon *flk1* knockdown (Lai et al., 2014). As in mice, the downstream effectors of Wnk1 in angiogenesis remain to be determined. Although it is not known if WNK1 acts downstream of Vegfr in mice, based on the fish and mouse data Wnk1 may have a conserved role in vascular development in humans (Lai et al., 2014; Xie et al., 2009, 2013). Such a function would also be consistent with recent data showing a requirement for WNK1 in human umbilical vein endothelial cell and human dermal microvascular endothelial cell models of in vitro angiogenesis (Dbouk et al., 2014).

WNK kinases may also have a function in the nervous system (see also Section 3.3 for *Drosophila* Wnk). For example, although no loss-of-function data are available, WNK2 is strongly expressed in the mouse brain, where it is found in a phospho-protein complex with SPAK and may regulate GABAergic signaling (Rinehart et al., 2011; see also Alessi et al., 2014 for a review of potential involvement of ion transporters). Whole-genome exome sequencing also identified rare variants in WNK1 in patients affected by Charcot–Marie–Tooth (CMT), a form of peripheral neuropathy (Gonzaga-Jauregui et al., 2015). Most interestingly though, stop codon mutations in an extra neuron-specific exon between exons 8 and 9 in WNK1 have been identified in hereditary sensory and autonomic neuropathy type II (HSANII) patients (*WNK1/HSN2* isoform; note that *WNK1/HSN2* mutations can occur in trans to an allele truncating *WNK1*) (Lafreniere et al., 2004; Shekarabi et al., 2008). HSANII is a recessive disease characterized by an early onset of lack of peripheral sensory functions (Auer-Grumbach, Mauko, Auer-Grumbach, & Pieber, 2006). As the disease is nonprogressive and nerves of affected individuals show fewer fibers without signs of degeneration, HSANII is thought to have developmental roots, which is supported by an elegant zebrafish model of HSANII developed by Bercier (2013) and Bercier et al. (2013). Sequence comparison showed that only zebrafish *wnk1b* has the ability to encode a HSN2 exon and antibody stainings confirmed its expression in the neuromasts of the posterior lateral line (PLL), a peripheral sensory organ responsive to water pressure. Knockdown of the HSN2 isoform of *wnk1b* using MOs targeting corresponding splice sites causes a strong reduction of neuromasts (Fig. 5E and F) and their hair cells that is partially rescued by injecting mRNA coding for the HSANII-type of *wnk1b* (Bercier et al., 2013). Interestingly, loss of neuromast hair cells coincides with a transcriptional upregulation of *kcc2* (distinct from regulating KCCs through phosphorylation). Indeed, overexpression of human *KCC2* mRNA mimics the PLL defects and the reduced size of neuromast precursor area of *wnk1b* knockdown. Consistent with this, combined knockdown of *kcc2* with *wnk1b* partially suppresses the defects caused by loss of *wnk1b*. Unexpectedly, although the phenotype was weaker than with overexpression of wild-type human *KCC2*, expression of a KCl transport-incompetent mutant of KCC2, KCC2^{C568A} (Reynolds et al., 2008), also prevented proper PLL formation, suggesting that loss of Wnk1b/HSN2 causes a transcriptional upregulation of KCC2, in turn preventing correct development of this sensory neuronal system in a potentially (at least partially) transport-independent manner (Bercier et al., 2013). Whether the effects of Wnk1 on KCC2 are mediated by SPAK/OSR1 was not examined in this study.

HSANII patients lose peripheral nerve fibers concomitant with reduced pain sensation (Lafreniere et al., 2004). *Wnk1* mutant mice specifically lacking the HSN2 exon (*Wnk1ΔHsn2*) have a somewhat different phenotype, with normal peripheral sensory neuron morphology and distribution (Kahle et al., 2016). Nevertheless, these mice were less susceptible to pain hypersensitivity resulting from peripheral nerve injury. Interestingly, in mice, loss of *Wnk1ΔHsn2* led to a reduced phosphorylation of KCC2 and thus a more active transporter (Kahle et al., 2016). Thus, the WNK1 HSN2 isoform may be a target for treatment of pain syndromes resulting from peripheral nerve injury. Future experiments will have to address the mechanistic differences between WNK1 and KCC2 in mice and fish in the peripheral sensory nervous system, and how these relate to the human HSANII phenotype.

The WNK1–KCC2 axis also plays a role in the developmental maturation of neurons. Neuronal intracellular chloride concentration in many cases decreases with postnatal development, due to an increase in KCC2 expression and activity (reviewed in Kaila, Price, Payne, Puskarjov, & Voipio, 2014). One potential role for maintaining lower KCC2 activity at earlier developmental timepoints is to allow normal neuronal migration (Inoue et al., 2012), while later increases in KCC2 activity allow the lowering of intraneuronal chloride concentration that allows GABA and glycine neurotransmission to result in a hyperpolarizing or inhibitory effect by opening ligand-gated chloride channels (Kaila et al., 2014). In cultured hippocampal neurons, the developmental shift in KCC2 activity is mimicked by WNK1 knockdown. Expression of a dominant-negative (DN), kinase-dead WNK1 in immature neurons has a similar effect that is reversed by simultaneous knockdown of KCC2. These results were also recapitulated with chemical inhibition of WNK1 (Friedel et al., 2015), suggesting that WNK1 inhibition of KCC2 in immature neurons maintains a higher intracellular chloride concentration. Indeed, WNK1 inhibits the activity of all mammalian KCCs when coexpressed in *Xenopus* oocytes (Fig. 3A) (Mercado et al., 2016). In mouse brain (Rinehart et al., 2009) or in cultured hippocampal and cortical neurons (Friedel et al., 2015), KCC2 phosphorylation, which results in transporter inactivation (Rinehart et al., 2009), decreases with maturation. In cells in which KCC2 is inactivated by introducing phospho-mimicking threonine-to-glutamate mutations, WNK1 inhibition has no effect on intracellular chloride. Together, the results suggest that increased WNK1 activity in immature neurons maintains KCC in a phosphorylated, inactive state that allows for higher intracellular chloride at that developmental timepoint (Friedel et al., 2015). How is higher WNK1 activity maintained in immature neurons despite the higher intracellular chloride, which is inhibitory toward WNK1? One possible mechanism is through activation of WNK1 by intracellular taurine, which is high in the fetal brain, though additional mechanisms may also play a role (Inoue et al., 2012). Additional roles for WNK signaling in nervous system physiology and disease are reviewed in Alessi et al. (2014) and Tang (2016).

3.3 Insights from the *Drosophila* Wnk-Frayed Axis

Over recent years, unexpected WNK functions critical for organismal development have been discovered using the fruit fly *D. melanogaster* model, and at least some of these are conserved in mice or human cells. As discussed earlier, the genome of *D. melanogaster* encodes one *wnk* gene and one homolog of OSR1/SPAK, *fray* (Fig. 2). *Drosophila wnk* was first identified in a genetic mosaic screen for axon pathfinding in the eye, a classical model system used by geneticists as the eye is dispensable for viability under lab conditions. Based on the identification of mutations within the kinase domain, it was suggested that the axon targeting function of Wnk required kinase activity, but no further functional studies were performed (Berger et al., 2008).

More recently, it was shown that *Drosophila* Wnk has additional important functions regulating Wnt signaling during wing development and regulating the LIM-homeobox transcription factor Arrowhead (Curtiss & Heilig, 1995, 1997) during development of the adult cuticle and likely the embryonic nervous system (Sato & Shibuya, 2013; Serysheva et al., 2013; Serysheva, Mlodzik, & Jenny, 2014).

3.3.1 A Conserved Role of Wnk in the Activation of Arrowhead/Lhx8

Like *fray* mutations, *wnk* mutations are embryonic or larval homozygous lethal (Leiserson, Harkins, & Keshishian, 2000; Sato & Shibuya, 2013; Serysheva et al., 2013). Homozygous mutant abdominal tissue in mosaic animals or abdominal tissue overexpressing a kinase dead, DN form of Wnk (Wnk^{D420A}) fails to form cuticle (Fig. 5G and H and not shown), a phenotype that can be suppressed by co-overexpression of constitutively active Fray (Fray^{S347D}; Fig. 5J) (Sato & Shibuya, 2013). Similarly, the peripheral axon growth phenotype caused by DN Wnk in *Drosophila* embryos can be suppressed by Fray^{S347D}. In addition, the formation of ectopic wing vein tissue in the posterior wing compartment caused by overexpression of Wnk is dominantly suppressed by the removal of one gene dose of *fray*, while Fray or human OSR1 overexpression causes similar wing vein defects (Sato & Shibuya, 2013). Thus, as is the case in the control of transepithelial ion flux in the *Drosophila* renal tubule (Wu et al., 2014), these data indicate that Wnk can act through Fray in *Drosophila*.

The adult *Drosophila* abdominal epidermal cuticle develops from histoblast cells set aside as nests in the embryo that only divide and migrate over the forming abdomen after metamorphosis (Madhavan & Madhavan, 1980). The mutant abdomen phenotype of *wnk* is highly reminiscent of the one caused by the loss of the LIM-Homeobox gene *arrowhead* (*Awh*), the homolog of vertebrate *Lhx8* (Curtiss & Heilig, 1995, 1997). Indeed, Sato et al. showed that histoblast nest-specific expression of *Awh* is lost in *wnk* mutant embryos and that overexpression of Awh in *wnk* mutant mosaics (Fig. 5I), or in the background of DN Wnk^{D420A}, can suppress the cuticle phenotype due to loss of *wnk* function (Sato & Shibuya, 2013).

Intriguingly, the functional axis from WNK to LHX8 is conserved in mice: expression of *Lhx8* mRNA is strongly reduced in E9.5 *Wnk1* mutant mouse embryos. Additionally, overexpression of *Wnk1* and *Wnk4* in NIH3T3 cells induces *Lhx8* expression in the presence of cycloheximide and in an *Osr1*-dependent manner. *Lhx8* mRNA and protein are also induced upon hypertonic stimulation, a known activator of WNK1 (Lenertz et al., 2005; Moriguchi et al., 2005), which is prevented by siRNA-mediated knockdown of *Wnk1* or *Wnk4*. LHX8 was known to be involved in the determination of cholinergic neurons in the forebrain (Zhao et al., 2003). Therefore, Sato et al. further tested, if differentiation of Neuro2A cells in culture was mediated by WNK. Indeed, induction of differentiation by retinoic acid treatment stimulated OSR1 phosphorylation and neurite outgrowth. Additionally, WNK1 and WNK4 were required in a redundant manner for the induction of the neural differentiation marker choline acetyl transferase (ChAT) in Neuro2A cells. However, while constitutively active OSR1^{S325D} expression was sufficient to suppress the knockdown of *Wnk1* and *Wnk4* with respect to neurite outgrowth and ChAT induction, overexpression of *Lhx8* was not (Sato & Shibuya, 2013). This parallels Wnk function in the fly nervous system, where only overexpression of constitutively active Fray (see earlier), but not Awh was able to suppress the DN Wnk^{D420A} phenotype. This suggests that in addition to LHX8, there are other pathways downstream of Wnk–Fray/OSR1 required for neurite outgrowth. Clearly, these experiments demonstrate a strong conservation between flies and vertebrates of a novel function of the Wnk–Fray/OSR1 axis in regulating the transcription factor Awh/LHX8. It will be critical to identify missing pathway components that link OSR1/Fray to the transcriptional induction of *Lhx8/Awh* and the additional factors under control of WNK required in addition to LHX8/Awh for neurite outgrowth.

It is worth noting that while Wnk is required for axon growth of photoreceptor neurons in the fly eye (Berger et al., 2008), and DN Wnk prevents correct axon outgrowth in the embryo, homozygous mutant *wnk* embryos do not show a clear axon outgrowth phenotype in the embryonic peripheral nervous system. Neither have axon outgrowth problems been reported for *fray* mutants. A very likely explanation for this discrepancy is that *wnk* and *fray* mRNAs are maternally deposited (Attrill et al., 2016), and identification of a peripheral axon outgrowth phenotype thus will require analysis of maternal-zygotic mutants (see also later). As mentioned, *fray* mutants show a defasciculation and axon bulging phenotype in embryos due to a function of Fray in ensheathing glia cells (Leiserson et al., 2000). In *fray* mutant embryos, subperineural glia cells (SPGs) that form the paracellular nerve blood barrier fail to completely wrap axons and cause fluid-filled bulges between axons and glia (Leiserson et al., 2000). These phenotypes can be rescued by glial specific reexpression of Fray or its rat homolog SPAK. Further characterization of the mechanism showed that this function of Fray is mediated by Ncc69, the fly homolog of NKCC1 (Leiserson et al., 2011). *ncc69* null mutants are homozygous viable and give rise to apparently normal adults, but mutant embryos show similar nerve bulging as *fray* mutants without affecting action potentials propagated by the nerves (Leiserson et al., 2011). Therefore, failure by SPGs to remove KCl from the space between neuron and glia likely draws H_2O into the intercellular space via osmosis, causing nerve bulging (Leiserson et al., 2011).

WNK kinases may also influence the etiology of spinocerebellar ataxia type 1 (SCA1), a neurodegenerative disease caused by the extension of a polyglutamine repeat in ataxin 1, as human WNK4 and *Drosophila* Wnk were identified in kinome-wide screens for SCA1 (Park et al., 2013). siRNA-mediated reduction of WNK4 destabilized ATXN1(82Q) in culture, while knockdown of *Drosophila wnk* in vivo suppressed the photoreceptor neuron degeneration induced by ATXN1(82Q) overexpression. Both screens also identified several components of MAP kinase cascades, suggesting that Wnks may affect ATXN1 stability via their effect on MAPK signaling, reduction of which in turn was shown to ameliorate phenotypes of a mouse SCA1 model (Park et al., 2013).

3.3.2 Wnk in Wnt Signaling

An additional function for WNK in *Drosophila* has been identified as a regulator of canonical Wnt signaling during wing development (Serysheva et al., 2013, 2014). The Wnt pathway (Fig. 3D) is a major and conserved signaling pathway regulating embryonic axis establishment in vertebrates and segmentation and patterning in *Drosophila*. Aberrant Wnt signaling not only causes strong developmental defects but also various diseases including cancer (see also Section 4) (Clevers, 2006; Clevers & Nusse, 2012; Swarup & Verheyen, 2012). In the absence of Wnt signaling, its central transcriptional cofactor, β-catenin, is targeted for degradation by a destruction complex consisting of Axin, GSK3, and APC. GSK3 phosphorylation of β-catenin marks it for ubiquitination and subsequent degradation by the proteasome. Signaling is activated by binding of a Wnt ligand (Wingless [Wg] in *Drosophila*) to a seven-pass transmembrane receptor of the Frizzled (Fz) family and a LRP5/6 coreceptor (Arrow [Arr] in *Drosophila*; Fig. 3D). Wnt binding induces recruitment of the Dishevelled (Dsh) adapter protein and formation of the LRP signalosome consisting of a complex of Fz, LRP5/6, Dsh, Axin, and GSK3 (Bilic et al., 2007). This ultimately leads to the inactivation of the destruction complex and concomitant stabilization of β-catenin (Kim et al., 2013; Li et al., 2012), allowing it to translocate to the nucleus to activate Wnt target genes by binding the transcription factor TCF/LEF (Pangolin in *Drosophila*) and additional cofactors (Brunner, Peter, Schweizer, & Basler, 1997; Kramps et al., 2002).

In *Drosophila* in vivo, Wnt signaling can be assessed by its function during wing formation. Adult fly wings develop from epithelial cells, the so-called wing imaginal discs that are set aside in the embryo and proliferate and differentiate during larval and pupal stages (reviewed in Swarup & Verheyen,

2012). In 3rd instar imaginal discs, Wg is expressed in a line of cells along the dorsoventral boundary in the wing pouch, the part of the wing disc that will give rise to the adult wing blade (Fig. 5K) (Swarup & Verheyen, 2012). There, Wg induces the bHLH transcription factor Senseless (Sens) and Distalless (Dll), the fly member of the Dlx family of homeobox genes, both of which are required for patterning of the wing. In particular, Sens specifies margin bristles, which are thus structures that depend on a high level of Wnt signaling (Jafar-Nejad, Tien, Acar, & Bellen, 2006). A loss of Wnt signaling can be detected as a loss of Sens or Dll during development and wing margin defects in adults. In vitro, Wnt signaling is commonly assessed using transcriptional reporters (TOPFLASH assays) (Korinek et al., 1997) or by monitoring the phosphorylation state of Dsh in gel shift assays, which correlates with Wnt pathway activation (Lee, Ishimoto, & Yanagawa, 1999; Matsubayashi et al., 2004; Yanagawa, van Leeuwen, Wodarz, Klingensmith, & Nusse, 1995; Yanfeng et al., 2011).

Wnk kinase has recently been identified in two kinome-wide screens as a regulator of Wnt signaling. First, in a cell culture based screen, knockdown of *wnk* was found to reduce the level of Dsh phosphorylation (Serysheva et al., 2013). The second screen was an in vivo knockdown screen monitoring Wnt targets by immunohistochemistry (Swarup, Pradhan-Sundd, & Verheyen, 2015). Phenotypically, overexpression of DN Wnk, in vivo RNAi-mediated knockdown of *wnk* in the wing, or loss of *wnk* function in mutant clones (patches of homozygous mutant *wnk* cells in a heterozygous background) causes lack of sensory bristles and margin defects in the adult wing (Sato & Shibuya, 2013; Serysheva et al., 2013). On the molecular level, staining of 3rd instar wing imaginal discs showed that Sens and Dll expression are lost after RNAi-mediated knockdown of *wnk* (Serysheva et al., 2013; Swarup et al., 2015), and Sens is lost in a cell autonomous manner in *wnk* mutant clones (Fig. 5L) (Serysheva et al., 2013). As the expression of the Wg ligand itself is not affected (Fig. 5L'), Wnk likely acts downstream of the Wnt ligand, but upstream or at the level of Dsh (the phosphorylation of which it affects). This is further supported by genetic interaction experiments: the cell death induced by overexpression of Dsh in the fly eye is dominantly suppressed by removal of one gene dose of *wnk*. Analogously, the ectopic Sens expression and margin bristles formed upon overexpression of the major Wnt receptor, dFz2, are dominantly suppressed by removal of one gene dose of *wnk* (the latter is also enhanced by concomitant overexpression of Wnk) (Serysheva et al., 2013). Consistent with this, knockdown of GSK3β to inhibit the destruction complex or expression of stable β-catenin in *wnk* mutant cells does not prevent constitutive Wnt pathway activation.

Significantly, the function of Wnk in regulating Wnt signaling is conserved in human HEK293T cells. siRNA-mediated knockdown of WNK1 or WNK2, the two WNKs expressed in the HEK293T cells tested, reduces Wnt3a induced Wnt signaling as measured by a decrease in soluble (active) β-catenin or by using a transcriptional luciferase reporter (TOPFLASH) (Serysheva et al., 2013). In contrast, overexpression of WNK2, but not a catalytically inactive version (WNK2^{K207A}) promotes Wnt3a-induced stabilization of β-catenin and TOPFLASH reporter activity, suggesting that the kinase activity is required for WNK to promote Wnt signaling. In line with these data, human WNK2/4 kinases were also identified in a genome-wide screen as candidate positive regulators of Wnt/β-catenin signaling in A375 melanoma cells (Biechele et al., 2012). However, for reasons not assessed, WNK1 in the same screen and Wnk in *Drosophila* Clone-8 cells antagonized Wnt signaling (Biechele et al., 2012; DasGupta et al., 2007).

Using purified *Drosophila* fusion proteins, direct phosphorylation of upstream Wnt signaling components, including Dsh, was not observed. As expected, however, *Drosophila* Wnk directly phosphorylates Fray (Sato & Shibuya, 2013; Serysheva et al., 2013). Consistent with this, knockdown of OSR1 and SPAK in HEK293T cells or RNAi-mediated knockdown of *fray* in *Drosophila* wing discs inhibits Wnt signaling, as assessed by reduction in TOPFLASH reporter activity and reduced Sens staining, respectively (Serysheva et al., 2013). This suggests that the conserved role of Wnk in Wnt signaling is mediated by OSR1/SPAK/Fray.

3.3.3 Intriguing Issues to Be Addressed

As reviewed earlier, the mechanistic link(s) of the WNK–Fray/OSR1/SPAK signaling axis to various downstream effectors in various developmental pathways is unknown, but may well be conserved between flies and vertebrates. In particular, it is unknown whether OSR1/SPAK/Fray can phosphorylate Wnt pathway components such as Dsh or the Fz receptors or coreceptors, or whether Wnt signaling is influenced by CCCs via an unknown mechanism. The *D. melanogaster* genome encodes one KCC (*kcc/CG5594*) and two putative NKCCs (*Ncc69/CG4357* and *Ncc83/CG31547*; NKCC transport activity has only been demonstrated for Ncc69) (Leiserson et al., 2011; Sun et al., 2010). *ncc69* null mutants are viable and show no obvious developmental defects (such as of wing margins). Cold-sensitive, hypomorphic *kcc* alleles are bang-sensitive and show seizures likely due to a function of KCC in neurons of the mushroom and ellipsoid bodies in the brain, but also appear externally normal (Hekmat-Scafe, Lundy, Ranga, & Tanouye, 2006; Hekmat-Scafe et al., 2010). Because strong loss-of-function *kcc* alleles are homozygous lethal, mosaics will have to be analyzed for Wnt-related defects, likely also in combination with *ncc69* and/or *ncc83*.

To this end, it is worth noting that there is no indication that Wnt signaling regulates Awh function in flies, and Wnk may thus independently affect Wnt signaling and the Awh expression required for cuticle formation and axon growth (Sato & Shibuya, 2013). In mice, however, the enhancer of *Lhx8*, which is required for craniofacial development, contains a conserved binding site for TCF/LEF that is critical for direct control of *Lhx8* expression in primary maxillary arch cells by Wnt signaling (Landin Malt, Cesario, Tang, Brown, & Jeong, 2014). Whether the activation of this enhancer is mediated by WNKs via effects on Wnt signaling remains to be determined.

An additional protein that may have functional relevance with respect to WNK signaling in development is Mo25, two paralogs of which exist in mice (see also Section 2.4). Mo25 proteins in various species from yeast to mammals have been shown to interact with Ste20-like kinases, stimulating their activity either directly, as for OSR1 and SPAK, or indirectly, via the induction of a trimeric complex with the pseudokinase STRAD in the case of the tumor suppressor LKB1 (Boudeau et al., 2003; Filippi et al., 2011; Mendoza, Redemann, & Brunner, 2005; Nozaki, Onishi, Togashi, & Miyamoto, 1996). In *Drosophila*, Mo25 is implicated in asymmetric division of embryonic neuroblasts (Yamamoto et al., 2008). Neuroblasts are stem cells in the embryo that asymmetrically divide to give rise to a neuroblast (self-renewal) and a ganglion mother cell (GMC) that later divides and differentiates into neurons or glia cells (reviewed in Gonczy, 2008). During neuroblast division, the cell fate determinants Prospero, a homeobox transcription factor, the phosphotyrosine-binding protein Numb, and their respective adaptor proteins Miranda and Pons (Partner of Numb) localize basally and are thus asymmetrically inherited, specifying the basal daughter cell as the GMC. Loss of maternal and zygotic *mo25* or *fray* in germ line clones causes identical phenotypes: localization of Miranda in the cytoplasm and at the mitotic spindle instead of at the basal cortex (Yamamoto et al., 2008). The originally described requirement of Lkb1 for asymmetric neuroblast division is controversial, as it was later reported that *lkb1* maternal-zygotic mutants show phenotypes during cellularization preventing the embryos to reach the stage of neurogenesis (Bonaccorsi et al., 2007;

Yamamoto et al., 2008). On the other hand, Lkb1 overexpression causes a similar phenotype as loss of *fray* or *mo25* and recruits Fray and Mo25 that are normally diffusely localized to the neuroblast cortex. Overexpression of both Mo25 and Fray are required to revert this effect, suggesting that levels of cytoplasmic Fray and Mo25 are critical for proper function in this context (Yamamoto et al., 2008). While Lkb1 can interact with Mo25 (in vivo co-IP) and Fray (in culture), it is not entirely clear whether the role of Lkb1 in neuroblast division is physiological. In particular, Lkb1 may not form a trimer with Fray and Mo25, as a Fray dimer formed via domain exchange will likely occur in a complex with Mo25 as discussed earlier (Ponce-Coria et al., 2012). Whether *Drosophila* Wnk has a role in regulation of Mo25/Fray in asymmetric neuroblast division has not been determined.

4. FUNCTIONS OF WNKs IN CANCER

WNK kinases have been linked to various cancers in recent years, although the mechanism by which they act is not well understood (for reviews, see also McCormick & Ellison, 2011; Moniz & Jordan, 2010; Tang, 2016). Since WNKs can positively regulate Wnt signaling in *Drosophila* and cultured human cells, it is conceivable that WNKs affect Wnt-related cancers such as the ones of the colon that are frequently caused by increased Wnt signaling (Clevers & Nusse, 2012). To this end, it was recently shown that β-catenin is an (indirect) transcriptional target of WNK1 and that the proliferation of certain cancer cell lines with high β-catenin activity is dependent on WNK1 (Dbouk et al., 2014; Rosenbluh et al., 2012). Furthermore, reduction of WNK1 also lowers levels of the epithelial–mesenchymal transition transcription factor Slug, thereby possibly having the potential to favor metastasis formation (Dbouk et al., 2014). In addition, WNK1 and WNK4 can phosphorylate Smad2, and silencing of WNK1 reduces Smad2 protein levels in HeLa cells, suggesting that WNKs have complex effects on TGFβ signaling (Lee, Chen, Stippec, & Cobb, 2007), which itself can promote cancer or act in a tumor suppressing manner (Derynck, Akhurst, & Balmain, 2001). To delineate the mechanism of tumor-promoting and -antagonizing effects of WNKs will be a demanding process, as it has also been shown that RNAi-mediated knockdown of *wnk* in cultured *Drosophila* cells reduces the level of mTORC1 activity (as measured by S6-kinase phosphorylation), and Wnks thus likely also can alter cellular metabolism and growth (Lindquist et al., 2011).

Recently, it was shown that overexpression of FoxF1 upregulates WNK1 and its target ERK5-MAPK in a mouse transgenic adenocarcinoma prostate cancer model (Fulford et al., 2016). Furthermore, FOXF1 expression correlated positively with that of MAP3K2 and WNK1 in human tumors. *Wnk1* is a direct target of FoxF1 and its knockdown in Myc-CaP cells, a prostate carcinoma cell line from Myc overexpressing mice (Watson et al., 2005), reduced primary tumor size, and metastasis formation in an orthotopic transplantation model. This effect was enhanced by simultaneous knockdown of MAP3K2 and is recapitulated by knockdown of their common downstream component ERK5, suggesting a role of WNK1 in the formation of prostate cancer (Fulford et al., 2016).

The situation for WNK2 is surprisingly different. In HeLa cells, WNK2 depletion reduces RhoA activation but increases Rac-GTP levels, thus causing stimulation of the p21-Cdc42-Rac1 activated kinase PAK1 and subsequent activation of MEK1 and ERK1/2 (Moniz, Matos, & Jordan, 2008), explaining WNK2's antiproliferative roles (Moniz et al., 2007). Promoter methylation studies showed that *WNK2* is often silenced in gliomas and meningiomas (tumors originating from glia or the membranous layers surrounding the CNS, respectively) (Hong et al., 2007; Jun et al., 2009; Moniz et al., 2013). Knockdown of *WNK2* in SW1088 cells, a "nonsilenced" glioblastoma cell line, promoted soft agar colony formation and increased cell migration in wound-healing (scratch) assays and invasion in Matrigel/membrane assays (Moniz et al., 2013). Conversely, reexpression of *WNK2* in "silenced" A172H glioblastoma cells inhibited their ability to form colonies in soft agar and tumor growth in a chick chorioallantoic membrane tumor model. Nevertheless, although *WNK2* promoter methylation correlated with protein expression in cell lines, a statistically significant association between methylation-dependent silencing of WNK2 and disease progression has not been found (Moniz et al., 2013). However, lower *WNK2* mRNA levels correlated with shorter survival time, suggesting that additional mechanisms other than promoter methylation must exist to lower *WNK2* levels in gliomas (Moniz et al., 2013). Taken together, these studies thus provide strong evidence for WNK2 acting as a tumor suppressor.

5. CONCLUSIONS

WNK kinases have multiple and expanding roles in numerous physiological and pathophysiological processes. Best characterized to date is the role of WNKs in the kidney in the regulation of ion transport and the maintenance of blood pressure and electrolyte homeostasis. This role of WNKs, as well as a role in osmoregulation, is evolutionary conserved in invertebrate organisms. Intriguingly, in recent years, WNKs also have been shown to function during development, including vascular and neuronal development, and in the Wnt signaling pathway. Additionally, studies support roles for WNKs in cancer and the immune system. Ongoing study of these fascinating kinases is sure to uncover more.

ACKNOWLEDGMENTS

The authors would like to thank Betsy Goldsmith, Xiaoshan Min, Chou-Long Huang, Chiou-Hwa (Cathy) Yuh, Pierre Drapeau, Atsushi Sato, and Jian Xie for sharing original figures, and Kevin Strange for helpful discussion. A.R.R. is supported by NIDDK (DK091316 and DK106350) and the AHA (16CSA28530002). A.J. is supported by AHA (13GRNT14680002) and NIGMS (R01GM115646).

REFERENCES

Aburto, N. J., Hanson, S., Gutierrez, H., Hooper, L., Elliott, P., & Cappuccio, F. P. (2013). Effect of increased potassium intake on cardiovascular risk factors and disease: Systematic review and meta-analyses. *BMJ*, *346*, f1378.

Alessi, D. R., Zhang, J., Khanna, A., Hochdorfer, T., Shang, Y., & Kahle, K. T. (2014). The WNK-SPAK/OSR1 pathway: Master regulator of cation-chloride cotransporters. *Science Signaling*, 7, re3.

An, S. W., Cha, S. K., Yoon, J., Chang, S., Ross, E. M., & Huang, C. L. (2011). WNK1 promotes PIP(2) synthesis to coordinate growth factor and GPCR-Gq signaling. *Current Biology*, *21*, 1979–1987.

Andrini, O., Keck, M., Briones, R., Lourdel, S., Vargas-Poussou, R., & Teulon, J. (2015). ClC-K chloride channels: Emerging pathophysiology of Bartter syndrome type 3. *American Journal of Physiology. Renal Physiology*, *308*, F1324–F1334.

Anselmo, A. N., Earnest, S., Chen, W., Juang, Y. C., Kim, S. C., Zhao, Y., et al. (2006). WNK1 and OSR1 regulate the Na+, K+, 2Cl− cotransporter in HeLa cells. *Proceedings of the National Academy of Sciences of the United States of America*, *103*, 10883–10888.

Arroyo, J. P., Kahle, K. T., & Gamba, G. (2013). The SLC12 family of electroneutral cation-coupled chloride cotransporters. *Molecular Aspects of Medicine*, *34*, 288–298.

Attrill, H., Falls, K., Goodman, J. L., Millburn, G. H., Antonazzo, G., Rey, A. J., et al. (2016). FlyBase: Establishing a gene group resource for *Drosophila melanogaster*. *Nucleic Acids Research*, *44*, D786–D792.

Auer-Grumbach, M., Mauko, B., Auer-Grumbach, P., & Pieber, T. R. (2006). Molecular genetics of hereditary sensory neuropathies. *Neuromolecular Medicine*, *8*, 147–158.

Bardutzky, J., & Schwab, S. (2007). Antiedema therapy in ischemic stroke. *Stroke*, *38*, 3084–3094.

Barker, M. (1932). Edema as influenced by a low ratio of sodium to potassium intake: Clinical observations. *Journal of American Medical Association*, *98*, 2193–2197.

Battilana, C. A., Dobyan, D. C., Lacy, F. B., Bhattacharya, J., Johnston, P. A., & Jamison, R. L. (1978). Effect of chronic potassium loading on potassium secretion by the pars recta or descending limb of the juxtamedullary nephron in the rat. *The Journal of Clinical Investigation*, *62*, 1093–1103.

Bazua-Valenti, S., Chavez-Canales, M., Rojas-Vega, L., Gonzalez-Rodriguez, X., Vazquez, N., Rodriguez-Gama, A., et al. (2015). The effect of WNK4 on the Na+-Cl− cotransporter is modulated by intracellular chloride. *Journal of the American Society of Nephrology*, *26*, 1781–1786.

Begum, G., Yuan, H., Kahle, K. T., Li, L., Wang, S., Shi, Y., et al. (2015). Inhibition of WNK3 kinase signaling reduces brain damage and accelerates neurological recovery after stroke. *Stroke, 46*, 1956–1965.
Bercier, V. (2013). WNK1/HSN2 isoform and the regulation of KCC2 activity. *Rare Diseases, 1*, e26537.
Bercier, V., Brustein, E., Liao, M., Dion, P. A., Lafreniere, R. G., Rouleau, G. A., et al. (2013). WNK1/HSN2 mutation in human peripheral neuropathy deregulates KCC2 expression and posterior lateral line development in zebrafish (*Danio rerio*). *PLoS Genetics, 9*, e1003124.
Bergaya, S., Faure, S., Baudrie, V., Rio, M., Escoubet, B., Bonnin, P., et al. (2011). WNK1 regulates vasoconstriction and blood pressure response to alpha 1-adrenergic stimulation in mice. *Hypertension, 58*, 439–445.
Berger, J., Senti, K. A., Senti, G., Newsome, T. P., Asling, B., Dickson, B. J., et al. (2008). Systematic identification of genes that regulate neuronal wiring in the Drosophila visual system. *PLoS Genetics, 4*, e1000085.
Biechele, T. L., Kulikauskas, R. M., Toroni, R. A., Lucero, O. M., Swift, R. D., James, R. G., et al. (2012). Wnt/beta-catenin signaling and AXIN1 regulate apoptosis triggered by inhibition of the mutant kinase BRAFV600E in human melanoma. *Science Signaling, 5*, ra3.
Bilic, J., Huang, Y. L., Davidson, G., Zimmermann, T., Cruciat, C. M., Bienz, M., et al. (2007). Wnt induces LRP6 signalosomes and promotes dishevelled-dependent LRP6 phosphorylation. *Science, 316*, 1619–1622.
Blumenthal, E. M. (2005). Modulation of tyramine signaling by osmolality in an insect secretory epithelium. *American Journal of Physiology. Cell Physiology, 289*, C1261–C1267.
Bonaccorsi, S., Mottier, V., Giansanti, M. G., Bolkan, B. J., Williams, B., Goldberg, M. L., et al. (2007). The Drosophila Lkb1 kinase is required for spindle formation and asymmetric neuroblast division. *Development, 134*, 2183–2193.
Boudeau, J., Baas, A. F., Deak, M., Morrice, N. A., Kieloch, A., Schutkowski, M., et al. (2003). MO25alpha/beta interact with STRADalpha/beta enhancing their ability to bind, activate and localize LKB1 in the cytoplasm. *The EMBO Journal, 22*, 5102–5114.
Boyden, L. M., Choi, M., Choate, K. A., Nelson-Williams, C. J., Farhi, A., Toka, H. R., et al. (2012). Mutations in kelch-like 3 and cullin 3 cause hypertension and electrolyte abnormalities. *Nature, 482*, 98–102.
Brandis, M., Keyes, J., & Windhager, E. E. (1972). Potassium-induced inhibition of proximal tubular fluid reabsorption in rats. *The American Journal of Physiology, 222*, 421–427.
Brunner, E., Peter, O., Schweizer, L., & Basler, K. (1997). pangolin encodes a Lef-1 homologue that acts downstream of Armadillo to transduce the Wingless signal in *Drosophila*. *Nature, 385*, 829–833.
Cabrero, P., Terhzaz, S., Romero, M. F., Davies, S. A., Blumenthal, E. M., & Dow, J. A. (2014). Chloride channels in stellate cells are essential for uniquely high secretion rates in neuropeptide-stimulated Drosophila diuresis. *Proceedings of the National Academy of Sciences of the United States of America, 111*, 14301–14306.
Cai, Q., & Brooks, H. L. (2011). Phosphorylation of eIF2alpha via the general control kinase, GCN2, modulates the ability of renal medullary cells to survive high urea stress. *American Journal of Physiology. Renal Physiology, 301*, F1202–F1207.
Carrisoza-Gaytan, R., Carattino, M. D., Kleyman, T. R., & Satlin, L. M. (2016). An unexpected journey: Conceptual evolution of mechanoregulated potassium transport in the distal nephron. *American Journal of Physiology. Cell Physiology, 310*, C243–C259.
Castaneda-Bueno, M., Cervantes-Perez, L. G., Rojas-Vega, L., Arroyo-Garza, I., Vazquez, N., Moreno, E., et al. (2014). Modulation of NCC activity by low and high K(+) intake: Insights into the signaling pathways involved. *American Journal of Physiology. Renal Physiology, 306*, F1507–F1519.

Castaneda-Bueno, M., Cervantes-Perez, L. G., Vazquez, N., Uribe, N., Kantesaria, S., Morla, L., et al. (2012). Activation of the renal Na+:Cl− cotransporter by angiotensin II is a WNK4-dependent process. *Proceedings of the National Academy of Sciences of the United States of America, 109*, 7929–7934.

Chen, W., Yazicioglu, M., & Cobb, M. H. (2004). Characterization of OSR1, a member of the mammalian Ste20p/germinal center kinase subfamily. *The Journal of Biological Chemistry, 279*, 11129–11136.

Cheng, C. J., Truong, T., Baum, M., & Huang, C. L. (2012). Kidney-specific WNK1 inhibits sodium reabsorption in the cortical thick ascending limb. *American Journal of Physiology. Renal Physiology, 303*, F667–F673.

Cheng, C. J., Yoon, J., Baum, M., & Huang, C. L. (2015). STE20/SPS1-related proline/alanine-rich kinase (SPAK) is critical for sodium reabsorption in isolated, perfused thick ascending limb. *American Journal of Physiology. Renal Physiology, 308*, F437–F443.

Choe, K. P., & Strange, K. (2007). Evolutionarily conserved WNK and Ste20 kinases are essential for acute volume recovery and survival after hypertonic shrinkage in Caenorhabditis elegans. *American Journal of Physiology. Cell Physiology, 293*, C915–C927.

Clevers, H. (2006). Wnt/beta-catenin signaling in development and disease. *Cell, 127*, 469–480.

Clevers, H., & Nusse, R. (2012). Wnt/beta-catenin signaling and disease. *Cell, 149*, 1192–1205.

Cogswell, M. E., Zhang, Z., Carriquiry, A. L., Gunn, J. P., Kuklina, E. V., Saydah, S. H., et al. (2012). Sodium and potassium intakes among US adults: NHANES 2003-2008. *The American Journal of Clinical Nutrition, 96*, 647–657.

Cruz-Rangel, S., Gamba, G., Ramos-Mandujano, G., & Pasantes-Morales, H. (2012). Influence of WNK3 on intracellular chloride concentration and volume regulation in HEK293 cells. *Pflügers Archiv, 464*, 317–330.

Curtiss, J., & Heilig, J. S. (1995). Establishment of Drosophila imaginal precursor cells is controlled by the Arrowhead gene. *Development, 121*, 3819–3828.

Curtiss, J., & Heilig, J. S. (1997). Arrowhead encodes a LIM homeodomain protein that distinguishes subsets of Drosophila imaginal cells. *Developmental Biology, 190*, 129–141.

DasGupta, R., Nybakken, K., Booker, M., Mathey-Prevot, B., Gonsalves, F., Changkakoty, B., et al. (2007). A case study of the reproducibility of transcriptional reporter cell-based RNAi screens in Drosophila. *Genome Biology, 8*, R203.

Dbouk, H. A., Weil, L. M., Perera, G. K., Dellinger, M. T., Pearson, G., Brekken, R. A., et al. (2014). Actions of the protein kinase WNK1 on endothelial cells are differentially mediated by its substrate kinases OSR1 and SPAK. *Proceedings of the National Academy of Sciences of the United States of America, 111*, 15999–16004.

de Los Heros, P., Alessi, D. R., Gourlay, R., Campbell, D. G., Deak, M., Macartney, T. J., et al. (2014). The WNK-regulated SPAK/OSR1 kinases directly phosphorylate and inhibit the K+-Cl− co-transporters. *The Biochemical Journal, 458*, 559–573.

Delpire, E., & Gagnon, K. B. (2008). SPAK and OSR1: STE20 kinases involved in the regulation of ion homoeostasis and volume control in mammalian cells. *The Biochemical Journal, 409*, 321–331.

Delpire, E., & Mount, D. B. (2002). Human and murine phenotypes associated with defects in cation-chloride cotransport. *Annual Review of Physiology, 64*, 803–843.

Denton, J., Nehrke, K., Yin, X., Morrison, R., & Strange, K. (2005). GCK-3, a newly identified Ste20 kinase, binds to and regulates the activity of a cell cycle-dependent ClC anion channel. *The Journal of General Physiology, 125*, 113–125.

Derynck, R., Akhurst, R. J., & Balmain, A. (2001). TGF-beta signaling in tumor suppression and cancer progression. *Nature Genetics, 29*, 117–129.

Donovan, M. D., Schellekens, H., Boylan, G. B., Cryan, J. F., & Griffin, B. T. (2016). In vitro bidirectional permeability studies identify pharmacokinetic limitations of NKCC1 inhibitor bumetanide. *European Journal of Pharmacology, 770*, 117–125.

Dow, J. A., Maddrell, S. H., Gortz, A., Skaer, N. J., Brogan, S., & Kaiser, K. (1994). The malpighian tubules of *Drosophila melanogaster*: A novel phenotype for studies of fluid secretion and its control. *The Journal of Experimental Biology, 197*, 421–428.

Dowd, B. F., & Forbush, B. (2003). PASK (proline-alanine-rich STE20-related kinase), a regulatory kinase of the Na-K-Cl cotransporter (NKCC1). *The Journal of Biological Chemistry, 278*, 27347–27353.

Falin, R. A., Miyazaki, H., & Strange, K. (2011). *C. elegans* STK39/SPAK ortholog-mediated inhibition of ClC anion channel activity is regulated by WNK-independent-ERK kinase signaling. *American Journal of Physiology. Cell Physiology, 300*, C624–C635.

Falin, R. A., Morrison, R., Ham, A. J., & Strange, K. (2009). Identification of regulatory phosphorylation sites in a cell volume- and Ste20 kinase-dependent ClC anion channel. *The Journal of General Physiology, 133*, 29–42.

Farkaš, R., Pecenova, L., Mentelova, L., Beno, M., Benova-Liszekova, D., Mahmoodova, S., et al. (2016). Massive excretion of calcium oxalate from late prepupal salivary glands of *Drosophila melanogaster* demonstrates active nephridial-like anion transport. *Development, Growth & Differentiation, 58*, 562–574.

Ferdaus, M. Z., Barber, K. W., Lopez-Cayuqueo, K. I., Terker, A. S., Argaiz, E. R., Gassaway, B. M., et al. (2016). SPAK and OSR1 play essential roles in potassium homeostasis through actions on the distal convoluted tubule. *The Journal of Physiology, 594*, 4945–4966.

Ferdaus, M. Z., & McCormick, J. A. (2016). The CUL3/KLHL3-WNK-SPAK/OSR1 pathway as a target for antihypertensive therapy. *American Journal of Physiology. Renal Physiology, 310*, F1389–F1396.

Filippi, B. M., de los Heros, P., Mehellou, Y., Navratilova, I., Gourlay, R., Deak, M., et al. (2011). MO25 is a master regulator of SPAK/OSR1 and MST3/MST4/YSK1 protein kinases. *The EMBO Journal, 30*, 1730–1741.

Friedel, P., Kahle, K. T., Zhang, J., Hertz, N., Pisella, L. I., Buhler, E., et al. (2015). WNK1-regulated inhibitory phosphorylation of the KCC2 cotransporter maintains the depolarizing action of GABA in immature neurons. *Science Signaling, 8*, ra65.

Frindt, G., Houde, V., & Palmer, L. G. (2011). Conservation of Na+ vs. K+ by the rat cortical collecting duct. *American Journal of Physiology. Renal Physiology, 301*, F14–F20.

Fulford, L., Milewski, D., Ustiyan, V., Ravishankar, N., Cai, Y., Le, T., et al. (2016). The transcription factor FOXF1 promotes prostate cancer by stimulating the mitogen-activated protein kinase ERK5. *Science Signaling, 9*, ra48.

Gagnon, K. B., & Delpire, E. (2010). On the substrate recognition and negative regulation of SPAK, a kinase modulating Na+-K+-2Cl− cotransport activity. *American Journal of Physiology. Cell Physiology, 299*, C614–C620.

Gagnon, K. B., England, R., & Delpire, E. (2006). Characterization of SPAK and OSR1, regulatory kinases of the Na-K-2Cl cotransporter. *Molecular and Cellular Biology, 26*, 689–698.

Gagnon, K. B., Rios, K., & Delpire, E. (2011). Functional insights into the activation mechanism of Ste20-related kinases. *Cellular Physiology and Biochemistry, 28*, 1219–1230.

Gamba, G. (2005). Molecular physiology and pathophysiology of electroneutral cation-chloride cotransporters. *Physiological Reviews, 85*, 423–493.

Gonczy, P. (2008). Mechanisms of asymmetric cell division: Flies and worms pave the way. *Nature Reviews. Molecular Cell Biology, 9*, 355–366.

Gonzaga-Jauregui, C., Harel, T., Gambin, T., Kousi, M., Griffin, L. B., Francescatto, L., et al. (2015). Exome sequence analysis suggests that genetic burden contributes to phenotypic variability and complex neuropathy. *Cell Reports, 12*, 1169–1183.

Grimm, P. R., Taneja, T. K., Liu, J., Coleman, R., Chen, Y. Y., Delpire, E., et al. (2012). SPAK isoforms and OSR1 regulate sodium-chloride co-transporters in a nephron-specific manner. *The Journal of Biological Chemistry, 287*, 37673–37690.

Haas, B. R., Cuddapah, V. A., Watkins, S., Rohn, K. J., Dy, T. E., & Sontheimer, H. (2011). With-No-Lysine Kinase 3 (WNK3) stimulates glioma invasion by regulating cell volume. *American Journal of Physiology. Cell Physiology, 301,* C1150–C1160.

Hadchouel, J., Ellison, D. H., & Gamba, G. (2016). Regulation of renal electrolyte transport by WNK and SPAK-OSR1 kinases. *Annual Review of Physiology, 78,* 367–389.

Hekmat-Scafe, D. S., Lundy, M. Y., Ranga, R., & Tanouye, M. A. (2006). Mutations in the K+/Cl− cotransporter gene kazachoc (kcc) increase seizure susceptibility in Drosophila. *The Journal of Neuroscience, 26,* 8943–8954.

Hekmat-Scafe, D. S., Mercado, A., Fajilan, A. A., Lee, A. W., Hsu, R., Mount, D. B., et al. (2010). Seizure sensitivity is ameliorated by targeted expression of K+-Cl− cotransporter function in the mushroom body of the Drosophila brain. *Genetics, 184,* 171–183.

Higashihara, E., & Kokko, J. P. (1985). Effects of aldosterone on potassium recycling in the kidney of adrenalectomized rats. *The American Journal of Physiology, 248,* F219–F227.

Hirata, T., Cabrero, P., Berkholz, D. S., Bondeson, D. P., Ritman, E. L., Thompson, J. R., et al. (2012). In vivo Drosophila genetic model for calcium oxalate nephrolithiasis. *American Journal of Physiology. Renal Physiology, 303,* F1555–F1562.

Hirata, T., Czapar, A., Brin, L., Haritonova, A., Bondeson, D. P., Linser, P., et al. (2012). Ion and solute transport by Prestin in Drosophila and Anopheles. *Journal of Insect Physiology, 58,* 563–569.

Hisamoto, N., Moriguchi, T., Urushiyama, S., Mitani, S., Shibuya, H., & Matsumoto, K. (2008). Caenorhabditis elegans WNK-STE20 pathway regulates tube formation by modulating ClC channel activity. *EMBO Reports, 9,* 70–75.

Hoffmann, E. K., Lambert, I. H., & Pedersen, S. F. (2009). Physiology of cell volume regulation in vertebrates. *Physiological Reviews, 89,* 193–277.

Hong, C., Moorefield, K. S., Jun, P., Aldape, K. D., Kharbanda, S., Phillips, H. S., et al. (2007). Epigenome scans and cancer genome sequencing converge on WNK2, a kinase-independent suppressor of cell growth. *Proceedings of the National Academy of Sciences of the United States of America, 104,* 10974–10979.

Howe, D. G., Bradford, Y. M., Conlin, T., Eagle, A. E., Fashena, D., Frazer, K., et al. (2013). ZFIN, the Zebrafish Model Organism Database: Increased support for mutants and transgenics. *Nucleic Acids Research, 41,* D854–D860.

Huang, C. L., & Cheng, C. J. (2015). A unifying mechanism for WNK kinase regulation of sodium-chloride cotransporter. *Pflügers Archiv, 467,* 2235–2241.

Inoue, K., Furukawa, T., Kumada, T., Yamada, J., Wang, T., Inoue, R., et al. (2012). Taurine inhibits K+-Cl− cotransporter KCC2 to regulate embryonic Cl− homeostasis via with-no-lysine (WNK) protein kinase signaling pathway. *The Journal of Biological Chemistry, 287,* 20839–20850.

Jafar-Nejad, H., Tien, A. C., Acar, M., & Bellen, H. J. (2006). Senseless and Daughterless confer neuronal identity to epithelial cells in the Drosophila wing margin. *Development, 133,* 1683–1692.

Johnson, M., Zaretskaya, I., Raytselis, Y., Merezhuk, Y., McGinnis, S., & Madden, T. L. (2008). NCBI BLAST: A better web interface. *Nucleic Acids Research, 36,* W5–W9.

Jun, P., Hong, C., Lal, A., Wong, J. M., McDermott, M. W., Bollen, A. W., et al. (2009). Epigenetic silencing of the kinase tumor suppressor WNK2 is tumor-type and tumor-grade specific. *Neuro-Oncology, 11,* 414–422.

Kahle, K. T., Rinehart, J., Ring, A., Gimenez, I., Gamba, G., Hebert, S. C., et al. (2006). WNK protein kinases modulate cellular Cl− flux by altering the phosphorylation state of the Na-K-Cl and K-Cl cotransporters. *Physiology (Bethesda), 21,* 326–335.

Kahle, K. T., Schmouth, J. F., Lavastre, V., Latremoliere, A., Zhang, J., Andrews, N., et al. (2016). Inhibition of the kinase WNK1/HSN2 ameliorates neuropathic pain by restoring GABA inhibition. *Science Signaling, 9,* ra32.

Kaila, K., Price, T. J., Payne, J. A., Puskarjov, M., & Voipio, J. (2014). Cation-chloride cotransporters in neuronal development, plasticity and disease. *Nature Reviews. Neuroscience, 15*, 637–654.

Keith, N. M., & Binger, M. W. (1935). Diuretic action of potassium. *Journal of American Medical Association, 105*, 1584–1591.

Kim, S. E., Huang, H., Zhao, M., Zhang, X., Zhang, A., Semonov, M. V., et al. (2013). Wnt stabilization of beta-catenin reveals principles for morphogen receptor-scaffold assemblies. *Science, 340*, 867–870.

Kochl, R., Thelen, F., Vanes, L., Brazao, T. F., Fountain, K., Xie, J., et al. (2016). WNK1 kinase balances T cell adhesion versus migration in vivo. *Nature Immunology, 17*(9), 1075–1083.

Kok, F. O., Shin, M., Ni, C. W., Gupta, A., Grosse, A. S., van Impel, A., et al. (2015). Reverse genetic screening reveals poor correlation between morpholino-induced and mutant phenotypes in zebrafish. *Developmental Cell, 32*, 97–108.

Kolotuev, I., Hyenne, V., Schwab, Y., Rodriguez, D., & Labouesse, M. (2013). A pathway for unicellular tube extension depending on the lymphatic vessel determinant Prox1 and on osmoregulation. *Nature Cell Biology, 15*, 157–168.

Korinek, V., Barker, N., Morin, P. J., van Wichen, D., de Weger, R., Kinzler, K. W., et al. (1997). Constitutive transcriptional activation by a beta-catenin-Tcf complex in APC $-/-$ colon carcinoma. *Science, 275*, 1784–1787.

Kramps, T., Peter, O., Brunner, E., Nellen, D., Froesch, B., Chatterjee, S., et al. (2002). Wnt/wingless signaling requires BCL9/legless-mediated recruitment of pygopus to the nuclear beta-catenin-TCF complex. *Cell, 109*, 47–60.

Krishna, G. G., Miller, E., & Kapoor, S. (1989). Increased blood pressure during potassium depletion in normotensive men. *The New England Journal of Medicine, 320*, 1177–1182.

Kupinski, A. P., Muller-Reichert, T., & Eckmann, C. R. (2010). The *Caenorhabditis elegans* Ste20 kinase, GCK-3, is essential for postembryonic developmental timing and regulates meiotic chromosome segregation. *Developmental Biology, 344*, 758–771.

Lafreniere, R. G., MacDonald, M. L., Dube, M. P., MacFarlane, J., O'Driscoll, M., Brais, B., et al. (2004). Identification of a novel gene (HSN2) causing hereditary sensory and autonomic neuropathy type II through the Study of Canadian Genetic Isolates. *The American Journal of Human Genetics, 74*, 1064–1073.

Lai, J. G., Tsai, S. M., Tu, H. C., Chen, W. C., Kou, F. J., Lu, J. W., et al. (2014). Zebrafish WNK lysine deficient protein kinase 1 (wnk1) affects angiogenesis associated with VEGF signaling. *PloS One, 9*, e106129.

Landin Malt, A., Cesario, J. M., Tang, Z., Brown, S., & Jeong, J. (2014). Identification of a face enhancer reveals direct regulation of LIM homeobox 8 (Lhx8) by wingless-int (WNT)/beta-catenin signaling. *The Journal of Biological Chemistry, 289*, 30289–30301.

Landry, G. M., Hirata, T., Anderson, J. B., Cabrero, P., Gallo, C. J., Dow, J. A., et al. (2016). Sulfate and thiosulfate inhibit oxalate transport via a dPrestin (Slc26a6)-dependent mechanism in an insect model of calcium oxalate nephrolithiasis. *American Journal of Physiology. Renal Physiology, 310*, F152–F159.

Larsen, E. H., Deaton, L. E., Onken, H., O'Donnell, M., Grosell, M., Dantzler, W. H., et al. (2014). Osmoregulation and excretion. *Comprehensive Physiology, 4*, 405–573.

Lee, B. H., Chen, W., Stippec, S., & Cobb, M. H. (2007). Biological cross-talk between WNK1 and the transforming growth factor beta-Smad signaling pathway. *The Journal of Biological Chemistry, 282*, 17985–17996.

Lee, S. J., Cobb, M. H., & Goldsmith, E. J. (2009). Crystal structure of domain-swapped STE20 OSR1 kinase domain. *Protein Sciences, 18*, 304–313.

Lee, J. S., Ishimoto, A., & Yanagawa, S. (1999). Characterization of mouse dishevelled (Dvl) proteins in Wnt/Wingless signaling pathway. *The Journal of Biological Chemistry, 274*, 21464–21470.

Lee, E. C., & Strange, K. (2012). GCN-2 dependent inhibition of protein synthesis activates osmosensitive gene transcription via WNK and Ste20 kinase signaling. *American Journal of Physiology. Cell Physiology*, *303*, C1269–C1277.

Leiserson, W. M., Forbush, B., & Keshishian, H. (2011). Drosophila glia use a conserved cotransporter mechanism to regulate extracellular volume. *Glia*, *59*, 320–332.

Leiserson, W. M., Harkins, E. W., & Keshishian, H. (2000). Fray, a Drosophila serine/threonine kinase homologous to mammalian PASK, is required for axonal ensheathment. *Neuron*, *28*, 793–806.

Lenertz, L. Y., Lee, B. H., Min, X., Xu, B. E., Wedin, K., Earnest, S., et al. (2005). Properties of WNK1 and implications for other family members. *The Journal of Biological Chemistry*, *280*, 26653–26658.

Li, V. S., Ng, S. S., Boersema, P. J., Low, T. Y., Karthaus, W. R., Gerlach, J. P., et al. (2012). Wnt signaling through inhibition of beta-catenin degradation in an intact Axin1 complex. *Cell*, *149*, 1245–1256.

Lindquist, R. A., Ottina, K. A., Wheeler, D. B., Hsu, P. P., Thoreen, C. C., Guertin, D. A., et al. (2011). Genome-scale RNAi on living-cell microarrays identifies novel regulators of Drosophila melanogaster TORC1-S6K pathway signaling. *Genome Research*, *21*, 433–446.

Linton, S. M., & O'Donnell, M. J. (1999). Contributions of K+:Cl− cotransport and Na+/K+-ATPase to basolateral ion transport in malpighian tubules of Drosophila melanogaster. *The Journal of Experimental Biology*, *202*, 1561–1570.

Louis-Dit-Picard, H., Barc, J., Trujillano, D., Miserey-Lenkei, S., Bouatia-Naji, N., Pylypenko, O., et al. (2012). KLHL3 mutations cause familial hyperkalemic hypertension by impairing ion transport in the distal nephron. *Nature Genetics*, *44*, 456–460. S1–S3.

Lykke, K., Tollner, K., Feit, P. W., Erker, T., MacAulay, N., & Loscher, W. (2016). The search for NKCC1-selective drugs for the treatment of epilepsy: Structure-function relationship of bumetanide and various bumetanide derivatives in inhibiting the human cation-chloride cotransporter NKCC1A. *Epilepsy & Behavior*, *59*, 42–49.

Madhavan, M. M., & Madhavan, K. (1980). Morphogenesis of the epidermis of adult abdomen of Drosophila. *Journal of Embryology and Experimental Morphology*, *60*, 1–31.

Mandai, S., Mori, T., Sohara, E., Rai, T., & Uchida, S. (2015). Generation of hypertension-associated STK39 polymorphism knockin cell lines with the clustered regularly interspaced short palindromic repeats/Cas9 system. *Hypertension*, *66*, 1199–1206.

Maruyama, J., Kobayashi, Y., Umeda, T., Vandewalle, A., Takeda, K., Ichijo, H., et al. (2016). Osmotic stress induces the phosphorylation of WNK4 Ser575 via the p38MAPK-MK pathway. *Scientific Reports*, *6*, 18710.

Matsubayashi, H., Sese, S., Lee, J. S., Shirakawa, T., Iwatsubo, T., Tomita, T., et al. (2004). Biochemical characterization of the Drosophila wingless signaling pathway based on RNA interference. *Molecular and Cellular Biology*, *24*, 2012–2024.

McCormick, J. A., & Ellison, D. H. (2011). The WNKs: Atypical protein kinases with pleiotropic actions. *Physiological Reviews*, *91*, 177–219.

Mederle, K., Mutig, K., Paliege, A., Carota, I., Bachmann, S., Castrop, H., et al. (2013). Loss of WNK3 is compensated for by the WNK1/SPAK axis in the kidney of the mouse. *American Journal of Physiology. Renal Physiology*, *304*, F1198–F1209.

Melo, Z., de los Heros, P., Cruz-Rangel, S., Vazquez, N., Bobadilla, N. A., Pasantes-Morales, H., et al. (2013). N-terminal serine dephosphorylation is required for KCC3 cotransporter full activation by cell swelling. *The Journal of Biological Chemistry*, *288*, 31468–31476.

Mendoza, M., Redemann, S., & Brunner, D. (2005). The fission yeast MO25 protein functions in polar growth and cell separation. *European Journal of Cell Biology*, *84*, 915–926.

Mente, A., O'Donnell, M. J., Rangarajan, S., McQueen, M. J., Poirier, P., Wielgosz, A., et al. (2014). Association of urinary sodium and potassium excretion with blood pressure. *The New England Journal of Medicine, 371*, 601–611.
Mercado, A., de Los Heros, P., Melo, Z., Chavez-Canales, M., Murillo-de-Ozores, A. R., Moreno, E., et al. (2016). With no lysine L-WNK1 isoforms are negative regulators of the K+:Cl− cotransporters. *American Journal of Physiology. Cell Physiology, 311*, C54–C66. http://dx.doi.org/10.1152/ajpcell.00193.2015.
Miyazaki, H., & Strange, K. (2012). Differential regulation of a CLC anion channel by SPAK kinase ortholog-mediated multisite phosphorylation. *American Journal of Physiology. Cell Physiology, 302*, C1702–C1712.
Miyazaki, H., Yamada, T., Parton, A., Morrison, R., Kim, S., Beth, A. H., et al. (2012). CLC anion channel regulatory phosphorylation and conserved signal transduction domains. *Biophysical Journal, 103*, 1706–1718.
Moniz, S., & Jordan, P. (2010). Emerging roles for WNK kinases in cancer. *Cellular and Molecular Life Sciences, 67*, 1265–1276.
Moniz, S., Martinho, O., Pinto, F., Sousa, B., Loureiro, C., Oliveira, M. J., et al. (2013). Loss of WNK2 expression by promoter gene methylation occurs in adult gliomas and triggers Rac1-mediated tumour cell invasiveness. *Human Molecular Genetics, 22*, 84–95.
Moniz, S., Matos, P., & Jordan, P. (2008). WNK2 modulates MEK1 activity through the Rho GTPase pathway. *Cellular Signalling, 20*, 1762–1768.
Moniz, S., Verissimo, F., Matos, P., Brazao, R., Silva, E., Kotelevets, L., et al. (2007). Protein kinase WNK2 inhibits cell proliferation by negatively modulating the activation of MEK1/ERK1/2. *Oncogene, 26*, 6071–6081.
Moriguchi, T., Urushiyama, S., Hisamoto, N., Iemura, S., Uchida, S., Natsume, T., et al. (2005). WNK1 regulates phosphorylation of cation-chloride-coupled cotransporters via the STE20-related kinases, SPAK and OSR1. *The Journal of Biological Chemistry, 280*, 42685–42693.
Naguro, I., Umeda, T., Kobayashi, Y., Maruyama, J., Hattori, K., Shimizu, Y., et al. (2012). ASK3 responds to osmotic stress and regulates blood pressure by suppressing WNK1-SPAK/OSR1 signaling in the kidney. *Nature Communications, 3*, 1285.
Naito, S., Ohta, A., Sohara, E., Ohta, E., Rai, T., Sasaki, S., et al. (2011). Regulation of WNK1 kinase by extracellular potassium. *Clinical and Experimental Nephrology, 15*, 195–202.
Nozaki, M., Onishi, Y., Togashi, S., & Miyamoto, H. (1996). Molecular characterization of the Drosophila Mo25 gene, which is conserved among Drosophila, mouse, and yeast. *DNA and Cell Biology, 15*, 505–509.
O'Donnell, M. J., Dow, J. A., Huesmann, G. R., Tublitz, N. J., & Maddrell, S. H. (1996). Separate control of anion and cation transport in malpighian tubules of Drosophila melanogaster. *The Journal of Experimental Biology, 199*, 1163–1175.
Oi, K., Sohara, E., Rai, T., Misawa, M., Chiga, M., Alessi, D. R., et al. (2012). A minor role of WNK3 in regulating phosphorylation of renal NKCC2 and NCC co-transporters in vivo. *Biology Open, 1*, 120–127.
Pacheco-Alvarez, D., Vazquez, N., Castaneda-Bueno, M., de-Los-Heros, P., Cortes-Gonzalez, C., Moreno, E., et al. (2012). WNK3-SPAK interaction is required for the modulation of NCC and other members of the SLC12 family. *Cellular Physiology and Biochemistry, 29*, 291–302.
Park, J., Al-Ramahi, I., Tan, Q., Mollema, N., Diaz-Garcia, J. R., Gallego-Flores, T., et al. (2013). RAS-MAPK-MSK1 pathway modulates ataxin 1 protein levels and toxicity in SCA1. *Nature, 498*, 325–331.
Piala, A. T., Moon, T. M., Akella, R., He, H., Cobb, M. H., & Goldsmith, E. J. (2014). Chloride sensing by WNK1 involves inhibition of autophosphorylation. *Science Signaling, 7*, ra41.

Ponce-Coria, J., Gagnon, K. B., & Delpire, E. (2012). Calcium-binding protein 39 facilitates molecular interaction between Ste20p proline alanine-rich kinase and oxidative stress response 1 monomers. *American Journal of Physiology. Cell Physiology, 303,* C1198–C1205.

Ponce-Coria, J., Markadieu, N., Austin, T. M., Flammang, L., Rios, K., Welling, P. A., et al. (2014). A novel Ste20-related proline/alanine-rich kinase (SPAK)-independent pathway involving calcium-binding protein 39 (Cab39) and serine threonine kinase with no lysine member 4 (WNK4) in the activation of Na-K-Cl cotransporters. *The Journal of Biological Chemistry, 289,* 17680–17688.

Ponce-Coria, J., San-Cristobal, P., Kahle, K. T., Vazquez, N., Pacheco-Alvarez, D., de Los Heros, P., et al. (2008). Regulation of NKCC2 by a chloride-sensing mechanism involving the WNK3 and SPAK kinases. *Proceedings of the National Academy of Sciences of the United States of America, 105,* 8458–8463.

Pressler, R. M., Boylan, G. B., Marlow, N., Blennow, M., Chiron, C., Cross, J. H., et al. (2015). Bumetanide for the treatment of seizures in newborn babies with hypoxic ischaemic encephalopathy (NEMO): An open-label, dose finding, and feasibility phase 1/2 trial. *The Lancet. Neurology, 14,* 469–477.

Rafiqi, F. H., Zuber, A. M., Glover, M., Richardson, C., Fleming, S., Jovanovic, S., et al. (2010). Role of the WNK-activated SPAK kinase in regulating blood pressure. *EMBO Molecular Medicine, 2,* 63–75.

Rengarajan, S., Lee, D. H., Oh, Y. T., Delpire, E., Youn, J. H., & McDonough, A. A. (2014). Increasing plasma [K+] by intravenous potassium infusion reduces NCC phosphorylation and drives kaliuresis and natriuresis. *American Journal of Physiology. Renal Physiology, 306,* F1059–F1068.

Reynolds, A., Brustein, E., Liao, M., Mercado, A., Babilonia, E., Mount, D. B., et al. (2008). Neurogenic role of the depolarizing chloride gradient revealed by global overexpression of KCC2 from the onset of development. *The Journal of Neuroscience, 28,* 1588–1597.

Rheault, M. R., & O'Donnell, M. J. (2001). Analysis of epithelial K(+) transport in Malpighian tubules of *Drosophila melanogaster*: Evidence for spatial and temporal heterogeneity. *The Journal of Experimental Biology, 204,* 2289–2299.

Richardson, C., Rafiqi, F. H., Karlsson, H. K., Moleleki, N., Vandewalle, A., Campbell, D. G., et al. (2008). Activation of the thiazide-sensitive Na+-Cl− cotransporter by the WNK-regulated kinases SPAK and OSR1. *Journal of Cell Science, 121,* 675–684.

Richardson, C., Sakamoto, K., de los Heros, P., Deak, M., Campbell, D. G., Prescott, A. R., et al. (2011). Regulation of the NKCC2 ion cotransporter by SPAK-OSR1-dependent and -independent pathways. *Journal of Cell Science, 124,* 789–800.

Rinehart, J., Maksimova, Y. D., Tanis, J. E., Stone, K. L., Hodson, C. A., Zhang, J., et al. (2009). Sites of regulated phosphorylation that control K-Cl cotransporter activity. *Cell, 138,* 525–536.

Rinehart, J., Vazquez, N., Kahle, K. T., Hodson, C. A., Ring, A. M., Gulcicek, E. E., et al. (2011). WNK2 is a novel regulator of essential neuronal cation-chloride cotransporters. *The Journal of Biological Chemistry, 286,* 30171–30180.

Rodan, A. R., Baum, M., & Huang, C. L. (2012). The Drosophila NKCC Ncc69 is required for normal renal tubule function. *American Journal of Physiology. Cell Physiology, 303,* C883–C894.

Romero, M. F., Henry, D., Nelson, S., Harte, P. J., Dillon, A. K., & Sciortino, C. M. (2000). Cloning and characterization of a Na+-driven anion exchanger (NDAE1). A new bicarbonate transporter. *The Journal of Biological Chemistry, 275,* 24552–24559.

Rosenbluh, J., Nijhawan, D., Cox, A. G., Li, X., Neal, J. T., Schafer, E. J., et al. (2012). beta-Catenin-driven cancers require a YAP1 transcriptional complex for survival and tumorigenesis. *Cell, 151,* 1457–1473.

Roy, A., Goodman, J. H., Begum, G., Donnelly, B. F., Pittman, G., Weinman, E. J., et al. (2015). Generation of WNK1 knockout cell lines by CRISPR/Cas-mediated genome editing. *American Journal of Physiology. Renal Physiology, 308*, F366–F376.

Sato, A., & Shibuya, H. (2013). WNK signaling is involved in neural development via Lhx8/Awh expression. *PloS One, 8*, e55301.

Sciortino, C. M., Shrode, L. D., Fletcher, B. R., Harte, P. J., & Romero, M. F. (2001). Localization of endogenous and recombinant Na(+)-driven anion exchanger protein NDAE1 from *Drosophila melanogaster*. *American Journal of Physiology. Cell Physiology, 281*, C449–C463.

Sengupta, S., Tu, S. W., Wedin, K., Earnest, S., Stippec, S., Luby-Phelps, K., et al. (2012). Interactions with WNK (with no lysine) family members regulate oxidative stress response 1 and ion co-transporter activity. *The Journal of Biological Chemistry, 287*, 37868–37879.

Serysheva, E., Berhane, H., Grumolato, L., Demir, K., Balmer, S., Bodak, M., et al. (2013). Wnk kinases are positive regulators of canonical Wnt/beta-catenin signalling. *EMBO Reports, 14*, 718–725.

Serysheva, E., Mlodzik, M., & Jenny, A. (2014). WNKs in Wnt/beta-catenin signaling. *Cell Cycle, 13*, 173–174.

Shaharabany, M., Holtzman, E. J., Mayan, H., Hirschberg, K., Seger, R., & Farfel, Z. (2008). Distinct pathways for the involvement of WNK4 in the signaling of hypertonicity and EGF. *The FEBS Journal, 275*, 1631–1642.

Shekarabi, M., Girard, N., Riviere, J. B., Dion, P., Houle, M., Toulouse, A., et al. (2008). Mutations in the nervous system-specific HSN2 exon of WNK1 cause hereditary sensory neuropathy type II. *The Journal of Clinical Investigation, 118*, 2496–2505.

Shi, Z., Jiao, S., Zhang, Z., Ma, M., Zhang, Z., Chen, C., et al. (2013). Structure of the MST4 in complex with MO25 provides insights into its activation mechanism. *Structure, 21*, 449–461.

Sorensen, M. V., Grossmann, S., Roesinger, M., Gresko, N., Todkar, A. P., Barmettler, G., et al. (2013). Rapid dephosphorylation of the renal sodium chloride cotransporter in response to oral potassium intake in mice. *Kidney International, 83*, 811–824.

Stokes, J. B. (1982). Consequences of potassium recycling in the renal medulla. Effects of ion transport by the medullary thick ascending limb of Henle's loop. *The Journal of Clinical Investigation, 70*, 219–229.

Stokum, J. A., Gerzanich, V., & Simard, J. M. (2016). Molecular pathophysiology of cerebral edema. *Journal of Cerebral Blood Flow and Metabolism, 36*, 513–538.

Sumanas, S., & Lin, S. (2006). Ets1-related protein is a key regulator of vasculogenesis in zebrafish. *PLoS Biology, 4*, e10.

Sun, Q., Tian, E., Turner, R. J., & Ten Hagen, K. G. (2010). Developmental and functional studies of the SLC12 gene family members from *Drosophila melanogaster*. *American Journal of Physiology. Cell Physiology, 298*, C26–C37.

Susa, K., Kita, S., Iwamoto, T., Yang, S. S., Lin, S. H., Ohta, A., et al. (2012). Effect of heterozygous deletion of WNK1 on the WNK-OSR1/SPAK-NCC/NKCC1/NKCC2 signal cascade in the kidney and blood vessels. *Clinical and Experimental Nephrology, 16*, 530–538.

Swarup, S., Pradhan-Sundd, T., & Verheyen, E. M. (2015). Genome-wide identification of phospho-regulators of Wnt signaling in Drosophila. *Development, 142*, 1502–1515.

Swarup, S., & Verheyen, E. M. (2012). Wnt/Wingless signaling in Drosophila. *Cold Spring Harbor Perspectives in Biology. 4*, pii:a007930, http://dx.doi.org/10.1101/cshperspect.a007930.

Takahashi, D., Mori, T., Nomura, N., Khan, M. Z., Araki, Y., Zeniya, M., et al. (2014). WNK4 is the major WNK positively regulating NCC in the mouse kidney. *Bioscience Reports. 34*, pii:e00107, http://dx.doi.org/10.1042/BSR20140047.

Tang, B. L. (2016). (WNK)ing at death: With-no-lysine (Wnk) kinases in neuropathies and neuronal survival. *Brain Research Bulletin, 125,* 92–98.

Taylor, C. A., 4th, Juang, Y. C., Earnest, S., Sengupta, S., Goldsmith, E. J., & Cobb, M. H. (2015). Domain-swapping switch point in Ste20 protein kinase SPAK. *Biochemistry, 54,* 5063–5071.

Terker, A. S., Zhang, C., Erspamer, K. J., Gamba, G., Yang, C. L., & Ellison, D. H. (2016). Unique chloride-sensing properties of WNK4 permit the distal nephron to modulate potassium homeostasis. *Kidney International, 89,* 127–134.

Terker, A. S., Zhang, C., McCormick, J. A., Lazelle, R. A., Zhang, C., Meermeier, N. P., et al. (2015). Potassium modulates electrolyte balance and blood pressure through effects on distal cell voltage and chloride. *Cell Metabolism, 21,* 39–50.

Vallon, V., Schroth, J., Lang, F., Kuhl, D., & Uchida, S. (2009). Expression and phosphorylation of the Na+-Cl− cotransporter NCC in vivo is regulated by dietary salt, potassium, and SGK1. *American Journal of Physiology. Renal Physiology, 297,* F704–F712.

van der Lubbe, N., Moes, A. D., Rosenbaek, L. L., Schoep, S., Meima, M. E., Danser, A. H., et al. (2013). K+-induced natriuresis is preserved during Na+ depletion and accompanied by inhibition of the Na+-Cl− cotransporter. *American Journal of Physiology. Renal Physiology, 305,* F1177–F1188.

Velazquez, H., & Silva, T. (2003). Cloning and localization of KCC4 in rabbit kidney: Expression in distal convoluted tubule. *American Journal of Physiology. Renal Physiology, 285,* F49–F58.

Verissimo, F., & Jordan, P. (2001). WNK kinases, a novel protein kinase subfamily in multicellular organisms. *Oncogene, 20,* 5562–5569.

Vitari, A. C., Deak, M., Collins, B. J., Morrice, N., Prescott, A. R., Phelan, A., et al. (2004). WNK1, the kinase mutated in an inherited high-blood-pressure syndrome, is a novel PKB (protein kinase B)/Akt substrate. *The Biochemical Journal, 378,* 257–268.

Vitari, A. C., Deak, M., Morrice, N. A., & Alessi, D. R. (2005). The WNK1 and WNK4 protein kinases that are mutated in Gordon's hypertension syndrome phosphorylate and activate SPAK and OSR1 protein kinases. *The Biochemical Journal, 391,* 17–24.

Wade, J. B., Fang, L., Coleman, R. A., Liu, J., Grimm, P. R., Wang, T., et al. (2011). Differential regulation of ROMK (Kir1.1) in distal nephron segments by dietary potassium. *American Journal of Physiology. Renal Physiology, 300,* F1385–F1393.

Wade, J. B., Liu, J., Coleman, R., Grimm, P. R., Delpire, E., & Welling, P. A. (2015). SPAK-mediated NCC regulation in response to low-K+ diet. *American Journal of Physiology. Renal Physiology, 308,* F923–F931.

Warsi, J., Hosseinzadeh, Z., Elvira, B., Bissinger, R., Shumilina, E., & Lang, F. (2014). Regulation of ClC-2 activity by SPAK and OSR1. *Kidney & Blood Pressure Research, 39,* 378–387.

Watson, P. A., Ellwood-Yen, K., King, J. C., Wongvipat, J., Lebeau, M. M., & Sawyers, C. L. (2005). Context-dependent hormone-refractory progression revealed through characterization of a novel murine prostate cancer cell line. *Cancer Research, 65,* 11565–11571.

Welling, P. A. (2013). Regulation of potassium channel trafficking in the distal nephron. *Current Opinion in Nephrology and Hypertension, 22,* 559–565.

Wilson, F. H., Disse-Nicodeme, S., Choate, K. A., Ishikawa, K., Nelson-Williams, C., Desitter, I., et al. (2001). Human hypertension caused by mutations in WNK kinases. *Science, 293,* 1107–1112.

Womersley, R. A., & Darragh, J. H. (1955). Potassium and sodium restriction in the normal human. *The Journal of Clinical Investigation, 34,* 456–461.

Wu, Y., Baum, M., Huang, C. L., & Rodan, A. R. (2015). Two inwardly rectifying potassium channels, Irk1 and Irk2, play redundant roles in Drosophila renal tubule function. *American Journal of Physiology. Regulatory, Integrative and Comparative Physiology, 309,* R747–R756.

Wu, Y., Schellinger, J. N., Huang, C. L., & Rodan, A. R. (2014). Hypotonicity stimulates potassium flux through the WNK-SPAK/OSR1 kinase cascade and the Ncc69 sodium-potassium-2-chloride cotransporter in the Drosophila renal tubule. *The Journal of Biological Chemistry, 289*, 26131–26142.

Xi, B., Chen, M., Chandak, G. R., Shen, Y., Yan, L., He, J., et al. (2013). STK39 polymorphism is associated with essential hypertension: A systematic review and meta-analysis. *PloS One, 8*, e59584.

Xie, J., Wu, T., Xu, K., Huang, I. K., Cleaver, O., & Huang, C. L. (2009). Endothelial-specific expression of WNK1 kinase is essential for angiogenesis and heart development in mice. *The American Journal of Pathology, 175*, 1315–1327.

Xie, J., Yoon, J., Yang, S. S., Lin, S. H., & Huang, C. L. (2013). WNK1 protein kinase regulates embryonic cardiovascular development through the OSR1 signaling cascade. *The Journal of Biological Chemistry, 288*, 8566–8574.

Xu, B., English, J. M., Wilsbacher, J. L., Stippec, S., Goldsmith, E. J., & Cobb, M. H. (2000). WNK1, a novel mammalian serine/threonine protein kinase lacking the catalytic lysine in subdomain II. *The Journal of Biological Chemistry, 275*, 16795–16801.

Xu, B. E., Stippec, S., Chu, P. Y., Lazrak, A., Li, X. J., Lee, B. H., et al. (2005a). WNK1 activates SGK1 to regulate the epithelial sodium channel. *Proceedings of the National Academy of Sciences of the United States of America, 102*, 10315–10320.

Xu, B. E., Stippec, S., Lazrak, A., Huang, C. L., & Cobb, M. H. (2005b). WNK1 activates SGK1 by a phosphatidylinositol 3-kinase-dependent and non-catalytic mechanism. *The Journal of Biological Chemistry, 280*, 34218–34223.

Yamada, T., Bhate, M. P., & Strange, K. (2013). Regulatory phosphorylation induces extracellular conformational changes in a CLC anion channel. *Biophysical Journal, 104*, 1893–1904.

Yamamoto, Y., Izumi, Y., & Matsuzaki, F. (2008). The GC kinase Fray and Mo25 regulate Drosophila asymmetric divisions. *Biochemical and Biophysical Research Communications, 366*, 212–218.

Yanagawa, S., van Leeuwen, F., Wodarz, A., Klingensmith, J., & Nusse, R. (1995). The Dishevelled protein is modified by Wingless signalling in Drosophila. *Genes & Development, 9*, 1087–1097.

Yanfeng, W. A., Berhane, H., Mola, M., Singh, J., Jenny, A., & Mlodzik, M. (2011). Functional dissection of phosphorylation of Disheveled in Drosophila. *Developmental Biology, 360*, 132–142.

Yang, S. S., Lo, Y. F., Wu, C. C., Lin, S. W., Yeh, C. J., Chu, P., et al. (2010). SPAK-knockout mice manifest Gitelman syndrome and impaired vasoconstriction. *Journal of the American Society of Nephrology, 21*, 1868–1877.

Zagorska, A., Pozo-Guisado, E., Boudeau, J., Vitari, A. C., Rafiqi, F. H., Thastrup, J., et al. (2007). Regulation of activity and localization of the WNK1 protein kinase by hyperosmotic stress. *The Journal of Cell Biology, 176*, 89–100.

Zaika, O., Tomilin, V., Mamenko, M., Bhalla, V., & Pochynyuk, O. (2016). New perspective of ClC-Kb/2 Cl− channel physiology in the distal renal tubule. *American Journal of Physiology. Renal Physiology, 310*, F923–F930.

Zambrowicz, B. P., Abuin, A., Ramirez-Solis, R., Richter, L. J., Piggott, J., BeltrandelRio, H., et al. (2003). Wnk1 kinase deficiency lowers blood pressure in mice: A gene-trap screen to identify potential targets for therapeutic intervention. *Proceedings of the National Academy of Sciences of the United States of America, 100*, 14109–14114.

Zeniya, M., Sohara, E., Kita, S., Iwamoto, T., Susa, K., Mori, T., et al. (2013). Dietary salt intake regulates WNK3-SPAK-NKCC1 phosphorylation cascade in mouse aorta through angiotensin II. *Hypertension, 62*, 872–878.

Zeqiraj, E., Filippi, B. M., Goldie, S., Navratilova, I., Boudeau, J., Deak, M., et al. (2009). ATP and MO25alpha regulate the conformational state of the STRADalpha pseudokinase and activation of the LKB1 tumour suppressor. *PLoS Biology, 7*, e1000126.

Zhang, C., Wang, L., Zhang, J., Su, X. T., Lin, D. H., Scholl, U. I., et al. (2014). KCNJ10 determines the expression of the apical Na-Cl cotransporter (NCC) in the early distal convoluted tubule (DCT1). *Proceedings of the National Academy of Sciences of the United States of America, 111*, 11864–11869.

Zhao, Y., Marin, O., Hermesz, E., Powell, A., Flames, N., Palkovits, M., et al. (2003). The LIM-homeobox gene Lhx8 is required for the development of many cholinergic neurons in the mouse forebrain. *Proceedings of the National Academy of Sciences of the United States of America, 100*, 9005–9010.

Zhao, H., Nepomuceno, R., Gao, X., Foley, L. M., Wang, S., Begum, G., et al. (2016). Deletion of the WNK3-SPAK kinase complex in mice improves radiographic and clinical outcomes in malignant cerebral edema after ischemic stroke. *Journal of Cerebral Blood Flow and Metabolism*, pii:0271678X16631561 [Epub ahead of print].

Zhu, W., Begum, G., Pointer, K., Clark, P. A., Yang, S. S., Lin, S. H., et al. (2014). WNK1-OSR1 kinase-mediated phospho-activation of Na+-K+-2Cl− cotransporter facilitates glioma migration. *Molecular Cancer, 13*, 31.

CHAPTER TWO

SGK1: The Dark Side of PI3K Signaling

Antonio Di Cristofano[1]

Albert Einstein College of Medicine, Bronx, NY, United States
[1]Corresponding author: e-mail address: antonio.dicristofano@einstein.yu.edu

Contents

1. Introduction: The Glucocorticoid-Regulated Kinase Family	50
2. SGK1: Expression and Stability Control	50
3. SGK1 Is Activated in a PI3K-Dependent Manner	51
4. AKT and SGK1: Target Overlap and Selectivity	52
5. Roadblocks to Defining Specific SGK1 Functions	56
6. SGK1 in Development and Differentiation	57
6.1 Ion Transport	58
6.2 Implantation and Pregnancy	58
6.3 Myocardial Injury	59
6.4 Muscle Mass	59
6.5 T Cell Activation	60
6.6 Macrophage Motility and Function	60
6.7 Insulin Sensitivity	61
7. SGK1 and Cancer	61
8. Concluding Remarks	64
Acknowledgments	64
References	65

Abstract

Activation of the PI3K pathway is central to a variety of physiological and pathological processes. In these contexts, AKT is classically considered the de facto mediator of PI3K-dependent signaling. However, in recent years, accumulating data point to the existence of additional effectors of PI3K activity, parallel to and independent of AKT, that play critical and unique roles in mediating different developmental, homeostatic, and pathological processes.

In this review, I summarize and discuss our current understanding of the function of the serine/threonine kinase SGK1 as a downstream effector of PI3K, and try to separate targets and pathways validated as uniquely SGK1-dependent from those shared with AKT.

1. INTRODUCTION: THE GLUCOCORTICOID-REGULATED KINASE FAMILY

Serum- and glucocorticoid-regulated kinases (SGKs) are members of the AGC (PK**A**-, PK**G**-, PK**C**-related) family of serine/threonine kinases, one of the most evolutionarily conserved groups of protein kinases, represented in most eukaryotic organisms (Arencibia, Pastor-Flores, Bauer, Schulze, & Biondi, 2013). Well-known members of the AGC family are AKT, PDK1, S6K, PKC, and RSK. SGK kinases share greatest sequence homology with the AKT family (Pearce, Komander, & Alessi, 2010). The SGK family consists of three distinct but highly homologous isoforms (SGK1, SGK2, and SGK3) that are produced from three distinct genes localized on different chromosomes (Lang & Cohen, 2001).

Structurally, SGK kinases, as most AGC kinases, consist of three domains: an N-terminal variable region, a catalytic domain, and the C-terminal tail. SGKs are subject to tight spatial and temporal regulation, mainly through phosphorylation of two conserved residues, one in the activation loop contained in the kinase domain, and one in the hydrophobic motif within the C-tail, which is indispensable for full kinase activation (Pearce et al., 2010).

While the N-terminal region of some AGC kinases, such as AKT and PDK1, contains a phosphoinositide-binding pleckstrin homology (PH) domain, essential for kinase recruitment to membrane-bound phosphatidylinositol-3-phosphate, SGK1 and SGK2 have no recognizable N-terminal functional domain. On the other hand, unique in the family, SGK3 possesses an N-terminal phosphoinositide-binding Phox homology (PX) domain, which interacts with phosphatidylinositol-3-phosphate to mediate the endosomal association of SGK3, essential for its phosphorylation and activation (Tessier & Woodgett, 2006).

2. SGK1: EXPRESSION AND STABILITY CONTROL

SGK isoforms are not equally expressed in all tissues. SGK2 expression is constitutive but restricted to the liver, pancreas, brain, and kidney proximal tubules (Kobayashi, Deak, Morrice, & Cohen, 1999; Pao et al., 2010). SGK3 is also constitutively expressed, but its expression is ubiquitous (Kobayashi et al., 1999).

On the other hand, expression of SGK1, while found in all tissues examined, is strictly transcriptionally and posttranscriptionally regulated. In fact, SGK1 was discovered as an immediate early gene, transcriptionally induced in rat mammary cancer cells by glucocorticoids and serum (Webster, Goya, Ge, Maiyar, & Firestone, 1993).

A multitude of stimuli, including growth factors (Mizuno & Nishida, 2001; Waldegger et al., 1999), mineralocorticoids (Naray-Fejes-Toth, Canessa, Cleaveland, Aldrich, & Fejes-Toth, 1999), cytokines (Fagerli et al., 2011), as well as various cellular stresses such as hyperosmotic cell shrinkage (Waldegger, Barth, Raber, & Lang, 1997), heat shock, ultraviolet irradiation, and oxidative stress (Leong, Maiyar, Kim, O'Keeffe, & Firestone, 2003), have been shown to induce SGK1 gene transcription. In addition, SGK1 mRNA has a short half-life, disappearing within 20 min from transcription (Waldegger et al., 1997).

A second level of tight control over SGK1 levels is represented by protein stability. SGK1 is polyubiquitinated and rapidly turned over, with a half-life of approximately 30 min (Brickley, Mikosz, Hagan, & Conzen, 2002).

The signals required for SGK1 degradation reside in the first 60 amino acids (Brickley et al., 2002). More specifically, a six amino acid motif devoid of lysines is required for polyubiquitination and rapid degradation by the 26S proteasome (Bogusz, Brickley, Pew, & Conzen, 2006). This process appears to involve different E3 ubiquitin ligases: SGK1 has been in fact reported to associate with the stress-associated, chaperone-dependent, U-box E3 ubiquitin ligase CHIP (Belova et al., 2006), with the ER-associated, transmembrane E3 ubiquitin ligase HRD1 (Arteaga, Wang, Ravid, Hochstrasser, & Canessa, 2006), with the HECT domain E3 ubiquitin ligase NEDD4L (Zhou & Snyder, 2005), and more recently, with a new E3 complex that includes Rictor, Cullin-1, and Rbx1 (Gao et al., 2010).

3. SGK1 IS ACTIVATED IN A PI3K-DEPENDENT MANNER

It was not until several years after SGK1 identification and characterization that a number of studies reported that SGK1 phosphorylation and activation was controlled by the PI3K signaling cascade (Kobayashi & Cohen, 1999; Park et al., 1999). These studies stemmed from the observation that the catalytic and C-terminal domains of SGK1 are highly homologous to those of other AGC kinases such as AKT, PKC, and S6K1, which had just been discovered to be phosphorylated and activated by PDK1 on a conserved residue in the activation loop. In fact, the PI3K inhibitor

LY294002 was found to completely abolish insulin- and IGF-1-induced SGK1 activity in HEK293 cells. Furthermore, these studies demonstrated that PDK1 is directly responsible for SGK1 phosphorylation on Thr256, and that Ser422 in the hydrophobic domain is phosphorylated in response to PI3K activation, likely by the same kinase that phosphorylates and activates AKT on Ser473 (Kobayashi & Cohen, 1999; Park et al., 1999).

Strikingly, these experiments also showed that, contrary to what is observed with AKT, the Ser422-to-Asp and Thr256-to-Asp SGK1 double mutant did not gain constitutive activity, a still unexplained feature shared with S6K1, which has significantly hindered progress on the delineation of SGK1 activity and role in cell models.

The identity of the kinase responsible for the PI3K-dependent phosphorylation of SGK1 remained mysterious for several years, until 2008, when the Alessi group, using genetic and biochemical approaches, identified mTORC2 as the complex phosphorylating SGK1 on Ser422 (Garcia-Martinez & Alessi, 2008).

While this finding provided clear evidence that the rapamycin–insensitive mTORC2 complex, already associated with AKT phosphorylation on Ser473, is the bona fide kinase for SGK1 Ser422, the question of the mechanism governing its dependence on PI3K activity still remained open. Very recently, however, the Wei laboratory has proposed a mechanism of mTORC2 activation that depends on the availability of phosphatidylinositol (3,4,5)-triphosphate, PIP3, the product of PI3K enzymatic activity (Liu et al., 2015). The Wei lab data support a model in which the PH domain of the mTORC2 essential component, SIN1, binds to and inhibits mTOR in the absence of PIP3. However, when growth factor-receptor stimulation leads to PI3K activation and accumulation of PIP3, the binding of SIN1 PH domain to PIP3 relieves the inhibition on mTOR, leading to complex activation, and recruits mTORC2 to the membrane, in proximity of many of its substrates.

4. AKT AND SGK1: TARGET OVERLAP AND SELECTIVITY

As described earlier, AKT, the prototypical effector of the PI3K signaling cascade, and SGK1 share the essential features of their activation mechanism, with mTORC2 and PDK1 phosphorylating the hydrophobic motif and the activation loop, respectively, of these two kinases. There are, however, some important differences. One is that AKT activation requires binding of its PH domain to PIP3, which induces a conformational change facilitating phosphorylation on Thr308 (Calleja et al., 2007; Milburn et al., 2003). A second critical difference is that, in the case of AKT, the two activating phosphorylations, both necessary for full kinase activation, are largely independent of each other (Biondi, Kieloch, Currie, Deak, & Alessi, 2001). On the other hand, in the case of SGK1, mTORC2-mediated phosphorylation on Ser422 has an essential priming function for the subsequent PDK1-mediated phosphorylation on Thr256, since it allows the PIF pocket of PDK1 to bind to phospho-Ser422 on SGK1 (Biondi et al., 2001; Collins, Deak, Arthur, Armit, & Alessi, 2003).

A key feature of these two kinases is that they share the same optimal target motif, Arg-X-Arg-X-X-Ser/Thr (Alessi, Caudwell, Andjelkovic, Hemmings, & Cohen, 1996; Kobayashi et al., 1999; Park et al., 1999). While AKT prefers a bulky hydrophobic residue immediately following the phosphorylation site, this characteristic does not seem to be shared by SGK1 (Kobayashi et al., 1999; Murray, Cummings, Bloomberg, & Cohen, 2005; Park et al., 1999). Thus, although the identity of surrounding amino acids, or the kinase subcellular localization, or even interacting adapter proteins might dictate some level of target specificity, AKT and SGK1 appear to display a high level of promiscuity, often complicating the attribution of specific biological functions to either of them (Fig. 1).

Nevertheless, to date, few of the classical AKT targets have been convincingly shown to be phosphorylated by SGK1 in physiological conditions. As an example, some of AKT best-known bona fide targets, such as GSK3β and FOXO3, were initially shown to be phosphorylated by SGK1 in vitro or in transfection-based systems (Brunet et al., 2001; Kobayashi & Cohen, 1999). However, the analysis of cells in which SGK1 cannot be activated because of a point mutation that was introduced in the PIF pocket of PDK1 failed to show changes in GSK3α/β and FOXO3 phosphorylation levels, strongly suggesting that SGK1 is not a primary driver of GSK3α/β and FOXO3 phosphorylation in physiological conditions (Collins et al., 2003).

Along the same line, early reports showing that both AKT (Gratton et al., 2001; Guan et al., 2000) and SGK1 (Chun et al., 2003; Zhang et al., 2001) could phosphorylate and negatively regulate BRAF and MEKK3 have never been conclusively validated.

On the other hand, a large amount of data from independent groups strongly supports the notion that the E3 ubiquitin ligase NEDD4L, which regulates internalization and turnover of a number of membrane proteins, can be negatively regulated by both AKT and SGK1 (Boehmer, Okur,

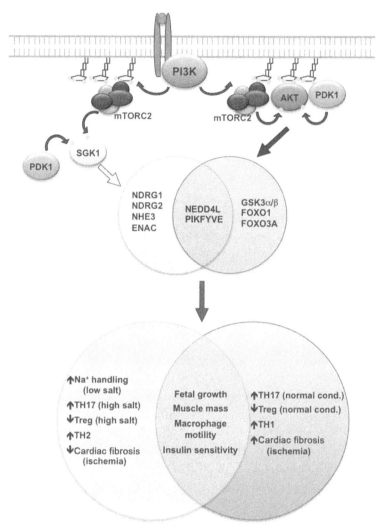

Fig. 1 Schematic representation of the pathways leading to AKT and SGK1 activation, and of the bona fide targets and physiological processes specifically associated with each kinase or coregulated by AKT and SGK1. See text for details.

Setiawan, Broer, & Lang, 2003; Caohuy et al., 2014; Lee, Dinudom, Sanchez-Perez, Kumar, & Cook, 2007), although gain- and loss-of-function experiments support SGK1 as more relevant for this function (Andersen et al., 2013).

Another shared target with related biological functions appears to be PIKFYVE (phosphatidylinositol-3-phosphate 5-kinase type III), which controls endomembrane homeostasis and protein trafficking by facilitating the progression of early endosomes toward multivesicular bodies (Shisheva, 2012). Both AKT (Berwick et al., 2004) and SGK1 (Seebohm et al., 2007) appear to be able to phosphorylate and activate PIKFYVE, thus regulating the localization and activity of diverse transporters, including the K^+ channel complex KCNQ1/KCNE1 (Seebohm et al., 2007), the Na^+/ monocarboxylate transporter SLC5A1 (Shojaiefard, Strutz-Seebohm, Tavare, Seebohm, & Lang, 2007), the creatine transporter SLC6A8 (Lopez-Barradas et al., 2016), the Ca^{2+}-permeable cation channel TRPV6 (Sopjani et al., 2010), as well as the amino acid transporters EAAT2, -3, -4 (Gehring et al., 2009). In all these cases, experimental validation of the relative role and relevance of AKT and SGK1 is still lacking.

The list of validated SGK1-specific phosphorylation targets is rather limited. NDRG1 and -2 (see later) are the best-known and most credible SGK1-specific targets identified so far (Murray et al., 2004). Notably, they are inefficiently phosphorylated by AKT (Murray et al., 2005). Although SGK1-mediated phosphorylation appears to prime them for subsequent hyperphosphorylation by GSK3 (Murray et al., 2004), the role of these modifications is still elusive. It has been proposed that SGK3-mediated phosphorylation of NDRG1 leads to its ubiquitination and degradation (Gasser et al., 2014); thus, it is possible that this function is also shared by SGK1 and AKT indirectly, although experimental validation of this inference is still lacking.

Two additional bona fide exclusive targets of SGK1 activity are NHE3 (Sandu et al., 2006; Wang, Sun, Lang, & Yun, 2005) and ENAC (Arteaga & Canessa, 2005; Lu et al., 2010). SGK-mediated phosphorylation of NHE3 (Na^+/H^+ exchanger 3) increases its activity toward the transcellular reabsorption of Na^+ and $HCO3^-$ in the kidney and intestine (Wang et al., 2005). Phosphorylation of the epithelial sodium (Na^+) channel (ENaC), instead, activates Na^+ transport in the distal nephron.

One particularly relevant feature of SGK1 and AKT is the ability to engage in opportunistic compensation when one of the two kinases is genetically repressed or pharmacologically inhibited. This notion might explain, for example, the reported ability of SGK1 to phosphorylate, in vitro or in overexpression conditions, AKT substrates such as GSK3β and FOXO3 (Brunet et al., 2001; Kobayashi & Cohen, 1999), and that of AKT to phosphorylate NDRG2 (Burchfield et al., 2004).

More importantly, breast cancer cells with high Akt activity and low SGK1 expression display significant NDRG1 phosphorylation that is suppressed by Akt inhibitors (Sommer et al., 2013). Furthermore, in the presence of a PI3K inhibitor that completely blocks AKT activity, SGK1 can directly phosphorylate TSC2, a bona fide AKT substrate, thus activating mTORC1 (Castel et al., 2016) (see later).

5. ROADBLOCKS TO DEFINING SPECIFIC SGK1 FUNCTIONS

The reported target promiscuity of AKT and SGK1 and the presence of three SGK genes, which could theoretically partially compensate each other in genetic ablation or depletion approaches, have represented major barriers to defining and validating SGK1 molecular targets and biological functions.

Highly selective inhibitors would be invaluable tools to mitigate these problems. However, only a handful of SGK inhibitors have been described, and most of them have fallen short of showing the required specificity, bioavailability, or cellular permeability.

The best-known and most used SGK1 inhibitor, GSK650394 (Sherk et al., 2008), has an IC_{50} of 64 nM for SGK1. Off-target effects are notable, with less than 10-fold selectivity for SGK1 over Aurora kinase, JNK1, and JNK3. This inhibitor has rather poor cell permeability (Zapf et al., 2016), and a 10 μM concentration is necessary to observe reduced phosphorylation of NEDD4L (Sherk et al., 2008) and NDRG1 (Mansley & Wilson, 2010).

Another commonly used inhibitor is EMD638683 (Ackermann et al., 2011). While its in vitro IC_{50} toward SGK1 was not specified, it showed better selectivity than GSK650394 (>30-fold for PKA and MSK1). However, once again, at least 10 μM was needed to abolish NDRG1 phosphorylation, underlining its poor cell permeability. Along the same line, when tested in vivo in mice, the effective dose was very high (600 mg/kg) (Ackermann et al., 2011).

The most recently described inhibitor, SGK-inh (Castel et al., 2016; Halland et al., 2015), has an IC_{50} of 4.8 nM for SGK1. Interestingly, the only other kinase inhibited with selectivity lower than 10-fold is p70S6K. Again, limited permeability resulted in the need of at least a 10 μM concentration to achieve significant inhibition of NDRG1 phosphorylation (Castel et al., 2016).

These data underline the critical need for better SGK1 inhibitors, in particular with improved pharmacodynamic and pharmacokinetic characteristics that allow in vivo application.

6. SGK1 IN DEVELOPMENT AND DIFFERENTIATION

Cell line-based data linking SGK1 activation to specific developmental or differentiation pathways are rather limited and often subjected to a number of caveats, including the use of overexpression systems and the absence of a physiological multicellular context. Using these approaches, for example, it has been proposed that SGK1 might regulate adipocyte differentiation, at least in part through the phosphorylation and relocalization of FOXO1, a canonical AKT target (Di Pietro et al., 2010).

On the other hand, some of these cell-based analyses have been designed to overcome most common caveats by using shRNA-mediated approaches as well as cells with targeted inactivation of *Sgk1*. As an example, a number of recent studies have uncovered a key role for SGK1 in promoting differentiation of specific T cell subsets under high salt conditions, which induce an increase of SGK1 expression. Two studies have found that high salt-induced SGK1 leads to increased differentiation of IL17-producing helper T cells, a $CD4^+$ population involved in inflammation and autoimmunity (Kleinewietfeld et al., 2013; Wu et al., 2013). Importantly, the involvement of PI3K in the differentiation of TH17 cells was already well established (Kim et al., 2005), and an AKT-mTORC1 pathway has been identified as essential for the differentiation of TH17 cells in normal conditions (Lee et al., 2010).

Similarly, the differentiation of $FOXP3^+$ Treg cells is the default outcome of T cell activation in mTORC1- and mTORC2-deficient mice under physiological conditions (Delgoffe et al., 2011), but not in T cell-specific $Sgk1^{-/-}$ mice, indicating a critical suppressive role for a linear AKT-mTORC1 pathway in physiological conditions. However, under high salt conditions, SGK1 can inhibit Treg generation (Hernandez et al., 2015).

Thus, in physiological conditions, AKT appears to be the primary driver of differentiation of activated T cells. However, under stress conditions such as high salt-induced *Sgk1* upregulation (conditions that do not alter AKT expression and activity), SGK1 appears to directly control the development of TH17 and the repression of Treg cells. Consequently, it is not implausible to extend this notion to other acute or nonphysiological stimuli that can increase SGK1 levels and activity.

The generation of isoform-specific *Sgk* mutant mice has been a critical driver in clarifying SGK1 roles in a physiological context.

Given the high degree of homology, it is theoretically conceivable that the other two SGK isoforms, which are constitutively expressed, could compensate for SGK1 absence, at least under physiological conditions. The same notion could apply to additional kinases with overlapping target specificity, such as AKT, which could engage in opportunistic compensation.

6.1 Ion Transport

The initial characterization of whole-body *Sgk1* knockout mice suggested, unexpectedly, that SGK1 is completely dispensable for basic, steady-state physiological functions (Wulff et al., 2002). Mutant mice and organs appeared physiologically normal, and only under conditions of sodium deprivation they showed a reduced ability to maintain Na^+ levels by activating distal-tubular resorption, despite the presence of appropriately increased aldosterone levels. Along the same line, *Sgk1* mutant mice did not display alterations in Na^+-coupled intestinal glucose transport under normal conditions, but, contrary to wild-type mice, they were unable to increase glucose transport upon glucocorticoid stimulation (Grahammer, Henke, et al., 2006).

Thus, based on in vivo data, it seems likely that, in normal conditions, SGK1 is not the primary regulator of those physiological processes to which it has been associated by in vitro studies, but is instead involved in their fine-tuning upon specific acute stresses.

Unexpectedly, the use of compound mutants has shown that functional compensation between isoforms is very limited, and that several functions ascribed to the whole SGK family appear to be quite isoform-specific, as clearly exemplified by renal Na^+ handling. Compound *Sgk1* and *Sgk3* mutants, in fact, show simple coexistence of the different phenotypes observed in single mutants (Na^+ handling in $Sgk1^{-/-}$ mice, hair growth delay in $Sgk3^{-/-}$ mice), without any exacerbation of these defects (Grahammer, Artunc, et al., 2006). Along the same line, while $Sgk2^{-/-}$ mice do not display any Na^+ handling problem in normal or even salt-deprivation conditions, compound *Sgk1*, *Sgk2* mutants further escalate the Na^+ resorption deficit seen in *Sgk1* knockouts upon NaCl deprivation (Schnackenberg et al., 2007), suggesting that SGK2 controls some aspects of Na^+ handling under acute stress, but in lesser measure compared to SGK1.

6.2 Implantation and Pregnancy

Following up on the finding that endometrial SGK1 levels rapidly increase upon progesterone stimulation, it recently has been shown that SGK1 is involved in implantation and pregnancy maintenance (Salker et al., 2011). Forced expression of SGK1 in the uterine luminal epithelium after conception in a time window that is normally characterized by a transient loss of SGK1 expression, dramatically impairs the rate of embryo implantation. This finding strikingly correlates with the reduced litter size and impaired trophoblast invasion in mice with decidua-specific loss of *Pten*, which constitutively activates PI3K in this compartment (Lague et al., 2010). Thus, it is likely that tight regulation of the PI3K-PTEN-SGK1 axis is essential during embryo implantation.

$Sgk1^{-/-}$ female mice experience a high rate (>30%) of pregnancy loss, associated with uterine bleeding, dramatic fetal growth restriction, and deregulation of reactive oxygen species scavengers in the decidua (Salker et al., 2011), further underlining a critical role for SGK1 in the maintenance of pregnancy. A similar growth restriction and pregnancy loss phenotype was also observed in $Akt1^{-/-}$ mice (Kent, Ohboshi, & Soares, 2012; Yang et al., 2003), suggesting that both AKT and SGK contribute to placental function.

6.3 Myocardial Injury

Recent in depth analysis of $Sgk1^{-/-}$ mice has revealed subtle cardiac phenotypes, including a smaller heart, reduced heart rate, and reduced cardiomyocyte size, compared to wild-type controls (Zarrinpashneh et al., 2013). $Sgk1^{-/-}$ endothelial cells display defective tube formation, which translates into increased cardiac fibrosis after ischemic insult in vivo due to inefficient capillary formation (Zarrinpashneh et al., 2013), a phenotype that contrasts the protection from fibrosis observed in $Akt1^{-/-}$ mice subjected to the same ischemic insult (Ma, Kerr, Naga Prasad, Byzova, & Somanath, 2014). These results underline the presence of clearly differential functions of SGK1 and AKT downstream of PI3K in the heart.

6.4 Muscle Mass

PI3K activity has long been known to regulate skeletal muscle atrophy and hypertrophy (Glass, 2010). Recent studies have shown that skeletal muscle-specific *Sgk1* inactivation leads to moderate atrophy and decreased muscle strength, despite a small increase in activated AKT (Andres-Mateos et al., 2013). Correspondingly, overexpression of SGK1 protects against muscle atrophy induced by disuse and starvation. Interestingly, also $Akt1^{-/-}$ and $Akt2^{-/-}$ mice display a similar muscle phenotype (Goncalves et al., 2010), and SGK1 expression is slightly increased in the muscle of $Akt1^{-/-}$ mice (Andres-Mateos et al., 2013). The similar phenotypes of Sgk and Akt mutant muscles, and the fact that loss of one gene induces increased expression of the other in the absence of a complete rescue strongly suggest that SGK1 and AKT cannot completely compensate for the loss of each other, and that both AKT and SGK1 are necessary to ensure normal muscle size and function.

6.5 T Cell Activation

Earlier genetic data had linked mTOR activity to the differentiation of $CD4^+$ T cells into both TH1 and TH2 populations, and had specifically identified a linear AKT-mTORC1 pathway leading to the TH1 phenotype (Delgoffe et al., 2011; Lee et al., 2010). More specifically, constitutively active AKT was shown to rescue the TH1 but not the TH2-impairment in *rictor* mutant mice, which cannot activate mTORC2 (Lee et al., 2010). By using $CD4^+$ T cell-specific *Sgk1* mutants, SGK1 has recently been identified as essential for the differentiation of $CD4^+$ T cells into TH2 helper cells, while being dispensable for TH1 differentiation (Heikamp et al., 2014). Mechanistically, SGK1-dependent inhibition of NEDD4L relieved posttranscriptional control of JunB, whose activity is essential for TH2 differentiation (Li, Tournier, Davis, & Flavell, 1999). The fact that AKT and SGK isoforms are coexpressed in $CD4^+$ T cells and that they are both activated upon TH1 and TH2 conditions underlines the notion that that the differential requirement of AKT and SGK1 for TH1 vs TH2 differentiation is likely determined by specific substrate availability, which might be dictated by the polarizing stimulus.

6.6 Macrophage Motility and Function

The importance of PI3K-dependent signaling, in particular PI3K-γ, in the response of monocytes and macrophages to atherogenic mediators is well established (Chang et al., 2007).

Deletion of *Sgk1* in a mouse model of atherosclerosis led to a 50% reduction in both lesion size and macrophage infiltration (Borst et al., 2015). The defect in macrophage migration was further validated in a mouse model of thioglycollate-induced peritoneal inflammation. *Sgk1* mutant macrophages displayed reduced NFkB-dependent MMP9 expression, which likely contributes to the migration defect. A different study examined the effect of SGK1 deficiency on hypertensive cardiac remodeling associated with activation of the renin-angiotensin system and found that *Sgk1* mutant mice were protected from cardiac inflammation and fibrosis after angiotensin II infusion (Yang et al., 2012). Attenuation of the infiltration of proinflammatory cells in the heart was associated with the prevention of activation of STAT3 and reduced polarization to M2 macrophages.

In a similar mouse model of cholesterol-induced atherosclerosis, specific deletion of *Akt1* in hematopoietic cells did not affect the size and macrophage content of atherosclerotic lesions (Babaev et al., 2014). Conversely, deletion of *Akt2* drastically reduced atherosclerotic lesion size and macrophage content. Interestingly, $Akt1^{-/-}$ blood monocytes and macrophages were found to be skewed to the M1 phenotype, while $Akt2^{-/-}$ monocyte and macrophages displayed the M2 phenotype (Arranz et al., 2012; Babaev et al., 2014).

Thus, it appears that SGK1 and AKT2 exert similar positive effects on macrophage recruitment, while SGK1 regulates macrophage polarization in a manner more similar to AKT1.

6.7 Insulin Sensitivity

Analysis of liver-specific *Sgk1* mutant mice has revealed a role for SGK1 in the control of insulin sensitivity. Mutant mice displayed a slight elevation in fasting blood glucose as well as a mild reduction in both glucose clearance and insulin sensitivity (Liu et al., 2014). The mechanisms for this impaired response to insulin seem to include reduced insulin-induced phosphorylation of IRS1. Notably, phosphorylation of the insulin receptor was not altered upon loss of SGK1, which suggests that SGK1 might negatively regulate a phosphatase that controls IRS1 phosphorylation.

It must be underlined that the insulin resistance observed in $Sgk1^{-/-}$ mice is significantly less severe than that observed in the overtly diabetic $Akt2^{-/-}$ mice (Cho et al., 2001). On the other hand, $Akt1^{-/-}$ mice do not display any insulin sensitivity phenotype (Chen et al., 2001). Thus, it is likely that SGK1 mediates only a minor part of the insulin-induced cellular uptake of glucose, which is instead mainly controlled by AKT2.

7. SGK1 AND CANCER

As discussed earlier, SGK1 expression is promptly induced by a variety of stimuli, especially stress-related ones. Furthermore, while being often dispensable for basal pathway activity, SGK1 is instead involved in signaling under acute stimulation. Taken together, these notions support the possibility that SGK1 plays a critical role in mediating neoplastic transformation as an effector of PI3K.

Lack of highly specific reagents (antibodies and small molecule inhibitors), its originally postulated exclusive role in ion transport, and the apparent promiscuity with AKT isoforms have historically limited the interest in SGK1 as a mediator of the transformed phenotype. However, the past few years have witnessed a significant increase in the number of reports showing that indeed SGK1 has a defined AKT-independent role in cellular transformation, downstream of activated PI3K signaling.

Similar to AKT isoforms, SGK1 does not appear to be frequently mutated in human tumors, although a recent study has found frequent SGK1 mutations (approximately 50% of samples) in nodular lymphocyte predominant Hodgkin lymphoma (NLPHL) (Hartmann et al., 2016). The functional significance of these alterations remains anyway to be defined. *SGK1* amplification has been observed in 31% of breast cancer patients in one study (Eirew et al., 2015) and overexpression in 48% of breast cancer patients in another study (Sahoo, Brickley, Kocherginsky, & Conzen, 2005).

The most compelling data linking SGK1 and tumor development come from studies employing genetically targeted *Sgk1* alleles.

The first in vivo demonstration that activation of SGK1 is important for neoplastic transformation came from the analysis of intestinal carcinogenesis in $Sgk1^{-/-}$ mice (Nasir et al., 2009). Wild-type mice subjected to a chemical carcinogenesis protocol developed on average 12 colonic tumors per mouse, while $Sgk1^{-/-}$ mice had a reduction of over 50% in the tumor load. Mechanistically, $Sgk1^{-/-}$ tissues displayed increased levels of FOXO3A and BIM, two molecules that may affect tumor growth by increasing apoptosis levels.

Along the same line, generation of $Apc^{Min/+}/Sgk1^{-/-}$ mice identified a clear protumorigenic role of SGK1 (Wang et al., 2010). $Apc^{Min/+}$ mice develop colonic adenomatous polyps, which eventually progress to colon carcinoma. Deletion of *Sgk1* resulted in approximately 75% reduction in the average number of tumors per mouse.

It is important to underline that these studies, performed with whole-body *Sgk1* mutants, do not clarify whether the observed reduction in tumor development is cell autonomous.

In fact, T cell-specific deletion of *Sgk1* resulted in a dramatic reduction in the number of lung metastases from mouse melanoma cells injected in the tail vein (Heikamp et al., 2014), strongly suggesting that deletion of *Sgk1* in T cells leads to enhanced antitumor immunity.

The recent availability of a conditional *Sgk1* allele will be instrumental in defining cell autonomous vs nonautonomous SGK1 functions associated with neoplastic transformation and tumor progression.

A critical issue is the identity of the pathways through which SGK1 activation contributes to PI3K-dependent tumorigenesis. While it is possible that increased SGK1 expression and activity impinge upon classical AKT targets, contributing to their phosphorylation, other proteins known to be primarily controlled by SGK1 might also play an important role.

One class of possible SGK-controlled players in neoplastic transformation is represented by the Ca^{2+} release-activated Ca^{2+} channel (I_{CRAC}), composed of the pore-forming subunits ORAI1, -2, and -3 and their regulatory subunits STIM1 and -2. SGK1 can strongly upregulate these channels through two mechanisms, repression of the ubiquitin ligase NEDD4L, leading to increased protein levels of ORAI1, and transcriptional induction of both *ORA1* and *STIM1*, by increasing the activity of NFkB, which directly interacts with their promoter (Lang & Shumilina, 2013). Store-operated calcium channels such as the ORAI/STIM complexes regulate key aspects of the cancer cell phenotype, including tumor growth, angiogenesis, and metastasis formation (Xie, Pan, Yao, Zhou, & Han, 2016).

At the same time, SGK1-controlled potassium channels, such as Kv1.3, mediate plasma membrane hyperpolarization, which provides increased driving force for Ca^{2+} entry, thus further fostering Ca^{2+}-dependent tumor growth (Urrego, Tomczak, Zahed, Stühmer, & Pardo, 2014).

Additionally, SGK1-controlled proton exchangers such as NHE3 have been associated with the control of lysosome trafficking in prostate cancer cells (Steffan, Williams, Welbourne, & Cardelli, 2010), and ENAC activity has been shown to mediate a variety of processes, including cancer cell proliferation and migration (Liu, Zhu, Xu, Ji, & Li, 2016).

Finally, it is important to underline that SGK1 is likely to phosphorylate a number of still unknown targets, some of which may have critical roles in neoplastic transformation either directly, or by controlling additional downstream players. In fact, novel SGK1-mediated pathways are being discovered on a regular basis, although the specific mechanisms at play are still undefined. For example, SGK1 has been recently shown to regulate *RANBP1* gene transcription in colon carcinoma cells, thus affecting mitotic spindle function as well as cell sensitivity to taxanes (Amato et al., 2013). Furthermore, a more recent study has proposed that SGK1 mediates an increase in rRNA synthesis under conditions of PI3K activation by regulating the nucleolar localization of the histone demethylase KDM4A (Salifou et al., 2016).

As mentioned earlier, an additional mechanism through which SGK1 contributes to transformation is through opportunistic compensation under conditions of PI3K/AKT pharmacological inhibition. As an example, in breast cancer cells expressing high levels of SGK1 and treated with a PI3K inhibitor, SGK1 stimulates residual mTORC1 activity through direct phosphorylation and inhibition of TSC2, thus contributing to resistance to PI3K inhibition (Castel et al., 2016). These data confirm and extend those presented in a previous report, in which high levels of SGK1 were found to predict resistance of breast cancer cells to AKT inhibitors (Sommer et al., 2013). A key corollary to this notion is that a certain level of SGK1 activation is still possible under condition of PI3K inhibition. Whether it is mediated by residual PI3K activity, or by a PI3K-independent pool of activated mTORC2, remains to be addressed.

8. CONCLUDING REMARKS

After spending two decades under the shadow of AKT, SGK1 is finally claiming its own spot as an important mediator of PI3K signaling activity, in particular in conditions where growth factor- or stress-related inputs increase SGK1 levels. Recent data support a major role for SGK1 that extends well beyond what has always been considered SGKs primary function, the control of ion transport across membranes.

Current and future efforts, in particular those employing genetically engineered mouse models and tightly controlled shRNA-based cell systems, will need to clarify the extent of both physiological and opportunistic target overlap between AKT and SGK1, and to define the signaling molecules and pathways that are directly and specifically controlled by SGK1.

Finally, the notions (i) that SGK1 is actively involved in the control of pathways that are altered during inflammation, neoplastic transformation, and tumor response to cytotoxic and targeted therapy, and (ii) that SGK1 loss, in vivo, has minimal consequences for normal homeostasis should drive a renewed effort to develop selective SGK1 inhibitors with pharmacodynamic and pharmacokinetic properties allowing in vivo preclinical and clinical applications.

ACKNOWLEDGMENTS

Every effort was made to include all relevant studies, and I apologize to those whose work was not referenced either due to space limitations or our oversight. Our research related to this chapter has received funding from NIH Grants CA172012, CA128943, and CA167839.

REFERENCES

Ackermann, T. F., Boini, K. M., Beier, N., Scholz, W., Fuchss, T., & Lang, F. (2011). EMD638683, a novel SGK inhibitor with antihypertensive potency. *Cellular Physiology and Biochemistry, 28*, 137–146.

Alessi, D. R., Caudwell, F. B., Andjelkovic, M., Hemmings, B. A., & Cohen, P. (1996). Molecular basis for the substrate specificity of protein kinase B; comparison with MAPKAP kinase-1 and p70 S6 kinase. *FEBS Letters, 399*, 333–338.

Amato, R., Scumaci, D., D'Antona, L., Iuliano, R., Menniti, M., Di Sanzo, M., et al. (2013). Sgk1 enhances RANBP1 transcript levels and decreases taxol sensitivity in RKO colon carcinoma cells. *Oncogene, 32*, 4572–4578.

Andersen, M. N., Krzystanek, K., Petersen, F., Bomholtz, S. H., Olesen, S. P., Abriel, H., et al. (2013). A phosphoinositide 3-kinase (PI3K)-serum- and glucocorticoid-inducible kinase 1 (SGK1) pathway promotes Kv7.1 channel surface expression by inhibiting Nedd4-2 protein. *The Journal of Biological Chemistry, 288*, 36841–36854.

Andres-Mateos, E., Brinkmeier, H., Burks, T. N., Mejias, R., Files, D. C., Steinberger, M., et al. (2013). Activation of serum/glucocorticoid-induced kinase 1 (SGK1) is important to maintain skeletal muscle homeostasis and prevent atrophy. *EMBO Molecular Medicine, 5*, 80–91.

Arencibia, J. M., Pastor-Flores, D., Bauer, A. F., Schulze, J. O., & Biondi, R. M. (2013). AGC protein kinases: From structural mechanism of regulation to allosteric drug development for the treatment of human diseases. *Biochimica et Biophysica Acta, 1834*, 1302–1321.

Arranz, A., Doxaki, C., Vergadi, E., Martinez de la Torre, Y., Vaporidi, K., Lagoudaki, E. D., et al. (2012). Akt1 and Akt2 protein kinases differentially contribute to macrophage polarization. *Proceedings of the National Academy of Sciences of the United States of America, 109*, 9517–9522.

Arteaga, M. F., & Canessa, C. M. (2005). Functional specificity of Sgk1 and Akt1 on ENaC activity. *American Journal of Physiology. Renal Physiology, 289*, F90–F96.

Arteaga, M. F., Wang, L., Ravid, T., Hochstrasser, M., & Canessa, C. M. (2006). An amphipathic helix targets serum and glucocorticoid-induced kinase 1 to the endoplasmic reticulum-associated ubiquitin-conjugation machinery. *Proceedings of the National Academy of Sciences of the United States of America, 103*, 11178–11183.

Babaev, V. R., Hebron, K. E., Wiese, C. B., Toth, C. L., Ding, L., Zhang, Y., et al. (2014). Macrophage deficiency of Akt2 reduces atherosclerosis in Ldlr null mice. *Journal of Lipid Research, 55*, 2296–2308.

Belova, L., Sharma, S., Brickley, D. R., Nicolarsen, J. R., Patterson, C., & Conzen, S. D. (2006). Ubiquitin-proteasome degradation of serum- and glucocorticoid-regulated kinase-1 (SGK-1) is mediated by the chaperone-dependent E3 ligase CHIP. *The Biochemical Journal, 400*, 235–244.

Berwick, D. C., Dell, G. C., Welsh, G. I., Heesom, K. J., Hers, I., Fletcher, L. M., et al. (2004). Protein kinase B phosphorylation of PIKfyve regulates the trafficking of GLUT4 vesicles. *Journal of Cell Science, 117*, 5985–5993.

Biondi, R. M., Kieloch, A., Currie, R. A., Deak, M., & Alessi, D. R. (2001). The PIF-binding pocket in PDK1 is essential for activation of S6K and SGK, but not PKB. *The EMBO Journal, 20*, 4380–4390.

Boehmer, C., Okur, F., Setiawan, I., Broer, S., & Lang, F. (2003). Properties and regulation of glutamine transporter SN1 by protein kinases SGK and PKB. *Biochemical and Biophysical Research Communications, 306*, 156–162.

Bogusz, A. M., Brickley, D. R., Pew, T., & Conzen, S. D. (2006). A novel N-terminal hydrophobic motif mediates constitutive degradation of serum- and glucocorticoid-induced kinase-1 by the ubiquitin-proteasome pathway. *The FEBS Journal, 273*, 2913–2928.

Borst, O., Schaub, M., Walker, B., Schmid, E., Munzer, P., Voelkl, J., et al. (2015). Pivotal role of serum- and glucocorticoid-inducible kinase 1 in vascular inflammation and atherogenesis. *Arteriosclerosis, Thrombosis, and Vascular Biology, 35,* 547–557.

Brickley, D. R., Mikosz, C. A., Hagan, C. R., & Conzen, S. D. (2002). Ubiquitin modification of serum and glucocorticoid-induced protein kinase-1 (SGK-1). *The Journal of Biological Chemistry, 277,* 43064–43070.

Brunet, A., Park, J., Tran, H., Hu, L. S., Hemmings, B. A., & Greenberg, M. E. (2001). Protein kinase SGK mediates survival signals by phosphorylating the forkhead transcription factor FKHRL1 (FOXO3a). *Molecular and Cellular Biology, 21,* 952–965.

Burchfield, J. G., Lennard, A. J., Narasimhan, S., Hughes, W. E., Wasinger, V. C., Corthals, G. L., et al. (2004). Akt mediates insulin-stimulated phosphorylation of Ndrg2: Evidence for cross-talk with protein kinase C theta. *The Journal of Biological Chemistry, 279,* 18623–18632.

Calleja, V., Alcor, D., Laguerre, M., Park, J., Vojnovic, B., Hemmings, B. A., et al. (2007). Intramolecular and intermolecular interactions of protein kinase B define its activation in vivo. *PLoS Biology, 5,* e95.

Caohuy, H., Yang, Q., Eudy, Y., Ha, T. A., Xu, A. E., Glover, M., et al. (2014). Activation of 3-phosphoinositide-dependent kinase 1 (PDK1) and serum- and glucocorticoid-induced protein kinase 1 (SGK1) by short-chain sphingolipid C4-ceramide rescues the trafficking defect of DeltaF508-cystic fibrosis transmembrane conductance regulator (DeltaF508-CFTR). *The Journal of Biological Chemistry, 289,* 35953–35968.

Castel, P., Ellis, H., Bago, R., Toska, E., Razavi, P., Carmona, F. J., et al. (2016). PDK1-SGK1 signaling sustains AKT-independent mTORC1 activation and confers resistance to PI3Kalpha inhibition. *Cancer Cell, 30,* 229–242.

Chang, J. D., Sukhova, G. K., Libby, P., Schvartz, E., Lichtenstein, A. H., Field, S. J., et al. (2007). Deletion of the phosphoinositide 3-kinase p110gamma gene attenuates murine atherosclerosis. *Proceedings of the National Academy of Sciences of the United States of America, 104,* 8077–8082.

Chen, W. S., Xu, P. Z., Gottlob, K., Chen, M. L., Sokol, K., Shiyanova, T., et al. (2001). Growth retardation and increased apoptosis in mice with homozygous disruption of the Akt1 gene. *Genes & Development, 15,* 2203–2208.

Cho, H., Mu, J., Kim, J. K., Thorvaldsen, J. L., Chu, Q., Crenshaw, E. B., 3rd, et al. (2001). Insulin resistance and a diabetes mellitus-like syndrome in mice lacking the protein kinase Akt2 (PKB beta). *Science, 292,* 1728–1731.

Chun, J., Kwon, T., Kim, D. J., Park, I., Chung, G., Lee, E. J., et al. (2003). Inhibition of mitogen-activated protein kinase kinase kinase 3 activity through phosphorylation by the serum- and glucocorticoid-induced kinase 1. *Journal of Biochemistry, 133,* 103–108.

Collins, B. J., Deak, M., Arthur, J. S., Armit, L. J., & Alessi, D. R. (2003). In vivo role of the PIF-binding docking site of PDK1 defined by knock-in mutation. *The EMBO Journal, 22,* 4202–4211.

Delgoffe, G. M., Pollizzi, K. N., Waickman, A. T., Heikamp, E., Meyers, D. J., Horton, M. R., et al. (2011). The kinase mTOR regulates the differentiation of helper T cells through the selective activation of signaling by mTORC1 and mTORC2. *Nature Immunology, 12,* 295–303.

Di Pietro, N., Panel, V., Hayes, S., Bagattin, A., Meruvu, S., Pandolfi, A., et al. (2010). Serum- and glucocorticoid-inducible kinase 1 (SGK1) regulates adipocyte differentiation via forkhead box O1. *Molecular Endocrinology, 24,* 370–380.

Eirew, P., Steif, A., Khattra, J., Ha, G., Yap, D., Farahani, H., et al. (2015). Dynamics of genomic clones in breast cancer patient xenografts at single-cell resolution. *Nature, 518,* 422–426.

Fagerli, U. M., Ullrich, K., Stuhmer, T., Holien, T., Kochert, K., Holt, R. U., et al. (2011). Serum/glucocorticoid-regulated kinase 1 (SGK1) is a prominent target gene of the transcriptional response to cytokines in multiple myeloma and supports the growth of myeloma cells. *Oncogene, 30*, 3198–3206.

Gao, D., Wan, L., Inuzuka, H., Berg, A. H., Tseng, A., Zhai, B., et al. (2010). Rictor forms a complex with Cullin-1 to promote SGK1 ubiquitination and destruction. *Molecular Cell, 39*, 797–808.

Garcia-Martinez, J. M., & Alessi, D. R. (2008). mTOR complex 2 (mTORC2) controls hydrophobic motif phosphorylation and activation of serum- and glucocorticoid-induced protein kinase 1 (SGK1). *The Biochemical Journal, 416*, 375–385.

Gasser, J. A., Inuzuka, H., Lau, A. W., Wei, W., Beroukhim, R., & Toker, A. (2014). SGK3 mediates INPP4B-dependent PI3K signaling in breast cancer. *Molecular Cell, 56*, 595–607.

Gehring, E. M., Zurn, A., Klaus, F., Laufer, J., Sopjani, M., Lindner, R., et al. (2009). Regulation of the glutamate transporter EAAT2 by PIKfyve. *Cellular Physiology and Biochemistry, 24*, 361–368.

Glass, D. J. (2010). PI3 kinase regulation of skeletal muscle hypertrophy and atrophy. *Current Topics in Microbiology and Immunology, 346*, 267–278.

Goncalves, M. D., Pistilli, E. E., Balduzzi, A., Birnbaum, M. J., Lachey, J., Khurana, T. S., et al. (2010). Akt deficiency attenuates muscle size and function but not the response to ActRIIB inhibition. *PloS One, 5*, e12707.

Grahammer, F., Artunc, F., Sandulache, D., Rexhepaj, R., Friedrich, B., Risler, T., et al. (2006). Renal function of gene-targeted mice lacking both SGK1 and SGK3. *American Journal of Physiology. Regulatory, Integrative and Comparative Physiology, 290*, R945–R950.

Grahammer, F., Henke, G., Sandu, C., Rexhepaj, R., Hussain, A., Friedrich, B., et al. (2006). Intestinal function of gene-targeted mice lacking serum- and glucocorticoid-inducible kinase 1. *American Journal of Physiology. Gastrointestinal and Liver Physiology, 290*, G1114–G1123.

Gratton, J. P., Morales-Ruiz, M., Kureishi, Y., Fulton, D., Walsh, K., & Sessa, W. C. (2001). Akt down-regulation of p38 signaling provides a novel mechanism of vascular endothelial growth factor-mediated cytoprotection in endothelial cells. *The Journal of Biological Chemistry, 276*, 30359–30365.

Guan, K. L., Figueroa, C., Brtva, T. R., Zhu, T., Taylor, J., Barber, T. D., et al. (2000). Negative regulation of the serine/threonine kinase B-Raf by Akt. *The Journal of Biological Chemistry, 275*, 27354–27359.

Halland, N., Schmidt, F., Weiss, T., Saas, J., Li, Z., Czech, J., et al. (2015). Discovery of N-[4-(1H-pyrazolo[3,4-b]pyrazin-6-yl)-phenyl]-sulfonamides as highly active and selective SGK1 inhibitors. *ACS Medicinal Chemistry Letters, 6*, 73–78.

Hartmann, S., Schuhmacher, B., Rausch, T., Fuller, L., Doring, C., Weniger, M., et al. (2016). Highly recurrent mutations of SGK1, DUSP2 and JUNB in nodular lymphocyte predominant Hodgkin lymphoma. *Leukemia, 30*, 844–853.

Heikamp, E. B., Patel, C. H., Collins, S., Waickman, A., Oh, M. H., Sun, I. H., et al. (2014). The AGC kinase SGK1 regulates TH1 and TH2 differentiation downstream of the mTORC2 complex. *Nature Immunology, 15*, 457–464.

Hernandez, A. L., Kitz, A., Wu, C., Lowther, D. E., Rodriguez, D. M., Vudattu, N., et al. (2015). Sodium chloride inhibits the suppressive function of FOXP3+ regulatory T cells. *The Journal of Clinical Investigation, 125*, 4212–4222.

Kent, L. N., Ohboshi, S., & Soares, M. J. (2012). Akt1 and insulin-like growth factor 2 (Igf2) regulate placentation and fetal/postnatal development. *The International Journal of Developmental Biology, 56*, 255–261.

Kim, K. W., Cho, M. L., Park, M. K., Yoon, C. H., Park, S. H., Lee, S. H., et al. (2005). Increased interleukin-17 production via a phosphoinositide 3-kinase/Akt and nuclear factor kappaB-dependent pathway in patients with rheumatoid arthritis. *Arthritis Research & Therapy, 7*, R139–R148.

Kleinewietfeld, M., Manzel, A., Titze, J., Kvakan, H., Yosef, N., Linker, R. A., et al. (2013). Sodium chloride drives autoimmune disease by the induction of pathogenic TH17 cells. *Nature, 496,* 518–522.

Kobayashi, T., & Cohen, P. (1999). Activation of serum- and glucocorticoid-regulated protein kinase by agonists that activate phosphatidylinositide 3-kinase is mediated by 3-phosphoinositide-dependent protein kinase-1 (PDK1) and PDK2. *The Biochemical Journal, 339*(Pt. 2), 319–328.

Kobayashi, T., Deak, M., Morrice, N., & Cohen, P. (1999). Characterization of the structure and regulation of two novel isoforms of serum- and glucocorticoid-induced protein kinase. *The Biochemical Journal, 344*(Pt. 1), 189–197.

Lague, M. N., Detmar, J., Paquet, M., Boyer, A., Richards, J. S., Adamson, S. L., et al. (2010). Decidual PTEN expression is required for trophoblast invasion in the mouse. *American Journal of Physiology. Endocrinology and Metabolism, 299,* E936–E946.

Lang, F., & Cohen, P. (2001). Regulation and physiological roles of serum- and glucocorticoid-induced protein kinase isoforms. *Science's STKE, 2001,* re17.

Lang, F., & Shumilina, E. (2013). Regulation of ion channels by the serum- and glucocorticoid-inducible kinase SGK1. *The FASEB Journal, 27,* 3–12.

Lee, I. H., Dinudom, A., Sanchez-Perez, A., Kumar, S., & Cook, D. I. (2007). Akt mediates the effect of insulin on epithelial sodium channels by inhibiting Nedd4-2. *The Journal of Biological Chemistry, 282,* 29866–29873.

Lee, K., Gudapati, P., Dragovic, S., Spencer, C., Joyce, S., Killeen, N., et al. (2010). Mammalian target of rapamycin protein complex 2 regulates differentiation of Th1 and Th2 cell subsets via distinct signaling pathways. *Immunity, 32,* 743–753.

Leong, M. L., Maiyar, A. C., Kim, B., O'Keeffe, B. A., & Firestone, G. L. (2003). Expression of the serum- and glucocorticoid-inducible protein kinase, Sgk, is a cell survival response to multiple types of environmental stress stimuli in mammary epithelial cells. *The Journal of Biological Chemistry, 278,* 5871–5882.

Li, B., Tournier, C., Davis, R. J., & Flavell, R. A. (1999). Regulation of IL-4 expression by the transcription factor JunB during T helper cell differentiation. *The EMBO Journal, 18,* 420–432.

Liu, P., Gan, W., Chin, Y. R., Ogura, K., Guo, J., Zhang, J., et al. (2015). PtdIns(3,4,5)P3-dependent activation of the mTORC2 kinase complex. *Cancer Discovery, 5,* 1194–1209.

Liu, H., Yu, J., Xia, T., Xiao, Y., Zhang, Q., Liu, B., et al. (2014). Hepatic serum- and glucocorticoid-regulated protein kinase 1 (SGK1) regulates insulin sensitivity in mice via extracellular-signal-regulated kinase 1/2 (ERK1/2). *The Biochemical Journal, 464,* 281–289.

Liu, C., Zhu, L. L., Xu, S. G., Ji, H. L., & Li, X. M. (2016). ENaC/DEG in tumor development and progression. *Journal of Cancer, 7,* 1888–1891.

Lopez-Barradas, A., Gonzalez-Cid, T., Vazquez, N., Gavi-Maza, M., Reyes-Camacho, A., Velazquez-Villegas, L. A., et al. (2016). Insulin and SGK1 reduce the function of Na^+/monocarboxylate transporter 1 (SMCT1/SLC5A8). *American Journal of Physiology. Cell Physiology, 311,* C720–C734.

Lu, M., Wang, J., Jones, K. T., Ives, H. E., Feldman, M. E., Yao, L. J., et al. (2010). mTOR complex-2 activates ENaC by phosphorylating SGK1. *Journal of the American Society of Nephrology, 21,* 811–818.

Ma, L., Kerr, B. A., Naga Prasad, S. V., Byzova, T. V., & Somanath, P. R. (2014). Differential effects of Akt1 signaling on short- versus long-term consequences of myocardial infarction and reperfusion injury. *Laboratory Investigation, 94,* 1083–1091.

Mansley, M. K., & Wilson, S. M. (2010). Effects of nominally selective inhibitors of the kinases PI3K, SGK1 and PKB on the insulin-dependent control of epithelial Na+ absorption. *British Journal of Pharmacology, 161,* 571–588.

Milburn, C. C., Deak, M., Kelly, S. M., Price, N. C., Alessi, D. R., & Van Aalten, D. M. (2003). Binding of phosphatidylinositol 3,4,5-trisphosphate to the pleckstrin homology domain of protein kinase B induces a conformational change. *The Biochemical Journal*, *375*, 531–538.

Mizuno, H., & Nishida, E. (2001). The ERK MAP kinase pathway mediates induction of SGK (serum- and glucocorticoid-inducible kinase) by growth factors. *Genes to Cells*, *6*, 261–268.

Murray, J. T., Campbell, D. G., Morrice, N., Auld, G. C., Shpiro, N., Marquez, R., et al. (2004). Exploitation of KESTREL to identify NDRG family members as physiological substrates for SGK1 and GSK3. *The Biochemical Journal*, *384*, 477–488.

Murray, J. T., Cummings, L. A., Bloomberg, G. B., & Cohen, P. (2005). Identification of different specificity requirements between SGK1 and PKBalpha. *FEBS Letters*, *579*, 991–994.

Naray-Fejes-Toth, A., Canessa, C., Cleaveland, E. S., Aldrich, G., & Fejes-Toth, G. (1999). sgk is an aldosterone-induced kinase in the renal collecting duct. Effects on epithelial na+ channels. *The Journal of Biological Chemistry*, *274*, 16973–16978.

Nasir, O., Wang, K., Foller, M., Gu, S., Bhandaru, M., Ackermann, T. F., et al. (2009). Relative resistance of SGK1 knockout mice against chemical carcinogenesis. *IUBMB Life*, *61*, 768–776.

Pao, A. C., Bhargava, A., Di Sole, F., Quigley, R., Shao, X., Wang, J., et al. (2010). Expression and role of serum and glucocorticoid-regulated kinase 2 in the regulation of Na+/H+ exchanger 3 in the mammalian kidney. *American Journal of Physiology. Renal Physiology*, *299*, F1496–F1506.

Park, J., Leong, M. L., Buse, P., Maiyar, A. C., Firestone, G. L., & Hemmings, B. A. (1999). Serum and glucocorticoid-inducible kinase (SGK) is a target of the PI 3-kinase-stimulated signaling pathway. *The EMBO Journal*, *18*, 3024–3033.

Pearce, L. R., Komander, D., & Alessi, D. R. (2010). The nuts and bolts of AGC protein kinases. *Nature Reviews. Molecular Cell Biology*, *11*, 9–22.

Sahoo, S., Brickley, D. R., Kocherginsky, M., & Conzen, S. D. (2005). Coordinate expression of the PI3-kinase downstream effectors serum and glucocorticoid-induced kinase (SGK-1) and Akt-1 in human breast cancer. *European Journal of Cancer*, *41*, 2754–2759.

Salifou, K., Ray, S., Verrier, L., Aguirrebengoa, M., Trouche, D., Panov, K. I., et al. (2016). The histone demethylase JMJD2A/KDM4A links ribosomal RNA transcription to nutrients and growth factors availability. *Nature Communications*, *7*, 10174.

Salker, M. S., Christian, M., Steel, J. H., Nautiyal, J., Lavery, S., Trew, G., et al. (2011). Deregulation of the serum- and glucocorticoid-inducible kinase SGK1 in the endometrium causes reproductive failure. *Nature Medicine*, *17*, 1509–1513.

Sandu, C., Artunc, F., Palmada, M., Rexhepaj, R., Grahammer, F., Hussain, A., et al. (2006). Impaired intestinal NHE3 activity in the PDK1 hypomorphic mouse. *American Journal of Physiology. Gastrointestinal and Liver Physiology*, *291*, G868–G876.

Schnackenberg, C. G., Costell, M. H., Bernard, R. E., Minuti, K. K., Grygielko, E. T., Parsons, M. J., et al. (2007). Compensatory role for Sgk2 mediated sodium reabsorption during salt deprivation in Sgk1 knockout mice. *The FASEB Journal*, *21*, A508.

Seebohm, G., Strutz-Seebohm, N., Birkin, R., Dell, G., Bucci, C., Spinosa, M. R., et al. (2007). Regulation of endocytic recycling of KCNQ1/KCNE1 potassium channels. *Circulation Research*, *100*, 686–692.

Sherk, A. B., Frigo, D. E., Schnackenberg, C. G., Bray, J. D., Laping, N. J., Trizna, W., et al. (2008). Development of a small-molecule serum- and glucocorticoid-regulated kinase-1 antagonist and its evaluation as a prostate cancer therapeutic. *Cancer Research*, *68*, 7475–7483.

Shisheva, A. (2012). PIKfyve and its Lipid products in health and in sickness. *Current Topics in Microbiology and Immunology*, *362*, 127–162.

Shojaiefard, M., Strutz-Seebohm, N., Tavare, J. M., Seebohm, G., & Lang, F. (2007). Regulation of the Na(+), glucose cotransporter by PIKfyve and the serum and glucocorticoid inducible kinase SGK1. *Biochemical and Biophysical Research Communications, 359,* 843–847.

Sommer, E. M., Dry, H., Cross, D., Guichard, S., Davies, B. R., & Alessi, D. R. (2013). Elevated SGK1 predicts resistance of breast cancer cells to Akt inhibitors. *The Biochemical Journal, 452,* 499–508.

Sopjani, M., Kunert, A., Czarkowski, K., Klaus, F., Laufer, J., Foller, M., et al. (2010). Regulation of the Ca(2+) channel TRPV6 by the kinases SGK1, PKB/Akt, and PIKfyve. *The Journal of Membrane Biology, 233,* 35–41.

Steffan, J. J., Williams, B. C., Welbourne, T., & Cardelli, J. A. (2010). HGF-induced invasion by prostate tumor cells requires anterograde lysosome trafficking and activity of Na^+-H^+ exchangers. *Journal of Cell Science, 123,* 1151–1159.

Tessier, M., & Woodgett, J. R. (2006). Role of the Phox homology domain and phosphorylation in activation of serum and glucocorticoid-regulated kinase-3. *The Journal of Biological Chemistry, 281,* 23978–23989.

Urrego, D., Tomczak, A. P., Zahed, F., Stühmer, W., & Pardo, L. A. (2014). Potassium channels in cell cycle and cell proliferation. *Philosophical Transactions of the Royal Society, B: Biological Sciences, 369,* 20130094.

Waldegger, S., Barth, P., Raber, G., & Lang, F. (1997). Cloning and characterization of a putative human serine/threonine protein kinase transcriptionally modified during anisotonic and isotonic alterations of cell volume. *Proceedings of the National Academy of Sciences of the United States of America, 94,* 4440–4445.

Waldegger, S., Klingel, K., Barth, P., Sauter, M., Rfer, M. L., Kandolf, R., et al. (1999). h-sgk serine-threonine protein kinase gene as transcriptional target of transforming growth factor beta in human intestine. *Gastroenterology, 116,* 1081–1088.

Wang, K., Gu, S., Nasir, O., Foller, M., Ackermann, T. F., Klingel, K., et al. (2010). SGK1-dependent intestinal tumor growth in APC-deficient mice. *Cellular Physiology and Biochemistry, 25,* 271–278.

Wang, D., Sun, H., Lang, F., & Yun, C. C. (2005). Activation of NHE3 by dexamethasone requires phosphorylation of NHE3 at Ser663 by SGK1. *American Journal of Physiology. Cell Physiology, 289,* C802–C810.

Webster, M. K., Goya, L., Ge, Y., Maiyar, A. C., & Firestone, G. L. (1993). Characterization of sgk, a novel member of the serine/threonine protein kinase gene family which is transcriptionally induced by glucocorticoids and serum. *Molecular and Cellular Biology, 13,* 2031–2040.

Wu, C., Yosef, N., Thalhamer, T., Zhu, C., Xiao, S., Kishi, Y., et al. (2013). Induction of pathogenic TH17 cells by inducible salt-sensing kinase SGK1. *Nature, 496,* 513–517.

Wulff, P., Vallon, V., Huang, D. Y., Volkl, H., Yu, F., Richter, K., et al. (2002). Impaired renal Na(+) retention in the sgk1-knockout mouse. *The Journal of Clinical Investigation, 110,* 1263–1268.

Xie, J., Pan, H., Yao, J., Zhou, Y., & Han, W. (2016). SOCE and cancer: Recent progress and new perspectives. *International Journal of Cancer, 138,* 2067–2077.

Yang, Z. Z., Tschopp, O., Hemmings-Mieszczak, M., Feng, J., Brodbeck, D., Perentes, E., et al. (2003). Protein kinase B alpha/Akt1 regulates placental development and fetal growth. *The Journal of Biological Chemistry, 278,* 32124–32131.

Yang, M., Zheng, J., Miao, Y., Wang, Y., Cui, W., Guo, J., et al. (2012). Serum-glucocorticoid regulated kinase 1 regulates alternatively activated macrophage polarization contributing to angiotensin II-induced inflammation and cardiac fibrosis. *Arteriosclerosis, Thrombosis, and Vascular Biology, 32,* 1675–1686.

Zapf, J., Meyer, T., Wade, W., Lingardo, L., Batova, A., Alton, G., et al. (2016). Abstract 2180: Drug-like inhibitors of SGK1: Discovery and optimization of low molecular weight fragment leads. *Cancer Research, 76*, 2180.

Zarrinpashneh, E., Poggioli, T., Sarathchandra, P., Lexow, J., Monassier, L., Terracciano, C., et al. (2013). Ablation of SGK1 impairs endothelial cell migration and tube formation leading to decreased neo-angiogenesis following myocardial infarction. *PloS One, 8*, e80268.

Zhang, B. H., Tang, E. D., Zhu, T., Greenberg, M. E., Vojtek, A. B., & Guan, K. L. (2001). Serum- and glucocorticoid-inducible kinase SGK phosphorylates and negatively regulates B-Raf. *The Journal of Biological Chemistry, 276*, 31620–31626.

Zhou, R., & Snyder, P. M. (2005). Nedd4-2 phosphorylation induces serum and glucocorticoid-regulated kinase (SGK) ubiquitination and degradation. *The Journal of Biological Chemistry, 280*, 4518–4523.

CHAPTER THREE

Homeodomain-Interacting Protein Kinases: Diverse and Complex Roles in Development and Disease

Jessica A. Blaquiere, Esther M. Verheyen[1]

Department of Molecular Biology and Biochemistry, Centre for Cell Biology, Development and Disease, Simon Fraser University, Burnaby, BC, Canada
[1]Corresponding author: e-mail address: everheye@sfu.ca

Contents

1. Introduction	74
1.1 Hipk Protein Family	74
1.2 Dynamic Tissue-Specific and Intracellular Localization of Hipk Proteins	76
2. Requirements for Hipk Proteins During Development	78
2.1 Regulation of Cell Proliferation and Cell Death	79
2.2 Neuronal Development and Survival	81
2.3 Angiogenesis and Hematopoiesis	81
3. Hipk Roles in Regulation of Diverse Signaling Pathways	82
3.1 Hipk Antagonizes the Global Corepressor Gro	82
3.2 Modulating the Cellular Response to DNA Damage Through p53	83
3.3 Regulation of Wnt Signaling by Hipks	85
3.4 Hipk Promotes Yki Activity in the Hippo Pathway	86
3.5 JNK Signaling Is Modulated Through Hipk	87
3.6 Hipk Is Required for JAK/STAT Activity in Normal Development and Tumorigenesis	87
4. Regulation of Hipk Activity	88
4.1 Phosphorylation	88
4.2 SUMO Modification	89
4.3 Ubiquitin and Proteasomal Degradation	90
4.4 Acetylation	90
4.5 Regulation of Hipk Gene Expression	91
4.6 Drug-Based Inhibition of Hipk	91
5. Hipks in Disease	91
5.1 Hipks as Tumor Suppressors in Cancer	92
5.2 Hipks as Oncogenic Factors in Cancer	92
5.3 Hipks and Fibrosis	93

6. Conclusions	93
Acknowledgments	94
References	94

Abstract

The Homeodomain-interacting protein kinase (Hipk) family of proteins plays diverse, and at times conflicting, biological roles in normal development and disease. In this review we will highlight developmental and cellular roles for Hipk proteins, with an emphasis on the pleiotropic and essential physiological roles revealed through genetic studies. We discuss the myriad ways of regulating Hipk protein function, and how these may contribute to the diverse cellular roles. Furthermore we will describe the context-specific activities of Hipk family members in diseases such as cancer and fibrosis, including seemingly contradictory tumor-suppressive and oncogenic activities. Given the diverse signaling pathways regulated by Hipk proteins, it is likely that Hipks act to fine-tune signaling and may mediate cross talk in certain contexts. Such regulation is emerging as vital for development and in disease.

1. INTRODUCTION
1.1 Hipk Protein Family

The Homeodomain-interacting protein kinase (Hipk) family of proteins was first identified in a yeast two-hybrid screen to identify cofactors for the NKx-1.2 family of homeoproteins (Kim, Choi, Lee, Conti, & Kim, 1998). Hipk1 and Hipk2, which are the most closely related sharing more than 93% amino acid identity in their kinase domains, were identified in this initial screen. Hipk3 is slightly less conserved, sharing roughly 87% identity with Hipk1 and 2 in the kinase domain (Van der Laden, Soppa, & Becker, 2015). The much smaller Hipk4 protein, which was identified later, is the most diverged member of the Hipk family and shares roughly 50% homology within the kinase domain (Arai et al., 2007; He et al., 2010). Hipk family members are conserved across species, with vertebrates possessing four orthologs, called Hipk1–4 and simpler organisms such as *Drosophila* (whose ortholog for clarity we will refer to as *dHipk*) and *C. elegans* (*hpk-1*) encoding one each. In this review we will refer to the family generically as Hipk, and use specific names (e.g., Hipk2, dHipk) when describing particular findings. Hipk2 is the best-characterized vertebrate paralog, and thus much of our understanding of the family comes from its study. Hipk proteins are dual specificity serine/threonine and tyrosine kinases which are roughly 40% identical to the Dyrk family of proteins within the kinase domain (Hofmann, Mincheva, Lichter, Droge, & Schmitz, 2000; Kentrup et al., 1996).

Hipk1 and Hipk3 share a number of functional domains with Hipk2 (Fig. 1). The N-terminus is composed of a highly conserved kinase domain, which is also found in Hipk4. The central portion of Hipk2 contains numerous protein interaction domains that bind homeodomain transcription factors (HID), as well as other proteins such as Siah-1, p53, and CtBP, the speckle retention sequence (SRS), and the SUMO-interacting motif (SIM) (Schmitz, Rodriguez-Gil, & Hornung, 2014; Sombroek & Hofmann, 2009). The HID domains show between 72% and 75% similarity between Hipk1–3 (Kim et al., 1998). Hipk2 has two or three nuclear localization sequences, while Hipk1 has just one, and the nuclear targeting sequence in Hipk3 remains unknown (Schmitz et al., 2014). The C-terminus contains a number of features that are referred to in varying ways. A PEST domain overlaps with the SRS region. The most C-terminal region is referred to either as a region rich in tyrosine and histidine residues (referred to as the YH domain in some studies (Kim et al., 1998)) or enriched for serine, glutamine, and alanine (referred to as the S, Q, A region in some studies (de la Vega et al., 2011)) and shows between 66% and 73% similarity between Hipks1–3.

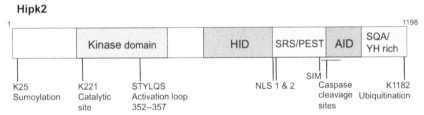

Fig. 1 A schematic diagram of Hipk2. The prototypical Hipk family member Hipk2 has domains conserved in Hipks1–3, including the kinase domain, homeoprotein-interacting domain (HID), PEST domain, speckle retention signal (SRS), and auto-inhibitory domain (AID). Also shown are sites that are regulated by phosphorylation, ubiquitination, and caspase cleavage. *Amino acid domain numbering adapted from Rinaldo, C., Prodosmo, A., Siepi, F., & Soddu, S. (2007). HIPK2: A multitalented partner for transcription factors in DNA damage response and development. Biochemistry and Cell Biology, 85, 411–418; and Van der Laden, J., Soppa, U., & Becker, W. (2015). Effect of tyrosine autophosphorylation on catalytic activity and subcellular localisation of homeodomain-interacting protein kinases (HIPK). Cell Communication and Signaling: CCS 13, 3 [Epub]. doi:10.1186/s12964-014-0082-6.*

This region can also serve as an autoinhibitory domain (AID), since its removal increases Hipk2's ability to stimulate p53 activity (Rui et al., 2004). Overall, Hipk proteins contain numerous sites of protein–protein interaction and regions targeted by posttranslational modifications (PTMs) (discussed below).

1.2 Dynamic Tissue-Specific and Intracellular Localization of Hipk Proteins

Hipk family members are expressed in diverse and dynamic temporal and spatial patterns, highlighting their important roles during development. Hipk protein levels are highly regulated by PTM and proteasomal degradation [for an excellent recent review, see Saul and Schmitz (2013)]. In this review we will highlight key features of protein modulation as they affect development and disease.

In mice, *Hipk2* has been found in multiple tissues including brain, kidney, heart, skin, and muscle (Pierantoni et al., 2002; Wei et al., 2007). In *X. laevis*, *Hipk1* is expressed throughout embryonic development, with a peak at gastrulation and in a number of tissues at later stages (Kondo et al., 2003; Louie et al., 2009). In *C. elegans Hpk-1* is expressed at low levels in somatic cell types (Berber et al., 2013). *Hipks* across species are expressed within multiple neuronal cell types, including the cerebral cortex, neural retina, neurofilament cells, and neural tube (Ciarapica et al., 2014; Inoue et al., 2010; Isono et al., 2006; Lee, Andrews, Faust, Walldorf, & Verheyen, 2009; Pierantoni et al., 2002; Rochat-Steiner et al., 2000; Wiggins et al., 2004).

Hipk proteins show dynamic intracellular localizations (Fig. 2). Endogenous Hipk proteins can be difficult to observe in certain contexts due to their low levels; thus, transfected epitope-tagged Hipk constructs are often utilized to track Hipk localization, with the caveat that they reflect exogenous proteins. Hipk proteins, in vertebrate cell culture, *C. elegans*, and *Drosophila* tissues, are primarily in nuclear structures termed speckles, as well as low to moderate levels in the nucleoplasm and cytoplasm (Raich et al., 2003) (Fig. 2), with the exception of Hipk4 which is found primarily in the cytoplasm (Arai et al., 2007). Kim et al. (1998) determined that this localization is due to the SRS, and deletion of the C-terminus of Hipk2 results in a diffuse nuclear stain (Wang, Hofmann, et al., 2001). A detailed mutational analysis has revealed a tightly regulated nuclear and speckle localization process (de la Vega et al., 2011). The Hipk2 drug inhibitor D-115893 causes a delocalization of the protein, with relatively uniform nuclear and cytoplasmic

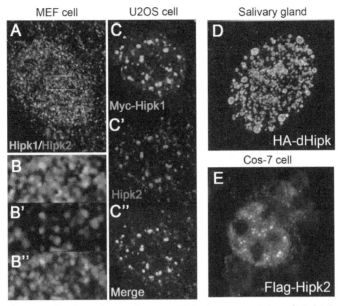

Fig. 2 Hipk expression in various cell types. (A) Endogenous stains of Hipk1 (*green*) and Hipk2 (*red*) in primary MEFs. (B–B″) Higher magnification images of the *boxed-in portion* seen in (A). (C–C″) A U2OS cell expressing Myc-Hipk1 and stained for endogenous Hipk2. (D) HA-dHipk expressed in the *Drosophila* salivary gland cell. (E) Flag-Hipk2 transfected into a Cos-7 cell. *Images (A–C″) courtesy of Isono, K., Nemoto, K., Li, Y., Takada, Y., Suzuki, R., Katsuki, M., et al. (2006). Overlapping roles for homeodomain-interacting protein kinases hipk1 and hipk2 in the mediation of cell growth in response to morphogenetic and genotoxic signals. Molecular and Cellular Biology, 26, 2758–2771, and images (D, E) (Joanna Chen, Jessica Blaquiere, and Esther Verheyen, unpublished).*

localization and reduced phosphorylation of p53 (de la Vega et al., 2011), suggesting altered localization impacts function. Localization is likely kinase independent as Hipk3 and kinase dead Hipk3 have the same nuclear and cytoplasmic distribution (Rochat-Steiner et al., 2000). Staining for Hipk1 and Hipk2 in mouse embryo fibroblasts reveals that they have largely non-overlapping speckle localizations, suggesting that they may carry out distinct roles (Fig. 2).

A subset of exogenous Hipk proteins colocalize with promyelocytic leukemia protein (PML) and p53-containing nuclear bodies (PML NBs) (D'Orazi et al., 2002; Engelhardt et al., 2003; Hofmann et al., 2002). Of note, endogenous Hipk2 is primarily localized to structures that do not contain PML, and Hipk2 can localize to speckles independent of p53. Nuclear speckles can also be interchromatin granule clusters which are enriched in pre-mRNA splicing factors (Spector & Lamond, 2011). Dyrk family members localize to such speckles and have been found to regulate splicing factors, although such colocalization has not been seen with Hipk2 (Alvarez, Estivill, & de la Luna, 2003; Kim, Choi, & Kim, 1999). Nuclear spots are also associated with antiviral defenses (Tavalai & Stamminger, 2008). Whether Hipk family members also carry out roles in splicing or antiviral defense is currently unknown.

While Hipks1–3 are primarily nuclear, in certain contexts Hipk proteins are found in nonnuclear compartments including at low levels in the cytoplasm and associated with the plasma membrane. The changes in Hipk localization are in part due to regulation by PTM, as has been seen in the case of sumoylation (see below), or through altered binding interactions and retention. Nuclear export or cytoplasmic retention could affect some Hipk functions, and may be important for others that are yet to be determined. Changes in localization are also likely to affect binding interactions, and may reflect cell type-specific roles. For example, the cytoskeletal protein Zyxin can colocalize with Hipk2 at cell edges in U2OS cells (Crone et al., 2011). Expression of a membrane-tethered dHipk can rescue some but not all aspects of *dhipk* mutant flies, suggesting that it has functions in both the nucleus and cytoplasm (J.A.B. and E.M.V., unpublished). A thorough understanding of the nonnuclear roles of Hipk family members will no doubt reveal additional functions for Hipk proteins.

2. REQUIREMENTS FOR HIPK PROTEINS DURING DEVELOPMENT

Analysis of loss-of-function phenotypes in diverse cell types and organisms has consistently shown that Hipk proteins can modulate numerous signaling pathways and that they are required for organismal survival (Fig. 3). Our goal in this review is to integrate and reconcile in vivo evidence with the plethora of cell culture work on Hipk family members and to highlight the context-dependent functions. We first describe biological effects of Hipk mutations, and then in subsequent sections describe molecular interactions and effects on signaling. These processes are highly intertwined, yet by using this approach we hope to emphasize the functional insights that can be gained through genetic studies, complemented by the biochemical and molecular evidence.

There is evidence of functional redundancy between the vertebrate Hipks (Inoue et al., 2010; Isono et al., 2006). While no single mouse ortholog is

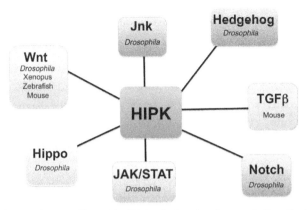

Fig. 3 Signaling pathways affected by Hipks during development. Genetically identified roles for Hipk proteins in development of model organisms. Loss-of-function and gain-of-function studies have revealed physiologically and developmentally relevant functions for Hipk family members, as described in the text. It is likely that many additional functions remain to be revealed after further study.

essential, loss of both *Hipk1* and *Hipk2* simultaneously results in lethality, caused in part by neural tube defects and homeotic transformations of the axial skeleton (Inoue et al., 2010; Isono et al., 2006; Kondo et al., 2003; Wiggins et al., 2004). *dHipk* and *hpk-1* are the only homologs in *Drosophila* and *C. elegans*, respectively, making them ideal for studying Hipk function. In *Drosophila dhipk* mutants die during larval and pupal stages, while loss of maternal and zygotic *dhipk* results in embryonic lethality (Lee, Andrews, et al., 2009; Lee, Swarup, Chen, Ishitani, & Verheyen, 2009).

2.1 Regulation of Cell Proliferation and Cell Death

Hipk family members are reported to exert distinct and contradictory effects on cell proliferation and tissue growth. Overexpressing dHipk causes dramatic overgrowth in the wing, eye, and leg tissues of *Drosophila* in a dose-dependent manner, due in part to promotion of Hippo pathway target gene expression (Chen & Verheyen, 2012; Lee, Andrews, et al., 2009; Lee, Swarup, et al., 2009; Poon, Zhang, Lin, Manning, & Harvey, 2012). In *C. elegans*, Hpk-1 nonautonomously promotes proliferation of the germline cells, and loss of *hpk-1* reduces the number of proliferating cells and size of the mitotic region (Berber et al., 2013). $Hipk2^{-/-}$ mice are born smaller than their littermates, have growth deficiencies and decreased amounts of adipose tissue, have difficulty gaining weight compared to their wild-type littermates, and 40% die prematurely (Chalazonitis et al., 2011; Sjölund,

Pelorosso, Quigley, DelRosario, & Balmain, 2014; Trapasso et al., 2009). $Hipk2^{-/-}$ mice are also less prone to become obese due to their increased insulin sensitivity. $Hipk3^{-/-}$ mice have disrupted glucose tolerance and insulin secretion, accompanied by decreased proliferation and increased apoptosis of pancreatic β-cells (Shojima et al., 2012). $Hipk3^{-/-}$ mice also show decreased proliferation that has been attributed to impaired Wnt signaling. Isono et al. (2006) found a decrease in mitotic cells in $Hipk1^{-/-} Hipk2^{-/-}$ mouse embryos. In normal human skin, Hipk2 protein expression is enriched in basal proliferating cells, while it is undetectable in nonproliferating cells (Iacovelli et al., 2009), and Hipk2 expression can be reactivated when cells are stimulated to proliferate, suggesting a close connection between Hipk protein function and cell proliferation.

Cell culture studies have also shown that Hipks are growth-promoting factors. Depletion of Hipk2 in peripheral blood lymphocytes causes cell cycle arrest (Iacovelli et al., 2009). Hipk2 depletion strongly reduced proliferation of MCF10A cells and human fibroblasts with no associated apoptosis. Further, $Hipk1^{-/-}$ mouse embryo fibroblasts (MEFs) show reduced proliferation and mice display decreased tumor formation after irradiation (Kondo et al., 2003). MEFs from $Hipk2^{-/-}$ knockout mice also show reduced proliferation associated with the G0/G1-specific cell cycle regulators CDK6 and Cyclin D (Trapasso et al., 2009), while another study claimed such cells proliferated more than wild type (Wei et al., 2007). Tetra- and polyploidization occurs in cultured $Hipk2^{-/-}$ cells, indicating that Hipk2 can promote cell proliferation by controlling the end stages of cytokinesis (Rinaldo et al., 2012).

Hipks can also inhibit cell proliferation in certain contexts. Loss of $Hipk2$ leads to increased proliferation of epidermal stem cells (Wei et al., 2007). Excessive endothelial cell growth was observed in the yolk sacs of $Hipk1^{-/-} Hipk2^{-/-}$ mice (Shang et al., 2013). We suggest that Hipk's functions are context dependent and may differ between homologs and paralogs. dHipk primarily acts as a positive regulator of growth, while vertebrate models have shown diverse roles for Hipks in growth. It will be interesting to see whether any of Hipk's effects are nonautonomous through compensatory proliferation mechanisms, since in C. elegans hpk-1 is required nonautonomously to promote germline proliferation (Berber et al., 2013).

In contrast to regulating proliferation, Hipks can play important roles in regulating apoptosis and programmed cell death (PCD) through various mechanisms during development and in particular cell stress. In Drosophila pupal development, dHipk is required for a form of collective cell death

(Link et al., 2007). Both dHipk overexpression and loss can induce caspase-related cell death, suggesting that the balance of Hipk levels is essential for overall homeostasis of tissues. In vertebrates, Hipk contributes to PCD that is utilized during different aspects of development to regulate neuronal survival and cell numbers (Doxakis, Huang, & Davies, 2004). The role of Hipk proteins in DNA damage-induced apoptosis is discussed below.

2.2 Neuronal Development and Survival

Hipks regulate the development of various neuronal subtypes in part through their role in inducing cell death. Flies mutant for *dHipk* possess small eyes with neuronal abnormalities (Blaquiere, Lee, & Verheyen, 2014; Lee, Andrews, et al., 2009, Lee, Swarup, et al., 2009). $Hipk1^{-/-}Hipk2^{-/-}$ mice have underdeveloped retinas and neural tube closure defects (Aikawa et al., 2006; Inoue et al., 2010; Isono et al., 2006). Hipk2 regulates the survival of sensory and sympathetic neurons as well as midbrain dopamine neurons in mice, and its expression peaks at stages where neuronal cell death occurs (Doxakis et al., 2004; Wiggins et al., 2004; Zhang et al., 2007). Hipk2 is a proapoptotic factor regulating cell death in the embryonic stages of mouse brain development and in cultured sensory neurons. In the mouse embryonic midbrain, Hipk2 seems to have an opposite effect such that loss of Hipk2 leads to a decrease in the number of neurons due to an increase in TGF-β-dependent apoptosis (Zhang et al., 2007), resulting in Parkinson-like psychomotor abnormalities. Postnatally, Hipk2 regulates the maintenance of enteric neurons and glia through BMP signaling (Chalazonitis et al., 2011). Blood vessels cannot grow into the neural tubes of $Hipk1^{-/-}Hipk2^{-/-}$ mice, which could contribute to neuronal deficiencies (Shang et al., 2013).

2.3 Angiogenesis and Hematopoiesis

Hipks can regulate angiogenesis, vasculogenesis, and hematopoiesis in numerous contexts (Aikawa et al., 2006; Li, Arai, et al., 2007; Ohtsu, Nobuhisa, Mochita, & Taga, 2007; Shang et al., 2013). $Hipk1^{-/-}Hipk2^{-/-}$ mice show severe abnormalities in the organization of blood vessel networks (Hattangadi, Burke, & Lodish, 2010). Hipks regulate angiogenesis through TGF-β signaling, and $Hipk1^{-/-}Hipk2^{-/-}$ mice have increased proliferation of endothelial cells. Perturbations in hematopoiesis occur in $Hipk1^{-/-}Hipk2^{-/-}$ mice which include decreased erythroid cell proliferation due to a dysfunctional cell cycle and downregulation of genes involved in proliferation and erythroid lineage specification (Hattangadi et al., 2010). $Hipk2^{-/-}$ mice show defects in thymic epithelial cells (Rattay et al., 2015), while $Hipk1^{-/-}$ mice have abnormal splenic B cell development (Guerra, Gommerman, Corfe, Paige, & Rottapel, 2012). dHipk also plays a role in the hemocyte lineage in *Drosophila* since knockdown of dHipk can reduce Jak/Stat-induced hematopoietic tumor burden and overexpression in blood cell precursors promotes tumors (Blaquiere, Wray, & Verheyen, 2016).

3. HIPK ROLES IN REGULATION OF DIVERSE SIGNALING PATHWAYS

Hipks were the first shown to act as transcriptional corepressors for homeodomain-containing transcription factors (Kim et al., 1998). Subsequent work in numerous systems and cell types has revealed their pleiotropic effects as modulators of diverse signaling pathways. Given the diverse pathways regulated by Hipk proteins, it is likely that Hipk function may mediate their cross talk and integration in certain contexts (recently reviewed in Schmitz et al., 2014). Hipks likely act to fine-tune signaling output. Such regulation is emerging as vital for development and tissue homeostasis, as well as in disease.

3.1 Hipk Antagonizes the Global Corepressor Gro

Groucho (Gro)/TLE proteins belong to a family of broadly expressed transcriptional corepressors that regulate multiple signaling pathways in different developmental contexts (Jennings & Ish-Horowicz, 2008). Gro, Hipk2, and the histone deacetylase complex (HDAC) act together as a corepressor complex and are recruited by NK-3 to regulate transcription (Choi, Kim, Kwon, & Kim, 1999). dHipk directly phosphorylates Gro on several residues to promote its dissociation from the corepressor complex (Choi et al., 2005).

Studies of Hipk and Gro revealed an interaction with the Pax6/Eyeless (Ey) master eye regulator proteins. dHipk can control Ey/Pax6 through Gro, but Hipk2 can also regulate Pax6 by phosphorylating it directly, enhancing its transactivation ability by accelerating the recruitment of the coactivator p300 (Kim et al., 2006). Notch signaling is important for growth of the *Drosophila* eye and dHipk promotes Notch target gene transcription by antagonizing Gro (Lee, Andrews, et al., 2009; Lee, Swarup, et al., 2009). dHipk's effect on Gro seems to be context dependent, since dHipk promotes eye imaginal disc growth by antagonizing Gro, while it promotes wing disc growth independently of this interaction (Lee, Andrews, et al., 2009; Lee, Swarup, et al., 2009). In the eye disc, dHipk does not affect other pathways in which Gro participates, such as Wnt/Wingless. Hipk2 can also promote neurogenic differentiation during mouse embryogenesis by antagonizing Gro/TLE (Ciarapica et al., 2014).

3.2 Modulating the Cellular Response to DNA Damage Through p53

The most extensively described role for Hipk2 is in regulating p53 function during genotoxic stress (see the following reviews for more detail: Bitomsky & Hofmann, 2009; Calzado, Renner, Roscic, & Schmitz, 2007; Puca, Nardinocchi, Givol, & D'Orazi, 2010; Rinaldo, Prodosmo, Siepi, & Soddu, 2007). p53 is a tumor suppressor and transcription factor that controls the cellular response to DNA damage. In normal, unstressed cells tight regulation of p53 by Mdm2 through continuous degradation prevents inappropriate cell cycle arrest or apoptosis (Haupt, Maya, Kazaz, & Oren, 1997; Momand, Zambetti, Olson, George, & Levine, 1992). Mdm2 (HDM2 in humans) is an E3-ubiquitin ligase that ubiquitinates p53, targeting it for proteasomal degradation. Depending on the severity of the DNA damage, p53 either triggers cell cycle arrest and DNA repair, or in cases of irreparable damage, drives senescence or apoptosis (Oda et al., 2000). Defects in this DNA damage response can lead to genomic instability and cancer.

Upon induction of severe lethal genotoxic stress, Hipks can phosphorylate p53 at Serine 46 (p53Ser46), which disrupts binding of Mdm2, allowing p53 to accumulate and activate transcription (D'Orazi et al., 2002; Hofmann et al., 2002; Rinaldo, Prodosmo, Mancini, et al., 2007; Rinaldo, Prodosmo, Siepi, et al., 2007). This interaction is stimulated by complex formation between Axin, p53, and Hipk2 (Li, Wang, et al., 2007; Rui et al., 2004). Phosphorylation of p53Ser46 by Hipk and numerous other kinases is proposed to affect the selectivity of apoptotic target genes, promoting a shift from cell cycle arrest genes to apoptosis-inducing targets (D'Orazi et al., 2002; Hofmann et al., 2002; Kodama et al., 2010; Mayo et al., 2005; Oda et al., 2000; Perfettini et al., 2005; Smeenk et al., 2011; Taira, Nihira, Yamaguchi, Miki, & Yoshida, 2007). Of note, Hipk2 does not affect all p53 transcriptional targets (D'Orazi et al., 2002; Wang, Debatin, & Hug, 2001). Recently the Ras effector NORE1A, which can bind Hipk1 directly, was shown to block phosphorylation of p53

(Donninger et al., 2015; Lee et al., 2012). Furthermore, Hipk2 can associate with the transcriptional activators and acetyltransferases p300/CBP in nuclear speckles. Death of UV-irradiated cells is enhanced by Hipk2 expression and reduced in cells with low-level Hipk2 expression (Hofmann et al., 2002). Hipk2 also interacts with the scaffold protein Axin and the death domain-associated protein Daxx to promote cell death through phosphorylation of p53 (Li, Wang, et al., 2007; Rui et al., 2004).

A complicated negative feedback loop exists between Hipk2, p53, and Mdm2. p53 induces *Mdm2* transcription and Hipk2 causes a posttranscriptional reduction of Mdm2 levels (Barak, Juven, Haffner, & Oren, 1993). In contrast, Mdm2 ubiquitinates Hipk2 and promotes its proteasome-mediated degradation upon nonlethal levels of DNA damage (Rinaldo, Prodosmo, Mancini, et al., 2007; Rinaldo, Prodosmo, Siepi, et al., 2007). Thus, Hipk2 protein stabilization is induced by lethal doses of genotoxic agents, and Hipk2 levels are repressed through degradation upon sublethal doses. In addition, p53-induced caspase-mediated cleavage of the C-terminus (containing the lysine targeted by Mdm2, Fig. 1) is required for activation of Hipk2's apoptotic function (Gresko et al., 2006). Thus, vertebrate Hipks play important and diverse roles in the DNA damage response. Whether this relationship exists in other organisms is unknown.

Some aspects of the Hipk–p53 relationship are still unclear. For example, PML is a proapoptotic target of p53, yet Hipk2 can also stabilize PML in a p53-independent mechanism following treatment with the DNA-damaging anticancer drug doxorubicin (Gresko et al., 2009). In contrast to the effect of Hipk2 on p53, it was shown that Hipk4 can phosphorylate p53Ser9 which enhances p53-mediated transcriptional repression (Arai et al., 2007). Furthermore, Mdm2 does not interact with Hipk2 in unstressed cells, nor in a yeast two-hybrid assay, suggesting the interaction may be limited to times of DNA damage (Wang, Debatin, et al., 2001). The association of Hipks and p53 does not appear to occur in normal resting cells or during differentiation of distinct cell types such as skeletal muscle, granulocytes, or macrophages (Iacovelli et al., 2009), and thus reflects a very context-specific function for Hipk proteins during DNA damage responses. Further, analysis of mouse Hipk2 and p53 function during skin carcinogenesis suggests that the proteins have distinct and nonoverlapping functions in cell growth and tumorigenesis (Wei et al., 2007).

Hipk family proteins also play roles in response to other stress inducers in addition to DNA damage, such as nutritional stress, reactive oxygen species, and hypoxia (reviewed in Schmitz et al., 2014). In *C. elegans*, loss of *hpk-1* causes sensitivity to heat and oxidative stress, and Hpk-1 can regulate expression of genes that affect aging and stress responses. In addition, Hpk-1 levels are increased following heat shock, suggesting a mechanism to overcome stress-related damage (Berber et al., 2016).

3.3 Regulation of Wnt Signaling by Hipks

Hipk proteins can modulate Wnt signaling in diverse and, at times, contradictory ways. It is likely that cell-type specificity and other signaling inputs will emerge as factors affecting how Hipks control Wnt signaling. A key regulatory event in the canonical Wnt signal transduction pathway is the stabilization of the transcriptional effector β-catenin (reviewed in Clevers & Nusse, 2012). β-Catenin levels are controlled by the destruction complex, which promotes β-catenin phosphorylation, ubiquitination, and subsequent proteasomal degradation. Multiple kinases act reiteratively in the Wnt pathway to both positively and negatively regulate its activity, including GSK3β and CK1 (Verheyen & Gottardi, 2010). β-Catenin and members of the TCF/LEF family of transcription factors activate Wnt target genes. TCF/LEF proteins act as transcriptional repressors and activators, depending on their binding partners. When bound with β-catenin they activate target genes, but when bound to Gro/TLE corepressors they act as transcriptional repressors.

Both *Drosophila* Hipk and murine Hipk2 can positively regulate the Wnt pathway by preventing the ubiquitination and degradation of β-catenin [*Drosophila* Armadillo (Arm)] (Lee, Andrews, et al., 2009; Lee, Swarup, et al., 2009; Swarup & Verheyen, 2011). In the absence of *dHipk*, cells fail to accumulate Arm, Wnt target genes are not induced, and wing phenotypes reminiscent of impaired Wg signaling are seen (Lee, Andrews, et al., 2009; Lee, Swarup, et al., 2009). dHipk and Hipk2 can both activate the Topflash TCF-responsive reporter in cell culture, and promote accumulation of Arm/β-catenin by inhibiting the negative regulation of Arm/β-catenin by the SCF^{Slimb}/β-TrCP E3 ubiquitin ligase (Swarup & Verheyen, 2011). SCF^{Slimb} is also involved in regulating the Hedgehog signaling pathway; thus, dHipk also can promote the Hedgehog pathway during tissue development (Swarup & Verheyen, 2011). Furthermore, dHipk and murine Hipk2 can increase the transcriptional activity of the β-catenin/TCF complex independent of β-catenin stabilization (Verheyen, Swarup, & Lee, 2012). Hipk2 also promotes the stability of β-catenin in HeLa cells (Shimizu et al., 2014). One study contradicts these finding, showing that Hipk2 in

HEK293T cells repressed Topflash expression in a kinase-independent manner, with no effect on β-catenin accumulation (Wei et al., 2007). The apparently conflicting results could be due to the levels of Hipk2 expression within cells, as there can be varied dose-dependent responses.

Hipks can also act directly on TCF/LEF family members to affect their activity. In response to Wnt signaling, both Hipk1 and Hipk2 can phosphorylate and inhibit *Xenopus* TCF3, which is a constitutive repressor of Wnt target genes (Hikasa et al., 2010; Hikasa & Sokol, 2011; Kuwahara et al., 2014; Louie et al., 2009; Wu et al., 2012). Knockdown of *Hipk1* during *Xenopus* axis formation causes derepressed Wnt targets. In contrast, *Hipk1* morphants showed dramatic loss of Wnt targets during gastrulation, demonstrating that at this developmental stage Hipk1 plays a role in promoting Wnt signaling. Thus, Hipks can either upregulate or repress target genes depending on which TCF/LEF family members they are regulating in diverse contexts.

Another point of intersection comes through the regulation of Dishevelled (Dsh/Dvl). Dsh is a modular component of numerous Wnt pathways and can promote β-catenin stabilization after Wnt signaling. In *Xenopus* extracts and HEK293T cells Dsh can interact with Hipk1, which modulates Wnt signaling output in a context-dependent way (Louie et al., 2009). Further, Hipk2 is proposed to regulate the stability of Dvl in Zebrafish in a kinase-independent manner (Shimizu et al., 2014). All these results combined suggest that Hipks can be versatile modulators of Wnt by affecting multiple pathway components, as is also seen for other critical Wnt regulatory kinases like GSK3 and CK1.

3.4 Hipk Promotes Yki Activity in the Hippo Pathway

Hipk also modulates the Hippo pathway, which is an essential conserved signaling pathway regulating tissue and organ growth (Heidary Arash & Attisano, 2013). The Hippo pathway comprises a cascade of kinases that act as tumor suppressors and negatively regulate the Yap/Taz/Yorkie (Yki) oncogene, which promotes the expression of growth-promoting and apoptosis-inhibiting genes. dHipk is required for Yki-induced overgrowths and increases in target gene expression (Chen & Verheyen, 2012; Poon et al., 2012). dHipk's kinase ability is necessary for the increase in Yki targets upon overexpression, and Hipk directly phosphorylates Yki. This effect is evolutionarily conserved since human Hipk2 can promote the activity and abundance of the Yki ortholog YAP (Poon et al., 2012).

3.5 JNK Signaling Is Modulated Through Hipk

Hipks have also been shown to regulate Jun N-terminal kinase (JNK) signaling in numerous contexts, although the mechanisms are generally indirect (Hofmann, Jaffray, Stollberg, Hay, & Will, 2005; Hofmann, Stollberg, Schmitz, & Will, 2003; Huang et al., 2011; Lan, Li, Lin, Lai, & Chung, 2007; Lan, Wu, Shih, & Chung, 2012; Rochat-Steiner et al., 2000; Song & Lee, 2003). The JNK pathway is evolutionarily conserved and affects processes ranging from apoptosis and tissue regeneration, to steroidogenesis (Weston & Davis, 2002). This pathway is kept at basal levels and becomes activated during moments of stress or morphogenesis. Hipk2 and the antiapoptotic protein Daxx can cooperatively promote JNK signaling. Furthermore, knocking down Hipk2 in Hep3B cells impairs TGF-β-induced JNK activation and apoptosis, suggesting Hipk2 is required for normal stimulation of apoptosis (Hofmann et al., 2003). Hipk2 also promotes neurodegeneration via JNK in ER stress-induced apoptosis (Lee et al., 2016). Hipk1 and Hipk3 appear to act in a manner similar to Hipk2 in these processes (Song & Lee, 2003), and dHipk participates in sumoylation-dependent regulation of JNK signaling (Huang et al., 2011).

3.6 Hipk Is Required for JAK/STAT Activity in Normal Development and Tumorigenesis

We have recently found that dHipk can affect JAK/STAT signaling under both normal and tumorigenic conditions (Blaquiere et al., 2016). The JAK/STAT cascade is a conserved pathway that plays a role in an extensive number of developmental processes and diseases (Amoyel, Anderson, & Bach, 2014; Chen, Giedt, Tang, & Harrison, 2014). Pathway activation causes the translocation of STAT/Stat92E into the nucleus where it activates target genes. Receptor-associated Janus kinase (JAK; Hop in *Drosophila*) activates Stat92E through phosphorylation, and gain-of-function mutations of JAK kinases (hop^{Tum-l} and *Jak2 V617F*) lead to the development of hematopoietic tumors, myeloproliferative neoplasms, and leukemia (Hanratty & Ryerse, 1981; Jones, Kreil, Zoi, & Waghorn, 2005; Lacronique, 1997). Reducing dHipk in fly blood cells suppresses the severity of the hop^{Tum-l} phenotype. Conversely, elevating dHipk within the hemocytes phenocopies melanotic tumors seen in hop^{Tum-l} flies in a kinase-dependent manner. dHipk and Stat92E physically interact in vivo, as shown by a proximity ligation assay, and dHipk is required for expression of a Stat92E-dependent reporter. Consistent with such a function, truncated, activated Hipk2 can interact with and phosphorylate Stat3, but the biological function of this finding is still unknown (Matsuo, Ochiai, Nakashima, & Taga, 2001).

This section has aimed to shed light on some of the functions of Hipk family proteins in modulating diverse signaling pathways during developmental processes in model organisms. We were unable to highlight some of the less well-characterized signaling roles. It is likely that new functions will be uncovered as in-depth analyses of all family members in multiple tissues are performed. An overarching theme that emerges is that Hipks likely act to integrate numerous signals and may mediate cross talk between signaling pathways. It is also likely that Hipk activity will be regulated through some of these pathways and may involve PTMs (see below).

4. REGULATION OF HIPK ACTIVITY

Peak activation of Hipk activity can be triggered by diverse morphogenetic signals, DNA damage, or cell stress. Unlike many mitogen-activated protein kinases (MAPKs), Hipk proteins seem to exist constitutively in a partially active state via autophosphorylation. Therefore, regulation of Hipk activity has evolved to focus on how to deactivate Hipk. Most notably, Hipk proteins are extensively PTM, and these modifications can be correlated with altered stability, localization, and consequently activity. In this next section, we briefly review these modes of regulation and how they affect the role of Hipk proteins in development and disease. For an excellent recent review on this topic, we also refer the reader to Saul and Schmitz (2013).

4.1 Phosphorylation

Hipks are dual specificity serine/threonine and tyrosine kinases. Hipk2 can autophosphorylate on conserved tyrosine residue in the kinase activation loop (Hipk2 Y354) (Saul et al., 2013; Siepi, Gatti, Camerini, Crescenzi, & Soddu, 2013). The activation loop is defined in earlier studies as the bold amino acids S**TY**LQS [S352, Y354] (Rinaldo, Prodosmo, Mancini, et al., 2007; Rinaldo, Prodosmo, Siepi, et al., 2007), while later studies propose that phosphorylation of the activation loop occurs on ST**Y**LQ**S** [Y354, S357] (Saul & Schmitz, 2013). Autophosphorylation is proposed to affect the degree of kinase activity and in turn the affinity for substrates, although a Y354F phospho-resistant mutant retains intermediate kinase activity (Saul et al., 2013; Siepi et al.,

2013) and Y354 phosphorylation alone is not sufficient for full Hipk2 activity (Saul & Schmitz, 2013; Schmitz et al., 2014). Hipk3 also undergoes autophosphorylation (Rochat-Steiner et al., 2000). Hipk2 can also be modified by transphosphorylation at other sites outside of the kinase domain. It is likely that differential phosphorylation of Hipk2 provides a mechanism for controlling the signal output of the kinase. Of particular interest are Hipk2 phosphorylation sites that are not autophosphorylated, suggesting that other, as yet unknown, kinases are responsible for phosphorylating these sites. One such kinase may be Tak1 which is proposed to activate Hipk2 (Kanei-Ishii et al., 2004).

4.2 SUMO Modification

Hipk2 can be modified through the addition of the small ubiquitin-like modifier (SUMO) by SUMO1 (Kim et al., 1999), and SUMO modification of Hipk2 at K25 was shown to regulate its stability and activity (Gresko, Moller, Roscic, & Schmitz, 2005; Hofmann et al., 2005; Sung et al., 2005). Hipk2 recruitment to PML nuclear bodies requires the SIM of Hipk2 (Sung et al., 2011). Mutation of these residues results in localization of Hipk2 to the entire cell and reduced interaction with PML, although Hipk2 retained the ability to phosphorylate PML (de la Vega et al., 2011). Deletion of the SIM in Hipk2ΔSIM impaired the ability of Hipk2 to induce p53-dependent transcription, while enhancing its ability to phosphorylate its cytosolic substrate Siah2 (de la Vega et al., 2011). The K25 residue is conserved across vertebrates (human, *Gallus gallus*, *Fugu rubripes*, *Xenopus*) and can also be SUMO-modified in Hipk1 and Hipk3 (Gresko et al., 2005; Li et al., 2005). Removal of the SUMO moiety by the SUMO-specific protease SENP1 causes Hipk2 dissociation from PML bodies (Kim et al., 2005). dHipk is also SUMOylated, though the exact site is not known (Huang et al., 2011). In *Drosophila* tissues lacking SUMO (*smt3* RNAi-expressing cells), JNK signaling becomes triggered, leading to ectopic apoptosis and compensatory proliferation. Loss of *smt3* also causes exogenous dHipk to become delocalized from the nucleus, and enriched in the cytoplasm which allows it to stimulate JNK signaling (Huang et al., 2011). It is possible that the sumoylation status of dHipk can regulate its interaction with nuclear vs cytoplasmic partners. While some studies observed that the nuclear localization of Hipk2 and dHipk is dependent on its SUMOylation state (Huang et al., 2011; Kim et al., 1999), others found that adding or removing SUMO from Hipk2 does not influence its localization to PML nuclear bodies (Hofmann et al., 2005). Changes to the SUMOylation state of Hipk's can also alter the binding between Gro and Hipk2 (Sung et al., 2005).

4.3 Ubiquitin and Proteasomal Degradation

Hipk2 levels in unstressed cells are kept low through ubiquitin-targeted degradation mediated by proteins such as the E3-ubiquitin ligases Mdm2, Siah-1, Siah-2, Fbx3, and WSB-1 (Calzado, de la Vega, Möller, Bowtell, & Schmitz, 2009; Calzado, De La Vega, Muñoz, & Schmitz, 2009; Choi et al., 2008; Rinaldo, Prodosmo, Mancini, et al., 2007; Rinaldo, Prodosmo, Siepi, et al., 2007; Shima et al., 2008; Winter et al., 2008). The role of Mdm2 has already been described above in the p53 signaling section. Following treatments that induce severe DNA damage, Hipk2 degradation induced by Siah1 and WSB-1 ceases, resulting in elevated Hipk2 levels. In contrast, mild genotoxic stress does not block Siah-1, and Hipk2 is rapidly degraded. The p53-inducible ligase Siah-1L targets Hipk2 and Hipk3 for degradation, thus preventing p53 activation and apoptosis (Calzado, de la Vega, Möller, et al., 2009; Calzado, De La Vega, Muñoz, et al., 2009). Zyxin promotes Hipk2 stability by interfering with Siah-1 function (Crone et al., 2011). The XIAP-associated factor 1 (XAF1) can also block Hipk2 degradation by interfering with the Siah2–Hipk2 interaction (Lee et al., 2014), while hypoxia can trigger the interaction between Hipk and Siah2 (Calzado, de la Vega, Möller, et al., 2009; Calzado, De La Vega, Muñoz, et al., 2009).

4.4 Acetylation

Hipks 1–3 can be acetylated on lysines in the presence of CBP, although this occurs at very low levels in normal cells (Aikawa et al., 2006; de la Vega et al., 2012; Hofmann et al., 2002). Acetylation causes delocalization of Hipk2 from speckles to the nucleoplasm and cytoplasm (de la Vega et al., 2012). Mutation of Hipk2 lysine residues does not affect its ability to phosphorylate its substrate Siah-2, suggesting that acetylation may not affect all cellular functions of Hipk2. The ability of Hipk2 to induce cell death is regulated by acetylation and levels of ROS (de la Vega et al., 2012). Inhibition of acetylation through mutation of lysine residues causes Hipk2 to have enhanced cell killing properties, suggesting acetylation normally protects against such death. SUMOylation of Hipk2 promotes its deacetylation by HDAC3 and favors cell survival. Under physiological conditions, acetylation provides a protective effect by delocalizing Hipk2 from sites where it interacts with p53, and could trigger cell death.

4.5 Regulation of Hipk Gene Expression

As described above, Hipk protein levels are tightly regulated by numerous PTMs that affect stability, localization, and function. Hipk gene expression levels are also controlled through microRNA-mediated posttranscriptional regulation (reviewed in Conte & Pierantoni, 2015). For example, in renal tubular epithelial cells, *miR-141* downregulates expression of Hipk2 (Huang et al., 2015). Currently, the transcriptional regulation of Hipk family genes is an understudied area, which may yet reveal additional modes of ensuring adequate Hipk function.

4.6 Drug-Based Inhibition of Hipk

While numerous generic protein kinase inhibitors can influence Hipk, only a few have been described to be specific to Hipk. The drug D-115893 has dramatic effects causing Hipk delocalization and distribution throughout the cell, in addition to blocking p53 serine 46 phosphorylation (de la Vega et al., 2011). Recently a very specific Hipk2 inhibitor named TBID was identified that can block phosphorylation of both p53 and a generic substrate by endogenous Hipk2 in human T-lymphoblastoid cells (Cozza et al., 2014). TBID is also capable of inhibiting Hipk1 and Hipk3 at a lower efficiency.

5. HIPKS IN DISEASE

Given the extensive role of Hipk proteins in signaling pathways and proliferation, it is not surprising that modulation of Hipk activity can also contribute to development of diverse disease states (Fig. 4). In this context

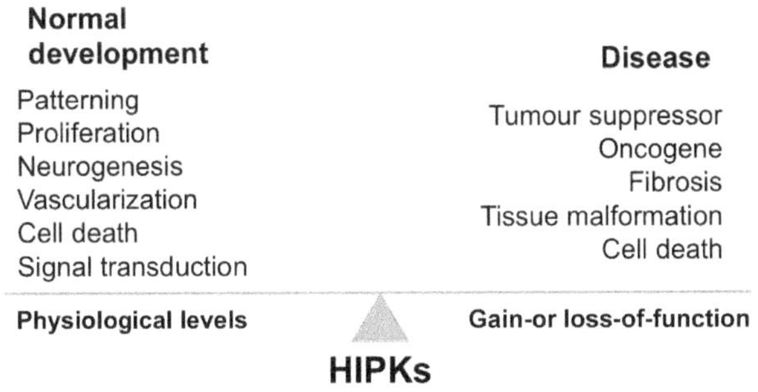

Normal development	Disease
Patterning	Tumour suppressor
Proliferation	Oncogene
Neurogenesis	Fibrosis
Vascularization	Tissue malformation
Cell death	Cell death
Signal transduction	
Physiological levels	**Gain-or loss-of-function**

HIPKs

Fig. 4 Hipk functions are maintained in a delicate balance in development and disease. Levels of Hipk proteins are tightly regulated. Numerous studies in various organisms and cell types reveal that a delicate balance is required to maintain normal function. Both gains and losses of function of Hipk proteins result in defects including cell death, hyperproliferation, tumor formation, and developmental abnormalities.

as well, there are conflicting models of Hipk protein family function that likely reflect tissue- and organism-specific effects, genetic background differences, and possibly diverged functions for the distinct vertebrate paralogs. It is difficult to determine at times whether certain disease states arise due to elevated Hipk levels, or whether Hipk overexpression occurs as a result of disease progression.

5.1 Hipks as Tumor Suppressors in Cancer

Since Hipks can both induce proliferation and trigger cell death, the roles in cell transformation will likely be complex and multifaceted. Given the extensive work on Hipks and p53, many studies have proposed a tumor-suppressive role for Hipks in response to ionizing radiation (D'Orazi, Rinaldo, & Soddu, 2012; Puca et al., 2010). Hipk2 cooperates with p53 to suppress gamma-ray radiation-induced mouse thymic lymphoma by triggering cell death (Mao et al., 2012). When Hipk is reduced due to mutation or downregulation, cancerous cells are capable of resisting apoptosis. Consistent with a tumor suppressor function, decreased expression of Hipk family members is found in several types of cancers including thyroid and breast carcinomas, idiopathic pulmonary fibrosis, thyroid carcinomas, and papillary and follicular thyroid carcinomas (Lavra et al., 2011; Pierantoni et al., 2002; Ricci et al., 2013) In gastric cancer stem cells *Hipk1* is targeted by miRNAs (Wu et al., 2013). Colon tumors with higher expression of Hipk responded better to treatment than tumors with low expression, suggesting that the presence of Hipk can repress the cancer severity, independent of p53 status (Soubeyran et al., 2011). In acute myeloid leukemia, two missense mutations in the Hipk2 speckle retention sequence led to altered protein localization and decreased induction of apoptosis by Hipk2 (Li, Arai, et al., 2007). In a two-step model of skin tumorigenesis, Hipk2 mutant mice were more susceptible to tumor induction (Wei et al., 2007). Hipk2 deficiency has also been linked to chromosomal instability via cytokinesis failure, increasing tumorigenicity in mouse embryonic fibroblasts (Valente et al., 2015). Thus, in these contexts, Hipk proteins act as tumor suppressors and their loss results in excessive proliferation or tumor survival.

5.2 Hipks as Oncogenic Factors in Cancer

In direct contrast to the proposed role of Hipks as tumor suppressors, it has been found that elevated Hipk levels are seen in cervical cancers, a pilocytic astrocytomas, idiopathic pulmonary fibrosis, colorectal cancer cells, and in several other proliferative diseases (Al-Beiti & Lu, 2008; Cheng et al., 2012; D'Orazi et al., 2006; Deshmukh et al., 2008; Saul & Schmitz, 2013; Yu et al., 2009). High Hipk2 protein levels are seen in thyroid follicular hyperplasia due to increased proliferation (Lavra et al., 2011). In astrocytomas, gene amplification leads to higher mRNA and protein levels of Hipk2 (Deshmukh et al., 2008). Further, overexpression of a wild-type Hipk2 in glioma cells confers a growth advantage. Cervical cancer and juvenile pilocytic astrocytomas frequently show amplification or elevated expression of Hipk2 (Al-Beiti & Lu, 2008; Jacob et al., 2009). Hipk1 is highly expressed in human breast cancer cell lines and oncogenically transformed MEFs, although it can also play a protective role in skin tumors (Kondo et al., 2003). Hipk1 is highly overexpressed in colorectal cancer samples, peaking in early stages (Rey et al., 2013).

In *Drosophila*, dHipk promotes numerous conserved oncogenic pathways (Notch, Wnt, Hh, Hippo), suggesting that it could contribute to development of cancers due to these pathways in humans (Chen & Verheyen, 2012; Lee, Andrews, et al., 2009; Lee, Swarup, et al., 2009; Poon et al., 2012; Swarup & Verheyen, 2011). Furthermore, dHipk is required for tumorigenesis of blood cells in a JAK/STAT leukemia model in *Drosophila* (Blaquiere et al., 2016). The evidence from gain-of-function studies, coupled with the role in proliferation demonstrated in loss-of-function mouse, fly, and worm models, strongly supports the model that Hipks can have oncogenic functions in certain contexts.

5.3 Hipks and Fibrosis

Fibrosis is a contributor to organ failure that occurs as a result of excessive fibroblast proliferation coupled with epithelial-to-mesenchymal transition (EMT) and deposition of extracellular matrix components. Hipks have been implicated in fibrosis which arises from dysregulation of numerous signaling pathways (Fan, Wang, Chuang, & He, 2014; Jin et al., 2012; Nugent, Lee, & He, 2015). EMT is also an important cellular process in development, wound healing, and cancer progression. Hipk2 expression in renal tubular epithelial cells triggers EMT, as shown by upregulation of vimentin and downregulation of E-cadherin (Huang et al., 2015).

6. CONCLUSIONS

Hipk proteins are multifunctional kinases that have profound and at times opposing effects on development and disease. Hipks likely act through context-dependent regulation of signaling pathways and gene expression. Given functional redundancy in vertebrates, the four orthologs have specialized functions, which in lower organisms such as *Drosophila* and *C. elegans* are carried out by single proteins. It is clear that through animal model studies coupled with investigations into disease states, we will continue to uncover important roles for Hipk proteins. Many unanswered questions remain. For example, how can Hipk proteins integrate so many signaling pathways effectively? How can Hipks promote both proliferation and cell death, and what decides how that balance is tipped? Future research will hopefully reveal insight into the complex role of Hipks and provide therapeutic targets for various diseases.

ACKNOWLEDGMENTS

We apologize to those authors whose work we were unable to cite. We are grateful to all Verheyen Lab members who have contributed to Hipk discussions, in particular Nathan Wray, Kenny Wong, Sharan Swarup, Wendy Lee, Joanna Chen, and Maryam Rahnama. We thank Nick Harden for comments on the review. We are grateful to Kyoichi Isono and Haruhiko Koseki for allowing us to reproduce images from Isono et al., 2006 in Fig. 2 in this review. Our work on Hipk has been supported by operating grants from the Canadian Institutes of Health Research.

REFERENCES

Aikawa, Y., Nguyen, L. A., Isono, K., Takakura, N., Tagata, Y., Schmitz, M. L., et al. (2006). Roles of HIPK1 and HIPK2 in AML1- and p300-dependent transcription, hematopoiesis and blood vessel formation. *The EMBO Journal, 25*, 3955–3965.

Al-Beiti, M. A. M., & Lu, X. (2008). Expression of HIPK2 in cervical cancer: Correlation with clinicopathology and prognosis. *The Australian & New Zealand Journal of Obstetrics & Gynaecology, 48*, 329–336. http://dx.doi.org/10.1111/j.1479-828X.2008.00874.x.

Alvarez, M., Estivill, X., & de la Luna, S. (2003). DYRK1A accumulates in splicing speckles through a novel targeting signal and induces speckle disassembly. *Journal of Cell Science, 116*, 3099–3107. http://dx.doi.org/10.1242/jcs.00618.

Amoyel, M., Anderson, A. M., & Bach, E. A. (2014). JAK/STAT pathway dysregulation in tumors: A Drosophila perspective. *Seminars in Cell & Developmental Biology, 28*, 96–103. http://dx.doi.org/10.1016/j.semcdb.2014.03.023.

Arai, S., Matsushita, A., Du, K., Yagi, K., Okazaki, Y., & Kurokawa, R. (2007). Novel homeodomain-interacting protein kinase family member, HIPK4, phosphorylates human p53 at serine 9. *FEBS Letters, 581*, 5649–5657. http://dx.doi.org/10.1016/j.febslet.2007.11.022.

Barak, Y., Juven, T., Haffner, R., & Oren, M. (1993). mdm2 expression is induced by wild type p53 activity. *The EMBO Journal, 12*, 461–468.

Berber, S., Llamosas, E., Thaivalappil, P., Boag, P. R., Crossley, M., & Nicholas, H. R. (2013). Homeodomain interacting protein kinase (HPK-1) is required in the soma for robust germline proliferation in C. elegans. *Developmental Dynamics, 242*, 1250–1261. http://dx.doi.org/10.1002/dvdy.24023.

Berber, S., Wood, M., Llamosas, E., Thaivalappil, P., Lee, K., Liao, B. M., et al. (2016). Homeodomain-interacting protein kinase (HPK-1) regulates stress responses and ageing in C. elegans. *Scientific Reports*, *6*, 19582. http://dx.doi.org/10.1038/srep19582.

Bitomsky, N., & Hofmann, T. G. (2009). Apoptosis and autophagy: Regulation of apoptosis by DNA damage signalling—Roles of p53, p73 and HIPK2. *The FEBS Journal*, *276*, 6074–6083. http://dx.doi.org/10.1111/j.1742-4658.2009.07331.x.

Blaquiere, J. A., Lee, W., & Verheyen, E. M. (2014). Hipk promotes photoreceptor differentiation through the repression of Twin of eyeless and Eyeless expression. *Developmental Biology*, *390*, 14–25.

Blaquiere, J. A., Wray, N. B., & Verheyen, E. M. (2016). Hipk is required for JAK/STAT activity and promotes hemocyte-derived tumorigenesis, bioRxiv. *Cold Spring Harbor Labs Journals*, http://dx.doi.org/10.1101/058156.

Calzado, M. A., de la Vega, L., Möller, A., Bowtell, D. D. L., & Schmitz, M. L. (2009a). An inducible autoregulatory loop between HIPK2 and Siah2 at the apex of the hypoxic response. *Nature Cell Biology*, *11*, 85–91. http://dx.doi.org/10.1038/ncb1816.

Calzado, M. A., De La Vega, L., Muñoz, E., & Schmitz, M. L. (2009b). Autoregulatory control of the p53 response by Siah-1 L-mediated HIPK2 degradation. *Biological Chemistry*, *390*, 1079–1083. http://dx.doi.org/10.1515/BC.2009.112.

Calzado, M. A., Renner, F., Roscic, A., & Schmitz, M. L. (2007). HIPK2: A versatile switchboard regulating the transcription machinery and cell death. *Cell Cycle*, *6*, 139–143. http://dx.doi.org/10.4161/cc.6.2.3788.

Chalazonitis, A., Tang, A. A., Shang, Y., Pham, T. D., Hsieh, I., Setlik, W., et al. (2011). Homeodomain interacting protein kinase 2 regulates postnatal development of enteric dopaminergic neurons and glia via BMP signaling. *The Journal of Neuroscience*, *31*, 13746–13757. http://dx.doi.org/10.1523/JNEUROSCI.1078-11.2011.

Chen, Q., Giedt, M., Tang, L., & Harrison, D. A. (2014). Tools and methods for studying the Drosophila JAK/STAT pathway. *Methods*, *68*, 160–172. http://dx.doi.org/10.1016/j.ymeth.2014.03.023.

Chen, J., & Verheyen, E. M. (2012). Homeodomain-interacting protein kinase regulates Yorkie activity to promote tissue growth. *Current Biology*, *22*, 1582–1586. http://dx.doi.org/10.1016/J.Cub.2012.06.074.

Cheng, Y., Al-Beiti, M. A. M., Wang, J., Wei, G., Li, J., Liang, S., et al. (2012). Correlation between homeodomain-interacting protein kinase 2 and apoptosis in cervical cancer. *Molecular Medicine Reports*, *5*, 1251–1255. http://dx.doi.org/10.3892/mmr.2012.810.

Choi, C. Y., Kim, Y. H., Kim, Y.-O., Park, S. J., Kim, E.-A., Riemenschneider, W., et al. (2005). Phosphorylation by the DHIPK2 protein kinase modulates the corepressor activity of Groucho. *The Journal of Biological Chemistry*, *280*, 21427–21436. http://dx.doi.org/10.1074/jbc.M500496200.

Choi, C. Y., Kim, Y. H., Kwon, H. J., & Kim, Y. (1999). The homeodomain protein NK-3 recruits Groucho and a histone deacetylase complex to repress transcription. *The Journal of Biological Chemistry*, *274*, 33194–33197.

Choi, D. W., Seo, Y.-M., Kim, E.-A., Sung, K. S., Ahn, J. W., Park, S.-J., et al. (2008). Ubiquitination and degradation of homeodomain-interacting protein kinase 2 by WD40 repeat/SOCS box protein WSB-1. *The Journal of Biological Chemistry*, *283*, 4682–4689. http://dx.doi.org/10.1074/jbc.M708873200.

Ciarapica, R., Methot, L., Tang, Y., Lo, R., Dali, R., Buscarlet, M., et al. (2014). Prolyl isomerase Pin1 and protein kinase HIPK2 cooperate to promote cortical neurogenesis by suppressing Groucho/TLE:Hes1-mediated inhibition of neuronal differentiation. *Cell Death and Differentiation*, *21*, 321–332. http://dx.doi.org/10.1038/cdd.2013.160.

Clevers, H., & Nusse, R. (2012). Wnt/β-catenin signaling and disease. *Cell*, *149*, 1192–1205. http://dx.doi.org/10.1016/j.cell.2012.05.012.

Conte, A., Pierantoni, G. M., Conte, A., & Pierantoni, G. M. (2015). Regulation of HIPK proteins by microRNAs. *MicroRNA (Shāriqah, United Arab Emirates), 4*, 148–157.
Cozza, G., Zanin, S., Determann, R., Ruzzene, M., Kunick, C., & Pinna, L. A. (2014). Synthesis and properties of a selective inhibitor of homeodomain-interacting protein kinase 2 (HIPK2). *PLoS One, 9*. e89176. http://dx.doi.org/10.1371/journal.pone.0089176.
Crone, J., Glas, C., Schultheiss, K., Moehlenbrink, J., Krieghoff-Henning, E., & Hofmann, T. G. (2011). Zyxin is a critical regulator of the apoptotic HIPK2-p53 signaling axis. *Cancer Research, 71*, 2350–2359. http://dx.doi.org/10.1158/0008-5472.CAN-10-3486.
D'Orazi, G., Cecchinelli, B., Bruno, T., Manni, I., Higashimoto, Y., Saito, S., et al. (2002). Homeodomain-interacting protein kinase-2 phosphorylates p53 at Ser 46 and mediates apoptosis. *Nature Cell Biology, 4*, 11–19. http://dx.doi.org/10.1038/ncb714.
D'Orazi, G., Rinaldo, C., & Soddu, S. (2012). Updates on HIPK2: A resourceful oncosuppressor for clearing cancer. *Journal of Experimental & Clinical Cancer Research, 31*, 63. http://dx.doi.org/10.1186/1756-9966-31-63.
D'Orazi, G., Sciulli, M. G., Di Stefano, V., Riccioni, S., Frattini, M., Falcioni, R., et al. (2006). Homeodomain-interacting protein kinase-2 restrains cytosolic phospholipase A2-dependent prostaglandin E2 generation in human colorectal cancer cells. *Clinical Cancer Research, 12*, 735–741.
de la Vega, L., Fröbius, K., Moreno, R., Calzado, M. A., Geng, H., & Schmitz, M. L. (2011). Control of nuclear HIPK2 localization and function by a SUMO interaction motif. *Biochimica et Biophysica Acta, 1813*, 283–297. http://dx.doi.org/10.1016/j.bbamcr.2010.11.022.
de la Vega, L., Grishina, I., Moreno, R., Krüger, M., Braun, T., & Schmitz, M. L. (2012). A redox-regulated SUMO/acetylation switch of HIPK2 controls the survival threshold to oxidative stress. *Molecular Cell, 46*, 472–483. http://dx.doi.org/10.1016/j.molcel.2012.03.003.
Deshmukh, H., Yeh, T. H., Yu, J., Sharma, M. K., Perry, A., Leonard, J. R., et al. (2008). High-resolution, dual-platform aCGH analysis reveals frequent HIPK2 amplification and increased expression in pilocytic astrocytomas. *Oncogene, 27*, 4745–4751. http://dx.doi.org/10.1038/onc.2008.110.
Donninger, H., Calvisi, D. F., Barnoud, T., Clark, J., Lee Schmidt, M., Vos, M. D., et al. (2015). NORE1A is a Ras senescence effector that controls the apoptotic/senescent balance of p53 via HIPK2. *The Journal of Cell Biology, 208*, 777–789. http://dx.doi.org/10.1083/jcb.201408087.
Doxakis, E., Huang, E. J., & Davies, A. M. (2004). Homeodomain-interacting protein kinase-2 regulates apoptosis in developing sensory and sympathetic neurons. *Current Biology, 14*, 1761–1765.
Engelhardt, O. G., Boutell, C., Orr, A., Ullrich, E., Haller, O., & Everett, R. D. (2003). The homeodomain-interacting kinase PKM (HIPK-2) modifies ND10 through both its kinase domain and a SUMO-1 interaction motif and alters the posttranslational modification of PML. *Experimental Cell Research, 283*, 36–50.
Fan, Y., Wang, N., Chuang, P., & He, J. C. (2014). Role of HIPK2 in kidney fibrosis. *Kidney International. Supplement, 4*, 97–101. http://dx.doi.org/10.1038/kisup.2014.18.
Gresko, E., Moller, A., Roscic, A., & Schmitz, M. L. (2005). Covalent modification of human homeodomain interacting protein kinase 2 by SUMO-1 at lysine 25 affects its stability. *Biochemical and Biophysical Research Communications, 329*, 1293–1299.
Gresko, E., Ritterhoff, S., Sevilla-Perez, J., Roscic, A., Fröbius, K., Kotevic, I., et al. (2009). PML tumor suppressor is regulated by HIPK2-mediated phosphorylation in response to DNA damage. *Oncogene, 28*, 698–708. http://dx.doi.org/10.1038/onc.2008.420.

Gresko, E., Roscic, A., Ritterhoff, S., Vichalkovski, A., del Sal, G., & Schmitz, M. L. (2006). Autoregulatory control of the p53 response by caspase-mediated processing of HIPK2. *The EMBO Journal, 25*, 1883–1894.

Guerra, F. M., Gommerman, J. L., Corfe, S. A., Paige, C. J., & Rottapel, R. (2012). Homeodomain-interacting protein kinase (HIPK)-1 is required for splenic B cell homeostasis and optimal T-independent type 2 humoral response. *PLoS One, 7*. e35533. http://dx.doi.org/10.1371/journal.pone.0035533.

Hanratty, W. P., & Ryerse, J. S. (1981). A genetic melanotic neoplasm of Drosophila melanogaster. *Developmental Biology, 83*, 238–249. http://dx.doi.org/10.1016/0012-1606(81)90470-X.

Hattangadi, S. M., Burke, K. a., & Lodish, H. F. (2010). Homeodomain-interacting protein kinase 2 plays an important role in normal terminal erythroid differentiation. *Blood, 115*, 4853–4861. http://dx.doi.org/10.1182/blood-2009-07-235093.

Haupt, Y., Maya, R., Kazaz, A., & Oren, M. (1997). Mdm2 promotes the rapid degradation of p53. *Nature, 387*, 296–299. http://dx.doi.org/10.1038/387296a0.

He, Q., Shi, J., Sun, H., An, J., Huang, Y., & Sheikh, M. S. (2010). Characterization of human homeodomain-interacting protein kinase 4 (HIPK4) as a unique member of the HIPK Family. *Molecular and Cellular Pharmacology, 2*, 61–68.

Heidary Arash, E., & Attisano, L. (2013). A role for Hipk in the Hippo pathway. *Science Signaling, 6*, pe18.

Hikasa, H., Ezan, J., Itoh, K., Li, X., Klymkowsky, M. W., & Sokol, S. Y. (2010). Regulation of TCF3 by Wnt-dependent phosphorylation during vertebrate axis specification. *Developmental Cell, 19*, 521–532. http://dx.doi.org/10.1016/j.devcel.2010.09.005.

Hikasa, H., & Sokol, S. Y. (2011). Phosphorylation of TCF proteins by homeodomain-interacting protein kinase 2. *The Journal of Biological Chemistry, 286*, 12093–12100. http://dx.doi.org/10.1074/jbc.M110.185280.

Hofmann, T. G., Jaffray, E., Stollberg, N., Hay, R. T., & Will, H. (2005). Regulation of homeodomain-interacting protein kinase 2 (HIPK2) effector function through dynamic small ubiquitin-related modifier-1 (SUMO-1) modification. *The Journal of Biological Chemistry, 280*, 29224–29232. http://dx.doi.org/10.1074/jbc.M503921200.

Hofmann, T. G., Mincheva, A., Lichter, P., Droge, W., & Schmitz, M. L. (2000). Human homeodomain-interacting protein kinase-2 (HIPK2) is a member of the DYRK family of protein kinases and maps to chromosome 7q32-q34. *Biochimie, 82*, 1123–1127.

Hofmann, T. G., Möller, A., Sirma, H., Zentgraf, H., Taya, Y., Dröge, W., et al. (2002). Regulation of p53 activity by its interaction with homeodomain-interacting protein kinase-2. *Nature Cell Biology, 4*, 1–10. http://dx.doi.org/10.1038/ncb715.

Hofmann, T. G., Stollberg, N., Schmitz, M. L., & Will, H. (2003). HIPK2 regulates transforming growth factor-beta-induced c-Jun NH(2)-terminal kinase activation and apoptosis in human hepatoma cells. *Cancer Research, 63*, 8271–8277.

Huang, H., Du, G., Chen, H., Liang, X., Li, C., Zhu, N., et al. (2011). Drosophila Smt3 negatively regulates JNK signaling through sequestering Hipk in the nucleus. *Development, 138*, 2477–2485. http://dx.doi.org/10.1242/dev.061770.

Huang, Y., Tong, J., He, F., Yu, X., Fan, L., Hu, J., et al. (2015). miR-141 regulates TGF-Fb1-induced epithelial-mesenchymal transition through repression of Hipk2 expression in renal tubular epithelial cells. *International Journal of Molecular Medicine, 35*, 311–318. http://dx.doi.org/10.3892/ijmm.2014.2008.

Iacovelli, S., Ciuffini, L., Lazzari, C., Bracaglia, G., Rinaldo, C., Prodosmo, A., et al. (2009). HIPK2 is involved in cell proliferation and its suppression promotes growth arrest independently of DNA damage. *Cell Proliferation, 42*, 373–384. http://dx.doi.org/10.1111/j.1365-2184.2009.00601.x.

Inoue, T., Kagawa, T., Inoue-Mochita, M., Isono, K., Ohtsu, N., Nobuhisa, I., et al. (2010). Involvement of the Hipk family in regulation of eyeball size, lens formation and retinal morphogenesis. *FEBS Letters*, *584*, 3233–3238. http://dx.doi.org/10.1016/j.febslet.2010.06.020.

Isono, K., Nemoto, K., Li, Y., Takada, Y., Suzuki, R., Katsuki, M., et al. (2006). Overlapping roles for homeodomain-interacting protein kinases hipk1 and hipk2 in the mediation of cell growth in response to morphogenetic and genotoxic signals. *Molecular and Cellular Biology*, *26*, 2758–2771.

Jacob, K., Albrecht, S., Sollier, C., Faury, D., Sader, E., Montpetit, A., et al. (2009). Duplication of 7q34 is specific to juvenile pilocytic astrocytomas and a hallmark of cerebellar and optic pathway tumours. *British Journal of Cancer*, *101*, 722–733. http://dx.doi.org/10.1038/sj.bjc.6605179.

Jennings, B. H., & Ish-Horowicz, D. (2008). The Groucho/TLE/Grg family of transcriptional co-repressors. *Genome Biology*, *9*, 205. http://dx.doi.org/10.1186/gb-2008-9-1-205.

Jin, Y., Ratnam, K., Chuang, P. Y., Fan, Y., Zhong, Y., Dai, Y., et al. (2012). A systems approach identifies HIPK2 as a key regulator of kidney fibrosis. *Nature Medicine*, *18*, 580–588. http://dx.doi.org/10.1038/nm.2685.

Jones, A., Kreil, S., Zoi, K., & Waghorn, K. (2005). Widespread occurrence of the JAK2 V617F mutation in chronic myeloproliferative disorders. *Blood*, *106*, 2162–2169. http://dx.doi.org/10.1182/blood-2005-03-1320.Supported.

Kanei-Ishii, C., Ninomiya-Tsuji, J., Tanikawa, J., Nomura, T., Ishitani, T., Kishida, S., et al. (2004). Wnt-1 signal induces phosphorylation and degradation of c-Myb protein via TAK1, HIPK2, and NLK. *Genes & Development*, *18*, 816–829.

Kentrup, H., Becker, W., Heukelbach, J., Wilmes, A., Schürmann, A., Huppertz, C., et al. (1996). Dyrk, a dual specificity protein kinase with unique structural features whose activity is dependent on tyrosine residues between subdomains VII and VIII. *The Journal of Biological Chemistry*, *271*, 3488–3495. http://dx.doi.org/10.1074/jbc.271.7.3488.

Kim, Y. H., Choi, C. Y., & Kim, Y. (1999). Covalent modification of the homeodomain-interacting protein kinase 2 (HIPK2) by the ubiquitin-like protein SUMO-1. *Proceedings of the National Academy of Sciences of the United States of America*, *96*, 12350–12355. http://dx.doi.org/10.1073/pnas.96.22.12350.

Kim, Y. H., Choi, C. Y., Lee, S. J., Conti, M. A., & Kim, Y. (1998). Homeodomain-interacting protein kinases, a novel family of co-repressors for homeodomain transcription factors. *The Journal of Biological Chemistry*, *273*, 25875–25879.

Kim, E. A., Noh, Y. T., Ryu, M. J., Kim, H. T., Lee, S. E., Kim, C. H., et al. (2006). Phosphorylation and transactivation of Pax6 by homeodomain-interacting protein kinase 2. *The Journal of Biological Chemistry*, *281*, 7489–7497.

Kim, Y. H., Sung, K. S., Lee, S.-J., Kim, Y.-O., Choi, C. Y., & Kim, Y. (2005). Desumoylation of homeodomain-interacting protein kinase 2 (HIPK2) through the cytoplasmic-nuclear shuttling of the SUMO-specific protease SENP1. *FEBS Letters*, *579*, 6272–6278. http://dx.doi.org/10.1016/j.febslet.2005.10.010.

Kodama, M., Otsubo, C., Hirota, T., Yokota, J., Enari, M., & Taya, Y. (2010). Requirement of ATM for rapid p53 phosphorylation at Ser46 without Ser/Thr-Gln sequences. *Molecular and Cellular Biology*, *30*, 1620–1633. http://dx.doi.org/10.1128/MCB.00810-09.

Kondo, S., Lu, Y., Debbas, M., Lin, A. W., Sarosi, I., Itie, A., et al. (2003). Characterization of cells and gene-targeted mice deficient for the p53-binding kinase homeodomain-interacting protein kinase 1 (HIPK1). *Proceedings of the National Academy of Sciences of the United States of America*, *100*, 5431–5436. http://dx.doi.org/10.1073/pnas.0530308100.

Kuwahara, A., Sakai, H., Xu, Y., Itoh, Y., Hirabayashi, Y., & Gotoh, Y. (2014). Tcf3 represses Wnt-β-catenin signaling and maintains neural stem cell population during neocortical development. *PLoS One*, 9. e94408. http://dx.doi.org/10.1371/journal. pone.0094408.

Lacronique, V. (1997). A TEL-JAK2 fusion protein with constitutive kinase activity in human leukemia. *Science*, *278*, 1309–1312. http://dx.doi.org/10.1126/science.278. 5341.1309.

Lan, H.-C., Li, H.-J., Lin, G., Lai, P.-Y., & Chung, B. (2007). Cyclic AMP stimulates SF-1-dependent CYP11A1 expression through homeodomain-interacting protein kinase 3-mediated Jun N-terminal kinase and c-Jun phosphorylation. *Molecular and Cellular Biology*, *27*, 2027–2036. http://dx.doi.org/10.1128/MCB.02253-06.

Lan, H.-C., Wu, C.-F., Shih, H.-M., & Chung, B.-C. (2012). Death-associated protein 6 (Daxx) mediates cAMP-dependent stimulation of Cyp11a1 (P450scc) transcription. *The Journal of Biological Chemistry*, *287*, 5910–5916. http://dx.doi.org/10.1074/jbc. M111.307603.

Lavra, L., Rinaldo, C., Ulivieri, A., Luciani, E., Fidanza, P., Giacomelli, L., et al. (2011). The loss of the p53 activator HIPK2 is responsible for galectin-3 overexpression in well differentiated thyroid carcinomas. *PLoS One*, *6*. e20665. http://dx.doi.org/10.1371/journal.pone.0020665.

Lee, W., Andrews, B. C., Faust, M., Walldorf, U., & Verheyen, E. M. (2009). Hipk is an essential protein that promotes Notch signal transduction in the Drosophila eye by inhibition of the global co-repressor Groucho. *Developmental Biology*, *325*, 263–272. http://dx.doi.org/10.1016/j.ydbio.2008.10.029.

Lee, M.-G., Han, J., Jeong, S.-I., Her, N.-G., Lee, J.-H., Ha, T.-K., et al. (2014). XAF1 directs apoptotic switch of p53 signaling through activation of HIPK2 and ZNF313. *Proceedings of the National Academy of Sciences*, *111*, 15532–15537. http://dx.doi.org/10.1073/pnas.1411746111.

Lee, D., Park, S.-J., Sung, K. S., Park, J., Lee, S. B., Park, S.-Y., et al. (2012). Mdm2 associates with Ras effector NORE1 to induce the degradation of oncoprotein HIPK1. *EMBO Reports*, *13*, 163–169. http://dx.doi.org/10.1038/embor.2011.235.

Lee, S., Shang, Y., Redmond, S. A., Urisman, A., Tang, A. A., Li, K. H., et al. (2016). Activation of HIPK2 promotes ER stress-mediated neurodegeneration in amyotrophic lateral sclerosis. *Neuron*, *91*, 41–55. http://dx.doi.org/10.1016/j.neuron. 2016.05.021.

Lee, W., Swarup, S., Chen, J., Ishitani, T., & Verheyen, E. M. (2009). Homeodomain-interacting protein kinases (Hipks) promote Wnt/Wg signaling through stabilization of beta-catenin/Arm and stimulation of target gene expression. *Development*, *136*, 241–251. http://dx.doi.org/10.1242/dev.025460.

Li, X.-L., Arai, Y., Harada, H., Shima, Y., Yoshida, H., Rokudai, S., et al. (2007). Mutations of the HIPK2 gene in acute myeloid leukemia and myelodysplastic syndrome impair AML1- and p53-mediated transcription. *Oncogene*, *26*, 7231–7239. http://dx.doi.org/10.1038/sj.onc.1210523.

Li, Q., Wang, X. X., Wu, X., Rui, Y., Liu, W., Wang, J., et al. (2007). Daxx cooperates with the Axin/HIPK2/p53 complex to induce cell death. *Cancer Research*, *67*, 66–74. http://dx.doi.org/10.1158/0008-5472.CAN-06-1671.

Li, X., Zhang, R., Luo, D., Park, S.-J., Wang, Q., Kim, Y., et al. (2005). Tumor necrosis factor alpha-induced desumoylation and cytoplasmic translocation of homeodomain-interacting protein kinase 1 are critical for apoptosis signal-regulating kinase 1-JNK/p38 activation. *The Journal of Biological Chemistry*, *280*, 15061–15070. http://dx.doi.org/10.1074/jbc.M414262200.

Link, N., Chen, P., Lu, W. J., Pogue, K., Chuong, A., Mata, M., et al. (2007). A collective form of cell death requires homeodomain interacting protein kinase. *The Journal of Cell Biology*, *178*, 567–574. http://dx.doi.org/10.1083/jcb.200702125.

Louie, S. H., Yang, X. Y., Conrad, W. H., Muster, J., Angers, S., Moon, R. T., et al. (2009). Modulation of the beta-catenin signaling pathway by the dishevelled-associated protein Hipk1. *PLoS One, 4*. e4310.

Mao, J.-H., Wu, D., Kim, I.-J., Kang, H. C., Wei, G., Climent, J., et al. (2012). Hipk2 cooperates with p53 to suppress γ-ray radiation-induced mouse thymic lymphoma. *Oncogene, 31*, 1176–1180. http://dx.doi.org/10.1038/onc.2011.306.

Matsuo, R., Ochiai, W., Nakashima, K., & Taga, T. (2001). A new expression cloning strategy for isolation of substrate-specific kinases by using phosphorylation site-specific antibody. *Journal of Immunological Methods, 247*, 141–151.

Mayo, L. D., Seo, Y. R., Jackson, M. W., Smith, M. L., Guzman, J. R., Korgaonkar, C. K., et al. (2005). Phosphorylation of human p53 at serine 46 determines promoter selection and whether apoptosis is attenuated or amplified. *The Journal of Biological Chemistry, 280*, 25953–25959. http://dx.doi.org/10.1074/jbc.M503026200.

Momand, J., Zambetti, G. P., Olson, D. C., George, D., & Levine, A. J. (1992). The mdm-2 oncogene product forms a complex with the p53 protein and inhibits p53-mediated transactivation. *Cell, 69*, 1237–1245.

Nugent, M. M., Lee, K., & He, J. C. (2015). HIPK2 is a new drug target for anti-fibrosis therapy in kidney disease. *Frontiers in Physiology, 6*, 132. http://dx.doi.org/10.3389/fphys.2015.00132.

Oda, K., Arakawa, H., Tanaka, T., Matsuda, K., Tanikawa, C., Mori, T., et al. (2000). p53AIP1, a potential mediator of p53-dependent apoptosis, and its regulation by Ser-46-phosphorylated p53. *Cell, 102*, 849–862. http://dx.doi.org/10.1016/S0092-8674(00)00073-8.

Ohtsu, N., Nobuhisa, I., Mochita, M., & Taga, T. (2007). Inhibitory effects of homeodomain-interacting protein kinase 2 on the aorta-gonad-mesonephros hematopoiesis. *Experimental Cell Research, 313*, 88–97.

Perfettini, J.-L., Castedo, M., Nardacci, R., Ciccosanti, F., Boya, P., Roumier, T., et al. (2005). Essential role of p53 phosphorylation by p38 MAPK in apoptosis induction by the HIV-1 envelope. *The Journal of Experimental Medicine, 201*, 279–289. http://dx.doi.org/10.1084/jem.20041502.

Pierantoni, G. M., Bulfone, A., Pentimalli, F., Fedele, M., Iuliano, R., Santoro, M., et al. (2002). The homeodomain-interacting protein kinase 2 gene is expressed late in embryogenesis and preferentially in retina, muscle, and neural tissues. *Biochemical and Biophysical Research Communications, 290*, 942–947.

Poon, C. L. C., Zhang, X., Lin, J. I., Manning, S. A., & Harvey, K. F. (2012). Homeodomain-interacting protein kinase regulates Hippo pathway-dependent tissue growth. *Current Biology, 22*, 1587–1594. http://dx.doi.org/10.1016/j.cub.2012.06.075.

Puca, R., Nardinocchi, L., Givol, D., & D'Orazi, G. (2010). Regulation of p53 activity by HIPK2: Molecular mechanisms and therapeutical implications in human cancer cells. *Oncogene, 29*, 4378–4387. http://dx.doi.org/10.1038/onc.2010.183.

Raich, W. B., Moorman, C., Lacefield, C. O., Lehrer, J., Bartsch, D., Plasterk, R. H. A., et al. (2003). Characterization of Caenorhabditis elegans homologs of the Down syndrome candidate gene DYRK1A. *Genetics, 163*, 571–580.

Rattay, K., Claude, J., Rezavandy, E., Matt, S., Hofmann, T. G., Kyewski, B., et al. (2015). Homeodomain-interacting protein kinase 2, a novel autoimmune regulator interaction partner, modulates promiscuous gene expression in medullary thymic epithelial cells. *Journal of Immunology, 194*, 921–928. http://dx.doi.org/10.4049/jimmunol.1402694.

Rey, C., Soubeyran, I., Mahouche, I., Pedeboscq, S., Bessede, A., Ichas, F., et al. (2013). HIPK1 drives p53 activation to limit colorectal cancer cell growth. *Cell Cycle, 12*, 1879–1891. http://dx.doi.org/10.4161/cc.24927.

Ricci, A., Cherubini, E., Ulivieri, A., Lavra, L., Sciacchitano, S., Scozzi, D., et al. (2013). Homeodomain-interacting protein kinase2 in human idiopathic pulmonary fibrosis. *Journal of Cellular Physiology, 228*, 235–241. http://dx.doi.org/10.1002/jcp.24129.

Rinaldo, C., Moncada, A., Gradi, A., Ciuffini, L., D'Eliseo, D., Siepi, F., et al. (2012). HIPK2 controls cytokinesis and prevents tetraploidization by phosphorylating histone H2B at the midbody. *Molecular Cell*, 47, 87–98. http://dx.doi.org/10.1016/j.molcel.2012.04.029.

Rinaldo, C., Prodosmo, A., Mancini, F., Iacovelli, S., Sacchi, A., Moretti, F., et al. (2007a). MDM2-regulated degradation of HIPK2 prevents p53Ser46 phosphorylation and DNA damage-induced apoptosis. *Molecular Cell*, 25, 739–750. http://dx.doi.org/10.1016/j.molcel.2007.02.008.

Rinaldo, C., Prodosmo, A., Siepi, F., & Soddu, S. (2007b). HIPK2: A multitalented partner for transcription factors in DNA damage response and development. *Biochemistry and Cell Biology*, 85, 411–418.

Rochat-Steiner, V., Becker, K., Micheau, O., Schneider, P., Burns, K., & Tschopp, J. (2000). FIST/HIPK3: A Fas/FADD-interacting serine/threonine kinase that induces FADD phosphorylation and inhibits fas-mediated Jun NH(2)-terminal kinase activation. *The Journal of Experimental Medicine*, 192, 1165–1174. http://dx.doi.org/10.1084/jem.192.8.1165.

Rui, Y., Xu, Z., Lin, S., Li, Q., Rui, H., Luo, W., et al. (2004). Axin stimulates p53 functions by activation of HIPK2 kinase through multimeric complex formation. *The EMBO Journal*, 23, 4583–4594. http://dx.doi.org/10.1038/sj.emboj.7600475.

Saul, V. V., de la Vega, L., Milanovic, M., Krüger, M., Braun, T., Fritz-Wolf, K., et al. (2013). HIPK2 kinase activity depends on cis-autophosphorylation of its activation loop. *Journal of Molecular Cell Biology*, 5, 27–38. http://dx.doi.org/10.1093/jmcb/mjs053.

Saul, V. V., & Schmitz, M. L. (2013). Posttranslational modifications regulate HIPK2, a driver of proliferative diseases. *Journal of Molecular Medicine (Berlin, Germany)*, 91, 1051–1058. http://dx.doi.org/10.1007/s00109-013-1042-0.

Schmitz, M. L., Rodriguez-Gil, A., & Hornung, J. (2014). Integration of stress signals by homeodomain interacting protein kinases. *Biological Chemistry*, 395, 375–386. http://dx.doi.org/10.1515/hsz-2013-0264.

Shang, Y., Doan, C. N., Arnold, T. D., Lee, S., Tang, A. A., Reichardt, L. F., et al. (2013). Transcriptional corepressors HIPK1 and HIPK2 control angiogenesis via TGF-β-TAK1-dependent mechanism. *PLoS Biology*, 11. http://dx.doi.org/10.1371/journal.pbio.1001527.

Shima, Y., Shima, T., Chiba, T., Irimura, T., Pandolfi, P. P., & Kitabayashi, I. (2008). PML activates transcription by protecting HIPK2 and p300 from SCFFbx3-mediated degradation. *Molecular and Cellular Biology*, 28, 7126–7138. http://dx.doi.org/10.1128/MCB.00897-08.

Shimizu, N., Ishitani, S., Sato, A., Shibuya, H., Ishitani, T., Angers, S., et al. (2014). Hipk1 and PP1c cooperate to maintain Dvl protein levels required for Wnt signal transduction. *Cell Reports*, 8, 1391–1404. http://dx.doi.org/10.1016/j.celrep.2014.07.040.

Shojima, N., Hara, K., Fujita, H., Horikoshi, M., Takahashi, N., Takamoto, I., et al. (2012). Depletion of homeodomain-interacting protein kinase 3 impairs insulin secretion and glucose tolerance in mice. *Diabetologia*, 55, 3318–3330. http://dx.doi.org/10.1007/s00125-012-2711-1.

Siepi, F., Gatti, V., Camerini, S., Crescenzi, M., & Soddu, S. (2013). HIPK2 catalytic activity and subcellular localization are regulated by activation-loop Y354 autophosphorylation. *Biochimica et Biophysica Acta*, 1833, 1443–1453. http://dx.doi.org/10.1016/j.bbamcr.2013.02.018.

Sjölund, J., Pelorosso, F. G., Quigley, D. A., DelRosario, R., & Balmain, A. (2014). Identification of Hipk2 as an essential regulator of white fat development. *Proceedings of the National Academy of Sciences of the United States of America*, 111, 7373–7378. http://dx.doi.org/10.1073/pnas.1322275111.

Smeenk, L., van Heeringen, S. J., Koeppel, M., Gilbert, B., Janssen-Megens, E., Stunnenberg, H. G., et al. (2011). Role of p53 Serine 46 in p53 target gene regulation. *PLoS One*, 6. http://dx.doi.org/10.1371/journal.pone.0017574.

Sombroek, D., & Hofmann, T. G. (2009). How cells switch HIPK2 on and off. *Cell Death and Differentiation*, *16*, 187–194. http://dx.doi.org/10.1038/cdd.2008.154.

Song, J. J., & Lee, Y. J. (2003). Role of the ASK1-SEK1-JNK1-HIPK1 signal in Daxx trafficking and ASK1 oligomerization. *The Journal of Biological Chemistry*, *278*, 47245–47252. http://dx.doi.org/10.1074/jbc.M213201200.

Soubeyran, I., Mahouche, I., Grigoletto, A., Leste-Lasserre, T., Drutel, G., Rey, C., et al. (2011). Tissue microarray cytometry reveals positive impact of homeodomain interacting protein kinase 2 in colon cancer survival irrespective of p53 function. *The American Journal of Pathology*, *178*, 1986–1998. http://dx.doi.org/10.1016/j.ajpath.2011.01.021.

Spector, D. L., & Lamond, A. I. (2011). Nuclear speckles. *Cold Spring Harbor Perspectives in Biology*, *3*. http://dx.doi.org/10.1101/cshperspect.a000646.

Sung, K. S., Go, Y. Y., Ahn, J. H., Kim, Y. H., Kim, Y., & Choi, C. Y. (2005). Differential interactions of the homeodomain-interacting protein kinase 2 (HIPK2) by phosphorylation-dependent sumoylation. *FEBS Letters*, *579*, 3001–3008.

Sung, K. S., Lee, Y. A., Kim, E. T., Lee, S. R., Ahn, J. H., & Choi, C. Y. (2011). Role of the SUMO-interacting motif in HIPK2 targeting to the PML nuclear bodies and regulation of p53. *Experimental Cell Research*, *317*, 1060–1070. http://dx.doi.org/10.1016/j.yexcr.2010.12.016.

Swarup, S., & Verheyen, E. M. (2011). Drosophila homeodomain-interacting protein kinase inhibits the Skp1-Cul1-F-box E3 ligase complex to dually promote Wingless and Hedgehog signaling. *Proceedings of the National Academy of Sciences of the United States of America*, *108*, 9887–9892. http://dx.doi.org/10.1073/pnas.1017548108.

Taira, N., Nihira, K., Yamaguchi, T., Miki, Y., & Yoshida, K. (2007). DYRK2 is targeted to the nucleus and controls p53 via Ser46 phosphorylation in the apoptotic response to DNA damage. *Molecular Cell*, *25*, 725–738. http://dx.doi.org/10.1016/j.molcel.2007.02.007.

Tavalai, N., & Stamminger, T. (2008). New insights into the role of the subnuclear structure ND10 for viral infection. *Biochimica et Biophysica Acta*, *1783*, 2207–2221. http://dx.doi.org/10.1016/j.bbamcr.2008.08.004.

Trapasso, F., Aqeilan, R. I., Iuliano, R., Visone, R., Gaudio, E., Ciuffini, L., et al. (2009). Targeted disruption of the murine homeodomain-interacting protein kinase-2 causes growth deficiency in vivo and cell cycle arrest in vitro. *DNA and Cell Biology*, *28*, 161–167. http://dx.doi.org/10.1089/dna.2008.0778.

Valente, D., Bossi, G., Moncada, A., Tornincasa, M., Indelicato, S., Piscuoglio, S., et al. (2015). HIPK2 deficiency causes chromosomal instability by cytokinesis failure and increases tumorigenicity. *Oncotarget*, *6*, 10320–10334. http://dx.doi.org/10.18632/oncotarget.3583.

Van der Laden, J., Soppa, U., & Becker, W. (2015). Effect of tyrosine autophosphorylation on catalytic activity and subcellular localisation of homeodomain-interacting protein kinases (HIPK). *Cell Communication and Signaling: CCS*, *13*, 3. http://dx.doi.org/10.1186/s12964-014-0082-6.

Verheyen, E. M., & Gottardi, C. J. (2010). Regulation of Wnt/beta-catenin signaling by protein kinases. *Developmental Dynamics*, *239*, 34–44. http://dx.doi.org/10.1002/dvdy.22019.

Verheyen, E. M., Swarup, S., & Lee, W. (2012). Hipk proteins dually regulate Wnt/Wingless signal transduction. *Fly*, *6*, 126–131. http://dx.doi.org/10.4161/fly.20143.

Wang, Y., Debatin, K.-M. M., & Hug, H. (2001). HIPK2 overexpression leads to stabilization of p53 protein and increased p53 transcriptional activity by decreasing Mdm2 protein levels. *BMC Molecular Biology*, *2*, 8. http://dx.doi.org/10.1186/1471-2199-2-8.

Wang, Y., Hofmann, T. G., Runkel, L., Haaf, T., Schaller, H., Debatin, K. M., et al. (2001). Isolation and characterization of cDNAs for the protein kinase HIPK2. *Biochimica et Biophysica Acta*, *1518*, 168–172. http://dx.doi.org/10.1016/S0167-4781(00)00308-0.

Wei, G., Ku, S., Ma, G. K., Saito, S., Tang, A. A., Zhang, J., et al. (2007). HIPK2 represses beta-catenin-mediated transcription, epidermal stem cell expansion, and skin tumorigenesis. *Proceedings of the National Academy of Sciences of the United States of America*, *104*, 13040–13045. http://dx.doi.org/10.1073/pnas.0703213104.

Weston, C. R., & Davis, R. J. (2002). The JNK signal transduction pathway. *Current Opinion in Genetics & Development*, *12*, 14–21. doi:S0959437X01002581 [pii].

Wiggins, A. K., Wei, G., Doxakis, E., Wong, C., Tang, A. A., Zang, K., et al. (2004). Interaction of Brn3a and HIPK2 mediates transcriptional repression of sensory neuron survival. *The Journal of Cell Biology*, *167*, 257–267.

Winter, M., Sombroek, D., Dauth, I., Moehlenbrink, J., Scheuermann, K., Crone, J., et al. (2008). Control of HIPK2 stability by ubiquitin ligase Siah-1 and checkpoint kinases ATM and ATR. *Nature Cell Biology*, *10*, 812–824. http://dx.doi.org/10.1038/ncb1743.

Wu, C.-I., Hoffman, J. A., Shy, B. R., Ford, E. M., Fuchs, E., Nguyen, H., et al. (2012). Function of Wnt/β-catenin in counteracting Tcf3 repression through the Tcf3-β-catenin interaction. *Development*, *139*, 2118–2129. http://dx.doi.org/10.1242/dev.076067.

Wu, Q., Yang, Z., Wang, F., Hu, S., Yang, L., Shi, Y., et al. (2013). MiR-19b/20a/92a regulates the self-renewal and proliferation of gastric cancer stem cells. *Journal of Cell Science*, *126*, 4220–4229. http://dx.doi.org/10.1242/jcs.127944.

Yu, J., Deshmukh, H., Gutmann, R. J., Emnett, R. J., Rodriguez, F. J., Watson, M. A., et al. (2009). Alterations of BRAF and HIPK2 loci predominate in sporadic pilocytic astrocytoma. *Neurology*, *73*, 1526–1531. http://dx.doi.org/10.1212/WNL.0b013e3181c0664a.

Zhang, J., Pho, V., Bonasera, S. J., Holtzman, J., Tang, A. T., Hellmuth, J., et al. (2007). Essential function of HIPK2 in TGFbeta-dependent survival of midbrain dopamine neurons. *Nature Neuroscience*, *10*, 77–86.

CHAPTER FOUR

ROR-Family Receptor Tyrosine Kinases

Sigmar Stricker*, Verena Rauschenberger[†], Alexandra Schambony[†,1]
*Institute for Chemistry and Biochemistry, Freie Universität Berlin, Berlin, Germany
[†]Developmental Biology, Friedrich-Alexander University Erlangen-Nuremberg, Erlangen, Germany
[1]Corresponding author: e-mail address: alexandra.schambony@fau.de

Contents

1. Domain Architecture and Expression Patterns   106
2. ROR Function as a WNT Receptor and Its Role in WNT Signal Transduction   110
3. Human Inheritable Syndromes Caused by *ROR2* Mutation   114
4. The Role of ROR-Family RTKs in Embryonic Development   119
   4.1 Body Axis Elongation and Axial Skeleton Formation: Gastrulation Movements and Mesoderm Development   119
   4.2 Neural Crest Development: Craniofacial and Cardiac Outflow Tract Defects   121
   4.3 Skeletal Development: Cartilage Growth Plate Defects in Robinow Syndrome   124
   4.4 Skeletal Development: Brachydactyly Caused by Defective Digit Elongation   128
   4.5 Kidney Development   131
References   133

Abstract

ROR-family receptor tyrosine kinases form a small subfamily of receptor tyrosine kinases (RTKs), characterized by a conserved, unique domain architecture. ROR RTKs are evolutionary conserved throughout the animal kingdom and act as alternative receptors and coreceptors of WNT ligands. The intracellular signaling cascades activated downstream of ROR receptors are diverse, including but not limited to ROR–Frizzled-mediated activation of planar cell polarity signaling, RTK-like signaling, and antagonistic regulation of WNT/β-Catenin signaling. In line with their diverse repertoire of signaling functions, ROR receptors are involved in the regulation of multiple processes in embryonic development such as development of the axial and paraxial mesoderm, the nervous system and the neural crest, the axial and appendicular skeleton, and the kidney. In humans, mutations in the *ROR2* gene cause two distinct developmental syndromes, recessive Robinow syndrome (RRS; MIM 268310) and dominant brachydactyly type B1 (BDB1; MIM 113000). In Robinow syndrome patients and animal models, the development of multiple organs is affected, whereas BDB1 results only in shortening of the distal phalanges of fingers and toes, reflecting the diversity of functions and signaling activities of ROR-family RTKs. In this chapter, we give an overview on ROR receptor structure and function. We discuss their signaling functions and role in vertebrate embryonic development with a focus on those developmental processes that are affected by mutations in the *ROR2* gene in human patients.

1. DOMAIN ARCHITECTURE AND EXPRESSION PATTERNS

Receptor tyrosine kinases (RTKs) regulate a multitude of cellular and tissue functions and play a crucial role in many events during embryonic development, in the adult organism as well as in disease like cancer. To date, the superfamily of transmembrane RTKs consists of about 58 known members in mammals which fall into 20 subfamilies (Robinson, Wu, & Lin, 2000), one of which is the ROR family.[a]

Initially, ROR RTKs have been identified based on their homology to the RTK family of neurotrophin receptors (NTRK), which play a role in development of the nervous system. In an attempt to find more members of the NTRK family, ROR1 and ROR2 were first isolated in 1992 from the human neuroblastoma cell line SH-SY5Y (Masiakowski & Carroll, 1992). Today, ROR RTKs form one family themselves that is characterized by its domain architecture and is comprised of two members, ROR1 and ROR2. ROR RTKs are single-pass type I transmembrane proteins that contain one immunoglobulin-like (Ig-like) domain, one Frizzled (FZD)-like cysteine-rich domain (CRD), and one Kringle domain in their extracellular part and a tyrosine kinase domain in their intracellular part (Fig. 1). The closest relatives to ROR RTKs are the MUSK-family RTKs (muscle, skeletal receptor tyrosine protein kinase) that play a role in neuromuscular synapse formation and show similar domain architecture. In contrast to ROR RTKs, MUSK RTKs contain three Ig-like domains and one CRD domain in their extracellular part, a single transmembrane domain and a cytoplasmic tyrosine kinase domain (DeChiara et al., 1996; Glass et al., 1996). Together with the DDR (epithelial discoidin domain-containing receptor) and NTRK family, ROR and MUSK RTKs form one branch of human tyrosine kinases (Robinson et al., 2000).

---

[a] For consistency, gene and protein nomenclature follows the guidelines of the HGNC (HUGO gene nomenclature committee; http://www.genenames.org/). When referring to homologous genes or proteins in other species, human nomenclature is used and preceded by a species identifier; the species-typic gene name is provided in parentheses.

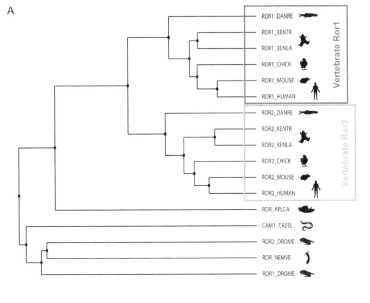

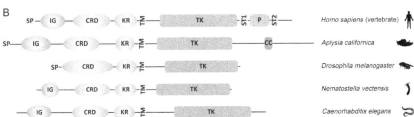

Fig. 1 Phylogeny and domain architecture of ROR-family RTKs. (A) Phylogenetic tree including selected vertebrate and invertebrate members of the ROR family. Species abbreviations according to uniprot.org: *DANRE*, *Danio rerio* (zebrafish); *XENTR*, *Xenopus tropicalis*; *XENLA*, *Xenopus laevis*; *CHICK*, *Gallus gallus*; *MOUSE*, *Mus musculus*; *HUMAN*, *Homo sapiens*; *APLCA*, *Aplysia californica* (sea hare); *CAEEL*, *Caenorhabditis elegans*; *DROME*, *Drosophila melanogaster* (fruit fly); *NEMVE*, *Nematostella vectensis* (sea anemone). (B) Domain architecture of selected ROR proteins. Human ROR2 is representative for all vertebrate ROR1 and ROR2 orthologs. Invertebrate ROR proteins share FZD-CRD, Kringle (KR), and tyrosine kinase (TK) domains and—except the *Drosophila melanogaster* orthologs—also the Ig-like domain. The region C-terminal of the TK domain is strictly conserved among vertebrates but highly variable in invertebrates. *ST*, serine/threonine-rich regions; *P*, proline-rich region; *CC*, coiled-coil domain.

ROR RTKs are evolutionary conserved in the animal kingdom. To date, orthologs of ROR-family RTKs have been identified in Cnidaria, in protostomes including *Drosophila melanogaster*, *Daphnia pulex*, *Caenorhabditis elegans*, and *Aplysia californica* as well as in all classes of vertebrates (Fig. 1A). Most ROR proteins share the abovementioned characteristic domain architecture of Ig-like, CRD, Kringle, and tyrosine kinase domain (Fig. 1B). An exception is found in the *D. melanogaster* orthologs *Drosophila* ROR1 (DRor) and *Drosophila* ROR2 (DNrk), which lack the Ig-like domain but show an extended, sometimes also called split or duplicated, CRD domain. Overall, *Drosophila* ROR1 shows higher similarity to human ROR1 and ROR2 than *Drosophila* ROR2. The most striking variations are observed in the intracellular part C-terminal of the tyrosine kinase domain. While such a C-terminal extension is absent in *Drosophila*, *C. elegans*, and *Nematostella* ROR proteins, *Aplysia* ROR shows a long C-terminal region that contains a coiled-coil domain. Vertebrate members of the ROR family, however, share a C-terminal domain of approximately 200 amino acids that contains two serine/threonine-rich and one proline-rich regions (Fig. 1B).

Despite the high structural homology of ROR proteins, the expression patterns of *ROR* genes during embryogenesis vary strikingly between invertebrates and vertebrates. Expression patterns and functional data of *ROR* orthologs have been reported for the mollusk *A. californica*, the nematode *C. elegans*, and *Drosophila*. In *Drosophila* but also in *Aplysia*, expression of *ROR* genes is restricted to neural tissue. *Drosophila ROR1* and *Drosophila ROR2* expression is first detectable at the extended germ band stage and is maintained in the neurogenic ectoderm during embryogenesis, suggesting a role in neural development and function (Oishi et al., 1997; Wilson, Goberdhan, & Steller, 1993). *Aplysia ROR* is expressed in developing neurons and clustered at neurites, indicating that ROR contributes to neuronal polarity; additionally, a role in the regulation of neuropeptides has been suggested (McKay et al., 2001). The *C. elegans ROR2* orthologous gene (*CAM-1*) shows a more diverse expression pattern and functionality. Similar to the other invertebrates, *C. elegans ROR2* is expressed in the nervous system where it plays a role in asymmetric cell division of neuronal precursors, neural cell migration, axon guidance, positioning of the anterior nerve ring (Forrester, Dell, Perens, & Garriga, 1999, Forrester, Kim, & Garriga, 2004; Kennerdell, Fetter, & Bargmann, 2009; Mentink et al., 2014), and is also required for neuronal polarity in mechanosensory neurons, synaptic function at the neuromuscular junction, and synaptic plasticity (Francis et al., 2005; Jensen et al., 2012). Interestingly, the intracellular domain is dispensable for polarity and migration of neural cells (Forrester et al., 1999) as well as for cholinergic neurotransmission (Francis et al., 2005; Jensen et al., 2012).

In addition, *C. elegans ROR2* is expressed in nonneural vulval precursor cells (VPCs) of *C. elegans*. In VPCs, *C. elegans ROR2* acts as WNT receptor and contributes to cell polarity and oriented cell division. Overexpression of *C. elegans ROR2* inhibits correct vulva development because VPCs fail to divide and arrange properly in these mutants (Green, Inoue, & Sternberg, 2007, 2008).

In vertebrates, the expression patterns of *ROR* genes have been analyzed in zebrafish, frog, chicken, and mouse, respectively. In chicken and mouse, *ROR1* and *ROR2* show a widespread expression during embryonic development with distinct but also overlapping domains. Most prominent expression domains were found in the central nervous system, heart, lung, mesonephros, the early limb bud, and also within cartilage condensations and the cartilage growth plate (DeChiara et al., 2000; Matsuda et al., 2001; Schwabe et al., 2004; Stricker et al., 2006). In accordance with their expression domains, both receptors exhibit partially redundant but also highly specific functions as discussed later.

In zebrafish and *Xenopus* to date only *ROR2* expression has been reported. Zebrafish *ROR2* is expressed maternally and throughout embryonic development; the expression levels are dynamically regulated with peaks at 2-h postfertilization (hpf), during gastrulation at 6–9 hpf and in organogenesis stages 24–48 hpf. Expression is ubiquitous in early development, whereas in organogenesis stages, expression is strongest in the nervous system (Bai et al., 2014). By contrast, *Xenopus ROR2* is only expressed zygotically and is first detected at the beginning of gastrulation (NF stage 10) in the dorsal marginal zone (DMZ). Expression levels rise continuously through gastrulation, peak at mid-to-late neurula stages (NF15–19), and persist at a slightly lower level till early tadpole stage NF28. Spatial expression is broad in gastrulation and becomes more restricted during neurulation when *ROR2* is expressed strongest in the neurogenic ectoderm and the notochord. In tadpole stages, *ROR2* transcripts are detected in the posterior neural tube, the migrating neural crest and later in branchial arches and cranial nerves, the otic placode, and the tailbud (Feike, Rachor, Gentzel, & Schambony, 2010; Hikasa, Shibata, Hiratani, & Taira, 2002) and also in the heart and epidermis (our unpublished data). Whereas temporal and spatial expression of *ROR2* in zebrafish embryos differs from chicken and mice, the expression patterns in *Xenopus* are highly similar to those in amniotes suggesting at least partially conserved roles in embryonic development.

2. ROR FUNCTION AS A WNT RECEPTOR AND ITS ROLE IN WNT SIGNAL TRANSDUCTION

Initially, no ligand for ROR RTKs was identified and they have been referred to as orphan receptors for about a decade. Today, it is well established that ROR RTKs act as receptors for WNT family ligands. Consistent with the high homology of ROR and FZD CRDs, the ROR CRD has been identified as the WNT-binding domain (Billiard et al., 2005; Oishi et al., 2003; Fig. 2A). ROR RTKs are able to bind multiple WNT ligands including WNT1, WNT2, WNT3 and WNT3A, WNT4, WNT5A and WNT5B, WNT6, WNT7A, WNT8, WNT11, as well as the *C. elegans* WNTs CWN-1, MOM-2, EGL-20, and LIN-44 (Billiard et al., 2005;

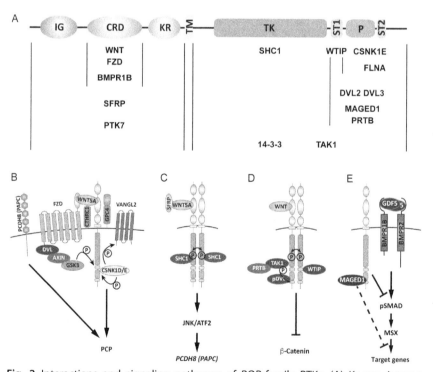

Fig. 2 Interactions and signaling pathways of ROR-family RTKs. (A) Known interacting proteins and binding regions in the ROR2 receptor (see text for details). (B) PCP signaling mediated by ROR2–FZD complexes; this signaling cascade likely depends on Clathrin-mediated endocytosis of the receptor complex, which is not shown in the schematic. (C) RTK-like signaling to JNK/ATF2. (D) Inhibition of β-Catenin signaling. (E) Antagonistic regulation of BMP signaling and MSX transcription factors.

Brinkmann et al., 2016; Green et al., 2007; Hikasa et al., 2002; Liu, Rubin, Bodine, & Billiard, 2008; Oishi et al., 2003; Paganoni, Bernstein, & Ferreira, 2010; Winkel et al., 2008).

WNT5A appears as the primary ligand of vertebrate ROR2. It has been shown that WNT5A is able to induce homodimerization and autophosphorylation of ROR2 (Feike et al., 2010; Liu, Ross, Bodine, & Billiard, 2007; Liu et al., 2008). Ligand-induced homodimerization and autophosphorylation is reminiscent to ligand-induced activation mechanisms described for other RTKs and consistent with a functional tyrosine kinase domain in ROR2 (Bainbridge et al., 2014; Feike et al., 2010; Liu et al., 2007; Mikels, Minami, & Nusse, 2009). In addition, the formation of heterodimers between ROR1 and ROR2 has been observed, where ROR2 seems to be the ligand-binding partner (Paganoni et al., 2010).

However, in other contexts, ROR2 acts as a coreceptor in a complex with FZD and potentially additional coreceptors rather than as autonomous RTK. The ROR2 and FZD CRDs are able to interact physically and both receptors form active, WNT-dependent receptor complexes (Brinkmann et al., 2016; Nishita et al., 2010; Oishi et al., 2003). FZD–ROR2 complexes may further include protein tyrosine kinase 7 (PTK7), collagen triple helix repeat containing 1 (CTHRC1, Fig. 2B), or the RTK RYK (Andre et al., 2012; Gao et al., 2011; Martinez et al., 2015; Yamamoto et al., 2008). In addition to WNT, FZD, and its coreceptors, a number of intracellular binding partners of vertebrate ROR2 have been described. Interestingly, most of those interactions have been mapped to the ROR2 C-terminus (Fig. 2A).

Signaling downstream of ROR2 is variable and context dependent. The major signaling cascade that is activated by ROR2–FZD complexes is the β-Catenin-independent WNT/planar cell polarity (PCP) pathway (Fig. 2B). PCP signaling has been first described in *Drosophila* wing epithelia and ommatidia where it is characterized by asymmetric, mutually exclusive localization of the core PCP complexes FZD/DVL and VANGL/PRICKLE at the plasma membrane. It should be noted that neither *Drosophila* ROR1 nor ROR2 is expressed in these tissues and so far no contribution of ROR to PCP in *Drosophila* has been reported. In vertebrates, the PCP pathway controls orientation of skin hair and stereocilia in the inner ear but also cell polarity and coordinated migration in gastrulation, neurulation, neural crest migration, gut elongation, and limb development (reviewed in Seifert & Mlodzik, 2007). PCP signaling is activated by a WNT ligand, mostly WNT5A, WNT11, or WNT4 in vertebrates (Andre, Song, Kim, Kispert, & Yang, 2015; Gros, Serralbo, & Marcelle, 2009; Heinonen, Vanegas, Lew, Krosl, &

Perreault, 2011; Heisenberg et al., 2000; Tanigawa et al., 2011) and wingless (wg) and WNT4 in *Drosophila* (Wu, Roman, Carvajal-Gonzalez, & Mlodzik, 2013). Downstream of the FZD receptor complex, the intracellular scaffold protein Disheveled (DVL) is phosphorylated and induces the activation of Rho-family GTPases RHOA and RAC1 and of the JNK cascade (Habas, Dawid, & He, 2003; Kim & Han, 2005). ROR2 interacts with phosphorylated DVL and is required for DVL-mediated AP1 and RAC1 activation (Nishita et al., 2010; Witte, Bernatik, et al., 2010). In addition, ROR2 has been found to promote VANGL2 phosphorylation by casein kinase 1 (CSNK1) and asymmetric localization of VANGL2 (Gao et al., 2011). Interestingly, ROR2 also interacts with and is itself phosphorylated by casein kinase 1 epsilon (CSNK1E) and glycogen synthase kinase 3 (GSK3), two kinases that also play a pivotal role in WNT/β-Catenin signaling, and similar mechanisms of receptor complex activation have been suggested for β-Catenin-dependent and -independent WNT pathways (Grumolato et al., 2010; Kani et al., 2004; Verkaar & Zaman, 2010; Yamamoto, Yoo, Nishita, Kikuchi, & Minami, 2007). Furthermore, ROR2–DVL interaction also contributes to cytoskeleton remodeling and JNK activation in migrating cells (Nishita et al., 2006; Nomachi et al., 2008; Oishi et al., 2003).

In addition to this direct role as part of the FZD receptor complex in WNT/PCP signaling, an additional mechanism of ROR2 contribution to PCP signaling has been described in *Xenopus* gastrulation, where ROR2 appears to activate a classic RTK pathway (Fig. 2C). ROR2 interacts with the RTK scaffold protein SHC1, which binds to a conserved phosphotyrosine-containing motif via its SH2 (SRC homology type 2) domain. Additional downstream effectors include phosphoinositide-3-kinase (PI3K/PIK3) and the RHO-family small GTPase CDC42, which activate the JNK cascade and regulate transcription of *paraxial protocadherin* (*PAPC*) via JUN and ATF2 (Feike et al., 2010; Schambony & Wedlich, 2007). PAPC is the *Xenopus* ortholog of human protocadherin 8 (PCDH8). In *Xenopus* gastrulation, PCDH8 (PAPC) is required for coordinated cell movements and tissue separation. Moreover, PCDH8 (PAPC) possesses signaling activity and has been found to activate RHOA and JNK likely by interaction with PCP pathway components (Kraft, Berger, Wallkamm, Steinbeisser, & Wedlich, 2012; Medina, Swain, Kuerner, & Steinbeisser, 2004; Unterseher et al., 2004; Fig. 2B).

ROR2 signaling to JNK and transcriptional regulation of *Xenopus PCDH8* (*PAPC*) is promoted by interactions between ROR2 and another RTK, PTK7 (Martinez et al., 2015), as well as by SFRPs (Brinkmann et al., 2016). SFRPs are secreted FZD-related proteins that contain an N-terminal

FZD-CRD and modulate WNT signaling by direct binding and sequestration of WNT ligands or by interaction with FZD receptors (reviewed in Kawano & Kypta, 2003). Interestingly, Brinkmann and colleagues suggest that SFRPs stabilize ROR2 complexes at the plasma membrane and promote *PCDH8* (*PAPC*) expression and at the same time inhibit FZD7 endocytosis (Brinkmann et al., 2016). It has been suggested that activation of PCP signaling requires Clathrin-mediated endocytosis of the FZD–ROR2 complex (Sakane, Yamamoto, Matsumoto, Sato, & Kikuchi, 2012; Yamamoto et al., 2008), accordingly, interaction with SFRPs would inhibit PCP signaling. By contrast, in the canonical model of RTK signaling, ligand-induced endocytosis results in receptor degradation and signal termination (reviewed in Goh & Sorkin, 2013), whereas signal transduction takes place at the plasma membrane. In this model and as shown for several RTKs, inhibition of endocytosis enhances and prolongs signaling, e.g., to SH2-domain adaptor proteins such as SHC1 and the lipid kinase PI3K. Consistent with ROR2 signaling through SHC1 and PI3K to regulate *PCDH8* (*PAPC*) transcription, this pathway is enhanced by SFRPs, suggesting that the activation of PCP vs canonical RTK signaling by ROR2 might indeed be distinct signaling events that are differentially regulated by SFRPs (Brinkmann et al., 2016; Feike et al., 2010; Schambony & Wedlich, 2007). Interestingly, recent work from Mentink and colleagues shows that the *C. elegans* ROR2 ortholog CAM-1 acts downstream of WNT but independent of the core PCP effectors VANGL and PRICKLE in the regulation of directional neuroblast migration (Mentink et al., 2014), further supporting the ability of ROR2 to signal in distinct β-Catenin-independent WNT pathways.

Different WNT signaling pathways form a complex signaling network where several effector proteins including WNT ligands, FZD, DVL, GSK3β, and CSNK1 play a role in distinct, even mutually antagonistic signaling cascades. In this network, ROR2 not only activates β-Catenin-independent WNT pathways but has also been found to act as an antagonist of WNT/β-Catenin signaling (Fig. 2D). The molecular mechanisms, however, are not fully understood. In *C. elegans* ROR2 apparently acts by sequestering WNT ligands, a function independent of its intracellular domain (Green et al., 2007, 2008), whereas in cellular systems intracellular signaling mechanisms involving tyrosine kinase activity and interaction with DVL have also been implicated (Mikels et al., 2009; Mikels & Nusse, 2006; Witte, Bernatik, et al., 2010). Further, the intracellular binding partners TAK1, PRTB, and WTIP have been found to bind to the C-terminal region of ROR2 and to enhance its inhibition of WNT/β-Catenin signaling (van Wijk et al., 2009; Winkel et al., 2008).

In addition to these well-characterized roles in WNT signaling, ROR2 apparently interacts with or modulates other signaling events. ROR2 binds to the type I bone morphogenetic protein (BMP) receptor BMPR1B, which inhibits SMAD phosphorylation and activation by BMPR1B (Sammar, Sieber, & Knaus, 2009; Sammar et al., 2004; Fig. 2E). SMADs regulate the expression of MSX homeobox transcription factors in vertebrate neural crest cells (Brugger et al., 2004; Tribulo, 2003). Interestingly, ROR2 also interacts with and sequesters melanoma-associated antigen 1 (MAGED1, DLXIN), a transcriptional cofactor of DLX and MSX transcription factors, and thereby inhibits MSX function (Matsuda et al., 2003). Although these interactions and their role in embryonic development have not been characterized in detail, these observations indicate that ROR2 might physically and functionally link and integrate multiple signaling events, which is further emphasized by the complexity of ROR2-associated pathologies.

3. HUMAN INHERITABLE SYNDROMES CAUSED BY *ROR2* MUTATION

Mutations in human *ROR2* cause two distinct developmental syndromes, recessive Robinow syndrome (RRS; MIM 268310) (Afzal et al., 2000; van Bokhoven et al., 2000) and brachydactyly type B1 (BDB1; MIM 113000) (Oldridge et al., 2000; Schwabe et al., 2000). RRS is characterized by a variety of phenotypic appearances including craniofacial malformations such as hypertelorism, midface hypoplasia, short upturned nose, cleft lip and/or cleft palate, triangular mouth, and micrognathia. These cause a specific appearance resembling a fetal face, which led to the original naming of the syndrome as "fetal facies" (later called "fetal face syndrome") by Robinow (Robinow, Silverman, & Smith, 1969). In addition to the craniofacial skeleton, the axial and appendicular skeleton is involved as well with deformations as short stature, mesomelic limb shortening (disproportionate shortening of the zeugopod bones, for a depiction of the skeletal structures of the limb; see Fig. 3), and hemivertebrae with fusion of the thoracic vertebrae. Apart from skeletal malformations RRS is characterized by genital hypoplasia, renal tract abnormalities, and occasional heart defects (Patton & Afzal, 2002). Brachydactylies on the other hand are congenital disorders altogether characterized by disproportionate shortening of skeletal structures specifically in the autopod (i.e., hands and feet; see Fig. 3). The term brachydactyly is of Greek origin (short = brachy; digit = dactylos). Brachydactyly can be a feature of complex malformation syndromes or appear as isolated trait. Isolated brachydactylies are categorized into types

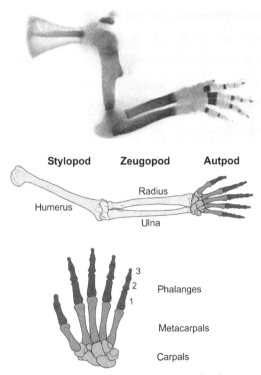

Fig. 3 Schematic depiction of limb skeletal structures. The three segments of the limb and their skeletal elements are depicted. The schematic drawing was kindly provided by S. Mundlos, Charité University Medicine Berlin, Berlin, Germany.

A–E (including subtypes), with each type and subtype exhibiting a specific pattern of digit malformation, i.e., specific phalanges and/or metacarpals in specific digits that are affected (see e.g., Mundlos, 2009; Stricker & Mundlos, 2011, for detailed depiction). Based on the highly overlapping phenotypic spectrum of the isolated brachydactylies, they were proposed a bona fide "molecular disease family" meaning that likely the common features seen are caused by mutations in genes involved in a common functional molecular network. BDB1 is characterized by hypoplasia/aplasia of distal phalanges in hands and feet as well as nail hypoplasia/aplasia (Mundlos, 2009).

Both syndromes, RRS and BDB1, are caused by a distinct set of mutations in *ROR2*. RRS mutations are found more or less scattered across the entire coding sequence; they include nonsense, missense, and frameshift mutations (Fig. 4). BDB1 on the other hand is caused by nonsense or frameshift mutations clustered at two hotspots that lead to the expression of a truncated protein that always consists of the extracellular part and the

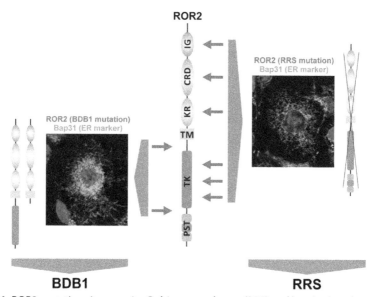

Fig. 4 ROR2 mutations in recessive Robinow syndrome (RRS) and brachydactyly type B1 (BDB1). Mutations found in human ROR2 causing either RRS or BDB1. *Arrows* indicate sites of mutations (type of mutation not depicted). RRS mutations (missense, nonsense, or frameshift) lead to retention of mutated protein in the endoplasmic reticulum (ER); BDB1 mutations (nonsense or frameshift) lead to expression of truncated proteins that localize to the cell membrane.

transmembrane domain but may or may not include the full tyrosine kinase domain; all BDB1 variants lack the proline–serine–threonine-rich C-terminal domains (Fig. 4).

Based on the nature of the mutations, it was speculated early on that RRS mutations may cause a loss of function, while BDB1 mutations may give rise to ROR2 variants with a dominant-negative or gain-of-function effect (Schwabe et al., 2000). This view was supported by the strong phenotypic overlap *ROR2*-null mice show with the features characteristic for RRS (Schwabe et al., 2004; Fig. 5A, C–E). Loss of *ROR2* in mice causes perinatal lethality due to respiratory dysfunction. In addition, *ROR2*-null mice show a cardiac septal defect and diverse skeletal defects, mainly prominent in severely shortened limbs and tail (DeChiara et al., 2000; Takeuchi et al., 2000). By contrast, loss of *ROR1* caused a similar respiratory defect leading to perinatal lethality but was reported to cause no or only very mild other defects (Lyashenko et al., 2010; Nomi et al., 2001).

It was later demonstrated that ROR2 proteins carrying RRS mutations are retained in the endoplasmic reticulum and also show decreased protein

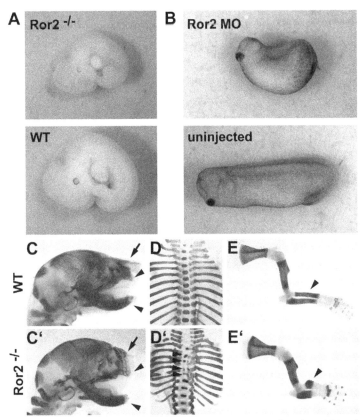

Fig. 5 Features of recessive Robinow syndrome are recapitulated in ROR2-deficient mouse and *Xenopus* embryos. (A and B) Overall shortened body axis and craniofacial malformations are recapitulated in ROR2-null mice (A) and in ROR2 knockdown *Xenopus* embryos (B); the corresponding control/WT embryos are shown below. In *Xenopus* embryos, ROR2 was knocked-down by injection of a translation-blocking morpholino oligonucleotide. (C–E) Alcian *blue*/Alizarin *red* stained skeletal preparations from E18.5 wild-type (WT) and Ror2-deficient (Ror2$^{-/-}$) mouse embryos are shown; cartilage stains *blue*, bone stains *red*. (C, C′) Show the craniofacial outgrowth defect, especially truncation of the nasal bone (*arrow*) and shortening of the mandible and maxilla (*arrowheads*) in Ror2$^{-/-}$ embryos. (D, D′) Vertebral malformations (*arrowheads*) in the Ror2$^{-/-}$ mutant. (E, E′) Mesomelic limb shortening in Ror2$^{-/-}$ mutants (*arrowheads*); note the concomitant thickening of skeletal elements.

stability (Chen, Bellamy, Seabra, Field, & Ali, 2005; Schwarzer, Witte, Rajab, Mundlos, & Stricker, 2009). While RRS mutations are found within conserved and presumably structurally highly ordered protein domains, BDB1 mutations are found in linker regions between such domains providing a potential explanation for the destabilizing nature of RRS mutations. Indeed, BDB1 mutations lead to the expression of stable proteins that are expressed on the cell membrane in accordance with a potential gain of function (Schwarzer et al., 2009).

RRS and BDB1 are clearly distinct syndromes; however, a small phenotypic overlap has to be noted. RRS is one example of a complex syndrome exhibiting mild brachydactyly as one of its features: RRS patients frequently show overall shortened phalanges with the distal phalanges predominantly affected. Notably, as opposed to BDB1, aplasia of phalanges is never observed. Also, occasional nail hypoplasia is seen in RRS. How can these overlaps be reconciled with the seemingly dichotomous nature of the mutations found in each syndrome? Interestingly, Schwabe et al. presented one patient that was homozygous for a frameshift mutation in *ROR2* (c.1321-25_del5/p.R441fsX15) truncating the protein before the TK domain. This mutation caused classical BDB1 in the heterozygous parents; however, the homozygous child exhibited severe brachydactyly (exacerbating the typical BDB1 phenotype) in addition to features overlapping the RS spectrum (Schwabe et al., 2000). Subsequently, Schwarzer et al. presented a patient homozygous for a truncation nonsense mutation at the identical position (c.1324C>T; p.R441X) that also showed severe brachydactyly and RS features; however, here the heterozygous parents were unaffected (Schwarzer et al., 2009). Using cell surface biotinylation in stable ROR2 variant-expressing cell lines, Schwarzer et al. (2009) carefully measured steady-state protein levels as well as cell surface localization. This provided evidence that the phenotypic outcome of *ROR2* mutation is not simply predicted by a gain- vs loss-of-function model, but that intermediate and overlapping phenotypes may be explained by the balance between intracellular retention (overall loss of function) vs surface expression (dominant-negative) of mutant protein. This would predict that a certain level of intracellular retention is needed for the loss-of-function effect thus causing RRS, while the expression of a truncated ROR2 protein at the cell surface in turn also needs to exceed a certain threshold to cause a recessive (as in the case of the c.1324C>T; p.R441X mutation), and a higher threshold to cause a dominant effect (as in the case of the c.1321-25_del5/p.R441fsX15 mutation). This hypothesis was further substantiated by an allelic series of mouse mutants using the *ROR2*-null allele and an allele carrying a BDB1 knock-in mutation, where an intermediate digit phenotype resembling digit involvement in RRS was found in mutants carrying both alleles (Schwarzer et al., 2009).

4. THE ROLE OF ROR-FAMILY RTKs IN EMBRYONIC DEVELOPMENT

4.1 Body Axis Elongation and Axial Skeleton Formation: Gastrulation Movements and Mesoderm Development

ROR2-null mice, as noted earlier, reproduce several RRS features such as craniofacial malformations, hemivertebrae/fused vertebrae, short stature, and mesomelic limb shortening (Fig. 5A, C–E). Shortening of the anterior–posterior body axis is also recapitulated in ROR2-deficient *Xenopus* embryos (Fig. 5B), and its role in primary axis elongation has been studied in more detail in *Xenopus*, in which early embryonic processes are more accessible due to entirely extracorporal development. *Xenopus ROR2* was first identified as a target gene of the organizer-specific LIM homeobox transcription factor XLIM-1 (Hikasa et al., 2002). The organizer's main function is the specification and patterning of dorsal tissues such as the notochord, presomitic mesoderm, and neuroectoderm as well as the initiation of gastrulation (Harland & Gerhart, 1997). During gastrulation and neurulation, coordinated mass cell movements of the mesoderm and neuroectoderm also referred to as convergent extension (CE) movements elongate and shape the embryonic dorsal body axis (Keller, 2002; Keller et al., 2000). The strongest CE movements are observed in the notochord and presomitic mesoderm that originates from the DMZ and in the neuroectoderm. ROR2 plays a crucial role in the regulation of CE movements during gastrulation where it acts as receptor for the "noncanonical" WNTs WNT5A and WNT11 in *Xenopus* and zebrafish (Bai et al., 2014; Hikasa et al., 2002; Schambony & Wedlich, 2007). Recently, it has been confirmed that the same WNT ligands are also required for anterior–posterior axis elongation and CE movements in the mouse (Andre et al., 2015). In *Xenopus*, ROR2 regulates gastrulation movements by an additional mechanism. A WNT5A/ROR2/PI3K pathway activates JUN/ATF2 and transcriptionally regulates *PCDH8* (also known as *PAPC*; Brinkmann et al., 2016; Feike et al., 2010; Ohkawara & Niehrs, 2010; Schambony & Wedlich, 2007). PCDH8 interacts with FZD7 and contributes to the coordination of cell polarity and migration during CE as a modulator of PCP signaling to the small GTPases RHOA and RAC1 (Kraft et al., 2012; Medina et al., 2004; Unterseher et al., 2004). As a consequence of ROR2 deficiency, CE movements are disrupted resulting in shortening of the notochord and presomitic mesoderm along the anterior–posterior axis. In vertebrates, the notochord is a transient structure that confers stability and rigidity during embryonic development and is replaced by the vertebral column later in development. The axial skeleton, i.e., the vertebral column, is derived from the paraxial, presomitic mesoderm. During development, cells from the presomitic mesoderm located at the caudal end of the embryo continuously generate mesodermal packages at either side of the neural tube called somites. The somites undergo epithelialization and differentiate into myotome, dermomyotome, and the sclerotome. The latter will give rise to the cartilage of the vertebrae and ribs. Vertebral malformations in ROR2-null mice were traced back to a reduced size of the presomitic mesoderm (Schwabe et al., 2004), reminiscent of the CE phenotype observed in *Xenopus* embryos, which indicates that similar processes might be controlled by ROR2 in early development of mice and frogs.

In addition to these early defects in murine *ROR2* KO embryos, smaller somites also have been observed that showed a perturbed segmentation and epithelialization process (Schwabe et al., 2004). Induction of the presomitic mesoderm and periodic segmentation of somites appears normal in ROR2-null embryos, but somites show reduced *mesp2* expression, indicating that fewer cells are available in the forming somite. Interestingly, in *WNT5A/WNT11* double-knockout mice, initial formation of the paraxial mesoderm is also normal, but proliferation and survival is compromised (Andre et al., 2015). Whether this might also account for the reduced cell numbers in the somites of ROR2-null mice has not been investigated. Subsequent differentiation into myotome, sclerotome, and dermomyotome takes place in the $ROR2^{-/-}$ embryos, but these tissues are improperly positioned, irregular, and occasionally fused due to perturbed epithelialization of somites (Schwabe et al., 2004). Interestingly, the ROR2 target gene *PCDH8* (*PAPC*) is also expressed in the developing somites and plays a role in somitogenesis in mouse and *Xenopus* embryos (Kim, Jen, De Robertis, & Kintner, 2000; Rhee, Takahashi, Saga, Wilson-Rawls, & Rawls, 2003). In early gastrula, *PCDH8* (*PAPC*) is directly regulated by ROR2 signaling and *PCHD8* (*PAPC*) expression is still downregulated in the presomitic mesoderm in ROR2-depleted *Xenopus* embryos just prior to the onset of somitogenesis (Feike et al., 2010; Schambony & Wedlich, 2007). *PCDH8* (*PAPC*) expression in the presomitic mesoderm is dependent on MESP2 in *Xenopus* and mouse and restricted to the anterior half of each somite (Kim et al., 2000; Rhee et al., 2003). Lower expression levels of MESP2 in ROR2-deficient embryos (Schwabe et al., 2004) would therefore also lead to reduced *PCDH8* (*PAPC*) expression. In *Xenopus*, it has been reported that PCDH8 (PAPC) regulates cell orientation and adhesion during somitogenesis (Kim et al., 2000), and consistently, tissue explants from mouse embryos fail to undergo epithelialization when treated with a secreted, dominant-negative form of PCDH8 (PAPC) (Rhee et al., 2003). Moreover, epithelialization of somites depends on a balanced activation of the small GTPases RAC1 and CDC42 (Nakaya, Kuroda, Katagiri, Kaibuchi, & Takahashi, 2004), which are regulated by PCDH8 (PAPC) and ROR2 (Medina et al., 2004; Schambony & Wedlich, 2007; Unterseher et al., 2004).

In summary, ROR2 and PCDH8 (PAPC)-mediated β-Catenin-independent signaling is required for anterior–posterior extension of the axial and presomitic mesoderm, epithelialization, and somitogenesis and consequently for the correct formation of the axial skeleton. In addition, ROR2 plays a role in chondrogenesis and ossification of skeletal elements as discussed later.

4.2 Neural Crest Development: Craniofacial and Cardiac Outflow Tract Defects

Craniofacial malformations including midface hypoplasia and cleft palate, summarized as "fetal face," are characteristic of RRS (Robinow et al., 1969) and are recapitulated in ROR2- and WNT5A-null mice (He et al., 2008; Nomi et al., 2001; Oishi et al., 2003; Takeuchi et al., 2000; Yamaguchi, Bradley, McMahon, & Jones, 1999). In addition, a subset of patients shows congenital heart defects including septal defects of atria or ventricles and cardiac outflow tract defects, which are also found in the corresponding mouse models. Craniofacial and cardiac defects are more severe in WNT5A-null mice as compared to ROR2-null mice; in some cases, the severity of malformations increases in *ROR1/ROR2*-double mutants, indicating partial redundancy of ROR1 and ROR2 (Nomi et al., 2001; Takeuchi et al., 2000; Yamaguchi et al., 1999).

The craniofacial skeleton and part of the cardiac outflow tract are formed by derivatives of the cranial neural crest. The neural crest is a vertebrate-specific cell population that arises from the neural plate border and is specified by the interaction of intermediate levels of BMP combined with WNT/β-Catenin, NOTCH, and fibroblast growth factor (FGF) signaling (for review, Milet & Monsoro-Burq, 2012). The neural crest is brought to the dorsal neural tube during neurulation, undergoes an epithelial-to-mesenchymal transition, and migrates on defined routes through the embryonic body (Fig. 6). The frontonasal, maxillary, and mandibulary prominences of the face are populated by cranial neural crest cells and form the nose, the palate, and Meckel's cartilage of the lower jaw (Bush & Jiang, 2012; Mayor & Theveneau, 2013).

A subpopulation of cranial neural crest cells migrates ventrally into the cardiac cushion and forms the cardiac outflow tracts with separated aortic and pulmonary trunks (Keyte, Alonzo-Johnsen, & Hutson, 2014). The cardiac neural crest marker *PLEXINA2* (*PLXNA2*) is downregulated in WNT5A and DVL2 mutants, and the animals show outflow tract defects similar to ROR2-null mice (Hamblet et al., 2002; Schleiffarth et al., 2007). In WNT5A-null mice the cardiac phenotype has been attributed to defective WNT/Ca^{2+} signaling to CAMK2 (Schleiffarth et al., 2007), which is intimately linked to WNT/PCP signaling in *Xenopus* gastrulation (Gentzel, Schille, Rauschenberger, & Schambony, 2015). In addition to the cardiac neural crest, WNT/PCP signaling by DVL2 and VANGL2 is required in the second heart field, which also contributes to the cardiac outflow tract (Henderson et al., 2001; Phillips, Murdoch, Chaudhry, Copp, & Henderson, 2005; Sinha, Wang, Evans, Wynshaw-Boris, & Wang, 2012).

Accordingly, the simultaneous occurrence of craniofacial and cardiac outflow tract malformations in RS patients and the corresponding mouse models indicate a pivotal role of WNT5A/ROR2 signaling in cranial neural crest development.

ROR2 is expressed at the neural plate border, in premigratory and migrating neural crest cells and in the neural crest-derived branchial arches in *Xenopus*, chicken, and mouse. In later stages of neural crest development, *ROR2* expression was detected in premigratory and migratory neural crest cells (Hikasa et al., 2002; Matsuda et al., 2001; Oishi et al., 1999; Schille, Bayerlová, Bleckmann, & Schambony, 2016; Stricker et al., 2006).

The role of WNT/PCP signaling in neural crest migration is well established (De Calisto, Araya, Marchant, Riaz, & Mayor, 2005; Mayor & Theveneau, 2014; Shnitsar & Borchers, 2008). PTK7 regulates neural crest migration via β-Catenin-independent WNT signaling, and it has been shown that ROR2 is capable to replace PTK7 function in this process (Podleschny, Grund, Berger, Rollwitz, & Borchers, 2015). It remains unclear, however, if these findings indicate a role of ROR2 in neural crest migration or simply reflects the functional redundancy of PTK7 and ROR2 in the WNT/PCP pathway. Notably and unexpectedly, instead of migration defects, impaired neural crest induction was the dominant phenotype in ROR2-depleted *Xenopus* embryos (Schille et al., 2016).

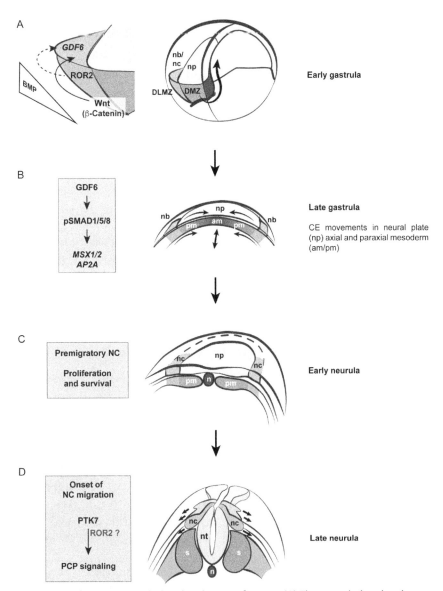

Fig. 6 Role of Ror2 in neural plate border specification. (A) The neural plate border territory is specified by intermediate levels of BMP signaling in combination with WNT/β-Catenin and FGF signals adjacent to the prospective neural plate during gastrulation. Inductive signals from the dorsolateral marginal zone (DLMZ) upregulate neural plate border-specific genes in the adjacent ectoderm, which will give rise to the neural plate border (nb) and neural crest (nc). ROR2 is required in the DLMZ to upregulate GDF6 in the nb territory. (B) The DLMZ and the dorsal marginal zone (DMZ) are mesodermal tissues that are internalized during gastrulation and placed underneath the nb and the

(Continued)

Evidence is accumulating that β-Catenin-independent WNT signaling is also required in neural crest induction (Ossipova & Sokol, 2011; Schille et al., 2016). In our own recent work, we have characterized the role of ROR2 in neural crest induction and found that ROR2 is required in the dorsolateral marginal zone of early gastrula-stage embryos to upregulate expression of the BMP ligand *GDF6* at the neural plate border (Schille et al., 2016). Expression of *GDF6* at the neural plate border in *Xenopus* and zebrafish is required for the local upregulation of BMP signaling at late gastrula stages that likely governs the further specification and maintenance of neural crest cell fates (Délot et al., 1999; Reichert, Randall, & Hill, 2013). Thereby, a role of ROR2 as positive regulator of BMP signaling activity in neural crest induction has been established (Schille et al., 2016).

Overall, ROR2 is required for cranial neural crest induction in *Xenopus* and possibly also contributes to WNT/PCP signaling that governs cranial and cardiac neural crest migration. Considering the craniofacial and cardiac malformations observed in RRS patients ROR2 deficiency seemingly affects predominantly mesenchymal derivatives of the cranial neural crest. Therefore, partial compensation by other receptors or an additional role of ROR2 in migration or differentiation of the mesenchymal lineage of cranial neural crest cannot be ruled out at this point.

4.3 Skeletal Development: Cartilage Growth Plate Defects in Robinow Syndrome

The axial and appendicular skeleton is formed by a process called endochondral ossification. Here, a cartilage template (condensation) is formed from mesenchymal progenitor cells. The prechondrogenic cells within this template expressing the cartilage master regulator SOX9 further differentiate to mature chondrocytes, which henceforth form the cartilage growth plate (Fig. 7A), a sophisticated anatomical structure that accomplishes growth

---

Fig. 6—Cont'd neural plate (np), respectively, by the end of gastrulation. At this stage, GDF6 locally upregulates BMP/SMAD1/5/8 signaling at the nb, which induces expression of BMP responsive genes such as *MSX1/2* and *AP2A*. In parallel, convergent extension (CE) movements in the mesoderm and neural plate shape these tissues. (C) Neural plate border genes cooperatively specify neural crest precursors and upregulate genes required for proliferation and survival as well as for initiation of migration in the neural crest cells. (D) At late neurula stages, the neural crest has been positioned dorsal of the neural tube in course of neural tube closure. The cells begin to emigrate from there and migrate on defined routes to their final destination. Neural crest migration depends among other signals on WNT/PCP activity mediated by PTK7 and putatively also by ROR2.

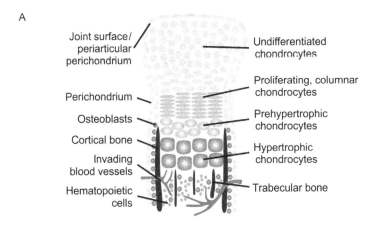
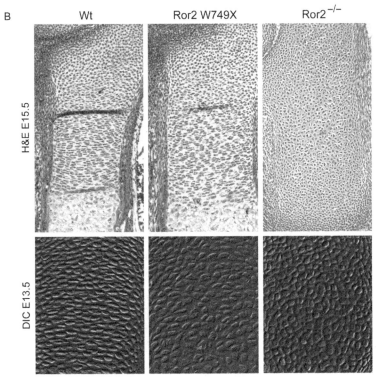

Fig. 7 Growth plate defects in ROR2 mutant mice. (A) Schematic depiction of the cartilage growth plate. Explanations see text. (B) Hematoxylin and Eosin (H&E; *top*) and differential interference contrast (DIC; *bottom*) images from sections of E15.5 (*top*) and E13.5 (*bottom*) ulnar growth plates of wild type, ROR2$^{W749X/W749X}$ (knock-in allele for a human BDB1 mutation) and ROR2$^{-/-}$ mice (RRS model) are shown. Note the apparent inability of ROR2-null growth plates to form columnar arrangement of chondrocytes; chondrocytes in ROR2-null growth plates remain rounded and unpolarized. ROR2$^{W749X/W749X}$ growth plates show an intermediate phenotype; polarization of chondrocytes at E13.5 is visibly impaired; however, this does not preclude the formation of an apparently intact growth plate at E15.5.

of the skeleton and with it body growth. Within the growth plate, chondrocytes undergo a stereotypic series of differentiation steps from small, rounded chondrocytes located near both ends of the skeletal element to the large so-called hypertrophic chondrocytes in the center of the skeletal element. The hypertrophic chondrocytes eventually undergo apoptosis and become replaced by bone. In between these stages chondrocytes proliferate, and, coupled to this, chondrocytes undergo a remarkable cellular change. Cells that have divided along a plane perpendicular to the longitudinal axis of the skeletal element flatten and shift on top of each other via the formation of an adhesive surface between the two daughter cells (Romereim, Conan, Chen, & Dudley, 2014). This process leads to the formation of long clonal columns of flattened chondrocytes reminiscent to stacks of coins (Fig. 7A).

In ROR2-null mice two prominent defects were observed. First, a delay in chondrocyte maturation shown by reduced expression of the prehypertrophic and hypertrophic markers Indian hedgehog (IHH) and collagen type 10 alpha1 (COL10A1) can obviously be seen as a cause for skeletal element shortening (Schwabe et al., 2004). IHH is a diffusible morphogen that coordinates the pace of cartilage maturation (Vortkamp et al., 1996). This is combined with an apparent inability to form a columnar arrangement of proliferating chondrocytes (Fig. 7B). The proliferation rate of chondrocytes was notably not changed (Schwabe et al., 2004). A prerequisite of chondrocyte column formation is the expression of cell adhesion molecules, specifically β1-integrins that mediate cell-matrix contacts at focal adhesions. ROR2 is involved in focal adhesion function and cell polarization/migration by interaction with FLNA (Nishita et al., 2006; Nomachi et al., 2008).

Interestingly, the chondrocyte stacking process has been compared to CE movements, fostered by the fact that it not only contributes to the elongation but also the narrowing (i.e., lateral constriction) of skeletal elements (Romereim & Dudley, 2011). In chicken embryos, disruption of the PCP pathway prevented chondrocyte polarization and column formation (Li & Dudley, 2009), indicating a critical involvement of WNT/PCP signaling in this process. We recently showed that HOXD13-induced expression of WNT5A is necessary for proper localization of PCP components DVL and PRICKLE1 in growth plate chondrocytes (Kuss et al., 2014). Failure of WNT5A induction or function in either HOXD13 or WNT5A mutants led to failure of chondrocyte polarization and stacking leading to the appearance of widened skeletal elements without oriented growth plates.

Intriguingly, autosomal-dominant Robinow syndrome (ADRS) can be caused by mutations in either WNT5A (Person et al., 2010; Roifman et al., 2014) or the central PCP components DVL1 (Bunn et al., 2015; White et al., 2015) or DVL3 (White et al., 2016). Further, *PRICKLE1* mouse mutants showing perturbed WNT5A signaling recapitulate Robinow syndrome features (Liu et al., 2014) altogether indicating a critical involvement of the PCP pathway in Robinow syndrome pathogenesis. Intriguingly, columnar arrangement is only mildly disturbed in mice carrying a BDB variant allele for Ror2 (Ror2$^{W749X/W749X}$; Fig. 7B). Thus, in the context of growth plate morphogenesis, the C-terminal proline–serine–threonine-rich domain of ROR2 appears partly dispensable.

Interestingly, ADRS-causing frameshift or truncating mutations in either DVL1 or DVL3 are found in the DVL C-terminal domain after the DEP domain (Bunn et al., 2015; White et al., 2016, 2015). The function of the DVL C-terminal domain is rather obscure. While in *Drosophila* it is apparently dispensable for both, canonical and PCP signaling (Yanfeng et al., 2011), its function has not been elucidated in vivo in mammals so far. The DVL C-terminal domain is hyperphosphorylated upon WNT stimulus (canonical or noncanonical), responsible for the characteristic gel size shift of DVL proteins (Bernatik et al., 2011, 2014; Yanfeng et al., 2011). Bernatik et al. (2011) proposed a mechanism, by which phosphorylation of the DVL C-terminus first is required for activation, but over time also for subsequent deactivation of DVL and the downstream signaling events (Bernatik et al., 2011). Functionally, a motif within the DVL C-terminus is required for binding to FZD receptors and hence for activation of β-Catenin signaling in vitro and in *Xenopus* embryos (Tauriello et al., 2012). Moreover, the DVL C-terminus interacts with ROR2 and it appears to be an intrinsic negative regulator of DVL function in the canonical pathway in this context (Witte, Bernatik, et al., 2010). Truncated DVL lacking the C-terminus shows exacerbate activity, while the C-terminus alone can inhibit β-Catenin signaling, a function that is dependent on the presence of ROR2 (Witte, Bernatik, et al., 2010). Notably, ROR2 interacts specifically with phosphorylated DVL (Witte, Bernatik, et al., 2010), implying ROR2 in a temporal feedback downregulation of canonical signaling as proposed by Bernatik et al. (2011). Furthermore, the DVL C-terminus appears to be as well involved in the feedback regulation of noncanonical WNT signaling (Witte, Bernatik, et al., 2010). The relationship of the DVL C-terminus to PCP signaling in vertebrates, and hence the RRS phenotype remains unresolved; however, it appears likely that

C-terminal truncation of DVL may affect the temporal balance of PCP signaling and thus dysregulate the pathway.

4.4 Skeletal Development: Brachydactyly Caused by Defective Digit Elongation

The appendicular skeleton originates from the lateral plate mesoderm. Limbs emerge from the body flank as a consequence of localized epithelial–mesenchymal interactions that result in rapid proliferation of mesenchymal cells leading to the formation of limb buds. At the most distal rim of the limb bud, separating the dorsal and ventral halves, the so-called apical ectodermal ridge (AER) is formed. The AER is a thickened ectodermal structure that produces several growth factors of the FGF family that diffuse to the underlying mesoderm and keep it undifferentiated and proliferating (Zeller, Lopez-Rios, & Zuniga, 2009). This distal proliferation leads to limb bud outgrowth; following the proximodistal expansion, the limb bud mesenchyme concomitantly differentiates in a proximodistal order. Thereby the lateral plate-derived limb bud mesenchyme gives rise to, e.g., the skeletal elements and the tendons of the limbs. Consequently, proximal skeletal elements in the stylopodes (humerus and femur) form first, followed by elements of the zeugopodes (radius/ulna, tibis/fibula). Finally, elements of the hands and feet form in the autopod (digits/toes; see Fig. 3). Notably, formation of the cartilage elements (i.e., condensation of cartilaginous mesenchyme followed by further differentiation as outlined earlier) occurs in a continuous fashion, where individual skeletal elements are subsequently divided by the intersection of synovial joints. Formation and shaping of the skeletal elements (but also the soft tissues) is under control of an intricate three-dimensional signaling system that provides a spatial and timely framework for coordinated cell differentiation. The mechanism by which this signaling interplay controls tissue patterning and differentiation is still under debate; in brief, cartilage differentiation passively follows distal outgrowth of the limb bud, whereby the most distal zone is kept undifferentiated via AER signals (Bénazet et al., 2009; Butterfield, McGlinn, & Wicking, 2010; Hasson, 2011). Moreover, the cartilaginous condensation is centered to the core of the limb bud by repressive action of ectoderm-expressed "canonical" WNT factors (Hill, Später, Taketo, Birchmeier, & Hartmann, 2005; Hill, Taketo, Birchmeier, & Hartmann, 2006; ten Berge, Brugmann, Helms, & Nusse, 2008). Formation of the digits, however, includes an additional mechanism. Once the initial digit condensations have formed, their distal elongation is driven by the continuous incorporation (recruitment) of undifferentiated mesenchymal cells from the distal pool (Gao et al., 2009; Suzuki, Hasso, & Fallon, 2008). Recruitment into the condensation is driven by a strong induction of BMP signaling via phosphorylation of their canonical downstream targets SMAD1/5/8, in a cell population immediately distal to the definitive cartilage, the so-called phalanx-forming region (Montero, Lorda-Diez, Gañan, Macias, & Hurle, 2008; Suzuki et al., 2008). Likely, correct dosage of BMP signaling is decisive for digit elongation since defects in BMP ligands or other BMP signaling components are underlying the majority of human brachydactylies, thus underscoring the molecular disease family hypothesis mentioned earlier (Stricker & Mundlos, 2011). This process is, however, not merely passive but is under influence by signals from the growing condensation itself. IHH, previously known for its role in the growth plate (see Section 4.3), is required for digit distal outgrowth in the mouse (Gao et al., 2009), and mutations in *IHH* cause human brachydactyly type A1 (Gao et al., 2001) showing a phenotypic overlap with BDB1. In knock-in mice homozygous for a human BDB1 mutation (p.W749X; Fig. 8), digit elongation via distal mesenchymal cell recruitment was severely impaired leading to digit shortening. This was in turn caused by disrupted BMP signaling in the phalanx-forming region altogether providing the first genetic proof for the concept of the phalanx-forming region and implicating it in the pathogenesis of human disease. $ROR2^{W749X/W749X}$ mice showed a perturbation of two independent signaling systems likely impacting in combination onto the phalanx-forming region leading to its shutdown. First, *IHH* expression was drastically decreased specifically in the distal cartilage condensation of $ROR2^{W749X/W749X}$ mice. This by itself affects the phalanx-forming region, since mice carrying a human BDA1 mutation in their *IHH* gene show reduced BMP signaling here. Second, an upregulation of canonical WNT signaling was observed in the mesenchyme distal to the condensation (Witte, Chan, Economides, Mundlos, & Stricker, 2010). In this context WNT5A, which is strongly expressed in the distal limb bud mesenchyme, is thought to signal via ROR2 to antagonize canonical WNT signaling, thus allowing phalanx-forming region establishment and/or maintenance. Indeed, loss of WNT5A produces severe brachydactyly in mice (Yamaguchi et al., 1999), causes upregulation of distal canonical WNT signaling (Topol et al., 2003), and WNT5A KO mice show no induction of a phalanx-forming region (S. Stricker, unpublished observation).

ROR1 and ROR2 are part of a WNT5A–ROR–DVL pathway in the limbs, where both receptors are in part redundant (Ho et al., 2012). It is clear, however, that a simple loss of WNT5A/ROR2 signaling cannot be the underlying cause for BDB1, since loss of ROR2 leads to the

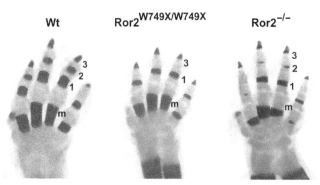

Fig. 8 Digit phenotypes of different ROR2 mutant mice. Alcian *blue*/Alizarin *red* stained skeletal preparations from E18.5 mouse embryos of the indicated genotypes are shown; cartilage stains *blue*, bone stains *red*. In the wild type, three phalanges are visible, in Ror2$^{W749X/W749X}$ mice (BDB1 model) only two phalanges form (best visible due to the absence of the second ossification center). In Ror2$^{-/-}$ embryos (RRS model), all three phalanges are present; however, they are hypoplastic as compared to the wild type, reflecting the mild brachydactyly seen in some RRS patients. *M*, metacarpal; *1-3*, phalanges 1-3. *Pictures are reproduced from Schwarzer, W., Witte, F., Rajab, A., Mundlos, S., Stricker, S., 2009. A gradient of ROR2 protein stability and membrane localization confers brachydactyly type B or Robinow syndrome phenotypes.* Human Molecular Genetics, 18, 4013–4021. http://dx.doi.org/10.1093/hmg/ddp345, *with kind permission from Oxford University Press.*

distinctive features seen in RRS. In digits, ROR2 is also involved in PCP signaling (Gao et al., 2011; Wang, Sinha, Jiao, Serra, & Wang, 2011); however, it remains unclear how this signaling contributes to the BDB1 phenotype and how it may be molecularly linked to the canonical pathway. It therefore appears likely that ROR receptors may have several (likely context dependent) functions. In cartilage ROR2 is likely part of a PCP signaling complex that is perturbed in its function in a differential way by either loss of ROR2 or expression of truncated ROR2. Moreover, the effect of truncated ROR2 on this complex is predicted to correlate with the truncated ROR2 protein dosage at the cell surface (Schwarzer et al., 2009). What causes this differential responsiveness remains to be elucidated. In summary it appears that deregulation of PCP signaling is at the center of the malformations seen in RRS. In digits it appears that ROR2 is required for two processes, enabling chondrogenic commitment of distal mesenchymal progenitors via inhibition of WNT/β-Catenin signaling (Witte, Chan, et al., 2010), and for nascent chondrocyte polarization and subsequent shaping of the condensation via PCP signaling (Gao et al., 2011).

4.5 Kidney Development

Some RRS patients show aberrant development of the renal tract such as hydronephrosis and cystic kidneys. Defects in urogenital tract development have also been associated with the more obvious genital abnormalities (Patton & Afzal, 2002).

The nephron epithelia and collecting ducts develop from intermediate mesoderm, which itself is specified and positioned during gastrulation. The dorsal intermediate mesoderm initially forms the pronephric duct, a primary tube that extends posteriorly during body axis elongation. Adjacent to the anterior pronephric duct, a primary set of tubules is induced in the intermediate mesoderm, the pronephros. The pronephros represents the functional kidney in fish and amphibian larvae. In mammals, it degenerates and is replaced by a second transient structure, the mesonephros and finally by the mammalian permanent kidney, the metanephros. The metanephros develops at the level of the hindlimb, where the nephric duct is in contact with the metanephric mesenchyme, a specialized population of intermediate mesoderm. The metanephric mesenchyme contains progenitor cells of the adult kidney and induces the initiation of bona fide kidney development: the outgrowth of a single ureteric bud. The ureteric bud branches extensively to form the highly branched collective duct system and induces the formation of nephrons from the metanephric mesenchyme (Costantini & Kopan, 2010; Dressler, 2006).

In WNT5A-null and ROR2-null mice positioning and patterning of the intermediate mesoderm relative to the body axis was not significantly altered (Nishita et al., 2014). However, mice in which WNT5A was selectively knocked-out in the mesoderm showed a shorter and wider nephric duct (Yun et al., 2014), and also a shortened body axis. Elongation of the nephric duct is driven primarily by cell rearrangements and migration, indicating a contribution of WNT5A and ROR2 to duct morphogenesis probably similar to the CE-like morphogenetic defects observed in the dorsal mesoderm.

In the mouse, *ROR2* is expressed in the metanephric mesenchyme during ureteric budding and *WNT5A* expression is detected not only in the metanephric mesenchyme but also in the adjacent tissues namely the caudal Wolffian duct and the ureteric bud (Nishita et al., 2014). Although to date no functional studies for ROR receptors in kidney development of lower vertebrates have been published, *ROR2* is expressed in the pronephros of *Xenopus* embryos, suggesting that some functions might be evolutionary conserved (Zhang, Tran, & Wessely, 2011).

The predominant phenotype observed in null and conditional Ror2 knockout mice was a duplication of the ureteric buds, which was preceded by duplication of the HoxB7-positive posterior nephric duct (Yun et al., 2014). Outgrowth of the ureteric bud is stimulated by GDNF derived from the metanephric mesenchyme and inhibited by BMP4; local expression of the BMP antagonist Gremlin1 allows ureteric bud outgrowth (Michos et al., 2007). Nishita and colleagues report that the metanephric mesenchyme was smaller and remained in a more ventral and caudal position relative to the nephric duct in WNT5A-null mice. In addition GDNF expression in the metanephric mesenchyme was mislocalized and spread rostrally in WNT5A- and ROR2-null embryos, resulting in the formation of an additional, ectopic ureteric bud rostral of the one localized at the normal position (Nishita et al., 2014). Recently it was shown that ROR1 also is involved in kidney development of the mouse. Although a loss of ROR1 alone does not cause kidney malformations, in *ROR1-; ROR2-* double-mutant mice kidney development is severely disturbed (Qi, Okinaka, Nishita, & Minami, 2016). Duplication of ureteric buds similar to WNT5A- and ROR2-null mice but also absence of the kidney has been observed. Kidney agenesis was traced back to a loss of the metanephric mesenchyme. The authors propose that separate ROR1-WNT5A and ROR2-WNT5A pathways interact in a cooperative way to position the metanephric mesenchyme around the ureteric bud (Qi et al., 2016); however, the underlying signaling pathways have not been studied. Therefore, it remains speculative albeit likely that WNT5A and ROR RTKs activate β-Catenin-independent pathways also in kidney development.

In summary, knockdown or knockout of ROR2 in *Xenopus* and mouse models recapitulates the clinical presentation of Robinow syndrome patients, which are distinct from patients bearing a BDB1 mutation in the ROR2 gene. Robinow syndrome phenotypes are reminiscent to WNT5A loss of function and partially overlapping to the phenotypes observed in knockout or mutant models of PCP genes such as VANGL2 or DVL2. Interestingly, ROR2 loss of function predominantly affects migratory mesodermal and mesenchymal tissues, suggesting that ROR2 plays a key role as activator of β-Catenin-independent signaling in migrating cells. By contrast, neural and sensory cells seem less affected by ROR2 deficiencies, which might be due to redundant functions with other RTKs, most likely ROR1 and PTK7.

REFERENCES

Afzal, A. R., Rajab, A., Fenske, C. D., Oldridge, M., Elanko, N., Ternes-Pereira, E., et al. (2000). Recessive Robinow syndrome, allelic to dominant brachydactyly type B, is caused by mutation of ROR2. *Nature Genetics, 25*, 419–422. http://dx.doi.org/10.1038/78107.

Andre, P., Song, H., Kim, W., Kispert, A., & Yang, Y. (2015). Wnt5a and Wnt11 regulate mammalian anterior-posterior axis elongation. *Development, 142*, 1516–1527. http://dx.doi.org/10.1242/dev.119065.

Andre, P., Wang, Q., Wang, N., Gao, B., Schilit, A., Halford, M. M., et al. (2012). The Wnt coreceptor Ryk regulates Wnt/planar cell polarity by modulating the degradation of the core planar cell polarity component Vangl2. *The Journal of Biological Chemistry, 287*, 44518–44525. http://dx.doi.org/10.1074/jbc.M112.414441.

Bai, Y., Tan, X., Zhang, H., Liu, C., Zhao, B., Li, Y., et al. (2014). Ror2 receptor mediates Wnt11 ligand signaling and affects convergence and extension movements in zebrafish. *The Journal of Biological Chemistry, 289*, 20664–20676. http://dx.doi.org/10.1074/jbc.M114.586099.

Bainbridge, T. W., DeAlmeida, V. I., Izrael-Tomasevic, A., Chalouni, C., Pan, B., Goldsmith, J., et al. (2014). Evolutionary divergence in the catalytic activity of the CAM-1, ROR1 and ROR2 kinase domains. *PloS One, 9*. e102695. http://dx.doi.org/10.1371/journal.pone.0102695.s004.

Bénazet, J.-D., Bischofberger, M., Tiecke, E., Gonçalves, A., Martin, J. F., Zuniga, A., et al. (2009). A self-regulatory system of interlinked signaling feedback loops controls mouse limb patterning. *Science, 323*, 1050–1053. http://dx.doi.org/10.1126/science.1168755.

Bernatik, O., Ganji, R. S., Dijksterhuis, J. P., Konik, P., Cervenka, I., Polonio, T., et al. (2011). Sequential activation and inactivation of Dishevelled in the Wnt/beta-catenin pathway by casein kinases. *The Journal of Biological Chemistry, 286*, 10396–10410. http://dx.doi.org/10.1074/jbc.M110.169870.

Bernatik, O., Sedová, K., Schille, C., Sri Ganji, R., Cervenka, I., Trantírek, L., et al. (2014). Functional analysis of Dishevelled-3 phosphorylation identifies distinct mechanisms driven by casein kinase 1ε and Frizzled5. *The Journal of Biological Chemistry, 289*, 23520–23533. http://dx.doi.org/10.1074/jbc.M114.590638.

Billiard, J., Way, D. S., Seestaller-Wehr, L. M., Moran, R. A., Mangine, A., & Bodine, P. V. N. (2005). The orphan receptor tyrosine kinase Ror2 modulates canonical Wnt signaling in osteoblastic cells. *Molecular Endocrinology, 19*, 90–101. http://dx.doi.org/10.1210/me.2004-0153.

Brinkmann, E.-M., Mattes, B., Kumar, R., Hagemann, A. I. H., Gradl, D., Scholpp, S., et al. (2016). Secreted Frizzled-related protein 2 (sFRP2) redirects non-canonical Wnt signaling from Fz7 to Ror2 during vertebrate gastrulation. *The Journal of Biological Chemistry, 291*, 13730–13742. http://dx.doi.org/10.1074/jbc.M116.733766.

Brugger, S. M., Merrill, A. E., Torres-Vazquez, J., Wu, N., Ting, M.-C., Cho, J. Y.-M., et al. (2004). A phylogenetically conserved cis-regulatory module in the Msx2 promoter is sufficient for BMP-dependent transcription in murine and Drosophila embryos. *Development, 131*, 5153–5165. http://dx.doi.org/10.1242/dev.01390.

Bunn, K. J., Daniel, P., Rösken, H. S., O'Neill, A. C., Cameron-Christie, S. R., Morgan, T., et al. (2015). Mutations in DVL1 cause an osteosclerotic form of Robinow syndrome. *American Journal of Human Genetics, 96*, 623–630. http://dx.doi.org/10.1016/j.ajhg.2015.02.010.

Bush, J. O., & Jiang, R. (2012). Palatogenesis: Morphogenetic and molecular mechanisms of secondary palate development. *Development, 139*, 231–243. http://dx.doi.org/10.1242/dev.067082.

Butterfield, N. C., McGlinn, E., & Wicking, C. (2010). The molecular regulation of vertebrate limb patterning. *Current Topics in Developmental Biology*, *90*, 319–341. http://dx.doi.org/10.1016/S0070-2153(10)90009-4.

Chen, Y., Bellamy, W. P., Seabra, M. C., Field, M. C., & Ali, B. R. (2005). ER-associated protein degradation is a common mechanism underpinning numerous monogenic diseases including Robinow syndrome. *Human Molecular Genetics*, *14*, 2559–2569. http://dx.doi.org/10.1093/hmg/ddi259.

Costantini, F., & Kopan, R. (2010). Patterning a complex organ: Branching morphogenesis and nephron segmentation in kidney development. *Developmental Cell*, *18*, 698–712. http://dx.doi.org/10.1016/j.devcel.2010.04.008.

De Calisto, J., Araya, C., Marchant, L., Riaz, C. F., & Mayor, R. (2005). Essential role of non-canonical Wnt signalling in neural crest migration. *Development*, *132*, 2587–2597. http://dx.doi.org/10.1242/dev.01857.

DeChiara, T. M., Bowen, D. C., Valenzuela, D. M., Simmons, M. V., Poueymirou, W. T., Thomas, S., et al. (1996). The receptor tyrosine kinase MuSK is required for neuromuscular junction formation in vivo. *Cell*, *85*, 501–512.

DeChiara, T. M., Kimble, R. B., Poueymirou, W. T., Rojas, J., Masiakowski, P., Valenzuela, D. M., et al. (2000). Ror2, encoding a receptor-like tyrosine kinase, is required for cartilage and growth plate development. *Nature Genetics*, *24*, 271–274. http://dx.doi.org/10.1038/73488.

Délot, E., Kataoka, H., Goutel, C., Yan, Y. L., Postlethwait, J., Wittbrodt, J., et al. (1999). The BMP-related protein radar: A maintenance factor for dorsal neuroectoderm cells? *Mechanisms of Development*, *85*, 15–25.

Dressler, G. R. (2006). The cellular basis of kidney development. *Annual Review of Cell and Developmental Biology*, *22*, 509–529. http://dx.doi.org/10.1146/annurev.cellbio.22.010305.104340.

Feike, A. C., Rachor, K., Gentzel, M., & Schambony, A. (2010). Wnt5a/Ror2-induced upregulation of xPAPC requires xShcA. *Biochemical and Biophysical Research Communications*, *400*, 500–506. http://dx.doi.org/10.1016/j.bbrc.2010.08.074.

Forrester, W. C., Dell, M., Perens, E., & Garriga, G. (1999). A C. elegans Ror receptor tyrosine kinase regulates cell motility and asymmetric cell division. *Nature*, *400*, 881–885. http://dx.doi.org/10.1038/23722.

Forrester, W. C., Kim, C., & Garriga, G. (2004). The Caenorhabditis elegans Ror RTK CAM-1 inhibits EGL-20/Wnt signaling in cell migration. *Genetics*, *168*, 1951–1962. http://dx.doi.org/10.1534/genetics.104.031781.

Francis, M. M., Evans, S. P., Jensen, M., Madsen, D. M., Mancuso, J., Norman, K. R., et al. (2005). The Ror receptor tyrosine kinase CAM-1 is required for ACR-16-mediated synaptic transmission at the C. elegans neuromuscular junction. *Neuron*, *46*, 581–594. http://dx.doi.org/10.1016/j.neuron.2005.04.010.

Gao, B., Guo, J., She, C., Shu, A., Yang, M., Tan, Z., et al. (2001). Mutations in IHH, encoding Indian hedgehog, cause brachydactyly type A-1. *Nature Genetics*, *28*, 386–388. http://dx.doi.org/10.1038/ng577.

Gao, B., Hu, J., Stricker, S., Cheung, M., Ma, G., Law, K. F., et al. (2009). A mutation in Ihh that causes digit abnormalities alters its signalling capacity and range. *Nature*, *458*, 1196–1200. http://dx.doi.org/10.1038/nature07862.

Gao, B., Song, H., Bishop, K., Elliot, G., Garrett, L., English, M. A., et al. (2011). Wnt signaling gradients establish planar cell polarity by inducing Vangl2 phosphorylation through Ror2. *Developmental Cell*, *20*, 163–176. http://dx.doi.org/10.1016/j.devcel.2011.01.001.

Gentzel, M., Schille, C., Rauschenberger, V., & Schambony, A. (2015). Distinct functionality of Dishevelled isoforms on Ca2+/calmodulin-dependent protein kinase 2 (CamKII) in Xenopus gastrulation. *Molecular Biology of the Cell*, *26*, 966–977. http://dx.doi.org/10.1091/mbc.E14-06-1089.

Glass, D. J., Bowen, D. C., Stitt, T. N., Radziejewski, C., Bruno, J., Ryan, T. E., et al. (1996). Agrin acts via a MuSK receptor complex. *Cell*, *85*, 513–523.
Goh, L. K., & Sorkin, A. (2013). Endocytosis of receptor tyrosine kinases. *Cold Spring Harbor Perspectives in Biology*, *5*, a017459. http://dx.doi.org/10.1101/cshperspect.a017459.
Green, J. L., Inoue, T., & Sternberg, P. W. (2007). The C. elegans ROR receptor tyrosine kinase, CAM-1, non-autonomously inhibits the Wnt pathway. *Development*, *134*, 4053–4062. http://dx.doi.org/10.1242/dev.005363.
Green, J. L., Inoue, T., & Sternberg, P. W. (2008). Opposing Wnt pathways orient cell polarity during organogenesis. *Cell*, *134*, 646–656. http://dx.doi.org/10.1016/j.cell.2008.06.026.
Gros, J., Serralbo, O., & Marcelle, C. (2009). WNT11 acts as a directional cue to organize the elongation of early muscle fibres. *Nature*, *457*, 589–593. http://dx.doi.org/10.1038/nature07564.
Grumolato, L., Liu, G., Mong, P., Mudbhary, R., Biswas, R., Arroyave, R., et al. (2010). Canonical and noncanonical Wnts use a common mechanism to activate completely unrelated coreceptors. *Genes & Development*, *24*, 2517–2530. http://dx.doi.org/10.1101/gad.1957710.
Habas, R., Dawid, I. B., & He, X. (2003). Coactivation of Rac and Rho by Wnt/Frizzled signaling is required for vertebrate gastrulation. *Genes & Development*, *17*, 295–309. http://dx.doi.org/10.1101/gad.1022203.
Hamblet, N. S., Lijam, N., Ruiz-Lozano, P., Wang, J., Yang, Y., Luo, Z., et al. (2002). Dishevelled 2 is essential for cardiac outflow tract development, somite segmentation and neural tube closure. *Development*, *129*, 5827–5838.
Harland, R., & Gerhart, J. (1997). Formation and function of Spemann's organizer. *Annual Review of Cell and Developmental Biology*, *13*, 611–667. http://dx.doi.org/10.1146/annurev.cellbio.13.1.611.
Hasson, P. (2011). "Soft" tissue patterning: Muscles and tendons of the limb take their form. *Developmental Dynamics*, *240*, 1100–1107. http://dx.doi.org/10.1002/dvdy.22608.
He, F., Xiong, W., Yu, X., Espinoza-Lewis, R., Liu, C., Gu, S., et al. (2008). Wnt5a regulates directional cell migration and cell proliferation via Ror2-mediated noncanonical pathway in mammalian palate development. *Development*, *135*, 3871–3879. http://dx.doi.org/10.1242/dev.025767.
Heinonen, K. M., Vanegas, J. R., Lew, D., Krosl, J., & Perreault, C. (2011). Wnt4 enhances murine hematopoietic progenitor cell expansion through a planar cell polarity-like pathway. *PloS One*. *6*. e19279. http://dx.doi.org/10.1371/journal.pone.0019279.t001.
Heisenberg, C. P., Tada, M., Rauch, G. J., Saúde, L., Concha, M. L., Geisler, R., et al. (2000). Silberblick/Wnt11 mediates convergent extension movements during zebrafish gastrulation. *Nature*, *405*, 76–81. http://dx.doi.org/10.1038/35011068.
Henderson, D. J., Conway, S. J., Greene, N. D., Gerrelli, D., Murdoch, J. N., Anderson, R. H., et al. (2001). Cardiovascular defects associated with abnormalities in midline development in the Loop-tail mouse mutant. *Circulation Research*, *89*, 6–12.
Hikasa, H., Shibata, M., Hiratani, I., & Taira, M. (2002). The Xenopus receptor tyrosine kinase Xror2 modulates morphogenetic movements of the axial mesoderm and neuroectoderm via Wnt signaling. *Development*, *129*, 5227–5239.
Hill, T. P., Später, D., Taketo, M. M., Birchmeier, W., & Hartmann, C. (2005). Canonical Wnt/beta-catenin signaling prevents osteoblasts from differentiating into chondrocytes. *Developmental Cell*, *8*, 727–738. http://dx.doi.org/10.1016/j.devcel.2005.02.013.
Hill, T. P., Taketo, M. M., Birchmeier, W., & Hartmann, C. (2006). Multiple roles of mesenchymal beta-catenin during murine limb patterning. *Development*, *133*, 1219–1229. http://dx.doi.org/10.1242/dev.02298.
Ho, H.-Y. H., Susman, M. W., Bikoff, J. B., Ryu, Y. K., Jonas, A. M., Hu, L., et al. (2012). Wnt5a-Ror-Dishevelled signaling constitutes a core developmental pathway that controls tissue morphogenesis. *Proceedings of the National Academy of Sciences*

of the United States of America, 109, 4044–4051. http://dx.doi.org/10.1073/pnas.1200421109.

Jensen, M., Hoerndli, F. J., Brockie, P. J., Wang, R., Johnson, E., Maxfield, D., et al. (2012). Wnt signaling regulates acetylcholine receptor translocation and synaptic plasticity in the adult nervous system. Cell, 149, 173–187. http://dx.doi.org/10.1016/j.cell.2011.12.038.

Kani, S., Oishi, I., Yamamoto, H., Yoda, A., Suzuki, H., Nomachi, A., et al. (2004). The receptor tyrosine kinase Ror2 associates with and is activated by casein kinase Iepsilon. The Journal of Biological Chemistry, 279, 50102–50109. http://dx.doi.org/10.1074/jbc.M409039200.

Kawano, Y., & Kypta, R. (2003). Secreted antagonists of the Wnt signalling pathway. Journal of Cell Science, 116, 2627–2634. http://dx.doi.org/10.1242/jcs.00623.

Keller, R. (2002). Shaping the vertebrate body plan by polarized embryonic cell movements. Science, 298, 1950–1954. http://dx.doi.org/10.1126/science.1079478.

Keller, R., Davidson, L., Edlund, A., Elul, T., Ezin, M., Shook, D., et al. (2000). Mechanisms of convergence and extension by cell intercalation. Philosophical Transactions of the Royal Society of London. Series B, Biological Sciences, 355, 897–922. http://dx.doi.org/10.1098/rstb.2000.0626.

Kennerdell, J. R., Fetter, R. D., & Bargmann, C. I. (2009). Wnt-Ror signaling to SIA and SIB neurons directs anterior axon guidance and nerve ring placement in C. elegans. Development, 136, 3801–3810. http://dx.doi.org/10.1242/dev.038109.

Keyte, A. L., Alonzo-Johnsen, M., & Hutson, M. R. (2014). Evolutionary and developmental origins of the cardiac neural crest: Building a divided outflow tract. Birth Defects Research. Part C, Embryo Today: Reviews, 102, 309–323. http://dx.doi.org/10.1002/bdrc.21076.

Kim, G.-H., & Han, J.-K. (2005). JNK and ROKalpha function in the noncanonical Wnt/RhoA signaling pathway to regulate Xenopus convergent extension movements. Developmental Dynamics, 232, 958–968. http://dx.doi.org/10.1002/dvdy.20262.

Kim, S. H., Jen, W. C., De Robertis, E. M., & Kintner, C. (2000). The protocadherin PAPC establishes segmental boundaries during somitogenesis in Xenopus embryos. Current Biology, 10, 821–830.

Kraft, B., Berger, C. D., Wallkamm, V., Steinbeisser, H., & Wedlich, D. (2012). Wnt-11 and Fz7 reduce cell adhesion in convergent extension by sequestration of PAPC and C-cadherin. The Journal of Cell Biology, 198, 695–709. http://dx.doi.org/10.1083/jcb.201110076.

Kuss, P., Kraft, K., Stumm, J., Ibrahim, D., Vallecillo-Garcia, P., Mundlos, S., et al. (2014). Regulation of cell polarity in the cartilage growth plate and perichondrium of metacarpal elements by HOXD13 and WNT5A. Developmental Biology, 385, 83–93. http://dx.doi.org/10.1016/j.ydbio.2013.10.013.

Li, Y., & Dudley, A. T. (2009). Noncanonical Frizzled signaling regulates cell polarity of growth plate chondrocytes. Development, 136, 1083–1092. http://dx.doi.org/10.1242/dev.023820.

Liu, C., Lin, C., Gao, C., May-Simera, H., Swaroop, A., & Li, T. (2014). Null and hypomorph Prickle1 alleles in mice phenocopy human Robinow syndrome and disrupt signaling downstream of Wnt5a. Biology Open, 3, 861–870. http://dx.doi.org/10.1242/bio.20148375.

Liu, Y., Ross, J. F., Bodine, P. V. N., & Billiard, J. (2007). Homodimerization of Ror2 tyrosine kinase receptor induces 14-3-3(beta) phosphorylation and promotes osteoblast differentiation and bone formation. Molecular Endocrinology, 21, 3050–3061. http://dx.doi.org/10.1210/me.2007-0323.

Liu, Y., Rubin, B., Bodine, P. V. N., & Billiard, J. (2008). Wnt5a induces homodimerization and activation of Ror2 receptor tyrosine kinase. Journal of Cellular Biochemistry, 105, 497–502. http://dx.doi.org/10.1002/jcb.21848.

Lyashenko, N., Weissenböck, M., Sharir, A., Erben, R. G., Minami, Y., & Hartmann, C. (2010). Mice lacking the orphan receptor ror1 have distinct skeletal abnormalities and are growth retarded. *Developmental Dynamics*, *239*, 2266–2277. http://dx.doi.org/10.1002/dvdy.22362.

Martinez, S., Scerbo, P., Giordano, M., Daulat, A. M., Lhoumeau, A.-C., Thomé, V., et al. (2015). The PTK7 and ROR2 protein receptors interact in the vertebrate WNT/planar cell polarity (PCP) pathway. *The Journal of Biological Chemistry*, *290*, 30562–30572. http://dx.doi.org/10.1074/jbc.M115.697615.

Masiakowski, P., & Carroll, R. D. (1992). A novel family of cell surface receptors with tyrosine kinase-like domain. *The Journal of Biological Chemistry*, *267*, 26181–26190.

Matsuda, T., Nomi, M., Ikeya, M., Kani, S., Oishi, I., Terashima, T., et al. (2001). Expression of the receptor tyrosine kinase genes, Ror1 and Ror2, during mouse development. *Mechanisms of Development*, *105*, 153–156.

Matsuda, T., Suzuki, H., Oishi, I., Kani, S., Kuroda, Y., Komori, T., et al. (2003). The receptor tyrosine kinase Ror2 associates with the melanoma-associated antigen (MAGE) family protein Dlxin-1 and regulates its intracellular distribution. *The Journal of Biological Chemistry*, *278*, 29057–29064. http://dx.doi.org/10.1074/jbc.M302199200.

Mayor, R., & Theveneau, E. (2013). The neural crest. *Development*, *140*, 2247–2251. http://dx.doi.org/10.1242/dev.091751.

Mayor, R., & Theveneau, E. (2014). The role of the non-canonical Wnt-planar cell polarity pathway in neural crest migration. *The Biochemical Journal*, *457*, 19–26. http://dx.doi.org/10.1042/BJ20131182.

McKay, S. E., Hislop, J., Scott, D., Bulloch, A. G., Kaczmarek, L. K., Carew, T. J., et al. (2001). Aplysia ror forms clusters on the surface of identified neuroendocrine cells. *Molecular and Cellular Neurosciences*, *17*, 821–841. http://dx.doi.org/10.1006/mcne.2001.0977.

Medina, A., Swain, R. K., Kuerner, K.-M., & Steinbeisser, H. (2004). Xenopus paraxial protocadherin has signaling functions and is involved in tissue separation. *The EMBO Journal*, *23*, 3249–3258. http://dx.doi.org/10.1038/sj.emboj.7600329.

Mentink, R. A., Middelkoop, T. C., Rella, L., Ji, N., Tang, C. Y., Betist, M. C., et al. (2014). Cell intrinsic modulation of Wnt signaling controls neuroblast migration in C. elegans. *Developmental Cell*, *31*, 188–201. http://dx.doi.org/10.1016/j.devcel.2014.08.008.

Michos, O., Gonçalves, A., Lopez-Rios, J., Tiecke, E., Naillat, F., Beier, K., et al. (2007). Reduction of BMP4 activity by gremlin 1 enables ureteric bud outgrowth and GDNF/WNT11 feedback signalling during kidney branching morphogenesis. *Development*, *134*, 2397–2405. http://dx.doi.org/10.1242/dev.02861.

Mikels, A., Minami, Y., & Nusse, R. (2009). Ror2 receptor requires tyrosine kinase activity to mediate Wnt5A signaling. *The Journal of Biological Chemistry*, *284*, 30167–30176. http://dx.doi.org/10.1074/jbc.M109.041715.

Mikels, A. J., & Nusse, R. (2006). Purified Wnt5a protein activates or inhibits beta-catenin-TCF signaling depending on receptor context. *PLoS Biology*. *4*, e115. http://dx.doi.org/10.1371/journal.pbio.0040115.

Milet, C., & Monsoro-Burq, A. H. (2012). Neural crest induction at the neural plate border in vertebrates. *Developmental Biology*, *366*, 22–33. http://dx.doi.org/10.1016/j.ydbio.2012.01.013.

Montero, J. A., Lorda-Diez, C. I., Gañan, Y., Macias, D., & Hurle, J. M. (2008). Activin/TGFbeta and BMP crosstalk determines digit chondrogenesis. *Developmental Biology*, *321*, 343–356. http://dx.doi.org/10.1016/j.ydbio.2008.06.022.

Mundlos, S. (2009). The brachydactylies: A molecular disease family. *Clinical Genetics*, *76*, 123–136.

Nakaya, Y., Kuroda, S., Katagiri, Y. T., Kaibuchi, K., & Takahashi, Y. (2004). Mesenchymal-epithelial transition during somitic segmentation is regulated by differential roles of Cdc42 and Rac1. *Developmental Cell, 7,* 425–438. http://dx.doi.org/10.1016/j.devcel.2004.08.003.

Nishita, M., Itsukushima, S., Nomachi, A., Endo, M., Wang, Z., Inaba, D., et al. (2010). Ror2/Frizzled complex mediates Wnt5a-induced AP-1 activation by regulating Dishevelled polymerization. *Molecular and Cellular Biology, 30,* 3610–3619. http://dx.doi.org/10.1128/MCB.00177-10.

Nishita, M., Qiao, S., Miyamoto, M., Okinaka, Y., Yamada, M., Hashimoto, R., et al. (2014). Role of Wnt5a-Ror2 signaling in morphogenesis of the metanephric mesenchyme during ureteric budding. *Molecular and Cellular Biology, 34,* 3096–3105. http://dx.doi.org/10.1128/MCB.00491-14.

Nishita, M., Yoo, S. K., Nomachi, A., Kani, S., Sougawa, N., Ohta, Y., et al. (2006). Filopodia formation mediated by receptor tyrosine kinase Ror2 is required for Wnt5a-induced cell migration. *The Journal of Cell Biology, 175,* 555–562. http://dx.doi.org/10.1083/jcb.200607127.

Nomachi, A., Nishita, M., Inaba, D., Enomoto, M., Hamasaki, M., & Minami, Y. (2008). Receptor tyrosine kinase Ror2 mediates Wnt5a-induced polarized cell migration by activating c-Jun N-terminal kinase via actin-binding protein filamin A. *The Journal of Biological Chemistry, 283,* 27973–27981. http://dx.doi.org/10.1074/jbc.M802325200.

Nomi, M., Oishi, I., Kani, S., Suzuki, H., Matsuda, T., Yoda, A., et al. (2001). Loss of mRor1 enhances the heart and skeletal abnormalities in mRor2-deficient mice: Redundant and pleiotropic functions of mRor1 and mRor2 receptor tyrosine kinases. *Molecular and Cellular Biology, 21,* 8329–8335. http://dx.doi.org/10.1128/MCB.21.24.8329-8335.2001.

Ohkawara, B., & Niehrs, C. (2010). An ATF2-based luciferase reporter to monitor non-canonical Wnt signaling in Xenopus embryos. *Developmental Dynamics, 240,* 188–194. http://dx.doi.org/10.1002/dvdy.22500.

Oishi, I., Sugiyama, S., Liu, Z. J., Yamamura, H., Nishida, Y., & Minami, Y. (1997). A novel Drosophila receptor tyrosine kinase expressed specifically in the nervous system. Unique structural features and implication in developmental signaling. *The Journal of Biological Chemistry, 272,* 11916–11923.

Oishi, I., Suzuki, H., Onishi, N., Takada, R., Kani, S., Ohkawara, B., et al. (2003). The receptor tyrosine kinase Ror2 is involved in non-canonical Wnt5a/JNK signalling pathway. *Genes to Cells, 8,* 645–654.

Oishi, I., Takeuchi, S., Hashimoto, R., Nagabukuro, A., Ueda, T., Liu, Z. J., et al. (1999). Spatio-temporally regulated expression of receptor tyrosine kinases, mRor1, mRor2, during mouse development: Implications in development and function of the nervous system. *Genes to Cells, 4,* 41–56.

Oldridge, M., Fortuna, A. M., Maringa, M., Propping, P., Mansour, S., Pollitt, C., et al. (2000). Dominant mutations in ROR2, encoding an orphan receptor tyrosine kinase, cause brachydactyly type B. *Nature Genetics, 24,* 275–278. http://dx.doi.org/10.1038/73495.

Ossipova, O., & Sokol, S. Y. (2011). Neural crest specification by noncanonical Wnt signaling and PAR-1. *Development, 138,* 5441–5450. http://dx.doi.org/10.1242/dev.067280.

Paganoni, S., Bernstein, J., & Ferreira, A. (2010). Ror1-Ror2 complexes modulate synapse formation in hippocampal neurons. *Neuroscience, 165,* 1261–1274. http://dx.doi.org/10.1016/j.neuroscience.2009.11.056.

Patton, M. A., & Afzal, A. R. (2002). Robinow syndrome. *Journal of Medical Genetics, 39,* 305–310.

Person, A. D., Beiraghi, S., Sieben, C. M., Hermanson, S., Neumann, A. N., Robu, M. E., et al. (2010). WNT5A mutations in patients with autosomal dominant Robinow syndrome. *Developmental Dynamics, 239,* 327–337. http://dx.doi.org/10.1002/dvdy.22156.

Phillips, H. M., Murdoch, J. N., Chaudhry, B., Copp, A. J., & Henderson, D. J. (2005). Vangl2 acts via RhoA signaling to regulate polarized cell movements during development of the proximal outflow tract. *Circulation Research, 96*, 292–299. http://dx.doi.org/10.1161/01.RES.0000154912.08695.88.

Podleschny, M., Grund, A., Berger, H., Rollwitz, E., & Borchers, A. (2015). A PTK7/Ror2 co-receptor complex affects Xenopus neural crest migration. *PloS One, 10*, e0145169.

Qi, X., Okinaka, Y., Nishita, M., & Minami, Y. (2016). Essential role of Wnt5a-Ror1/Ror2 signaling in metanephric mesenchyme and ureteric bud formation. *Genes to Cells, 21*, 325–334. http://dx.doi.org/10.1111/gtc.12342.

Reichert, S., Randall, R. A., & Hill, C. S. (2013). A BMP regulatory network controls ectodermal cell fate decisions at the neural plate border. *Development, 140*, 4435–4444. http://dx.doi.org/10.1242/dev.098707.

Rhee, J., Takahashi, Y., Saga, Y., Wilson-Rawls, J., & Rawls, A. (2003). The protocadherin papc is involved in the organization of the epithelium along the segmental border during mouse somitogenesis. *Developmental Biology, 254*, 248–261.

Robinow, M., Silverman, F. N., & Smith, H. D. (1969). A newly recognized dwarfing syndrome. *American Journal of Diseases of Children, 117*, 645–651.

Robinson, D. R., Wu, Y. M., & Lin, S. F. (2000). The protein tyrosine kinase family of the human genome. *Oncogene, 19*, 5548–5557. http://dx.doi.org/10.1038/sj.onc.1203957.

Roifman, M., Marcelis, C. L. M., Paton, T., Marshall, C., Silver, R., Lohr, J. L., et al. (2014). De novo WNT5A-associated autosomal dominant Robinow syndrome suggests specificity of genotype and phenotype. *Clinical Genetics, 87*, 34–41. http://dx.doi.org/10.1111/cge.12401.

Romereim, S. M., Conoan, N. H., Chen, B., & Dudley, A. T. (2014). A dynamic cell adhesion surface regulates tissue architecture in growth plate cartilage. *Development, 141*, 2085–2095. http://dx.doi.org/10.1242/dev.105452.

Romereim, S. M., & Dudley, A. T. (2011). Cell polarity: The missing link in skeletal morphogenesis? *Organogenesis, 7*, 217–228. http://dx.doi.org/10.4161/org.7.3.18583.

Sakane, H., Yamamoto, H., Matsumoto, S., Sato, A., & Kikuchi, A. (2012). Localization of glypican-4 in different membrane microdomains is involved in the regulation of Wnt signaling. *Journal of Cell Science, 125*, 449–460. http://dx.doi.org/10.1242/jcs.091876.

Sammar, M., Sieber, C., & Knaus, P. (2009). Biochemical and functional characterization of the Ror2/BRIb receptor complex. *Biochemical and Biophysical Research Communications, 381*, 1–6. http://dx.doi.org/10.1016/j.bbrc.2008.12.162.

Sammar, M., Stricker, S., Schwabe, G. C., Sieber, C., Hartung, A., Hanke, M., et al. (2004). Modulation of GDF5/BRI-b signalling through interaction with the tyrosine kinase receptor Ror2. *Genes to Cells, 9*, 1227–1238. http://dx.doi.org/10.1111/j.1365-2443.2004.00799.x.

Schambony, A., & Wedlich, D. (2007). Wnt-5A/Ror2 regulate expression of XPAPC through an alternative noncanonical signaling pathway. *Developmental Cell, 12*, 779–792. http://dx.doi.org/10.1016/j.devcel.2007.02.016.

Schille, C., Bayerlová, M., Bleckmann, A., & Schambony, A. (2016). Ror2 signaling is required for local upregulation of GFD6 and activation of BMP signaling at the neural plate border. *Development, 143*, 3182–3194. http://dx.doi.org/10.1242/dev.135426.

Schleiffarth, J. R., Person, A. D., Martinsen, B. J., Sukovich, D. J., Neumann, A., Baker, C. V. H., et al. (2007). Wnt5a is required for cardiac outflow tract septation in mice. *Pediatric Research, 61*, 386–391. http://dx.doi.org/10.1203/pdr.0b013e3180323810.

Schwabe, G. C., Tinschert, S., Buschow, C., Meinecke, P., Wolff, G., Gillessen-Kaesbach, G., et al. (2000). Distinct mutations in the receptor tyrosine kinase gene ROR2 cause brachydactyly type B. *American Journal of Human Genetics*, 67, 822–831. http://dx.doi.org/10.1086/303084.

Schwabe, G. C., Trepczik, B., Süring, K., Brieske, N., Tucker, A. S., Sharpe, P. T., et al. (2004). Ror2 knockout mouse as a model for the developmental pathology of autosomal recessive Robinow syndrome. *Developmental Dynamics*, 229, 400–410. http://dx.doi.org/10.1002/dvdy.10466.

Schwarzer, W., Witte, F., Rajab, A., Mundlos, S., & Stricker, S. (2009). A gradient of ROR2 protein stability and membrane localization confers brachydactyly type B or Robinow syndrome phenotypes. *Human Molecular Genetics*, 18, 4013–4021. http://dx.doi.org/10.1093/hmg/ddp345.

Seifert, J. R. K., & Mlodzik, M. (2007). Frizzled/PCP signalling: A conserved mechanism regulating cell polarity and directed motility. *Nature Reviews. Genetics*, 8, 126–138. http://dx.doi.org/10.1038/nrg2042.

Shnitsar, I., & Borchers, A. (2008). PTK7 recruits dsh to regulate neural crest migration. *Development*, 135, 4015–4024. http://dx.doi.org/10.1242/dev.023556.

Sinha, T., Wang, B., Evans, S., Wynshaw-Boris, A., & Wang, J. (2012). Dishevelled mediated planar cell polarity signaling is required in the second heart field lineage for outflow tract morphogenesis. *Developmental Biology*, 370, 135–144. http://dx.doi.org/10.1016/j.ydbio.2012.07.023.

Stricker, S., & Mundlos, S. (2011). Mechanisms of digit formation: Human malformation syndromes tell the story. *Developmental Dynamics*, 240, 990–1004. http://dx.doi.org/10.1002/dvdy.22565.

Stricker, S., Verhey Van Wijk, N., Witte, F., Brieske, N., Seidel, K., & Mundlos, S. (2006). Cloning and expression pattern of chicken Ror2 and functional characterization of truncating mutations in brachydactyly type B and Robinow syndrome. *Developmental Dynamics*, 235, 3456–3465. http://dx.doi.org/10.1002/dvdy.20993.

Suzuki, T., Hasso, S. M., & Fallon, J. F. (2008). Unique SMAD1/5/8 activity at the phalanx-forming region determines digit identity. *Proceedings of the National Academy of Sciences of the United States of America*, 105, 4185–4190. http://dx.doi.org/10.1073/pnas.0707899105.

Takeuchi, S., Takeda, K., Oishi, I., Nomi, M., Ikeya, M., Itoh, K., et al. (2000). Mouse Ror2 receptor tyrosine kinase is required for the heart development and limb formation. *Genes to Cells*, 5, 71–78.

Tanigawa, S., Wang, H., Yang, Y., Sharma, N., Tarasova, N., Ajima, R., et al. (2011). Wnt4 induces nephronic tubules in metanephric mesenchyme by a non-canonical mechanism. *Developmental Biology*, 352, 58–69. http://dx.doi.org/10.1016/j.ydbio.2011.01.012.

Tauriello, D. V. F., Jordens, I., Kirchner, K., Slootstra, J. W., Kruitwagen, T., Bouwman, B. A. M., et al. (2012). Wnt/β-catenin signaling requires interaction of the Dishevelled DEP domain and C terminus with a discontinuous motif in Frizzled. *Proceedings of the National Academy of Sciences of the United States of America*, 109, E812–E820. http://dx.doi.org/10.1073/pnas.1114802109.

ten Berge, D., Brugmann, S. A., Helms, J. A., & Nusse, R. (2008). Wnt and FGF signals interact to coordinate growth with cell fate specification during limb development. *Development*, 135, 3247–3257. http://dx.doi.org/10.1242/dev.023176.

Topol, L., Jiang, X., Choi, H., Garrett-Beal, L., Carolan, P. J., & Yang, Y. (2003). Wnt-5a inhibits the canonical Wnt pathway by promoting GSK-3-independent beta-catenin degradation. *The Journal of Cell Biology*, 162, 899–908. http://dx.doi.org/10.1083/jcb.200303158.

Tribulo, C. (2003). Regulation of Msx genes by a Bmp gradient is essential for neural crest specification. *Development*, 130, 6441–6452. http://dx.doi.org/10.1242/dev.00878.

Unterseher, F., Hefele, J. A., Giehl, K., De Robertis, E. M., Wedlich, D., & Schambony, A. (2004). Paraxial protocadherin coordinates cell polarity during convergent extension via

Rho A and JNK. *The EMBO Journal, 23*, 3259–3269. http://dx.doi.org/10.1038/sj.emboj.7600332.
van Bokhoven, H., Celli, J., Kayserili, H., van Beusekom, E., Balci, S., Brussel, W., et al. (2000). Mutation of the gene encoding the ROR2 tyrosine kinase causes autosomal recessive Robinow syndrome. *Nature Genetics, 25*, 423–426. http://dx.doi.org/10.1038/78113.
van Wijk, N. V., Witte, F., Feike, A. C., Schambony, A., Birchmeier, W., Mundlos, S., et al. (2009). The LIM domain protein Wtip interacts with the receptor tyrosine kinase Ror2 and inhibits canonical Wnt signalling. *Biochemical and Biophysical Research Communications, 390*, 211–216. http://dx.doi.org/10.1016/j.bbrc.2009.09.086.
Verkaar, F., & Zaman, G. J. R. (2010). A model for signaling specificity of Wnt/Frizzled combinations through co-receptor recruitment. *FEBS Letters, 584*, 3850–3854. http://dx.doi.org/10.1016/j.febslet.2010.08.030.
Vortkamp, A., Lee, K., Lanske, B., Segre, G. V., Kronenberg, H. M., & Tabin, C. J. (1996). Regulation of rate of cartilage differentiation by Indian hedgehog and PTH-related protein. *Science, 273*, 613–622.
Wang, B., Sinha, T., Jiao, K., Serra, R., & Wang, J. (2011). Disruption of PCP signaling causes limb morphogenesis and skeletal defects and may underlie Robinow syndrome and brachydactyly type B. *Human Molecular Genetics, 20*, 271–285. http://dx.doi.org/10.1093/hmg/ddq462.
White, J. J., Mazzeu, J. F., Hoischen, A., Bayram, Y., Withers, M., Gezdirici, A., et al. (2016). DVL3 alleles resulting in a -1 frameshift of the last exon mediate autosomal-dominant Robinow syndrome. *American Journal of Human Genetics, 98*, 553–561. http://dx.doi.org/10.1016/j.ajhg.2016.01.005.
White, J., Mazzeu, J. F., Hoischen, A., Jhangiani, S. N., Gambin, T., Alcino, M. C., et al. (2015). DVL1 frameshift mutations clustering in the penultimate exon cause autosomal-dominant Robinow syndrome. *American Journal of Human Genetics, 96*, 612–622. http://dx.doi.org/10.1016/j.ajhg.2015.02.015.
Wilson, C., Goberdhan, D. C., & Steller, H. (1993). Dror, a potential neurotrophic receptor gene, encodes a Drosophila homolog of the vertebrate Ror family of Trk-related receptor tyrosine kinases. *Proceedings of the National Academy of Sciences of the United States of America, 90*, 7109–7113.
Winkel, A., Stricker, S., Tylzanowski, P., Seiffart, V., Mundlos, S., Gross, G., et al. (2008). Wnt-ligand-dependent interaction of TAK1 (TGF-beta-activated kinase-1) with the receptor tyrosine kinase Ror2 modulates canonical Wnt-signalling. *Cellular Signalling, 20*, 2134–2144. http://dx.doi.org/10.1016/j.cellsig.2008.08.009.
Witte, F., Bernatik, O., Kirchner, K., Masek, J., Mahl, A., Krejci, P., et al. (2010). Negative regulation of Wnt signaling mediated by CK1-phosphorylated Dishevelled via Ror2. *The FASEB Journal, 24*, 2417–2426. http://dx.doi.org/10.1096/fj.09-150615.
Witte, F., Chan, D., Economides, A. N., Mundlos, S., & Stricker, S. (2010). Receptor tyrosine kinase-like orphan receptor 2 (ROR2) and Indian hedgehog regulate digit outgrowth mediated by the phalanx-forming region. *Proceedings of the National Academy of Sciences of the United States of America, 107*, 14211–14216. http://dx.doi.org/10.1073/pnas.1009314107.
Wu, J., Roman, A.-C., Carvajal-Gonzalez, J. M., & Mlodzik, M. (2013). Wg and Wnt4 provide long-range directional input to planar cell polarity orientation in Drosophila. *Nature Cell Biology, 15*, 1045–1055. http://dx.doi.org/10.1038/ncb2806.
Yamaguchi, T. P., Bradley, A., McMahon, A. P., & Jones, S. (1999). A Wnt5a pathway underlies outgrowth of multiple structures in the vertebrate embryo. *Development, 126*, 1211–1223.
Yamamoto, S., Nishimura, O., Misaki, K., Nishita, M., Minami, Y., Yonemura, S., et al. (2008). Cthrc1 selectively activates the planar cell polarity pathway of Wnt signaling

by stabilizing the Wnt-receptor complex. *Developmental Cell*, *15*, 23–36. http://dx.doi.org/10.1016/j.devcel.2008.05.007.

Yamamoto, H., Yoo, S. K., Nishita, M., Kikuchi, A., & Minami, Y. (2007). Wnt5a modulates glycogen synthase kinase 3 to induce phosphorylation of receptor tyrosine kinase Ror2. *Genes to Cells*, *12*, 1215–1223. http://dx.doi.org/10.1111/j.1365-2443.2007.01128.x.

Yanfeng, W. A., Berhane, H., Mola, M., Singh, J., Jenny, A., & Mlodzik, M. (2011). Functional dissection of phosphorylation of Disheveled in Drosophila. *Developmental Biology*, *360*, 132–142. http://dx.doi.org/10.1016/j.ydbio.2011.09.017.

Yun, K., Ajima, R., Sharma, N., Costantini, F., Mackem, S., Lewandoski, M., et al. (2014). Non-canonical Wnt5a/Ror2 signaling regulates kidney morphogenesis by controlling intermediate mesoderm extension. *Human Molecular Genetics*, *23*, 6807–6814. http://dx.doi.org/10.1093/hmg/ddu397.

Zeller, R., Lopez-Rios, J., & Zuniga, A. (2009). Vertebrate limb bud development: Moving towards integrative analysis of organogenesis. *Nature Reviews. Genetics*, *10*, 845–858. http://dx.doi.org/10.1038/nrg2681.

Zhang, B., Tran, U., & Wessely, O. (2011). Expression of Wnt signaling components during Xenopus pronephros development. *PloS One*, *6*, e26533. http://dx.doi.org/10.1371/journal.pone.0026533.

CHAPTER FIVE

Regulation of *Drosophila* Development by the Golgi Kinase Four-Jointed

Yoko Keira, Moe Wada, Hiroyuki O. Ishikawa[1]

Graduate School of Science, Chiba University, Chiba, Japan
[1]Corresponding author: e-mail address: ishikawaho@faculty.chiba-u.jp

Contents

1. Introduction	144
2. Genetic and Molecular Identification of *Drosophila* Four-Jointed	145
2.1 Phenotypes of *fj* Mutants	145
2.2 Structure of *fj* Gene	146
2.3 Localization and Posttranslational Modification of Fj	149
2.4 Transcription of *fj* During *Drosophila* Development	149
3. The Fat/Dachsous/Four-Jointed Pathway	149
3.1 Identification of the Fat Pathway	149
3.2 The Role of the Fat Pathway in Growth Regulation	152
3.3 The Role of the Fat Pathway in the Establishment of PCP	155
4. Biochemical Characterization of Fj	156
4.1 Fj Is a Golgi-Resident Kinase That Phosphorylates the Cadherin Domains of Fat and Ds	156
4.2 Substrate Specificity of Fj	157
5. Modulation of Fat-Ds Binding by Fj	158
5.1 Analysis of Fat-Ds Binding	158
5.2 Effect of Fj on Fat-Ds Localization	159
5.3 Phosphorylation Sites That Modulate Fat-Ds Binding	161
5.4 Fj Functions in the Golgi	163
6. Fj Polarizes Fat Activity	164
6.1 A Model for Polarization of Fat Activity in Response to a Fj Gradient	164
6.2 Computational Models of the Fat/Ds/Fj Pathway	165
7. Vertebrate Four-Jointed, Fat, Dachsous, and Other Kinases in the Secretory Pathway	165
7.1 Four-Jointed Box 1 (Fjx1)	165
7.2 Other Kinases in the Secretory Pathway	168
8. Conclusion	171
Acknowledgments	171
References	171

Current Topics in Developmental Biology, Volume 123
ISSN 0070-2153
http://dx.doi.org/10.1016/bs.ctdb.2016.11.003

© 2017 Elsevier Inc.
All rights reserved.

Abstract

Despite intensive research on kinases and protein phosphorylation, most studies focus on kinases localized to the cytosol and nucleus. Studies in *Drosophila* discovered a novel signaling pathway that regulates growth and planar cell polarity. In this pathway, the atypical cadherin Fat acts as a receptor, and the cadherin Dachsous (Ds) serves as its ligand. Genetic studies in *Drosophila* identified the *four-jointed* gene as a regulator of the Fat pathway. Four-jointed (Fj) resides in the Golgi and phosphorylates the cadherin domains of Fat and Ds. Fj-mediated phosphorylations promote the ability of Fat to bind to its ligand Ds and inhibit the ability of Ds to bind Fat, which is biased toward a stronger effect on Fat. Fj is expressed in a gradient in many developing tissues. The Fat-Ds-binding gradient can be explained by the graded activity of Fj that is sufficient to propagate the polarization of complexes across whole tissues. Recent studies revealed a new class of kinases that localize within the secretory pathway and the extracellular space, and phosphorylate proteins and sugar chains in the secretory pathway. Further, they appear to regulate extracellular processes. Mutations of the genes encoding these kinases cause human disease, thus underscoring the biological importance of phosphorylation events within the secretory pathway.

ABBREVIATIONS

ATP adenosine 5′-triphosphate
Da dalton
DNA deoxyribonucleic acid
Gal galactose
GFP green fluorescent protein
GlcA glucuronic acid
GST glutathione *S*-transferase
RNA ribonucleic acid
Xyl xylose

1. INTRODUCTION

Protein phosphorylation, which is a common posttranslational modification required for the regulation of protein function, was first described by Hammarsten, who detected phosphorous in the secreted milk protein casein (Hammarsten, 1883). Subsequently, the phosphorylated forms of numerous secreted proteins, peptide hormones, and proteoglycans were discovered. However, the molecular identities of the kinases responsible for these posttranslational modifications are unknown. Therefore, research on extracellular protein phosphorylation has not been pursued with sufficient vigor. A family of atypical kinases that localize within the secretory pathway was recently discovered (Ishikawa, Takeuchi, Haltiwanger, & Irvine, 2008; Ishikawa, Xu, Ogura, Manning, & Irvine, 2012; Tagliabracci et al., 2012). These enzymes have different substrate specificities and other biochemical characteristics compared with canonical protein kinases that localize to the cytosol and nucleus. Importantly, developmental genetic studies of *Drosophila* shed light on this class of kinases that function in the secretory pathway and phosphorylate extracellular proteins. This review focuses on these kinases with particular emphasis on Four-jointed (Fj), which represents the first molecularly identified Golgi-resident kinase that functions during the development of *Drosophila*.

2. GENETIC AND MOLECULAR IDENTIFICATION OF *DROSOPHILA* FOUR-JOINTED

2.1 Phenotypes of *fj* Mutants

The gene encoding *four-jointed* (*fj*) was discovered by Schultz in 1931 as a viable mutation of the fruit fly *Drosophila melanogaster*. Schultz was one of the last graduate students of Thomas Hunt Morgan to be awarded a degree in classical fly genetics (Fig. 1A and B; reviewed by Lindsley & Zimm, 1992). Subsequently, Waddington described the phenotype of *fj* flies (Waddington, 1940, 1943) characterized by short legs owing to the failure of joint formation. In homozygous *fj* mutants, the femur, tibia, and first three tarsal segments are truncated, and the T2 and T3 tarsal segments are fused (Fig. 1C and D; Tokunaga & Gerhart, 1976; Villano & Katz, 1995; Waddington, 1943). The wings of *fj* flies are slightly shorter compared to those of wild type, and the two crossveins of the mutants are closer than normal (Fig. 1E and F; Brodsky & Steller, 1996; Villano & Katz, 1995; Waddington, 1940). Further, defects in planar cell polarity (PCP), which is the coordinated alignment of cell polarity across the tissue plane, occur in the eye as well as the proximal–distal (PD) leg and the developing wing (Fig. 1G and H; Zeidler, Perrimon, & Strutt, 1999, 2000). The PCP defects observed in *fj* flies occasionally cause minor effects compared with those of mutations in genes encoding other PCP regulators such as *fat*, *dachsous* (*ds*), *dachs* (*d*), *frizzled* (*fz*), and *flamingo* (*fmi*). Further, growth is disturbed only in the longitudinal dimension of the leg, and mosaic analysis demonstrates that the *fj* mutation causes local nonautonomy at the joint, particularly joint failure in heterozygous tissue adjacent to mutant cells (Tokunaga & Gerhart, 1976). Together, these phenotypes suggest that the product of *fj* serves as a signal that regulates growth and patterning along the PD axis.

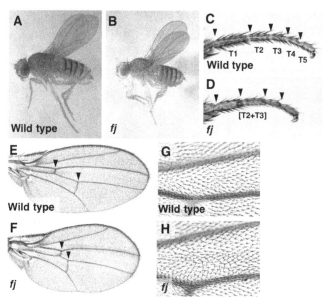

Fig. 1 Phenotypes of *four-jointed* (*fj*) adult flies. (A) Wild-type female. (B) Homozygotes of fj^{d1}, a null allele, exhibit a slightly smaller body size with characteristically shortened legs. (C) Tarsal region of the prothoracic leg of a female wild-type fly. *Arrowheads* mark the positions of the joints separating the five tarsal segments. (D) The prothoracic leg of fj^{d1} is shorter than normal, and the second and third tarsal segments are fused (T2/3). (E) Wild-type wing. *Arrowheads* indicate anterior and posterior crossveins. (F) The distance between crossveins is significantly reduced in fj^{d1} flies. (G) High magnification of a proximal region of the wild-type wing between veins 2 and 3. Wing hairs (trichomes) are arranged in a proximal to distal direction. (H) Wing-hair polarity is disrupted, with swirls of hairs forming in fj^{d1} flies.

2.2 Structure of *fj* Gene

Identification of the *fj* locus and the molecular characterization of *fj* were achieved independently by two groups (Brodsky & Steller, 1996; Villano & Katz, 1995) that performed enhancer-trap screens to identify genes that might play roles in the positional specification of imaginal discs (the larval primordia of adult cuticular structures composed of sheet or sac of epithelial cells). In this system, a *lacZ* gene lacking its own enhancer is carried on the P-element and may be regulated by genomic enhancers integrated near the transposon insertion site. As a result, an enhancer-trap line (termed *fj-lacZ*), in which the *lacZ* gene is expressed in larval imaginal discs, including those of the wing, eye, and leg, was isolated (Fig. 2A, C, and E; Brodsky & Steller, 1996; Villano & Katz, 1995). The *lacZ* gene is also

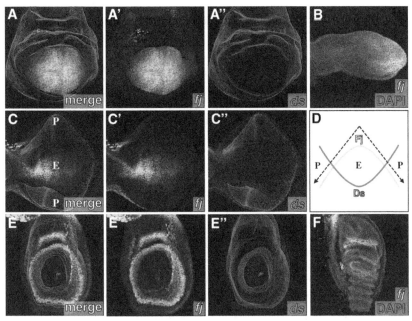

Fig. 2 Fj and Ds expression patterns in *Drosophila* imaginal discs. (A–A″) Wing disc from third instar larva. (A) Merged panel. (A′) *four-jointed* (*fj*) expression (*green*), revealed using an *fj-lacZ* enhancer trap, is low in peripheral regions and highest toward the center of the wing pouch that generates the distal adult wing (see Fig. 4B). (A″) Ds expression (*magenta*), revealed using an allele expressing the Ds endogenously tagged with GFP (Ds-GFP), is graded in a pattern complementary to *fj*. (B) 14–16 h after puparium formation (APF) pupal wing disc. (C–C″) Eye disc from third instar larva. (C) Merged panel. (C′) The levels of *fj* are high at the equator (E) and low at the poles (P). (C″) Ds expression is graded in a complementary pattern, with higher levels at the poles and lower levels at the equator. (D) Schematic: In wild-type flies, the arrangement of ommatidia is symmetrical with respect to the equator [E] of the eye, represented here by *arrows* pointing out toward the poles [P] (reviewed by Jenny, 2010). The vector of polarity can be considered as descending the Fj slope and ascending the Ds slope. Thus, in an eye with the clones either remove *fj* or *ds* (mutant clones), or augment *fj* or *ds* (overexpressing clones), reversals of polarity occur where the change in Fj or Ds expression causes a local reversal of the gradient. (E–E″) Leg disc from third instar larva. (E) Merged panel. (E′) The expression of *fj* is detected in concentric rings, corresponding to segmental boundaries. (E″) Ds expression occurs in pattern complementary to *fj*. (F) 4–6 h APF prepupal leg disc.

expressed in a subset of cells in the optic lobe of the larvae (Villano & Katz, 1995), and the P-element insertion site was determined using plasmid rescue, which was used to identify the *fj* locus (Brodsky & Steller, 1996; Villano & Katz, 1995).

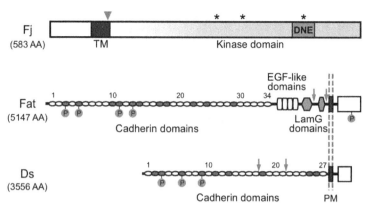

Fig. 3 Diagram of Four-jointed (Fj), Fat, and Dachsous (Ds). The *dark gray boxes* represent the transmembrane domain. Fj is a type II transmembrane protein (N-terminus in the cytosol). The *asterisks* denote the locations of predicted N-linked glycosylation sites. The *red arrowhead* indicates a predicted signal peptidase site. The *red box* shows the DNE motif in the extracellular domain, which is essential for the kinase activity of Fj. Extracellular cadherin domains of Fat and Ds are indicated as *ovals*. Shading *ovals* identify the cadherin domains with a Ser or Thr as the seventh amino acid residue, and "P" identifies cadherin domains that are phosphorylated by Fj. Intracellular domain of Fat is phosphorylated by Dco. The *red arrows* indicate cleavage sites.

The *fj* locus resides on right arm of *Drosophila* chromosome 2. A *fj* cDNA contains a single large open reading frame without introns, which encodes a predicted 65-kDa protein comprising 583 amino acid residues (Fig. 3; Brodsky & Steller, 1996; Villano & Katz, 1995). Amino acid residues 77–101 represent a potential transmembrane domain. The absence of an N-terminal region that resembles a signal sequence suggests that this hydrophobic domain serves as a transmembrane domain and an internal signal sequence, similar to those of type II transmembrane proteins.

Fj comprises a small N-terminal intracellular and a large C-terminal extracellular domain (Fig. 3; Brodsky & Steller, 1996; Villano & Katz, 1995). A potential signal peptidase cleavage site is located between amino acid residues 93 and 94 (red arrowhead in Fig. 3; Buckles, Rauskolb, Villano, & Katz, 2001). Further, the C-terminal region comprises several dibasic sites that, when cleaved, will produce a secreted protein. The C-terminal domain harbors three consensus sites for asparagine-linked glycosylation (asterisks in Fig. 3; Brodsky & Steller, 1996; Villano & Katz, 1995). The acidic amino acid sequence DNE, which is present at positions 490–492 in Fj as well as in its vertebrate homologs, is essential for kinase activity in vitro and biological activity in vivo (Fig. 3; Ishikawa et al., 2008; Simon, Xu, Ishikawa, & Irvine, 2010).

2.3 Localization and Posttranslational Modification of Fj

Immunohistochemical analysis revealed that endogenous Fj principally resides in the Golgi of imaginal eye and wing disc cells, although some spots of Fj expression do not colocalize with a Golgi marker in the basal regions of imaginal disc cells (Hale, Brittle, Fisher, Monk, & Strutt, 2015; Strutt, Mundy, Hofstra, & Strutt, 2004). Western blot analysis revealed that Fj is cleaved and secreted in vitro and in vivo, although a significant fraction of Fj remains membrane bound (Buckles et al., 2001; Villano & Katz, 1995).

2.4 Transcription of *fj* During *Drosophila* Development

The expression pattern of *fj* mRNA, visualized using in situ hybridization and *lacZ* expression in enhancer-trap lines, is consistent with the conclusion that Fj serves as a localized regional signal, consistent with the phenotypes of *fj* mutants (Brodsky & Steller, 1996; Villano & Katz, 1995). In the leg disc of third instar larvae, *fj* mRNA is expressed in concentric circles, similar to a subset of the concentric restrictions that mark the future segment boundaries of the leg (Fig. 2E; Brodsky & Steller, 1996; Villano & Katz, 1995). This expression pattern persists through the pupal stage (Fig. 2F; Buckles et al., 2001; Villano & Katz, 1995). In the wing disc, *fj* mRNA is expressed as a gradient in the wing pouch with low concentrations in the peripheral regions and with the highest concentrations toward the center of the pouch that develops into the distal adult wing (Fig. 2A and B; Brodsky & Steller, 1996; Strutt et al., 2004; Villano & Katz, 1995). Moreover, *fj* is expressed as a gradient from the equator toward the poles of the eye imaginal disc (Fig. 2C and D; Brodsky & Steller, 1996; Villano & Katz, 1995).

Immunohistochemical analysis of the imaginal discs of the eye, wing, and leg revealed that the pattern of Fj expression is similar to that of its mRNA (Hale et al., 2015; Strutt et al., 2004). Further, quantitative image analyses of immunohistochemical staining using Fj antibodies showed that the typical slope of a gradient of Fj between cells across the PD axis in the third instar larval wing discs of *Drosophila* is approximately 3% (Hale et al., 2015). Moreover, *fj* mRNA is expressed in the embryonic denticle field, larval brain, and adult abdomen (Casal, Struhl, & Lawrence, 2002; Donoughe & DiNardo, 2011; Villano & Katz, 1995).

3. THE FAT/DACHSOUS/FOUR-JOINTED PATHWAY

3.1 Identification of the Fat Pathway

Mutations of the *Drosophila* genes *dachsous* (*ds*), *fat*, and *dachs* (*d*) cause defects in the wing blade and leg growth similar to those of *fj*, as well as a reduction in the distance between crossveins (Fig. 1E and F; Brodsky & Steller, 1996; Villano & Katz, 1995; Waddington, 1940, 1943). Further, these mutants exhibit PCP phenotypes (Adler, Charlton, & Liu, 1998; Casal et al., 2002; Held, Duarte, & Derakhshanian, 1986; Mao et al., 2006; Rawls, Guinto, & Wolff, 2002; Strutt & Strutt, 2002; Yang, Axelrod, & Simon, 2002; Zeidler et al., 1999, 2000).

The *fat* gene, which encodes a large transmembrane protein (5147 amino acid residues) with 34 cadherin domains in its extracellular region (Fig. 3; Mahoney et al., 1991), was identified as a *Drosophila* tumor suppressor approximately 30 years ago (Bryant, Huettner, Held, Ryerse, & Szidonya, 1988). *fat* null alleles are lethal, and overgrown imaginal discs are present in mutants. However, viable mutants with weak alleles exhibit a broadening of the abdomen and wing, and this phenotype inspired the name *fat* (Mohr, 1923; Waddington, 1940).

The *ds* gene encodes a large transmembrane protein (3556 amino acid residues) with 27 cadherin domains (Fig. 3; Clark et al., 1995), and Dachs (the German word "badger") is an unconventional myosin (Mao et al., 2006). Genetic studies using *Drosophila* revealed that these four genes function together within a signaling pathway that influences growth, gene expression, and PCP (Fig. 4; Cho & Irvine, 2004; Fanto et al., 2003; Ma, Yang, McNeill, Simon, & Axelrod, 2003; Matakatsu & Blair, 2004; Yang et al., 2002). Genetics studies revealed that *fj* inhibits *fat* activity and promotes *ds* activity (Mao et al., 2006), and genetic epistasis experiments suggest that *fj* and *ds* act upstream of *fat* and that *dachs* is a downstream effector of *fat* (Fig. 4; Cho & Irvine, 2004; Yang et al., 2002). Fat and Ds are large atypical cadherins that bind each other through heterophilic interactions (Fig. 4). The results of cell aggregation and protein localization experiments provide direct support for the binding of Ds to Fat. Cultured *Drosophila* S2 cells do not normally aggregate, although they can be induced to aggregate when they express interacting proteins. Thus, cells that express Fat and Ds specifically bind each other (Matakatsu & Blair, 2004), and the cellular localizations of Fat and Ds in vivo suggest that they engage in heterophilic binding (Cho & Irvine, 2004; Ma et al., 2003; Mao et al., 2006; Strutt & Strutt, 2002). Ds levels are expressed in a mosaic pattern, such that a cell that expresses Fat is confronted with neighboring cells that differ in the levels of Ds expressed. Then, Fat concentrates at the interface with neighboring cells that express higher levels of Ds and is lost from interfaces of those that express lower levels (Ma et al., 2003; Mao et al., 2006; Strutt & Strutt, 2002). The localization of Ds can be similarly affected by manipulating Fat expression,

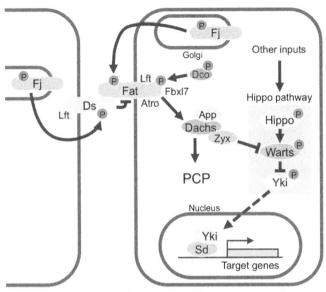

Fig. 4 Schematic representation of the *Drosophila* Fat-Hippo pathway. The atypical cadherin Fat acts as a receptor for a signaling pathway that regulates growth, gene expression, and PCP. Gene expression and growth mediated by the Fat pathway are regulated through the Hippo pathway. The linkage of the Fat and Hippo pathways depends on an unconventional myosin Dachs. Fj phosphorylates the extracellular cadherin domains of Fat and its ligand Ds.

and mutation or overexpression of Fj can modulate Fat and Ds localization (Cho & Irvine, 2004; Ma et al., 2003; Mao et al., 2006; Strutt & Strutt, 2002). For example, Ds staining is reduced in *fj* mutant clones, and Fat expression in a wing containing *fj* mutant clones is elevated (Ma et al., 2003). The hypothesis that *fj* and *ds* regulate Fat is supported by their influence on gene expression. The morphogen Wingless (Wg) is expressed in a ring of cells in the proximal *Drosophila* wing. In the absence of *fat* expression, Wg expression within the proximal wing is elevated and extended, and this effect is cell autonomous (Cho & Irvine, 2004). Manipulations of *fj* and *ds* expression influence Wg expression as well, but their effects are non-autonomous (Cho & Irvine, 2004).

Similarly, mutant clone analyses revealed that *fj* and *ds* exert non-autonomous effects on the expression of *fj*, *Serrate* (*Ser*), and *Death-associated inhibitor of apoptosis 1* (*Diap1*; Buckles et al., 2001; Cho et al., 2006; Rogulja, Rauskolb, & Irvine, 2008; Zeidler et al., 1999). In contrast, the expression of these genes is upregulated through a cell-autonomous mechanism within *fat* mutant clones (Cho et al., 2006; Mao et al., 2006; Yang et al., 2002).

The effects of *fj* and *ds* on *Diap1* expression and cell proliferation are suppressed in *dachs* mutants (Rogulja et al., 2008). Together, these observations indicate that Fat binds to Ds as a ligand–receptor pair for a pathway that influences gene expression, and further suggest that Fj influences this binding. Moreover, numerous observations implicate Ds as a Fat ligand, although *ds* mutants have weaker effects on growth compared with those of *fat* mutants. Thus, to a certain extent, Fat activity might be ligand independent. Further, recent studies revealed that Ds acts not only as a ligand for Fat but also as a receptor that signals through its cytoplasmic domain to regulate growth and PCP (Degoutin et al., 2013; Matakatsu & Blair, 2012; Willecke, Hamaratoglu, Sansores-Garcia, Tao, & Halder, 2008). The known components of the Fat pathway are presented in Fig. 4 and Table 1 (described in detail below).

3.2 The Role of the Fat Pathway in Growth Regulation

Gene expression and growth mediated by the Fat pathway are regulated through the Hippo pathway. Linkage of the Fat and Hippo pathways is indicated by their regulation of a common set of downstream target genes (Fig. 4; Bennett & Harvey, 2006; Cho et al., 2006; Silva, Tsatskis, Gardano, Tapon, & McNeill, 2006; Tyler & Baker, 2007; Willecke et al., 2006). This linkage depends on Dachs through the regulation of the turnover of Warts (Wts) (Fig. 4; Cho et al., 2006; Cho & Irvine, 2004; Mao et al., 2006; Rauskolb, Pan, Reddy, Oh, & Irvine, 2011).

An RNAi screen in *Drosophila* identified the LIM-domain protein Zyx102 (Zyx) as a Fat pathway component (Fig. 4; Rauskolb et al., 2011). Dachs stimulates Zyx-Wts binding, suggesting that regulated localization of Dachs could serve to regulate Zyx-Wts binding (Rauskolb et al., 2011). Hippo (Hpo) and Wts are Ser/Thr kinases, and their activity is regulated by phosphorylation and by their association with Salvador (Sav) and Mob as tumor suppressor (Mats). Hpo can be activated by intermolecular autophosphorylation (Glantschnig, Rodan, & Reszka, 2002; Lee & Yonehara, 2002), and activated Hpo then phosphorylates Wts, Sav, and Mats (Fig. 4; Wei, Shimizu, & Lai, 2007; Wu, Huang, Dong, & Pan, 2003). The activation of Wts is associated with autophosphorylation (Wei et al., 2007). Once activated, Wts then phosphorylates the transcriptional coactivator Yorkie (Yki) (Huang, Wu, Barrera, Matthews, & Pan, 2005). Yki does not have a DNA-binding domain and functions as transcriptional coactivator. DNA-binding proteins, such as Scalloped (Sd), Mothers against dpp (Mad), and homothorax (Hth), were

Table 1 Fat/Ds/Fj Pathway Components in *Drosophila*

Drosophila	Vertebrate Homologues	Protein Type or Motifs	Function
Fat (Ft)	Fat4	Atypical cadherin	Receptor
Dachsous (Ds)	Dchs1, 2	Atypical cadherin	Ligand for Fat
Four-jointed (Fj)	Fjx1	Golgi-localized kinase	Phosphorylates Fat and Ds cadherin domains
Dachs (D)	?	Unconventional myosin	Transduces signal from Fat to PCP and Hippo pathways
Discs overgrown (Dco)	CK1δ, ε	CK1 family kinase	Phosphorylates Fat cytoplasmic domain
Lowfat (Lft)	Lix1, Lix1L	Novel	Levels of Fat and Ds
Approximated (App)	ZDHHC9, 14, 18	Palmitoyltransferase	Dachs membrane localization
Zyx102 (Zyx)	Zyxin, Lpp, Trip6	LIM domains	Interacts with Wts
Fbxl7	Fbxl7	F-box and leucine-rich repeat protein	Regulates Dachs localization by binding to FatICD
Grunge (Gug)/ Atrophin (Atro)	RERE, ATN1	Transcriptional corepressor	Binds to FatICD
Wings apart (Wap)/ Riquiqui (Riq)	DCAF7	WD40 repeat protein	Binds to DsICD
Minibrain (Mnb)	DYRK1A	DYRK-family kinase	Binds to Wap/Riq
Hippo (Hpo)	Mst1,2	Sterile-20 family Ser/Thr kinase	Phosphorylates Wts, Hippo pathway component
Salvador (Sav)	WW45	WW, SARAH domains	Scaffolding protein, Hippo pathway component
Mob as tumor suppressor (Mats)	Mob1, 2	NDR kinase family cofactor	Promotes Wts activity, Hippo pathway component
Warts (Wts)	Lats1, 2	NDR family Ser/Thr kinase	Phosphorylates Yki, Hippo pathway component
Yorkie (Yki)	Yap, Taz	WW domains	Transcriptional coactivator, Hippo pathway component

Continued

Table 1 Fat/Ds/Fj Pathway Components in *Drosophila*—cont'd

Drosophila	Vertebrate Homologues	Protein Type or Motifs	Function
Wbp2	Wbp2	PPXY motifs	Transcriptional coactivator, binds to Yki, Hippo pathway component
Scalloped (Sd)	TEAD/TEF 1-4	TEA domain	DNA binding, binds to Yki, Hippo pathway component
Mothers against dpp (Mad)	Smad	MH domains	DNA binding, binds to Yki, Hippo pathway component
Homothorax (Hth)	Meis1-3	Homeodomain	DNA binding, binds to Yki, Hippo pathway component
Teashirt (Tsh)	Tshz1-3	Zn finger	DNA binding, binds to Yki, Hippo pathway component

identified as binding partners of Yki (Fig. 4; Goulev et al., 2008; Oh & Irvine, 2011; Peng, Slattery, & Mann, 2009; Wu, Liu, Zheng, Dong, & Pan, 2008; Zhang et al., 2008). Further, the transcriptional coactivator WW domain-binding protein-2 (Wbp2) binds Yki (Zhang, Milton, Poon, Hong, & Harvey, 2011). Wbp2 can enhance Yorkie's transcriptional coactivator properties (Zhang et al., 2011). Targets of Yki include genes that promote cellular growth (*Myc* and *bantam*), genes involved in cell cycle progression (*CyclinB*, *CyclinD*, *CyclinE*, and *E2F1*), and genes that inhibit apoptosis (*Diap1* and *cIap1*) (Dong et al., 2007; Goulev et al., 2008; Nicolay & Frolov, 2008; Nolo, Morrison, Tao, Zhang, & Halder, 2006; Shimizu, Ho, & Lai, 2008; Silva et al., 2006; Tapon et al., 2002; Thompson & Cohen, 2006; Tyler & Baker, 2007; Wu et al., 2003). The Hippo pathway regulates the expression of Fat pathway targets such as *wg*, *Ser*, and *fj* (Cho et al., 2006; Cho & Irvine, 2004). Discs overgrown (Dco, also called double-time), a CK1 homolog, phosphorylates the intracellular domain of Fat and induces Fat activity (Feng & Irvine, 2009; Sopko et al., 2009). The effects of Dco on Fat influence growth and gene expression, but do not influence PCP (Feng & Irvine, 2009; Sopko et al., 2009).

Although the Fat pathway is crucial for the regulation of the Hippo pathway, the latter receives multiple upstream signals, for example, from the Merlin (Mer) and Expanded (Ex) complex, the Lethal giant larvae (Lgl) complex, the levels of F-actin, and mechanical stress (Fig. 4; reviewed by Oh & Irvine, 2010; Staley & Irvine, 2012; Yu & Guan, 2013). Studies of

Drosophila showed that the Fat pathway regulates the Hippo pathway in numerous tissues, such as imaginal discs and the neuroepithelium (Bennett & Harvey, 2006; Cho et al., 2006; Meignin, Alvarez-Garcia, Davis, & Palacios, 2007; Reddy, Rauskolb, & Irvine, 2010; Silva et al., 2006; Willecke et al., 2006). In contrast, the Fat pathway is dispensable for the regulation of the Hippo pathway during the development of the *Drosophila* ovary (Polesello & Tapon, 2007).

3.3 The Role of the Fat Pathway in the Establishment of PCP

The Fat pathway affects PCP, which is the polarization of cells within the plane of a tissue, perpendicular to the apical–basal polarity of epithelial cells. Studies of PCP have focused on Frizzled-dependent PCP signaling, which involves a set of core PCP proteins, including Frizzled (Fz), Dishevelled (Dsh), Flamingo (Fmi, also called Starry night/Stan), and Prickle (Pk) (reviewed by Seifert & Mlodzik, 2007). However, *Drosophila fj*, *dachs*, *ds*, and *fat* mutants also exhibit PCP phenotypes (Adler et al., 1998; Casal et al., 2002; Held et al., 1986; Mao et al., 2006; Rawls et al., 2002; Strutt & Strutt, 2002; Yang et al., 2002; Zeidler et al., 1999, 2000). PCP phenotypes in *dachs* mutants are milder than the phenotypes of *fat* or *ds* mutants (Mao et al., 2006). The Fat/Ds/Fj pathway (global pathway) was proposed to link the direction of polarization to the tissue axes (Ma et al., 2003; Yang et al., 2002). The Fz/Fmi pathway (core pathway) functions locally to coordinate and amplify molecular asymmetry within and between cells, and core pathway proteins segregate within cells into two distinct spatially separated complexes (reviewed by Strutt & Strutt, 2009). Although both the global and core pathways regulate PCP, the relationship between the two pathways is unclear. For example, studies of the *Drosophila* eye and wing suggest that the global pathway acts upstream of the core pathway (Adler et al., 1998; Ma et al., 2003; Matakatsu & Blair, 2004; Yang et al., 2002). On the other hand, a detailed examination of the relationship between global pathway and core pathway in the abdomen indicates that these pathways can act in parallel to influence PCP (Casal, Lawrence, & Struhl, 2006). Recent studies showed that Ds and Dachs interact with an Fz/Fmi pathway component Spiny-legs and direct its localization in vivo. These explain a connection between the global pathway and the core pathway (Ambegaonkar & Irvine, 2015; Ayukawa et al., 2014).

The downstream targets of the global pathway are poorly understood, although certain genes are implicated. The conserved cytoplasmic protein Lowfat (Lft), which binds to the cytoplasmic domains of Fat and Ds, is required to maintain the normal levels of Fat and Ds (Fig. 4; Mao, Kucuk, & Irvine, 2009). Atrophin (Atro, also called Grunge) is linked to the global pathway according to observations that *Atro* mutant clones in the eye discs exhibit PCP phenotypes similar of *fat* mutant clones and that Atro binds the Fat cytoplasmic domain (Fig. 4; Fanto et al., 2003). Atro is a transcriptional corepressor that influences the expression of *fj* (Fanto et al., 2003), although to the best of our knowledge, there are no reports suggesting that it influences the growth or expression of other target genes of the Hippo pathway (Cho & Irvine, 2004; Fanto et al., 2003). Further, Atro regulates PCP only in regions near the equator of the eye (Sharma & McNeill, 2013).

Dachs is linked to the global pathway based on the observations of *dachs* mutants that partially suppress *fat* PCP phenotypes and findings that the subcellular localization of Dachs is polarized (Mao et al., 2006). The *approximated* (*app*) gene, encoding a DHHC palmitoyltransferase, was identified as a negative regulator of the Fat pathway. App, which is required for the localization of Dachs to the membrane, influences growth and PCP (Fig. 4; Matakatsu & Blair, 2008). Dachs is regulated by the ubiquitin ligase Fbxl7 (Fig. 4; Bosch et al., 2014; Rodrigues-Campos & Thompson, 2014) that binds to the cytoplasmic domain of Fat and regulates the level and localization of Dachs. Fbxl7 influences growth and PCP (Bosch et al., 2014; Rodrigues-Campos & Thompson, 2014).

4. BIOCHEMICAL CHARACTERIZATION OF FJ

4.1 Fj Is a Golgi-Resident Kinase That Phosphorylates the Cadherin Domains of Fat and Ds

Genetic studies identified the gene *fj* as a regulator of the Fat pathway that genetically acts upstream of *fat* and *ds*. However, the biochemical activity of Fj had been unknown. Fat and Ds are atypical cadherins, and Fj is a type II transmembrane protein that functions in the Golgi (Strutt et al., 2004; Villano & Katz, 1995). Thus, Fj might influence the Fat pathway by introducing posttranslational modifications in a component of the Fat pathway, similar to the activity of the Golgi-resident glycosyltransferase Fringe in Notch signaling (reviewed by Haines & Irvine, 2003).

When a subset of the cadherin domains of Ds and Fat (e.g., Ds2–3) are coexpressed with Fj in S2 cells, there is a shift in mobility (Ishikawa et al., 2008). Most glycosyltransferases contain the conserved sequence motif Asp-X-Asp (DXD; X, any amino acid), which is essential for their activity (Wiggins & Munro, 1998). A related sequence motif [Asp-Asn-Glu (DNE)] at amino acid residues 490–492 (Fig. 3) is present in Fj and its vertebrate homologs (Ashery-Padan, Alvarez-Bolado, Klamt, Gessler, & Gruss, 1999; Rock, Heinrich, Schumacher, & Gessler, 2005). A mutant form of Fj in which DNE was changed to GGG (Fj^{GGG}; G, glycine) was cloned, and its expression does not shift the mobility of Ds2–3, although the expression levels and Golgi localization of Fj^{GGG} appear normal (Ishikawa et al., 2008). The posttranslational modification associated with this mobility shift was identified by tandem mass spectrometry analysis. The Fj-dependent modification of Ds2–3 comprises an addition of 80 Da linked to Ser^{236} (Ishikawa et al., 2008), which does not correspond to that of known glycans, but does correspond to the mass of a phosphate group. Moreover, phosphatase treatment of Fj-modified cadherin fragments reversed the Fj-dependent mobility shifts, providing strong evidence that Ds and Fat cadherin domains undergo Fj-dependent phosphorylation (Ishikawa et al., 2008).

Bioinformatic analyses revealed a weak similarity between the amino acid sequences of Fj and the bacterial kinase HipA as well as that between Fj and the mammalian lipid kinase phosphatidylinositol 4-kinase II (Barylko et al., 2001; Correia et al., 2006). To determine whether Fj is a kinase, a secreted epitope-tagged Fj (sFj:V5) was purified from the medium of cultured S2 cells and incubated with purified Ds2–3 and γ-^{32}P-ATP. Transfer of ^{32}P to Ds2–3 was observed in the presence, but not absence, of sFj and was not catalyzed by sFj^{GGG} (Ishikawa et al., 2008). Bacterial recombinant Fj protein is also active. GST:Fj fusion protein was expressed in *Escherichia coli*, and purified GST:Fj also catalyzes the transfer of ^{32}P to Ds2–3 (Ishikawa et al., 2008). The generic kinase substrates myelin basic protein and casein were not detectably phosphorylated by sFj in vitro (Ishikawa et al., 2008). Thus, the substrate specificity of Fj appears to be high (see below). Fj autophosphorylates, although this activity is weak compared with the phosphorylation of Ds2–3. The autophosphorylation reaction is apparently unimolecular, because GST:Fj and sFj:V5 do not phosphorylate each other, and the fraction of Fj phosphorylated is independent of Fj concentration (Ishikawa et al., 2008). Unfortunately, positions and function of Fj autophosphorylation are unknown and must be addressed by future studies.

4.2 Substrate Specificity of Fj

A Ser residue at a specific location within the second of the two cadherin domains (e.g., Ser^{236} in Ds and Ser^{273} in Fat) is phosphorylated by Fj, although a Thr residue is compatible in vitro (Ishikawa et al., 2008). This Ser residue occupies the seventh amino acid in a structurally solved cadherin domain, and it is predicted to reside on the surface, near the middle of the cadherin domain (Patel et al., 2006). There are 9 and 11 potential phosphorylation sites in Ds and Fat, respectively (Fig. 3; Ishikawa et al., 2008). Cadherin domain 3 of Fat and domain 3 and 6 of Ds are the best substrates of Fj in vitro. However, cadherin domains 10 of Fat and 2, 11, 13, and 18 of Ds are not detectably phosphorylated, despite the presence of Ser or Thr at these positions, indicating that other structural features are also important for recognition by Fj. Fj exhibits a strong preference for Mn^{2+} as a cofactor, similar to glycosyltransferases present in the Golgi (Ishikawa et al., 2008). Asp residues play critical roles in catalysis and in the coordination of Mg^{2+} in other kinases (Correia et al., 2006). Uniform overexpression of *fj* reduces growth and leads to PCP defects (Casal et al., 2002; Cho & Irvine, 2004; Yang et al., 2002; Zeidler et al., 2000). However, overexpressed Fj^{GGG} is inactive, even though Fj:V5 and Fj^{GGG}:V5 exhibited Golgi localization in the cells of the wing imaginal disc (Ishikawa et al., 2008). Thus, the DNE motif of Fj is essential for the biological activity of Fj. In addition, the conserved aspartic acid residues D447 and D454 are essential for Fj kinase activity (Brittle, Repiso, Casal, Lawrence, & Strutt, 2010).

5. MODULATION OF FAT-Ds BINDING BY FJ

5.1 Analysis of Fat-Ds Binding

As mentioned above, genetic studies of *Drosophila* showed that *fj* acts upstream of *fat* in the PCP and Hippo pathways (Casal et al., 2006; Cho et al., 2006; Cho & Irvine, 2004; Strutt et al., 2004; Yang et al., 2002). Moreover, biochemical studies showed that Fj is a Golgi kinase that phosphorylates a subset of the cadherin domains of Fat and Ds (Ishikawa et al., 2008). These findings raised the possibility that Fj modulates Fat-Ds binding. A system for quantitatively measuring the effect of Fj on the Fat-Ds interaction was established to demonstrate direct binding between Fat and Ds and to establish an assay for characterizing the influence of Fj on this binding (Brittle et al., 2010; Simon et al., 2010). A secreted protein (Ds:AP) containing the entire Ds extracellular domain fused to human placental alkaline phosphatase (AP) was expressed in *Drosophila* S2 cells. The conditioned medium was incubated with S2 cells expressing full-length Fat (S2-Fat) or Fat and Fj (S2-Fat/Fj). Binding sites for the Ds:AP protein on the surface of the S2-Fat and S2-Fat/Fj cells were determined by measuring cell-associated AP activity after removal of unbound Ds:AP.

Fj expression strongly enhanced the ability of Fat-expressing cells to bind Ds:AP. Ds:AP binding by S2-Fat/Fj cells was readily detected compared with that of S2-Fat cells (Simon et al., 2010). A reversed binding assay was performed using S2 cells that secrete the Fat extracellular domain fused to AP (Fat:AP) with or without concomitant Fj expression. Medium conditioned by each of these two cell lines was incubated with S2 cells expressing full-length Ds (S2-Ds). Whereas binding of Fat:AP coproduced with Fj was readily detected, equivalent amounts of Fat:AP produced in the absence of Fj failed to detectably bind S2-Ds cells. These experiments demonstrate that Fj increases the ability of Fat to bind Ds (Fig. 5C; Simon et al., 2010). Expression of Fj^{GGG} failed to promote the ability of Fat to bind Ds, indicating that Fj acts on Fat via its kinase activity (Simon et al., 2010).

Next, S2-Ds:AP-expressing cells were transfected with an Fj expression vector. The Fat-binding activity of conditioned medium produced by S2-Ds:AP/Fj cells was then compared with that of S2-Ds:AP cells. Fj expression had a profound effect on the ability of Ds to bind Fat. When media containing equivalent amounts of Ds:AP were compared, binding of Ds:AP produced in the absence of Fj expression was detected. In contrast, Ds:AP produced in the presence of Fj had no detectable binding activity. These results indicate that Fj inhibits the ability of Ds to bind Fat (Fig. 5C; Simon et al., 2010). Similar to Fj's effect on Fat, the ability of Fj to regulate Ds depends on the Fj DNE motif (Simon et al., 2010). Fj does not enhance Fat-Ds binding by modulating the trafficking or stability of Fat and Ds in vivo. This indicates that Fj-mediated phosphorylation can act alone to directly modulate Fat-Ds binding, rather than acting as a precursor to subsequent posttranslational modifications or recruitment of cofactors (Simon et al., 2010).

5.2 Effect of Fj on Fat-Ds Localization

Genetic analyses were performed to test the effect of Fj on Fat-Ds localization in vivo. Clones of cells overexpressing Fj and lacking Fat function in pupal wing discs were generated. As the Fj-expressing cells lack Fat, the extent of Fat accumulation along the clone border reflects the ability of Ds produced in the presence of Fj overexpression to recruit Fat from neighboring wild-type cells. Fat failed to localize to the clone border (Fig. 5A; Simon et al., 2010). This result indicates that Fj overexpression reduces

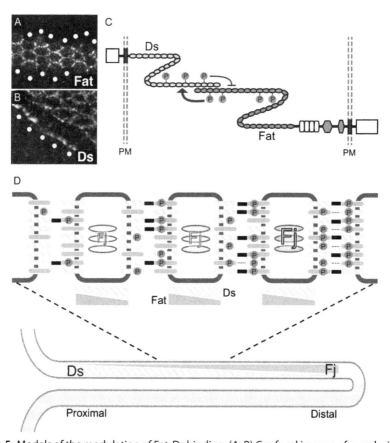

Fig. 5 Models of the modulation of Fat-Ds binding. (A, B) Confocal images of pupal wing discs stained with either anti-Fat or anti-Ds antibodies. The mutant cells (not stained) at the edge of the clone are indicated by the *white dots*. (A) Clones of *fat* mutant cells with Fj overexpression. Fat in the wild-type cells adjacent to the Fj-overexpressing cells fails to accumulate along the border with the Fj-overexpressing cells. (B) A clone of *ds* mutant cells with Fj overexpression. Ds in the adjacent wild-type cells is preferentially drawn to the border of the Fj-overexpressing cells. Note the reduced Ds staining at the cell–cell boundaries between the wild-type cells that border the Fj-overexpressing cells. (C) Fj-mediated phosphorylation of Fat increases the Ds-binding activity of Fat, and Fj-mediated phosphorylation of Ds decreases the Fat-binding activity of Ds, which is biased toward a stronger effect on Fat. Note that these phosphorylations are catalyzed in the Golgi, where Fj is localized. (D) A model of Fat-Ds binding across an epithelial tissue according to the Fj gradient during wing development. Fj acts cell autonomously to decrease the Fat-binding activity of Ds (*green* with *shaded lines*) and to increase the Ds-binding activity of Fat (*light blue*). Weak binding (*no line*) between Fat-Ds heterodimers can occur between all populations, intermediate binding (*dotted line*) can occur when phosphorylated Fat is available, and the strongest bond (*bold line*) occurs only between phosphorylated Fat and Ds. A gradient of tissue binding is produced with the left-most cell producing fewer intermediate bonds compared with

the ability of Ds to bind Fat in vivo (Simon et al., 2010). Further, the localization of Ds was examined in wing discs with clones of cells overexpressing Fj, which lacked Ds. Here, the ability of Ds to accumulate along the clone border is an indication of the ability of Fat produced by the Fj-expressing cells to bind Ds from neighboring wild-type cells (Fig. 5B; Simon et al., 2010). Examination with pupal wing discs showed that the Ds in the adjacent wild-type cells was drawn preferentially to the clone border, indicating that Fat produced in overexpression of Fj has enhanced ability to bind Ds. These results confirm the effects of Fj on Fat-Ds binding that were defined using the cell-based binding assays, which reflect the effects of Fj on Fat-Ds binding in vivo (Simon et al., 2010).

5.3 Phosphorylation Sites That Modulate Fat-Ds Binding

To determine the phosphorylation sites of Fat that modulate Fat-Ds binding, an in vitro Fat-Ds-binding assay was performed using N-terminal fragments of Fat and Ds, which comprise the first 10 cadherin domains of each protein (Fat1-10:FLAG and Ds1-10:AP) (Simon et al., 2010). Modest binding of Ds1-10:AP to Fat1-10:FLAG was detected, and the binding of Ds1-10:AP to Fat1-10:FLAG increased when Fat1-10:FLAG was purified from cells cotransfected with an Fj expression vector. These findings suggest that Fat and Ds binding is mediated by the N-terminal 10 of their extracellular cadherin domains (Fat and Ds contain 34 and 27 cadherin domains, respectively). Moreover, Fj modulates binding between Fat1-10 and Ds1-10 and stimulates binding when expressed on the Fat side and inhibits binding when expressed on the Ds side, indicating that sites sufficient for Fj-mediated modulation of binding reside within these 10 N-terminal cadherin domains. Binding of Ds1-10:AP to full-length Fat in a cell-based assay is lower compared with the binding of full-length Ds:AP, which may reflect an influence of C-terminal cadherin domains on the folding, structure, or stability of N-terminal cadherin domains (Simon et al., 2010). Fat1-10:FLAG was

---

the right-most cell. Phosphorylated Fat allows for stronger binding in cells with higher Fj, and therefore Ds in the adjacent left-most cell preferentially accumulates on the right cell edge, leading to the accumulation of Fat on the proximal side (closest to the body), and Ds accumulates on the distal side (closest to the wing tip). Thus, the Fj gradient generates cellular asymmetry. This pattern provides directional cues to the epithelial tissue. (A) and (B) were reproduced with permission from Simon, M. A., Xu, A., Ishikawa, H. O., & Irvine, K. D. (2010). Modulation of Fat:Dachsous binding by the cadherin domain kinase Four-jointed. Current Biology, 20, 811–817.

purified using anti-FLAG beads and then phosphorylated in vitro with affinity-purified Fj:V5 in the presence of ATP. Comparison of the Ds1-10:AP-binding activity of phosphorylated Fat1-10:FLAG with that of mock-treated Fat1-10:FLAG (incubated in the absence of Fj or ATP) established that in vitro phosphorylation of Fat1-10:FLAG enhances its binding to Ds. When Fat1-10:FLAG was purified from Fj-expressing cells and then incubated with phosphatase, its ability to bind Ds1-10:AP was reduced. Together, these observations establish that the presence or absence of phosphate groups covalently bound to the cadherin domains of Fat is sufficient to modulate its binding to Ds (Simon et al., 2010). Among the three Fj sites within the first 10 cadherin domains of Fat (Cad3, Cad5, and Cad10; Fig. 3), only the site in cadherin domain-3 is highly conserved among vertebrate and invertebrate Fat homologs. The Ser to Ala mutation of Cad3 of Fat1-10:FLAG (Fat1-10:FLAG^{S273A}) completely blocks the ability of Fj to enhance Fat-Ds binding in the in vitro assay. This observation indicates that the enhancement of Fat-Ds binding is dependent upon phosphorylation of this single Ser residue in cadherin domain 3 of Fat.

A cell-based binding assay using full-length Fat with S273A mutation was performed. The Fat^{S273A} reduced, but did not abolish, the ability of Fj to promote Fat-Ds binding. This confirms that Ser273 in the cadherin domain 3 of Fat contributes to Fj modulation of Fat-Ds binding and implies that binding interactions of full-length Fat and Ds may be more complex, with contributions from multiple Fj phosphorylation sites (Simon et al., 2010).

To test the effects of phosphorylation of Ds by Fj, the three conserved Ser residues in cadherin domains 3, 6, and 9 (Fig. 3) in Ds-GFP were mutated to Ala to prevent phosphorylation ($ds^{S>Ax3}$-GFP) (Brittle et al., 2010). These mutations did not alter the levels of Ds expression or its cell surface localization. In the absence of fj expression, $ds^{S>Ax3}$-GFP-expressing cells bind fat-expressing cells at levels similar to those of wild-type ds-GFP-expressing cells. However, in the cell aggregation assay, $ds^{S>Ax3}$-GFP-expressing cells do not respond to coexpression of Fj, unlike the control ds-GFP-expressing cells. These results suggest that these three Ds phosphorylation sites contribute to the modulation of Ft-Ds-binding affinity by Fj (Brittle et al., 2010).

A phosphomimetic form of Ds with mutations of these three Ser to Asp residues ($ds^{S>Dx3}$-GFP) was cloned (Brittle et al., 2010). The $ds^{S>Dx3}$-GFP-expressing cells exhibit significantly reduced levels of binding to Fat cells compared with wild-type ds-GFP- or $ds^{S>Ax3}$-GFP-expressing cells. Coexpression of Fj with $ds^{S>Dx3}$-GFP did not reduce binding further, indicating that if there were other potential phosphorylation sites in Ds, they do not contribute significantly to the regulation of Ft-Ds binding (Brittle et al., 2010). Moreover, these studies support the hypothesis that these three phosphorylation sites are important for Ds function in vivo (Brittle et al., 2010).

To measure the strength of Fat-Ds binding in vivo, the fluorescence recovery after photobleaching (FRAP) technique was employed (Hale et al., 2015). Transgenic flies expressing Ds endogenously tagged with GFP at the C terminus (Ds-GFP) and a strain in which GFP was inserted at the C terminus of the endogenous Fat coding region (Fat-GFP) were used (Brittle, Thomas, & Strutt, 2012; Hale et al., 2015). In the wing disc, Fat-GFP and Ds-GFP exhibit junctional populations with a punctate distribution, which is similar to the results of immunostaining of fixed tissue (Brittle et al., 2012; Hale et al., 2015; Ma et al., 2003). FRAP assays indicated that there are stable and unstable populations that express Fat-GFP and Ds-GFP at junctions, with stable expression concentrated into bright punctate regions (Hale et al., 2015). Further, Fat and Ds are required for junctional stability through mutual binding across cell membranes (Hale et al., 2015). The stability of the Fat-Ds dimer was reduced in the *fj* mutant, indicating that at least in the wing disc, the effect of Fj on Fat is dominant to that on Ds (Fig. 5C; Hale et al., 2015). FRAP experiments using the phosphomutants (Ser to Ala or Asp) of Fat and Ds revealed the in vivo effects of Fj phosphorylation on Fat and Ds in vitro (Hale et al., 2015).

5.4 Fj Functions in the Golgi

Fj is a type II transmembrane protein with a putative signal peptidase site in its transmembrane domain, and its C terminus is secreted (Fig. 3). Therefore, Fj was originally proposed to act as a secreted signaling molecule (Villano & Katz, 1995). Immunohistochemical analysis revealed that Fj resides in the Golgi in vivo and the Golgi of cultured Fj-expressing cells (Strutt et al., 2004). To test whether cleavage and secretion of Fj is required for its function in vivo, three mutant forms were generated (Strutt et al., 2004). The noncleavable form of Fj (Fj^{Un}) lacks signal peptidase cleavage sites. Further, a constitutively secreted Fj variant $CD2^{Signal}$-Fj (CD2-Fj) and a Golgi-retained form GalNAcT3-Fj (GNT-Fj) were generated by fusing either the rat CD2 signal sequence (Williams, Barclay, Clark, Paterson, & Willis, 1987) or the transmembrane domain and Golgi retention signal of GalNAc-T3 glycosyltransferase (Bennett, Hassan, & Clausen, 1996), respectively, to the C terminus of Fj downstream of its predicted cleavage sites. An *fj-14kb* transgene flanked by 7 and 4 kb of genomic upstream and downstream sequences, respectively, partially rescues the *fj* wing PD patterning defect (Strutt et al., 2004). The *CD2-fj-14kb* transgene only poorly rescues the PD growth defect caused by *fj* mutant. In contrast, rescue mediated by the *fj^{Un}-14kb* transgene is similar compared with that of the *fj-14kb* transgene. Further, *GNT-fj-14kb* fully rescues the *fj*-null wing phenotype, suggesting that the activity of GNT-Fj is higher compared with that of wild-type Fj (Strutt et al., 2004). Moreover, in the Fat-Ds-binding assay, the GNT-Fj form was at least as potent as wild-type Fj in inhibiting Ds binding to Fat, consistent with Fj mediating its effect in the Golgi and not at the cell surface (Brittle et al., 2010).

6. FJ POLARIZES FAT ACTIVITY

6.1 A Model for Polarization of Fat Activity in Response to a Fj Gradient

The induction of distinct cell fates in response to quantitatively distinct levels of morphogen signaling is a classic paradigm for developmental patterning (reviewed by Briscoe & Small, 2015). Further, evidence indicates that the vector and slope of morphogen gradients can be interpreted by cells, and they are used to direct PCP and growth (Casal et al., 2002; Ma et al., 2003; Matakatsu & Blair, 2004; Rogulja and Irvine, 2005; Rogulja et al., 2008; Simon, 2004; Willecke et al., 2008; Yang et al., 2002). The Fat/Ds/Fj pathway regulates growth and PCP, and Fat is regulated by the graded expression of Fj and Ds (Casal et al., 2002; Ma et al., 2003; Rogulja et al., 2008; Strutt & Strutt, 2002; Willecke et al., 2008; Yang et al., 2002). Moreover, planar polarization of Fat and Ds occurs across cells, and Ds and Ft accumulate distally and proximally, respectively, in wing imaginal discs (Fig. 5D; Ambegaonkar, Pan, Mani, Feng, & Irvine, 2012; Bosveld et al., 2012; Brittle et al., 2012).

Fat and Ds can bind each other when they are not phosphorylated, and Fj acts directly on Fat to promote its binding to Ds and acts on Ds to inhibit its binding to Fat in vitro and in vivo (Fig. 5C; Brittle et al., 2010; Hale et al., 2015; Simon et al., 2010). Moreover, the effect of Fj on Fat is stronger compared with the effect of Fj on Ds (Fig. 5C; Hale et al., 2015). These observations explain how the slope and vector of a gradient can be interpreted to establish polarity with high fidelity within cells (Fig. 5D; Brittle et al., 2010; Hale et al., 2015; Simon et al., 2010).

For any cell at any point within an Fj expression gradient, Fj acts cell autonomously to decrease the Fat-binding activity of Ds and to increase the Ds-binding activity of Fat. In a cell that expresses a low level of Fj, although more Ds is available, phosphorylated Fat is less abundant, indicating that fewer intermediate bonds will form. Phosphorylated Fat allows for stronger binding in cells with higher levels of Fj. As a result, Ds preferentially accumulates in the adjacent cells with lower Fj levels compared with cells that express higher levels of Fj at the cell edge. This indicates that the Fj gradient induces cellular asymmetry. A gradient of tissue binding is produced because cells in the lower Fj-expressing cell population produce fewer intermediate bonds compared with those in higher Fj-expressing cells. Therefore, a gradient of Fat-Ds binding will be observed across the tissue (Fig. 5D; Brittle et al., 2010; Hale et al., 2015; Simon et al., 2010).

Although the expression patterns of Fj and Ds in the gradient serve as important cues for specifying Ft-Ds asymmetry in the wing pouch, evidence suggests that the boundaries of Ds expression may act only as a patterning cue over a few cell diameters (Ambegaonkar et al., 2012; Brittle et al., 2012). Therefore, at least in the wing pouch, the Fj gradient is likely the dominant cue.

6.2 Computational Models of the Fat/Ds/Fj Pathway

Computational models of the underlying mechanism that generates the subcellular asymmetry of the Fat-Ds heterodimer on cell interfaces (Hale et al., 2015; Jolly, Rizvi, Kumar, & Sinha, 2014; Mani, Goyal, Irvine, & Shraiman, 2013; Yoshida, Bando, Mito, Ohuchi, & Noji, 2014) allow conclusions to be drawn to explain the observed phenotypes and to propose quantitative predictions. For example, RNAi experiments revealed that the Fat/Ds/Fj pathway is essential for leg regeneration in the two-spotted cricket *Gryllus bimaculatus* (Bando et al., 2009; Bando, Mito, Nakamura, Ohuchi, & Noji, 2011). Further, genetic studies of *Drosophila* showed that the Fat/Ds/Fj pathway controls looping and left-right asymmetry of the hindgut (González-Morales et al., 2015).

While elaborations to understand the significance are needed, these computational models may be able to help in understanding such experimental observations.

7. VERTEBRATE FOUR-JOINTED, FAT, DACHSOUS, AND OTHER KINASES IN THE SECRETORY PATHWAY

7.1 Four-Jointed Box 1 (Fjx1)

The *Drosophila* genome encodes single copies of the *fj* and *ds* and two copies of *fat* (*fat* and *kugelei*, known as *fat2* or *fat-like*). Mammalian genomes encode four Fat homologs (Fat1-4), two Ds homologs (Dchs1 and Dchs2), and one

Fj ortholog (Fjx1) (Ashery-Padan et al., 1999; Rock, Heinrich, et al., 2005; Rock, Schrauth, & Gessler, 2005). During development, mouse *Fjx1* is expressed in a subset of neuroepithelial cells, epithelial cells of multiple organs, and limbs (Rock, Heinrich, et al., 2005; Rock, Schrauth, et al., 2005). *Fjx1* is expressed in the adult brains of mice (Ashery-Padan et al., 1999; Probst, Rock, Gessler, Vortkamp, & Püschel, 2007). Similar to *Drosophila*, complementary expression of *Fjx1* and *Dchs1* occurs in developing organs such as kidney, lung, and intestine (Rock, Heinrich, et al., 2005; Rock, Schrauth, et al., 2005). These findings are therefore consistent with the conclusion that the *Drosophila* Fat/Ds/Fj pathway is conserved in higher vertebrates.

Homozygous *Fjx1* mutant mice are healthy and fertile and do not exhibit overt morphological or behavioral defects, although they exhibit abnormal morphology of dendritic arbors in the hippocampus (Probst et al., 2007). Further, there is an increase in dendrite extension and branching in cultured hippocampal neurons (Probst et al., 2007). In *Drosophila*, *fj* mutant clones show more severe PCP phenotypes than homozygotes of *fj* mutant because mutant *fj* clones reverse a gradient of Fat-Ds (Simon, 2004; Strutt et al., 2004; Yang et al., 2002). Therefore, it is possible that the manipulation of gene expression of *Fjx1* in a mosaic pattern shows severe or additional phenotypes even in mice. Overexpression of Fjx1 to cultures of dissociated hippocampal neurons causes an opposite effect, reduces the length of dendrites, and decreases dendritic branching (Probst et al., 2007). Thus, Fjx1 acts as an inhibitory factor that regulates dendrite extension in mice (Probst et al., 2007).

A study of *fat* knockout mice showed that loss of *Fat4* disrupts oriented cell division and tubule elongation during kidney development, leading to cystic kidney disease, indicating that *Fat4* is an essential gene in vertebrate PCP (Saburi et al., 2008). Mice with *Dchs1* mutations exhibit similar phenotypes compared with those of *Fat4* mutants, and comparison with the latter revealed that these two genes are required for PCP in multiple organs such as the brain, ear, kidney, skeleton, intestine, heart, and lung (Badouel et al., 2015; Bagherie-Lachidan et al., 2015; Cappello et al., 2013; Das et al., 2013; Durst et al., 2015; Kuta et al., 2016; Mao, Francis-West, & Irvine, 2015; Mao et al., 2016, 2011; Saburi et al., 2008; Zakaria et al., 2014). According to the conservation of certain functions among *Drosophila* and mammals, a model in which Fat4 and Dchs1 act as a receptor–ligand pair to regulate several developmental processes in mammals is proposed (Mao et al., 2011). Electron microscopy of the structure of

Fat4-Dchs1 complexes revealed that although their N-terminal regions are linear, the C-terminal regions of the extracellular domains are kinked with multiple hairpin-like bends (Fig. 5C; Tsukasaki et al., 2014). *Dchs1* and *Dchs2* function in a partially redundant fashion to regulate the number of nephron progenitors essential for kidney development, although *Dchs2* null mice are viable and fertile (Bagherie-Lachidan et al., 2015). Analysis of *Fat3* null mice found that Fat3 regulates the formation of the inner plexiform layer (Deans et al., 2011), and genetic studies of mice showed that loss of *Fjx1* enhances *Fat3* and *Fat4* phenotypes, indicating that the Fat cadherins of mammals may be modulated by Fjx1 (Deans et al., 2011; Saburi et al., 2008).

Western blot analysis detected phosphoserine and phosphothreonine residues in the extracellular domain of Fat1 in Fat1-expressing cultured human cells (Sadeqzadeh, de Bock, & Thorne, 2014). RNAi knockdown of *Fjx1* expression does not influence the levels of Fat1 phosphorylation, indicating that other mechanisms are likely responsible (Sadeqzadeh et al., 2014). Future studies are therefore required to characterize the biochemical activity of Fjx1 and to identify additional substrates. Fjx1 resides in the Golgi and is processed and secreted (Probst et al., 2007; Rock, Heinrich, et al., 2005; Rock, Schrauth, et al., 2005). Further, rescue experiments using cultured neurons showed that Fjx1 acts cell autonomously and non-autonomously (Probst et al., 2007). These observations indicate that Fjx1 may function as a secreted ligand.

Strong evidence indicates that Fat regulates PCP in mammals, although the link to the Hippo pathway is unknown. In the kidneys of fetal *Fat4* mutant mice, the nuclear localization of Yap (the mammalian homologue of *Drosophila yorkie*) is increased (Murphy et al., 2014). In contrast, *Fat4* regulates mesenchymal nephron progenitor cells independently of *Yap* (Bagherie-Lachidan et al., 2015). Moreover, Fat4-Dchs1 signaling mediates *Yap*-independent regulation of sclerotome proliferation (Kuta et al., 2016). These observations suggest that novel mechanisms of Fat4-Dchs1 signaling evolved to regulate cell proliferation, likely through a Hippo-independent pathway (Bagherie-Lachidan et al., 2015; Kuta et al., 2016).

Fat may contribute to oncogenesis in humans, because mutation or reduced expression of human *Fat* genes occurs, for example, in gastrointestinal and hematopoietic cancers (Cheng et al., 2015; Furukawa et al., 2015; Gao et al., 2014; Garg et al., 2015; Parry et al., 2013; Sadeqzadeh et al., 2014; Tenedini et al., 2014; Wadhwa et al., 2013). Fjx1 may be involved in the malignant transformation of human cells as well. For example, Fjx1 is overexpressed in various tumors (Buckanovich et al., 2007; Kakiuchi et al., 2003;

Lu et al., 2007), and the levels of *Fjx1* mRNA and protein correlate with poor survival of patients with colorectal cancer (Al-Greene et al., 2013). Further, Fjx1-specific peptides induce the secretion of cytokines that are cytotoxic to Fjx1-expressing cancer cells, indicating the possibility of tumor immunotherapy (Chai et al., 2015).

7.2 Other Kinases in the Secretory Pathway

Fj was the first molecularly characterized Golgi-localized protein kinase (Ishikawa et al., 2008). Advances in genome sequencing and genetics have paved the way for the discovery of a new class of kinases that localize within the ER, Golgi, and the extracellular space (Table 2; reviewed by Sreelatha, Kinch, & Tagliabracci, 2015). These novel kinases phosphorylate proteins and sugar chains of the secretory pathway and regulate extracellular processes. The homologs most closely related to Fj are encoded by *Family with sequence similarity 20* (*FAM20*) genes, which includes the human genes *FAM20A*, *FAM20B*, and *FAM20C* (Nalbant et al., 2005). FAM20C is the homolog most closely related to Fj among the three FAM20 family proteins. FAM20C was originally identified as a secretory calcium-binding protein named Dentin Matrix Protein 4 (DMP4) that modulates odontoblast differentiation (Hao, Narayanan, Muni, Ramachandran, & George, 2007). FAM20C resides in the Golgi, is the physiological Golgi casein kinase

Table 2 Mammalian Kinases in the Secretory Pathway

Kinase	Substrate	Disease	KO Mouse
Fjx1	Unknown		Increase in dendrite extension and branching
FAM20A	Pseudokinase	Amelogenesis imperfecta, hypophosphatemia	Amelogenesis imperfecta, hypophosphatemia
FAM20B	Xylose on proteoglycan		Embryonic lethal
FAM20C	Casein, SIBLING, FGF23, etc.	Raine syndrome, hypophosphatemia	Amelogenesis imperfecta, hypophosphatemia
VLK	MMP, collagen, etc.		Neonatal lethal
POMK	O-Mannose on α-dystroglycan	Congenital muscular dystrophy	

(G-CK), and phosphorylates Ser/Thr-X-Asp/Glu/pSer (S/T-X-D/E/pS) motifs of secreted proteins (Ishikawa et al., 2012; Lasa-Benito, Marin, Meggio, & Pinna, 1996; Tagliabracci et al., 2012). Similar to Fj, FAM20C prefers Mn^{2+} vs Mg^{2+} as a cofactor (Ishikawa et al., 2012; Tagliabracci et al., 2012).

FAM20C phosphorylates the small integrin-binding ligand N-linked glycoproteins (SIBLINGs), which modulate biomineralization in tooth and bone (Bellahcène, Castronovo, Ogbureke, Fisher, & Fedarko, 2008; Ishikawa et al., 2012; Qin, Baba, & Butler, 2004; Tagliabracci et al., 2012). Further, mass spectrometry analysis showed that FAM20C generates the majority of the secreted phosphoproteome, suggesting that FAM20C substrates contribute to a broad spectrum of biological processes (Tagliabracci et al., 2015). Loss-of-function mutations in *FAM20C* cause Raine syndrome, a rare disorder first described as a lethal osteosclerotic bone dysplasia (Raine, Winter, Davey, & Tucker, 1989; Rejjal, 1998). Patients with Raine syndrome and *FAM20C* null mice exhibit increased levels of circulating FGF23 (Vogel et al., 2012; Wang et al., 2012). FAM20C phosphorylates FGF23, which inhibits the O-glycosylation of FGF23 by the polypeptide GalNAc-T3, which then allows for proteolytic inactivation by furin (Tagliabracci et al., 2014).

FAM20C is the only secretory pathway kinase with a solved crystal structure (Xiao, Tagliabracci, Wen, Kim, & Dixon, 2013). Most protein kinases are activated by a phosphorylation event within an activation loop following the DFG motif (Kornev & Taylor, 2010). The DNE motif of Fj is considered a variant of this motif. Structural analysis of *Caenorhabditis elegans* FAM20C suggests that the analogous loop in the nematode FAM20C is not phosphorylated (Xiao et al., 2013). Thus, FAM20C appears to be constitutively active and does not require phosphorylation or significant conformational changes for catalysis.

FAM20B phosphorylates the 2-OH moiety of a xylose residue within the tetrasaccharide linkage region GlcA-β1,3-Gal-β1,3-Gal-β1,4-Xyl-β1-O-Ser of proteoglycans (Koike, Izumikawa, Tamura, & Kitagawa, 2009). Genetic and biochemical studies revealed that this xylose phosphorylation is essential for glycosaminoglycan formation and acts as a quality control system for proteoglycan biosynthesis in the Golgi (Eames et al., 2011; Nadanaka et al., 2013; Wen et al., 2014). *FAM20B* null mice show embryonic lethality (Table 2) and exhibit severe stunting, multisystem organ hypoplasia, and delayed development in the lung, eyes, liver, gastrointestinal tract, and skeletal system (Vogel et al., 2012). A tissue-specific conditional knockout of *FAM20B* causes chondrosarcoma in the joint as well as defects of postnatal ossification in the long bones (Ma et al., 2016).

FAM20A is a paralog of FAM20C, and both genes function in enamel formation (Vogel et al., 2012; Wang et al., 2013, 2014). *FAM20A* mutations are associated with an autosomal recessive type of amelogenesis imperfecta (AI; Table 2). Patients with AI with *FAM20A* mutations have major dental abnormalities such as generalized hypoplastic enamel, intrapulpal calcification, delayed tooth eruption, and fail to develop teeth (Cho et al., 2012; Kantaputra et al., 2014; O'Sullivan et al., 2011). FAM20A is a pseudokinase that forms a functional complex with FAM20C, and this complex enhances extracellular protein phosphorylation within the secretory pathway. FAM20A potentiates FAM20C kinase activity and promotes the phosphorylation of enamel matrix proteins in vitro and in cells (Cui et al., 2015). Further, binding of FAM20A to FAM20C enhances FAM20C secretion in osteoblasts (Ohyama et al., 2016).

Vertebrate lonesome kinase (VLK) is a secreted kinase that phosphorylates tyrosine residues of extracellular proteins (Bordoli et al., 2014). VLK phosphorylates proteins in the secretory pathway as well (Bordoli et al., 2014). Notably, VLK is the only known kinase that functions in the extracellular space using extracellular ATP (Bordoli et al., 2014). In contrast to Fj and FAM20C, Mg^{2+} and Ca^{2+} rather than Mn^{2+} enhance VLK kinase activity (Bordoli et al., 2014). Further, VLK phosphorylates a wide range of extracellular proteins, including matrix metalloproteinases (MMPs), with no recognizable consensus sequence (Bordoli et al., 2014). *Vlk* null mice show a defect in lung development, delayed ossification of endochondral bone, and neonatal lethality due to respiratory failure as well as a suckling defect arising from a cleft palate (Table 2; Kinoshita, Era, Jakt, & Nishikawa, 2009).

The gene encoding protein O-mannose kinase (POMK, also called SGK196) was identified in a haploid screen through its involvement in the posttranslational modification of α-dystroglycan (Jae et al., 2013). POMK phosphorylates O-mannose of N-acetylgalactosamine-β3-N-acetylglucosamine-β4-mannose-α-O-Ser/Thr on α-dystroglycan (Yoshida-Moriguchi et al., 2013). Phosphorylation as well as glycosylation of α-dystroglycan is required for α-dystroglycan to bind to ligands such as laminin (Hewitt, 2009; Yoshida-Moriguchi et al., 2010). Dystroglycan is a transmembrane glycoprotein whose interactions with the extracellular matrix are necessary for normal muscle and brain development. Therefore, *POMK* mutations lead to a spectrum of neuromuscular defects (Di Costanzo et al., 2014).

8. CONCLUSION

Fj is the first molecularly characterized Golgi-localized protein kinase. Genetics and biochemical studies revealed how Fj regulate growth and PCP in *Drosophila* development. The gradient expression of Fj is important for establishing PCP in the development of *Drosophila* imaginal discs. Studies on the regulation of gene expression of *fj* will help understand the mechanisms of conversion from morphogens gradients to PCP. Subsequently the identification of Fj, a new class of kinases that function in the secretory pathway, has been revealed. Mutations of these kinases lead to developmental defects and human diseases. Studies on how these kinases are regulated and the identification of their substrates will be important to understand animal development and human pathophysiology.

ACKNOWLEDGMENTS

Authors thank to Ken Irvine and members of our laboratory for comments and discussions about this manuscript.

REFERENCES

Adler, P. N., Charlton, J., & Liu, J. (1998). Mutations in the cadherin superfamily member gene *dachsous* cause a tissue polarity phenotype by altering *frizzled* signaling. *Development*, *125*, 959–968.

Al-Greene, N. T., Means, A. L., Lu, P., Jiang, A., Schmidt, C. R., Chakravarthy, A. B., et al. (2013). *Four jointed box 1* promotes angiogenesis and is associated with poor patient survival in colorectal carcinoma. *PLoS One*, *8*, e69660.

Ambegaonkar, A. A., & Irvine, K. D. (2015). Coordination of planar cell polarity pathways through Spiny-legs. *eLife*, *4*, e09946.

Ambegaonkar, A. A., Pan, G., Mani, M., Feng, Y., & Irvine, K. D. (2012). Propagation of Dachsous-Fat planar cell polarity. *Current Biology*, *22*, 1302–1308.

Ashery-Padan, R., Alvarez-Bolado, G., Klamt, B., Gessler, M., & Gruss, P. (1999). *Fjx1*, the murine homologue of the *Drosophila four-jointed* gene, codes for a putative secreted protein expressed in restricted domains of the developing and adult brain. *Mechanisms of Development*, *80*, 213–217.

Ayukawa, T., Akiyama, M., Mummery-Widmer, J. L., Stoeger, T., Sasaki, J., Knoblich, J. A., et al. (2014). Dachsous-dependent asymmetric localization of spiny-legs determines planar cell polarity orientation in *Drosophila*. *Cell Reports*, *8*, 610–621.

Badouel, C., Zander, M. A., Liscio, N., Bagherie-Lachidan, M., Sopko, R., Coyaud, E., et al. (2015). Fat1 interacts with Fat4 to regulate neural tube closure, neural progenitor proliferation and apical constriction during mouse brain development. *Development*, *142*, 2781–2791.

Bagherie-Lachidan, M., Reginensi, A., Pan, Q., Zaveri, H. P., Scott, D. A., Blencowe, B. J., et al. (2015). Stromal *Fat4* acts non-autonomously with *Dchs1/2* to restrict the nephron progenitor pool. *Development*, *142*, 2564–2573.

Bando, T., Mito, T., Maeda, Y., Nakamura, T., Ito, F., Watanabe, T., et al. (2009). Regulation of leg size and shape by the Dachsous/Fat signalling pathway during regeneration. *Development, 136*, 2235–2245.
Bando, T., Mito, T., Nakamura, T., Ohuchi, H., & Noji, S. (2011). Regulation of leg size and shape: Involvement of the Dachsous-fat signaling pathway. *Developmental Dynamics, 240*, 1028–1041.
Barylko, B., Gerber, S. H., Binns, D. D., Grichine, N., Khvotchev, M., Südhof, T. C., et al. (2001). A novel family of phosphatidylinositol 4-kinases conserved from yeast to humans. *The Journal of Biological Chemistry, 276*, 7705–7708.
Bellahcène, A., Castronovo, V., Ogbureke, K. U., Fisher, L. W., & Fedarko, N. S. (2008). Small integrin-binding ligand N-linked glycoproteins (SIBLINGs): Multifunctional proteins in cancer. *Nature Reviews. Cancer, 8*, 212–226.
Bennett, F. C., & Harvey, K. F. (2006). Fat cadherin modulates organ size in *Drosophila* via the Salvador/Warts/Hippo signaling pathway. *Current Biology, 16*, 2101–2110.
Bennett, E. P., Hassan, H., & Clausen, H. (1996). cDNA cloning and expression of a novel human UDP-N-acetyl-alpha-D-galactosamine. Polypeptide N-acetylgalactosaminyltransferase, GalNAc-T3. *The Journal of Biological Chemistry, 271*, 17006–17012.
Bordoli, M. R., Yum, J., Breitkopf, S. B., Thon, J. N., Italiano, J. E., Jr., Xiao, J., et al. (2014). A secreted tyrosine kinase acts in the extracellular environment. *Cell, 158*, 1033–1044.
Bosch, J. A., Sumabat, T. M., Hafezi, Y., Pellock, B. J., Gandhi, K. D., & Hariharan, I. K. (2014). The *Drosophila* F-box protein Fbxl7 binds to the protocadherin Fat and regulates Dachs localization and Hippo signaling. *eLife, 3*, e03383.
Bosveld, F., Bonnet, I., Guirao, B., Tlili, S., Wang, Z., Petitalot, A., et al. (2012). Mechanical control of morphogenesis by Fat/Dachsous/Four-jointed planar cell polarity pathway. *Science, 336*, 724–727.
Briscoe, J., & Small, S. (2015). Morphogen rules: Design principles of gradient-mediated embryo patterning. *Development, 142*, 3996–4009.
Brittle, A. L., Repiso, A., Casal, J., Lawrence, P. A., & Strutt, D. (2010). Four-jointed modulates growth and planar polarity by reducing the affinity of dachsous for fat. *Current Biology, 20*, 803–810.
Brittle, A., Thomas, C., & Strutt, D. (2012). Planar polarity specification through asymmetric subcellular localization of Fat and Dachsous. *Current Biology, 22*, 907–914.
Brodsky, M. H., & Steller, H. (1996). Positional information along the dorsal-ventral axis of the *Drosophila* eye: Graded expression of the *four-jointed* gene. *Developmental Biology, 173*, 428–446.
Bryant, P. J., Huettner, B., Held, L. I., Jr., Ryerse, J., & Szidonya, J. (1988). Mutations at the *fat* locus interfere with cell proliferation control and epithelial morphogenesis in *Drosophila*. *Developmental Biology, 129*, 541–554.
Buckanovich, R. J., Sasaroli, D., O'Brien-Jenkins, A., Botbyl, J., Hammond, R., Katsaros, D., et al. (2007). Tumor vascular proteins as biomarkers in ovarian cancer. *Journal of Clinical Oncology, 25*, 852–861.
Buckles, G. R., Rauskolb, C., Villano, J. L., & Katz, F. N. (2001). *four-jointed* interacts with *dachs*, *abelson* and *enabled* and feeds back onto the *Notch* pathway to affect growth and segmentation in the *Drosophila* leg. *Development, 128*, 3533–3542.
Cappello, S., Gray, M. J., Badouel, C., Lange, S., Einsiedler, M., Srour, M., et al. (2013). Mutations in genes encoding the cadherin receptor-ligand pair DCHS1 and FAT4 disrupt cerebral cortical development. *Nature Genetics, 45*, 1300–1308.
Casal, J., Lawrence, P. A., & Struhl, G. (2006). Two separate molecular systems, Dachsous/Fat and Starry night/Frizzled, act independently to confer planar cell polarity. *Development, 133*, 4561–4572.

Casal, J., Struhl, G., & Lawrence, P. A. (2002). Developmental compartments and planar polarity in *Drosophila*. *Current Biology, 12*, 1189–1198.

Chai, S. J., Yap, Y. Y., Foo, Y. C., Yap, L. F., Ponniah, S., Teo, S. H., et al. (2015). Identification of four-jointed box 1 (FJX1)-specific peptides for immunotherapy of nasopharyngeal carcinoma. *PLoS One, 10*, e0130464.

Cheng, F., Liu, C., Lin, C. C., Zhao, J., Jia, P., Li, W. H., et al. (2015). A gene gravity model for the evolution of cancer genomes: A study of 3,000 cancer genomes across 9 cancer types. *PLoS Computational Biology, 11*, e1004497.

Cho, E., Feng, Y., Rauskolb, C., Maitra, S., Fehon, R., & Irvine, K. D. (2006). Delineation of a Fat tumor suppressor pathway. *Nature Genetics, 38*, 1142–1150.

Cho, E., & Irvine, K. D. (2004). Action of *fat, four-jointed, dachsous* and *dachs* in distal-to-proximal wing signaling. *Development, 131*, 4489–4500.

Cho, S. H., Seymen, F., Lee, K. E., Lee, S. K., Kweon, Y. S., Kim, K. J., et al. (2012). Novel FAM20A mutations in hypoplastic amelogenesis imperfecta. *Human Mutation, 33*, 91–94.

Clark, H. F., Brentrup, D., Schneitz, K., Bieber, A., Goodman, C., & Noll, M. (1995). Dachsous encodes a member of the cadherin superfamily that controls imaginal disc morphogenesis in *Drosophila*. *Genes & Development, 9*, 1530–1542.

Correia, F. F., D'Onofrio, A., Rejtar, T., Li, L., Karger, B. L., Makarova, K., et al. (2006). Kinase activity of overexpressed HipA is required for growth arrest and multidrug tolerance in *Escherichia coli*. *Journal of Bacteriology, 188*, 8360–8367.

Cui, J., Xiao, J., Tagliabracci, V. S., Wen, J., Rahdar, M., & Dixon, J. E. (2015). A secretory kinase complex regulates extracellular protein phosphorylation. *eLife, 4*, e06120.

Das, A., Tanigawa, S., Karner, C. M., Xin, M., Lum, L., Chen, C., et al. (2013). Stromal-epithelial crosstalk regulates kidney progenitor cell differentiation. *Nature Cell Biology, 15*, 1035–1044.

Deans, M. R., Krol, A., Abraira, V. E., Copley, C. O., Tucker, A. F., & Goodrich, L. V. (2011). Control of neuronal morphology by the atypical cadherin Fat3. *Neuron, 71*, 820–832.

Degoutin, J. L., Milton, C. C., Yu, E., Tipping, M., Bosveld, F., Yang, L., et al. (2013). Riquiqui and minibrain are regulators of the hippo pathway downstream of Dachsous. *Nature Cell Biology, 15*, 1176–1185.

Di Costanzo, S., Balasubramanian, A., Pond, H. L., Rozkalne, A., Pantaleoni, C., Saredi, S., et al. (2014). POMK mutations disrupt muscle development leading to a spectrum of neuromuscular presentations. *Human Molecular Genetics, 23*, 5781–5792.

Dong, J., Feldmann, G., Huang, J., Wu, S., Zhang, N., Comerford, S. A., et al. (2007). Elucidation of a universal size-control mechanism in *Drosophila* and mammals. *Cell, 130*, 1120–1133.

Donoughe, S., & DiNardo, S. (2011). *dachsous* and *frizzled* contribute separately to planar polarity in the *Drosophila* ventral epidermis. *Development, 138*, 2751–2759.

Durst, R., Sauls, K., Peal, D. S., deVlaming, A., Toomer, K., Leyne, M., et al. (2015). Mutations in DCHS1 cause mitral valve prolapse. *Nature, 525*, 109–113.

Eames, B. F., Yan, Y. L., Swartz, M. E., Levic, D. S., Knapik, E. W., Postlethwait, J. H., et al. (2011). Mutations in *fam20b* and *xylt1* reveal that cartilage matrix controls timing of endochondral ossification by inhibiting chondrocyte maturation. *PLoS Genetics, 7*, e1002246.

Fanto, M., Clayton, L., Meredith, J., Hardiman, K., Charroux, B., Kerridge, S., et al. (2003). The tumor-suppressor and cell adhesion molecule Fat controls planar polarity via physical interactions with Atrophin, a transcriptional co-repressor. *Development, 130*, 763–774.

Feng, Y., & Irvine, K. D. (2009). Processing and phosphorylation of the Fat receptor. *Proceedings of the National Academy of Sciences of the United States of America, 106*, 11989–11994.

Furukawa, T., Sakamoto, H., Takeuchi, S., Ameri, M., Kuboki, Y., Yamamoto, T., et al. (2015). Whole exome sequencing reveals recurrent mutations in BRCA2 and FAT genes in acinar cell carcinomas of the pancreas. *Scientific Reports, 5*, 8829.

Gao, Y. B., Chen, Z. L., Li, J. G., Hu, X. D., Shi, X. J., Sun, Z. M., et al. (2014). Genetic landscape of esophageal squamous cell carcinoma. *Nature Genetics, 46*, 1097–1102.

Garg, M., Nagata, Y., Kanojia, D., Mayakonda, A., Yoshida, K., Haridas Keloth, S., et al. (2015). Profiling of somatic mutations in acute myeloid leukemia with FLT3-ITD at diagnosis and relapse. *Blood, 126*, 2491–2501.

Glantschnig, H., Rodan, G. A., & Reszka, A. A. (2002). Mapping of MST1 kinase sites of phosphorylation. Activation and autophosphorylation. *The Journal of Biological Chemistry, 277*, 42987–42996.

González-Morales, N., Géminard, C., Lebreton, G., Cerezo, D., Coutelis, J. B., & Noselli, S. (2015). The atypical cadherin dachsous controls left-right asymmetry in *Drosophila*. *Developmental Cell, 33*, 675–689.

Goulev, Y., Fauny, J. D., Gonzalez-Marti, B., Flagiello, D., Silber, J., & Zider, A. (2008). SCALLOPED interacts with YORKIE, the nuclear effector of the *hippo* tumor-suppressor pathway in *Drosophila*. *Current Biology, 18*, 435–441.

Haines, N., & Irvine, K. D. (2003). Glycosylation regulates Notch signalling. *Nature Reviews. Molecular Cell Biology, 4*, 786–797.

Hale, R., Brittle, A. L., Fisher, K. H., Monk, N. A., & Strutt, D. (2015). Cellular interpretation of the long-range gradient of Four-jointed activity in the *Drosophila* wing. *eLife, 4*, e05789.

Hammarsten, O. (1883). Zur Frage ob Caseïn ein einheitlicher Stoff sei. *Hoppe-Seyler's Zeitschrift Fur Physiologische Chemie, 7*, 227–273.

Hao, J., Narayanan, K., Muni, T., Ramachandran, A., & George, A. (2007). Dentin matrix protein 4, a novel secretory calcium-binding protein that modulates odontoblast differentiation. *The Journal of Biological Chemistry, 282*, 15357–15365.

Held, L. I., Jr., Duarte, C. M., & Derakhshanian, K. (1986). Extra joints and misoriented bristles on *Drosophila* legs. *Progress in Clinical and Biological Research, 217*, 293–296.

Hewitt, J. E. (2009). Abnormal glycosylation of dystroglycan in human genetic disease. *Biochimica et Biophysica Acta, 1792*, 853–861.

Huang, J., Wu, S., Barrera, J., Matthews, K., & Pan, D. (2005). The Hippo signaling pathway coordinately regulates cell proliferation and apoptosis by inactivating Yorkie, the *Drosophila* homolog of YAP. *Cell, 122*, 421–434.

Ishikawa, H. O., Takeuchi, H., Haltiwanger, R. S., & Irvine, K. D. (2008). Four-jointed is a Golgi kinase that phosphorylates a subset of cadherin domains. *Science, 321*, 401–404.

Ishikawa, H. O., Xu, A., Ogura, E., Manning, G., & Irvine, K. D. (2012). The Raine syndrome protein FAM20C is a Golgi kinase that phosphorylates bio-mineralization proteins. *PLoS One, 7*, e42988.

Jae, L. T., Raaben, M., Riemersma, M., van Beusekom, E., Blomen, V. A., Velds, A., et al. (2013). Deciphering the glycosylome of dystroglycanopathies using haploid screens for lassa virus entry. *Science, 340*, 479–483.

Jenny, A. (2010). Planar cell polarity signaling in the *Drosophila* eye. *Current Topics in Developmental Biology, 93*, 189–227.

Jolly, M. K., Rizvi, M. S., Kumar, A., & Sinha, P. (2014). Mathematical modeling of subcellular asymmetry of fat-dachsous heterodimer for generation of planar cell polarity. *PLoS One, 9*, e97641.

Kakiuchi, S., Daigo, Y., Tsunoda, T., Yano, S., Sone, S., & Nakamura, Y. (2003). Genome-wide analysis of organ-preferential metastasis of human small cell lung cancer in mice. *Molecular Cancer Research, 1*, 485–499.

Kantaputra, P. N., Kaewgahya, M., Khemaleelakul, U., Dejkhamron, P., Sutthimethakorn, S., Thongboonkerd, V., et al. (2014). Enamel-renal-gingival syndrome and FAM20A mutations. *American Journal of Medical Genetics. Part A, 164A*, 1–9.

Kinoshita, M., Era, T., Jakt, L. M., & Nishikawa, S. (2009). The novel protein kinase Vlk is essential for stromal function of mesenchymal cells. *Development, 136*, 2069–2079.

Koike, T., Izumikawa, T., Tamura, J., & Kitagawa, H. (2009). FAM20B is a kinase that phosphorylates xylose in the glycosaminoglycan-protein linkage region. *The Biochemical Journal, 421*, 157–162.

Kornev, A. P., & Taylor, S. S. (2010). Defining the conserved internal architecture of a protein kinase. *Biochimica et Biophysica Acta, 1804*, 440–444.

Kuta, A., Mao, Y., Martin, T., Ferreira de Sousa, C., Whiting, D., Zakaria, S., et al. (2016). Fat4-Dchs1 signalling controls cell proliferation in developing vertebrae. *Development, 143*, 2367–2375.

Lasa-Benito, M., Marin, O., Meggio, F., & Pinna, L. A. (1996). Golgi apparatus mammary gland casein kinase: Monitoring by a specific peptide substrate and definition of specificity determinants. *FEBS Letters, 382*, 149–152.

Lee, K. K., & Yonehara, S. (2002). Phosphorylation and dimerization regulate nucleocytoplasmic shuttling of mammalian STE20-like kinase (MST). *The Journal of Biological Chemistry, 277*, 12351–12358.

Lindsley, D. L., & Zimm, G. G. (1992). *The genome of Drosophila melanogaster* (p.215). San Diego, CA: Academic Press.

Lu, C., Bonome, T., Li, Y., Kamat, A. A., Han, L. Y., Schmandt, R., et al. (2007). Gene alterations identified by expression profiling in tumor-associated endothelial cells from invasive ovarian carcinoma. *Cancer Research, 67*, 1757–1768.

Ma, D., Yang, C. H., McNeill, H., Simon, M. A., & Axelrod, J. D. (2003). Fidelity in planar cell polarity signalling. *Nature, 421*, 543–547.

Ma, P., Yan, W., Tian, Y., Wang, J., Feng, J. Q., Qin, C., et al. (2016). Inactivation of Fam20B in joint cartilage leads to chondrosarcoma and postnatal ossification defects. *Sci Rep, 6*, 29814.

Mahoney, P. A., Weber, U., Onofrechuk, P., Biessmann, H., Bryant, P. J., & Goodman, C. S. (1991). The *fat* tumor suppressor gene in *Drosophila* encodes a novel member of the cadherin gene superfamily. *Cell, 67*, 853–868.

Mani, M., Goyal, S., Irvine, K. D., & Shraiman, B. I. (2013). Collective polarization model for gradient sensing via Dachsous-Fat intercellular signaling. *Proceedings of the National Academy of Sciences of the United States of America, 110*, 20420–20425.

Mao, Y., Francis-West, P., & Irvine, K. D. (2015). Fat4/Dchs1 signaling between stromal and cap mesenchyme cells influences nephrogenesis and ureteric bud branching. *Development, 142*, 2574–2585.

Mao, Y., Kucuk, B., & Irvine, K. D. (2009). *Drosophila lowfat*, a novel modulator of Fat signaling. *Development, 136*, 3223–3233.

Mao, Y., Kuta, A., Crespo-Enriquez, I., Whiting, D., Martin, T., Mulvaney, J., et al. (2016). Dchs1-Fat4 regulation of polarized cell behaviours during skeletal morphogenesis. *Nature Communications, 7*, 11469.

Mao, Y., Mulvaney, J., Zakaria, S., Yu, T., Morgan, K. M., Allen, S., et al. (2011). Characterization of a *Dchs1* mutant mouse reveals requirements for Dchs1-Fat4 signaling during mammalian development. *Development, 138*, 947–957.

Mao, Y., Rauskolb, C., Cho, E., Hu, W. L., Hayter, H., Minihan, G., et al. (2006). Dachs: An unconventional myosin that functions downstream of Fat to regulate growth, affinity and gene expression in *Drosophila*. *Development, 133*, 2539–2551.

Matakatsu, H., & Blair, S. S. (2004). Interactions between Fat and Dachsous and the regulation of planar cell polarity in the *Drosophila* wing. *Development, 131*, 3785–3794.

Matakatsu, H., & Blair, S. S. (2008). The DHHC palmitoyltransferase approximated regulates Fat signaling and Dachs localization and activity. *Current Biology, 18*, 1390–1395.

Matakatsu, H., & Blair, S. S. (2012). Separating planar cell polarity and Hippo pathway activities of the protocadherins Fat and Dachsous. *Development, 139*, 1498–1508.

Meignin, C., Alvarez-Garcia, I., Davis, I., & Palacios, I. M. (2007). The Salvador-Warts-Hippo pathway is required for epithelial proliferation and axis specification in *Drosophila*. *Current Biology*, *17*, 1871–1878.

Mohr, O. L. (1923). Modifications of the sex-ratio through a sex-linked semi-lethal in *Drosophila melanogaster* (besides notes on an autosomal section deficiency). In H. Iltis (Ed.), *Studia Mendeliana* (pp. 266–287). Brunn, Czechoslovakia: Apud Typos.

Murphy, A. J., Pierce, J., de Caestecker, C., Libes, J., Neblett, D., de Caestecker, M., et al. (2014). Aberrant activation, nuclear localization, and phosphorylation of Yes-associated protein-1 in the embryonic kidney and Wilms tumor. *Pediatric Blood & Cancer*, *61*, 198–205.

Nadanaka, S., Zhou, S., Kagiyama, S., Shoji, N., Sugahara, K., Sugihara, K., et al. (2013). EXTL2, a member of the EXT family of tumor suppressors, controls glycosaminoglycan biosynthesis in a xylose kinase-dependent manner. *The Journal of Biological Chemistry*, *288*, 9321–9333.

Nalbant, D., Youn, H., Nalbant, S. I., Sharma, S., Cobos, E., Beale, E. G., et al. (2005). FAM20: An evolutionarily conserved family of secreted proteins expressed in hematopoietic cells. *BMC Genomics*, *6*, 11.

Nicolay, B. N., & Frolov, M. V. (2008). Context-dependent requirement for dE2F during oncogenic proliferation. *PLoS Genetics*, *4*, e1000205.

Nolo, R., Morrison, C. M., Tao, C., Zhang, X., & Halder, G. (2006). The *bantam* microRNA is a target of the hippo tumor-suppressor pathway. *Current Biology*, *16*, 1895–1904.

Oh, H., & Irvine, K. D. (2010). Yorkie: The final destination of Hippo signaling. *Trends in Cell Biology*, *20*, 410–417.

Oh, H., & Irvine, K. D. (2011). Cooperative regulation of growth by Yorkie and Mad through *bantam*. *Developmental Cell*, *20*, 109–122.

Ohyama, Y., Lin, J. H., Govitvattana, N., Lin, I. P., Venkitapathi, S., Alamoudi, A., et al. (2016). FAM20A binds to and regulates FAM20C localization. *Scientific Reports*, *6*, 27784.

O'Sullivan, J., Bitu, C. C., Daly, S. B., Urquhart, J. E., Barron, M. J., Bhaskar, S. S., et al. (2011). Whole-exome sequencing identifies *FAM20A* mutations as a cause of amelogenesis imperfecta and gingival hyperplasia syndrome. *American Journal of Human Genetics*, *88*, 616–620.

Parry, M., Rose-Zerilli, M. J., Gibson, J., Ennis, S., Walewska, R., Forster, J., et al. (2013). Whole exome sequencing identifies novel recurrently mutated genes in patients with splenic marginal zone lymphoma. *PLoS One*, *8*, e83244.

Patel, S. D., Ciatto, C., Chen, C. P., Bahna, F., Rajebhosale, M., Arkus, N., et al. (2006). Type II cadherin ectodomain structures: Implications for classical cadherin specificity. *Cell*, *124*, 1255–1268.

Peng, H. W., Slattery, M., & Mann, R. S. (2009). Transcription factor choice in the Hippo signaling pathway: *homothorax* and *yorkie* regulation of the microRNA *bantam* in the progenitor domain of the *Drosophila* eye imaginal disc. *Genes & Development*, *23*, 2307–2319.

Polesello, C., & Tapon, N. (2007). Salvador-Warts-Hippo signaling promotes *Drosophila* posterior follicle cell maturation downstream of Notch. *Current Biology*, *17*, 1864–1870.

Probst, B., Rock, R., Gessler, M., Vortkamp, A., & Püschel, A. W. (2007). The rodent Four-jointed ortholog Fjx1 regulates dendrite extension. *Developmental Biology*, *312*, 461–470.

Qin, C., Baba, O., & Butler, W. T. (2004). Post-translational modifications of sibling proteins and their roles in osteogenesis and dentinogenesis. *Critical Reviews in Oral Biology and Medicine*, *15*, 126–136.

Raine, J., Winter, R. M., Davey, A., & Tucker, S. M. (1989). Unknown syndrome: Microcephaly, hypoplastic nose, exophthalmos, gum hyperplasia, cleft palate, low set ears, and osteosclerosis. *Journal of Medical Genetics*, *26*, 786–788.

Rauskolb, C., Pan, G., Reddy, B. V., Oh, H., & Irvine, K. D. (2011). Zyxin links fat signaling to the hippo pathway. *PLoS Biology, 9,* e1000624.
Rawls, A. S., Guinto, J. B., & Wolff, T. (2002). The cadherins fat and dachsous regulate dorsal/ventral signaling in the *Drosophila* eye. *Current Biology, 12,* 1021–1026.
Reddy, B. V., Rauskolb, C., & Irvine, K. D. (2010). Influence of Fat-Hippo and Notch signaling on the proliferation and differentiation of *Drosophila* optic neuroepithelia. *Development, 137,* 2397–2408.
Rejjal, A. (1998). Raine syndrome. *American Journal of Medical Genetics, 78,* 382–385.
Rock, R., Heinrich, A. C., Schumacher, N., & Gessler, M. (2005). Fjx1: A Notch-inducible secreted ligand with specific binding sites in developing mouse embryos and adult brain. *Developmental Dynamics, 234,* 602–612.
Rock, R., Schrauth, S., & Gessler, M. (2005). Expression of mouse *dchs1, fjx1,* and *fat-j* suggests conservation of the planar cell polarity pathway identified in *Drosophila. Developmental Dynamics, 234,* 747–755.
Rodrigues-Campos, M., & Thompson, B. J. (2014). The ubiquitin ligase FbxL7 regulates the Dachsous-Fat-Dachs system in *Drosophila. Development, 141,* 4098–4103.
Rogulja, D., & Irvine, K. D. (2005). Regulation of cell proliferation by a morphogen gradient. *Cell, 123,* 449–461.
Rogulja, D., Rauskolb, C., & Irvine, K. D. (2008). Morphogen control of wing growth through the Fat signaling pathway. *Developmental Cell, 15,* 309–321.
Saburi, S., Hester, I., Fischer, E., Pontoglio, M., Eremina, V., Gessler, M., et al. (2008). Loss of *Fat4* disrupts PCP signaling and oriented cell division and leads to cystic kidney disease. *Nature Genetics, 40,* 1010–1015.
Sadeqzadeh, E., de Bock, C. E., & Thorne, R. F. (2014). Sleeping giants: Emerging roles for the fat cadherins in health and disease. *Medicinal Research Reviews, 34,* 190–221.
Seifert, J. R., & Mlodzik, M. (2007). Frizzled/PCP signalling: A conserved mechanism regulating cell polarity and directed motility. *Nature Reviews. Genetics, 8,* 126–138.
Sharma, P., & McNeill, H. (2013). Regulation of long-range planar cell polarity by Fat-Dachsous signaling. *Development, 140,* 3869–3881.
Shimizu, T., Ho, L. L., & Lai, Z. C. (2008). The *mob* as tumor suppressor gene is essential for early development and regulates tissue growth in *Drosophila. Genetics, 178,* 957–965.
Silva, E., Tsatskis, Y., Gardano, L., Tapon, N., & McNeill, H. (2006). The tumor-suppressor gene *fat* controls tissue growth upstream of *expanded* in the hippo signaling pathway. *Current Biology, 16,* 2081–2089.
Simon, M. A. (2004). Planar cell polarity in the *Drosophila* eye is directed by graded Four-r-jointed and Dachsous expression. *Development, 131,* 6175–6184.
Simon, M. A., Xu, A., Ishikawa, H. O., & Irvine, K. D. (2010). Modulation of Fat:Dachsous binding by the cadherin domain kinase Four-jointed. *Current Biology, 20,* 811–817.
Sopko, R., Silva, E., Clayton, L., Gardano, L., Barrios-Rodiles, M., Wrana, J., et al. (2009). Phosphorylation of the tumor suppressor Fat is regulated by its ligand Dachsous and the kinase Discs overgrown. *Current Biology, 19,* 1112–1117.
Sreelatha, A., Kinch, L. N., & Tagliabracci, V. S. (2015). The secretory pathway kinases. *Biochimica et Biophysica Acta, 1854,* 1687–1693.
Staley, B. K., & Irvine, K. D. (2012). Hippo signaling in *Drosophila*: Recent advances and insights. *Developmental Dynamics, 241,* 3–15.
Strutt, H., Mundy, J., Hofstra, K., & Strutt, D. (2004). Cleavage and secretion is not required for Four-jointed function in *Drosophila* patterning. *Development, 131,* 881–890.
Strutt, H., & Strutt, D. (2002). Nonautonomous planar polarity patterning in *Drosophila*: *dishevelled*-independent functions of *frizzled. Developmental Cell, 3,* 851–863.
Strutt, H., & Strutt, D. (2009). Asymmetric localisation of planar polarity proteins: Mechanisms and consequences. *Seminars in Cell & Developmental Biology, 20,* 957–963.

Tagliabracci, V. S., Engel, J. L., Wen, J., Wiley, S. E., Worby, C. A., Kinch, L. N., et al. (2012). Secreted kinase phosphorylates extracellular proteins that regulate biomineralization. *Science*, *336*, 1150–1153.
Tagliabracci, V. S., Engel, J. L., Wiley, S. E., Xiao, J., Gonzalez, D. J., Nidumanda Appaiah, H., et al. (2014). Dynamic regulation of FGF23 by Fam20C phosphorylation, GalNAc-T3 glycosylation, and furin proteolysis. *Proceedings of the National Academy of Sciences of the United States of America*, *111*, 5520–5525.
Tagliabracci, V. S., Wiley, S. E., Guo, X., Kinch, L. N., Durrant, E., Wen, J., et al. (2015). A single kinase generates the majority of the secreted phosphoproteome. *Cell*, *161*, 1619–1632.
Tapon, N., Harvey, K. F., Bell, D. W., Wahrer, D. C. R., Schiripo, T. A., Haber, D. A., et al. (2002). Salvador promotes both cell cycle exit and apoptosis in Drosophila and is mutated in human cancer cell lines. *Cell*, *110*, 467–478.
Tenedini, E., Bernardis, I., Artusi, V., Artuso, L., Roncaglia, E., Guglielmelli, P., et al. (2014). Targeted cancer exome sequencing reveals recurrent mutations in myeloproliferative neoplasms. *Leukemia*, *28*, 1052–1059.
Thompson, B. J., & Cohen, S. M. (2006). The hippo pathway regulates the *bantam* microRNA to control cell proliferation and apoptosis in Drosophila. *Cell*, *126*, 767–774.
Tokunaga, C., & Gerhart, J. C. (1976). The effect of growth and joint formation on bristle pattern in D. melanogaster. *The Journal of Experimental Zoology*, *198*, 79–96.
Tsukasaki, Y., Miyazaki, N., Matsumoto, A., Nagae, S., Yonemura, S., Tanoue, T., et al. (2014). Giant cadherins Fat and Dachsous self-bend to organize properly spaced intercellular junctions. *Proceedings of the National Academy of Sciences of the United States of America*, *111*, 16011–16016.
Tyler, D. M., & Baker, N. E. (2007). Expanded and *fat* regulate growth and differentiation in the Drosophila eye through multiple signaling pathways. *Developmental Biology*, *305*, 187–201.
Villano, J. L., & Katz, F. N. (1995). Four-jointed is required for intermediate growth in the proximal-distal axis in Drosophila. *Development*, *121*, 2767–2777.
Vogel, P., Hansen, G. M., Read, R. W., Vance, R. B., Thiel, M., Liu, J., et al. (2012). Amelogenesis imperfecta and other biomineralization defects in *Fam20a* and *Fam20c* null mice. *Veterinary Pathology*, *49*, 998–1017.
Waddington, C. H. (1940). The genetic control of wing development in Drosophila. *Journal of Genetics*, *41*, 75–139.
Waddington, C. H. (1943). The development of some 'leg genes' in Drosophila. *Journal of Genetics*, *45*, 39–43.
Wadhwa, R., Song, S., Lee, J. S., Yao, Y., Wei, Q., & Ajani, J. A. (2013). Gastric cancer-molecular and clinical dimensions. *Nature Reviews. Clinical Oncology*, *10*, 643–655.
Wang, S. K., Aref, P., Hu, Y., Milkovich, R. N., Simmer, J. P., El-Khateeb, M., et al. (2013). *FAM20A* mutations can cause enamel-renal syndrome (ERS). *PLoS Genetics*, *9*, e1003302.
Wang, S. K., Reid, B. M., Dugan, S. L., Roggenbuck, J. A., Read, L., Aref, P., et al. (2014). *FAM20A* mutations associated with enamel renal syndrome. *Journal of Dental Research*, *93*, 42–48.
Wang, X., Wang, S., Li, C., Gao, T., Liu, T., Rangiani, A., et al. (2012). Inactivation of a novel FGF23 regulator, FAM20C, leads to hypophosphatemic rickets in mice. *PLoS Genetics*, *8*, e1002708.
Wen, J., Xiao, J., Rahdar, M., Choudhury, B. P., Cui, J., Taylor, G. S., et al. (2014). Xylose phosphorylation functions as a molecular switch to regulate proteoglycan biosynthesis. *Proceedings of the National Academy of Sciences of the United States of America*, *111*, 15723–15728.
Wei, X., Shimizu, T., & Lai, Z. C. (2007). Mob as tumor suppressor is activated by Hippo kinase for growth inhibition in Drosophila. *EMBO J*, *26*, 1772–1781.

Wiggins, C. A., & Munro, S. (1998). Activity of the yeast MNN1 alpha-1,3-mannosyltransferase requires a motif conserved in many other families of glycosyltransferases. *Proceedings of the National Academy of Sciences of the United States of America, 95*, 7945–7950.

Willecke, M., Hamaratoglu, F., Kango-Singh, M., Udan, R., Chen, C., Tao, C., et al. (2006). The fat cadherin acts through the hippo tumor-suppressor pathway to regulate tissue size. *Current Biology, 16*, 2090–2100.

Willecke, M., Hamaratoglu, F., Sansores-Garcia, L., Tao, C., & Halder, G. (2008). Boundaries of Dachsous Cadherin activity modulate the Hippo signaling pathway to induce cell proliferation. *Proceedings of the National Academy of Sciences of the United States of America, 105*, 14897–14902.

Williams, A. F., Barclay, A. N., Clark, S. J., Paterson, D. J., & Willis, A. C. (1987). Similarities in sequences and cellular expression between rat CD2 and CD4 antigens. *The Journal of Experimental Medicine, 165*, 368–380.

Wu, S., Huang, J., Dong, J., & Pan, D. (2003). *hippo* encodes a Ste-20 family protein kinase that restricts cell proliferation and promotes apoptosis in conjunction with *salvador* and *warts*. *Cell, 114*, 445–456.

Wu, S., Liu, Y., Zheng, Y., Dong, J., & Pan, D. (2008). The TEAD/TEF family protein scalloped mediates transcriptional output of the hippo growth-regulatory pathway. *Developmental Cell, 14*, 388–398.

Xiao, J., Tagliabracci, V. S., Wen, J., Kim, S. A., & Dixon, J. E. (2013). Crystal structure of the Golgi casein kinase. *Proceedings of the National Academy of Sciences of the United States of America, 110*, 10574–10579.

Yang, C. H., Axelrod, J. D., & Simon, M. A. (2002). Regulation of Frizzled by fat-like cadherins during planar polarity signaling in the *Drosophila* compound eye. *Cell, 108*, 675–688.

Yoshida, H., Bando, T., Mito, T., Ohuchi, H., & Noji, S. (2014). An extended steepness model for leg-size determination based on Dachsous/Fat trans-dimer system. *Scientific Reports, 4*, 4335.

Yoshida-Moriguchi, T., Willer, T., Anderson, M. E., Venzke, D., Whyte, T., Muntoni, F., et al. (2013). SGK196 is a glycosylation-specific O-mannose kinase required for dystroglycan function. *Science, 341*, 896–899.

Yoshida-Moriguchi, T., Yu, L., Stalnaker, S. H., Davis, S., Kunz, S., Madson, M., et al. (2010). O-mannosyl phosphorylation of alpha-dystroglycan is required for laminin binding. *Science, 327*, 88–92.

Yu, F. X., & Guan, K. L. (2013). The Hippo pathway: Regulators and regulations. *Genes & Development, 27*, 355–371.

Zakaria, S., Mao, Y., Kuta, A., Ferreira de Sousa, C., Gaufo, G. O., McNeill, H., et al. (2014). Regulation of neuronal migration by Dchs1-Fat4 planar cell polarity. *Current Biology, 24*, 1620–1627.

Zeidler, M. P., Perrimon, N., & Strutt, D. I. (1999). The *four-jointed* gene is required in the *Drosophila* eye for ommatidial polarity specification. *Current Biology, 9*, 1363–1372.

Zeidler, M. P., Perrimon, N., & Strutt, D. I. (2000). Multiple roles for *four-jointed* in planar polarity and limb patterning. *Developmental Biology, 228*, 181–196.

Zhang, X., Milton, C. C., Poon, C. L., Hong, W., & Harvey, K. F. (2011). Wbp2 cooperates with Yorkie to drive tissue growth downstream of the Salvador-Warts-Hippo pathway. *Cell Death and Differentiation, 18*, 1346–1355.

Zhang, L., Ren, F., Zhang, Q., Chen, Y., Wang, B., & Jiang, J. (2008). The TEAD/TEF family of transcription factor scalloped mediates Hippo signaling in organ size control. *Developmental Cell, 14*, 377–387.

CHAPTER SIX

The Hippo Pathway: A Master Regulatory Network Important in Development and Dysregulated in Disease

Cathie M. Pfleger[*,†,1]

[*]The Icahn School of Medicine at Mount Sinai, New York, NY, United States
[†]The Graduate School of Biomedical Sciences, The Icahn School of Medicine at Mount Sinai, New York, NY, United States
[1]Corresponding author: e-mail address: cathie.pfleger@mssm.edu

Contents

1. Introduction	182
2. The Highly Conserved Core Kinase Cassette	183
3. The Hippo Pathway Restricts Mass Accumulation and Proliferation	186
4. The Hippo Pathway Regulates Apoptosis	187
5. The Hippo Pathway Restricts Organ Size and Maintains Organ Homeostasis	187
6. Molecular Mechanisms and Additional Roles of the Hippo Pathway: Posttranslational Targets and Transcriptional Outputs	188
6.1 Wts/Lats Inhibits the Yki/YAP Oncogene to Regulate Proliferation, Growth, Apoptosis, Stemness and Differentiation, and Organ Size	189
6.2 The Hippo Pathway Core Kinase Cassette Phosphorylates DIAP1 to Promote Its Degradation	193
6.3 The Hippo Pathway Regulates the Actin Cytoskeleton	194
6.4 The Hippo Pathway Regulates Asymmetric Cell Division and Mitotic Spindle Orientation by Targeting Canoe/Afadin, Bazooka, and Mud	194
6.5 The Hippo Pathway Regulates Proliferation, the Mitotic Cyclins, and Organ Size by Targeting Rae1	195
6.6 Yki/YAP/TAZ-Independent Hippo Pathway-Regulated Processes With Unresolved Mechanisms	195
7. Upstream Signals and Regulators and Mechanisms of Pathway Homeostasis	197
7.1 FERM Domain Proteins Merlin and Expanded Regulate the Hippo Pathway	198
7.2 AMOT Family Members Regulate Lats1/2 and YAP/TAZ	201
7.3 The Actin, Spectrin, and Microtubule Cytoskeletal Networks Regulate the Hippo Pathway	203
7.4 Microtubule Regulator Rae1 Feeds Back to Control Mer, Hpo, and Wts	204
7.5 Myc Inhibits Yki Activity	205
7.6 Cell Adhesion and Cell Contact Inhibition Regulate the Hippo Pathway	205

7.7 Oxidative Stress and MST1 Engage a Positive Feedback Circuit   206
 7.8 A Bistable Feedback Loop Between Wts and Melted in *Drosophila*
     Photoreceptors   206
8. Dysregulation of the Hippo Pathway in Disease   207
 8.1 Eye Health and Disease   207
 8.2 Cardiovascular Health and Heart Disease   208
 8.3 Neuronal Disease   208
 8.4 Cancer and Chemotherapeutic Resistance   209
9. Conclusions and Open Questions   210
Acknowledgments   212
References   212

Abstract

The Hippo Pathway is a master regulatory network that regulates proliferation, cell growth, stemness, differentiation, and cell death. Coordination of these processes by the Hippo Pathway throughout development and in mature organisms in response to diverse external and internal cues plays a role in morphogenesis, in controlling organ size, and in maintaining organ homeostasis. Given the importance of these processes, the Hippo Pathway also plays an important role in organismal health and has been implicated in a variety of diseases including eye disease, cardiovascular disease, neurodegeneration, and cancer. This review will focus on *Drosophila* reports that identified the core components of the Hippo Pathway revealing specific downstream biological outputs of this complicated network. A brief description of mammalian reports will complement review of the *Drosophila* studies. This review will also survey upstream regulation of the core components with a focus on feedback mechanisms.

1. INTRODUCTION

Multicellular eukaryotes develop from a single cell to a patterned, functioning, reproductive adult through a series of important events including cycles of proliferation, cell fate decisions, morphogenesis, and programmed waves of cell death. During these crucial developmental processes, organisms must respond to a variety of stressors (both intracellular and external) at the cellular level and also at the tissue level, they must balance pools of stem cells with differentiated counterparts, and they must monitor the size of individual organs and the overall organisms to stop growth once a predetermined size has been reached. Mature organisms then maintain adult health by continuing surveillance and maintenance of these processes. The Hippo Pathway (also referred to as the Salvador–Warts–Hippo Pathway, the Hippo/Yap Pathway, and other names) plays a role in each of these processes. Our understanding of the Hippo Pathway is still in early stages compared to many of the classical developmental signaling pathways, having been revealed as a signaling network by genetic screens and subsequent genetic and biochemical studies as recently as 2002–03 (Harvey, Pfleger, & Hariharan, 2003; Jia, Zhang, Wang, Trinko, & Jiang, 2003; Kango-Singh et al., 2002; Pantalacci, Tapon, & Léopold, 2003; Tapon et al., 2002; Udan, Kango-Singh, Nolo, Tao, & Halder, 2003; Wu, Huang, Dong, & Pan, 2003). As a result, there are many exciting open questions and areas of ongoing controversy. Despite the youth of this field, the depth and breadth of research are already far too extensive for a single review. This review (1) will focus on the core components of the Hippo Pathway and both widely accepted and emerging roles of the pathway defined by regulation of multiple downstream targets, (2) will briefly survey modulation of the core cassette by upstream factors and contexts with an emphasis on feedback mechanisms, and (3) will review implications of the pathway in disease and open questions and existing controversies relevant to fundamental biology.

2. THE HIGHLY CONSERVED CORE KINASE CASSETTE

Core components of the Hippo Pathway had been described individually in reports from multiple systems prior to our understanding that these components physically associated to form a core cassette in the same signaling network (Creasy & Chernoff, 1995a, 1995b; Justice, Zilian, Woods, Noll, & Bryant, 1995; Kakeya, Onose, & Osada, 1998; Taylor, Wang, & Erikson, 1996; Xu, Wang, Zhang, Stewart, & Yu, 1995). The genetic and phenotypic characterization of mutations in these genes in *Drosophila* and accompanying biochemical analyses led to an understanding that these components worked together with a broad role to restrict proliferation, cell growth, and organ size, and to promote apoptosis. In 2002, mosaic *Drosophila* screens identified mutations in novel gene *salvador*, *sav* (also called *shar-pei*) (Kango-Singh et al., 2002; Tapon et al., 2002) and in the tumor suppressor gene *warts* (*wts*)/*Lats* (Tapon et al., 2002) which had been first described as a *Drosophila* tumor suppressor in 1995 (Justice et al., 1995; Xu et al., 1995).

In the following years, multiple independent efforts identified *hippo* (*hpo*), whose mutant phenotypes phenocopied loss of *sav* or *wts* in terms of proliferation, growth, cell death, and organ size phenotypes and which interacted genetically with *sav* and *wts* and biochemically with Sav and Wts proteins (Harvey et al., 2003; Jia et al., 2003; Pantalacci et al., 2003; Udan et al., 2003; Wu et al., 2003). Upon linking Sav, Wts, and Hpo together through physical association studies and also genetic interaction studies, a widely accepted view emerged that these components represented a new signaling pathway. Shortly thereafter, identification of *mob as tumor suppressor* (*mats*) as a Wts cofactor (Lai et al., 2005) led to our current model of the core components of the pathway (core components, Table 1; simplified pathway schematic, Fig. 1).

These central components are highly conserved. *hpo* and *wts* are conserved across metazoa including orthologs in yeast where they have been characterized with roles in the mitotic exit network (Frenz, Lee, Fesquet, & Johnston, 2000; Jaspersen, Charles, Tinker-Kulberg, & Morgan, 1998; Mah, Jang, & Deshaies, 2001). The evolutionary origins and divergence of core components and downstream targets in different species have been reported and discussed elsewhere (Bossuyt et al., 2014; Ikmi et al., 2014; Sebé-Pedrós, Zheng, Ruiz-Trillo, & Pan, 2012). The signaling cascade and/or individual components and downstream targets have been described in diverse multicellular eukaryotes including *Caenorhabditis elegans* (Cai et al., 2009), planaria (Lin & Pearson, 2014), and Xenopus (Hayashi et al., 2014; Nejigane, Takahashi, Haramoto, Michiue, & Asashima, 2013).

Hpo is represented by Mst1 and Mst2 in mammals; Wts by Lats1 and Lats2 in mammals; Sav by Sav1 (also called hWW45) in mammals, Mats by MOB1 in mammals. Within the field, the Hpo/MSt, Sav, Wts/Lats, and Mats/MOB1 components are often referred to as the Hippo Pathway's "core cassette" or "core kinase cassette." Hpo/Mst is the upstream serine/threonine kinase and phosphorylates Sav (Wu et al., 2003), Mats/Mob1 (Wei, Shimizu, & Lai, 2007), and downstream serine/threonine kinase Wts/Lats (Pantalacci et al., 2003; Wu et al., 2003). Sav has multiple protein–protein interaction domains and is presumed to act as a scaffold

Table 1 Isolation and Characterization of the Core Components of the Hippo Pathway in *Drosophila*

Core Component	References
Hippo (Hpo)	Wu et al. (2003), Jia et al. (2003), Pantalacci et al. (2003), Udan et al. (2003), and Harvey et al. (2003)
Salvador (Sav)	Tapon et al. (2002) and Kango-Singh et al. (2002)
Mob as tumor suppressor (Mats)	Lai et al. (2005)
Warts (Wts)	Xu et al. (1995), Justice et al. (1995), and Tapon et al. (2002)

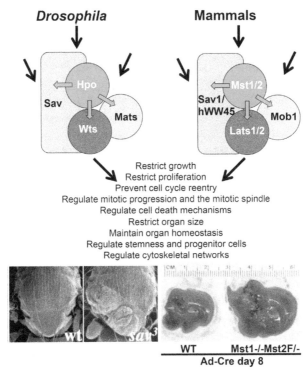

Fig. 1 The Hippo Pathway core kinase cassette. In flies (*left*) and mammals (*right*), a presumed scaffold protein (Sav in flies, Sav1/hWW45 in mammals) facilitates interaction between an upstream kinase (Hpo in flies, Mst1 and Mst2 in mammals) and a downstream kinase (Wts in flies, Lats1 and Lats2 in mammals) as described in Section 2. Hpo/MST phosphorylates and stabilizes Sav/hWW45, phosphorylates and activates Wts/Lats, and phosphorylates Wts/Lats coactivator Mats/Mob1. A variety of upstream signals (represented by *arrows*) promote activation or regulation of the core cassette or individual components in the core cassette. Upstream regulation and feedback are reviewed in Section 7 and Table 4. Active Wts/Lats phosphorylates a number of downstream targets to carry out the listed biological outputs. Specific targets and processes of the core cassette kinases Hpo/MST and Wts/Lats are reviewed in Section 6 and Tables 2 and 3. *Bottom panels* reflect overgrowth phenotypes in the fly (*left*) and mouse (*right*). In flies, wild-type clones (*left notum*) do not disrupt notum growth or architecture, but sav^3 mutant clones (*right notum*) produce dramatic, bulging overgrowth (fly images are from Tapon et al., 2002). In mice, a control liver (*left liver*) shows appropriate liver size. Inactivation of Mst2 in Mst1 null liver results in a substantially overgrown liver (liver images are from Zhou et al., 2009).

to facilitate interactions between Hpo/Mst and Wts/Lats. Normally, Wts/Lats activation is limiting and requires an activating phosphorylation from Hpo/Mst and association with the cofactor mats/Mob1 for full activity (Lai et al., 2005; Shimizu, Ho, & Lai, 2008). Once active, Wts/Lats acts as an effector kinase that phosphorylates a host of downstream targets to define the biological outputs of the pathway.

3. THE HIPPO PATHWAY RESTRICTS MASS ACCUMULATION AND PROLIFERATION

A number of roles for the Hippo Pathway were established by characterization of loss-of-function mutations in the core components *hpo/MST*, *sav*, *mats/MOB1*, and/or *wts/Lats*. These efforts revealed an important role for the pathway to restrict growth and proliferation. Developing multicellular organisms must increase their cell numbers and accumulate mass to create those cells. The Hippo Pathway plays distinct roles at different phases of the cell cycle to maintain proper control of growth and proliferation during development and then in proliferative tissues and stem compartments in mature adults. In *Drosophila*, the Hippo Pathway restricts cell cycle entry by inhibiting transcription of *cyclin E* (*cycE*) to restrain CycE protein levels. Loss of Hippo Pathway activity led to increases in CycE levels sufficient to cause cells to reenter the cell cycle and initiate proliferation (Harvey et al., 2003; Jia et al., 2003; Kango-Singh et al., 2002; Pantalacci et al., 2003; Tapon et al., 2002; Udan et al., 2003; Wu et al., 2003).

Modulating pathway components also revealed a role in mitotic exit and in restricting the protein levels of *cyclin A* (*cycA*) and *cyclin B* (*cycB*) (Jahanshahi, Hsiao, Jenny, & Pfleger, 2016; Pellock, Buff, White, & Hariharan, 2007; Shimizu et al., 2008). CycA and CycB proteins accumulated in a variety of Hippo Pathway mutants (Jahanshahi et al., 2016; Pellock et al., 2007; Shimizu et al., 2008). In addition to cyclin regulation, mitotic roles of the pathway revealed in diverse systems extend to regulation of the mitotic spindle (Bettencourt-Dias et al., 2004), regulation of chromosome segregation (Shimizu et al., 2008), and regulation of centrosomal dynamics (Hergovich, Schmitz, & Hemmings, 2006) and genome integrity (Iida et al., 2004). Loss of pathway components has resulted in multiple mitotic phenotypes including chromosome segregation defects, spindle defects, and alteration of centrosomal dynamics.

Proliferating cells must "grow" to accumulate sufficient mass to divide between daughter cells upon division. If mutant cells exhibit an increase in proliferation rate without a corresponding increase in mass accumulation, resulting daughter cells will be of smaller size than the mother cell. Surprisingly, mutations in the Hippo Pathway core components resulted in cells of wild-type size despite their decrease in doubling time indicating that the cells had not only an increase rate of proliferation but also an accompanying increased rate of cell growth (Harvey et al., 2003; Jia et al., 2003; Wu et al., 2003).

Curiously, faster growing cells eliminate their more slowly growing neighbors by a process known as "cell competition." *Drosophila* genetic screens investigating this phenomenon discovered that mutations in the core Hippo Pathway components allowed cells to evade mechanisms of cell competition (Tyler, Li, Zhuo, Pellock, & Baker, 2007).

4. THE HIPPO PATHWAY REGULATES APOPTOSIS

During development, many organisms employ waves of apoptosis at specific stages to eliminate extra cells created during earlier stages of rapid proliferation to finely pattern specific structures. In the *Drosophila* eye, a wave of apoptosis eliminates extra interommatidial cells in the pupal retina so that the final hexagonal lattice of the retina has a single layer of interommatidial cells. The Hippo Pathway plays an important role in coordinating this wave of apoptosis. Mutations in the core components led to increased levels of the inhibitor of apoptosis protein DIAP1 (Harvey et al., 2003; Jia et al., 2003; Pantalacci et al., 2003; Tapon et al., 2002; Udan et al., 2003; Wu et al., 2003) and decreased levels of the caspase Dronc (an ortholog of caspase 9) (Verghese, Bedi, & Kango-Singh, 2012) and corresponding decreased apoptosis in the pupal retina. A role in apoptosis is highly conserved. Originally, the Hippo homologs in mammalian systems were identified as stress response kinases Mst1 and Mst2 (also referred to as Krs1, Krs2 or Stk4 and Stk3) and were reported to play a role in responding to stress and in promoting apoptosis (Creasy & Chernoff, 1995a, 1995b; Kakeya et al., 1998; Taylor et al., 1996).

5. THE HIPPO PATHWAY RESTRICTS ORGAN SIZE AND MAINTAINS ORGAN HOMEOSTASIS

Developing organs know when to stop growing at the appropriate size; in contexts of changes in proliferation or developmental stressors, they employ mechanisms to achieve but not exceed that predetermined size in what is sometimes called an "organ size checkpoint" (Leevers & McNeill, 2005). Mature organs then continue to engage homeostatic mechanisms to maintain that size in a mature organism. How a predetermined organ size is set and then maintained remains a major unanswered question in developmental biology and is crucial in developing organisms and in maintaining health of mature organisms. Interestingly, mutations in the core cassette not only affected proliferation and cell death but also caused dramatic organ overgrowth (Harvey et al., 2003; Jia et al., 2003; Kango-Singh et al., 2002; Lai et al., 2005; Pantalacci et al., 2003; Tapon et al., 2002; Udan et al., 2003; Wu et al., 2003). Because these mutations managed to bypass organ size control mechanisms, these genes are involved in not only growth and proliferation regulation but also organ size control. A role in organ size and organ homeostasis was highly conserved; reducing core components resulted in dramatic overgrowth of the mouse liver (Lee et al., 2010; Lu et al., 2010; Zhou et al., 2009).

6. MOLECULAR MECHANISMS AND ADDITIONAL ROLES OF THE HIPPO PATHWAY: POSTTRANSLATIONAL TARGETS AND TRANSCRIPTIONAL OUTPUTS

As noted, before there was a clear understanding that the core components Hpo/MST, Sav, Mats/MOB1, and Wts/Lats worked in the same cascade, mammalian reports focusing on Mst1 and Mst2 had identified and characterized a role for these kinases in stress response and apoptosis (Creasy & Chernoff, 1995a, 1995b; Kakeya et al., 1998; Taylor et al., 1996). Mechanistic work to elucidate this role investigated changes in chromatin and chromatin-associated proteins during apoptosis and identified a role for Mst1 to phosphorylate histone H2B at serine 14 (Cheung et al., 2003). Other reports focusing on Mst1/2 in mammals have described additional Mst1/2 substrates involved in stress response, apoptosis, autophagy and other processes (Table 2).

The phenotypes of loss-of-function mutations or knockdown of the core components in the Hippo Pathway initially revealed roles for the pathway to restrict growth and proliferation, to promote apoptosis, and to control organ size. As more in-depth phenotypic analyses and mechanistic studies followed the discovery of the Hippo Pathway and extended to different tissues, developmental stages, and contexts, multiple efforts uncovered other important roles of the pathway and identified many of the downstream targets phosphorylated by the core complex or directly by the Wts/Lats effector kinase

Table 2 Reported Targets of the MST1 and MST2 Kinases

Substrate	References
Histone H2B	Cheung et al. (2003)
FOXO1	Lehtinen et al. (2006)
PDX1	Ardestani et al. (2014)
PRDX1	Rawat, Creasy, Peterson, and Chernoff (2013)
NDR1/2	Tang et al. (2015), Hergovich et al. (2009), and Vichalkovski et al. (2008)
Beclin 1	Maejima, Isobe, and Sadoshima (2016)
LC3	Wilkinson et al. (2015) and Wilkinson and Hansen (2015)
Cardiac troponin I	You et al. (2009)
H2AX	Wen et al. (2010)

Table 3 Reported Targets of the Wts/Lats Kinase

Substrate	References
Yki/YAP	Huang, Wu, Barrera, Matthews, and Pan (2005)
Canoe/Afadin	Keder et al. (2015)
Bazooka	Keder et al. (2015)
Mud/NuMA	Dewey, Sanchez, and Johnston (2015)
Ena/VASP	Lucas et al. (2013)
p21	Suzuki et al. (2013)
AMOT family members	Adler et al. (2013), Chan et al. (2013), and Dai et al. (2013)

to achieve these roles (Table 3). In some cases, the mechanisms of specific processes regulated downstream of Wts/Lats remain unresolved.

6.1 Wts/Lats Inhibits the Yki/YAP Oncogene to Regulate Proliferation, Growth, Apoptosis, Stemness and Differentiation, and Organ Size

The most characterized target of the Hippo Pathway is the transcriptional regulator Yorkie (Yki) in flies represented by YAP and TAZ in mammals

(Huang et al., 2005). Phosphorylation by Wts/Lats downregulates Yki/YAP/TAZ by two primary mechanisms: (1) phosphorylation of Yki/YAP/TAZ creates 14-3-3 binding sites that result in 14-3-3-mediated sequestration of Yki/YAP/TAZ in the cytoplasm (Basu, Totty, Irwin, Sudol, & Downward, 2003; Dong et al., 2007; Oh & Irvine, 2008; Ren, Zhang, & Jiang, 2010; Zhao et al., 2007) where it cannot promote transcription of target genes and (2) phosphorylation also increases Yki/YAP/TAZ degradation (Liu et al., 2010; Zhao, Li, Tumaneng, Wang, & Guan, 2010).

Yki/YAP/TAZ lacks a DNA-binding domain but has been found in complex with chromatin remodeling machinery including GAGA factor, Brahma complex, and mediator complex (Oh et al., 2013) and also pairs with specific DNA-binding partners to promote transcription of target genes. There is some controversy over the role of specific partners in mediating Yki/YAP/TAZ transcriptional outputs, but Yki/YAP/TAZ has been linked to a variety of binding partners including Mothers against dpp (Mad) (Oh & Irvine, 2011), Homothorax (Hth) (Peng, Slattery, & Mann, 2009), Teashirt (Tsh) (Peng et al., 2009), Lz/Runx (Jang et al., 2016; Zaidi et al., 2004), and p53 family member *p73* (Basu et al., 2003; Strano et al., 2005, 2001). The most well-studied and widely discussed binding partners for Yki/YAP/TAZ are members of the TEA domain/Transcription enhancer family (TEAD/TEF) family of transcription factors (represented by Scalloped, Sd, in flies and TEAD1-4 in humans) (Ota & Sasaki, 2008; Wu, Liu, Zheng, Dong, & Pan, 2008; Zhang et al., 2008; Zhao et al., 2008). The Sd/TEAD factors do not promote growth when overexpressed on their own but instead act as transcriptional repressors in the absence of Yki/YAP/TAZ interaction; transcriptional repression is relieved and they become transcriptionally active when in complex with Yki/YAP/TAZ. Sd/TEAD and Yki/YAP/TAZ act together to promote transcription of specific genes (Koontz et al., 2013).

Yki transcriptional targets described in flies include *cycE* and *Diap1* (Huang et al., 2005). Yki-mediated transcription of *cycE* in part explains the increased *cycE* transcription and CycE protein accumulation, cell cycle reentry, and increased proliferation phenotypes upon mutation in the core components. Yki-mediated *Diap1* transcription explains the increased *Diap1* transcription upon mutation of the core components (see also Section 6.2 for a transcription-independent regulation of *Diap1*). Additional Yki targets identified include growth and proliferation promoting genes *bantam*, a microRNA (Nolo, Morrison, Tao, Zhang, & Halder, 2006; Thompson & Cohen, 2006) and *myc* (Neto-Silva, de Beco, & Johnston,

2010; Ziosi et al., 2010). A host of other transcriptional targets of Yki and Yap/TAZ have been revealed by genome-wide transcriptional studies, microarray studies, RNA-SEQ, ChIP-SEQ, and ChIP-on-chip approaches and in individual reports including tumor suppressor genes *kibra*, *expanded* (*ex*), and *Merlin* (*mer*) (also called *NF2*) (Diepenbruck et al., 2014; Dong et al., 2007; Gregorieff, Liu, Inanlou, Khomchuk, & Wrana, 2015; Hamaratoglu et al., 2006; Huang et al., 2005; Ikmi et al., 2014; Lian et al., 2010; Zhang, Pasolli, & Fuchs, 2011; Zhao et al., 2008). YAP transcriptional targets include *CTGF1* which is implicated in YAP-mediated growth (Zhao et al., 2008).

Yki activity can be further modulated by binding to additional factors. The Ecdysone receptor coactivator, Taiman (Tai) can bind Yki and modulate its transcriptional activity (Zhang et al., 2015). Although Tai is not required for expression of the standard Yki targets, Tai binding to Yki enhanced its activity. Moreover, Tai interaction with Yki resulted in transcription of additional target genes not normally regulated by Yki alone. This sets up a paradigm where crosstalk with other pathways could modulate the expression of genes associated with Yki activity and also could reveal additional context-dependent transcriptional targets.

In addition to transcriptional repression mediated by the Sd/TEAD factors when not in complex with Yki/YAP/TAZ, transcriptional repression has been reported upon inactivation of the core components concurrent to and dependent on activation of Yki. Upon activation of the core complex, the caspase *Death regulator Nedd2-like caspase* (*Dronc*, the fly homolog of caspase 9) was transcribed. When the core complex was inactive, resulting in activation of Yki, *Dronc* transcription was repressed (Verghese, Bedi, & Kango-Singh, 2012). Hippo Pathway regulation of *Dronc* transcription thus is another means by which the Pathway promotes apoptosis. This Yki-dependent repression of *Dronc* is not fully understood, but might represent an exciting role for Yki in repressing transcription distinct from its role in transcriptional activation.

Mature multicellular organisms composed of a variety of cell types must establish a balance between populations of uncommitted, undifferentiated cells (stem cells or progenitor cells) capable of differentiating into a variety of cell types and differentiated cells specified for specific fates or tissue identities. Those cells that respond to differentiation cues and adopt a differentiated identity are influenced by a variety of inputs. Regulating the proliferation of a pool of stem and progenitor cells and establishing and maintaining the balance between stem and progenitor cells and their differentiated counterparts are crucial in development and in the response to injury in mature tissues. The role of Yki/YAP/TAZ in promoting proliferation is consistent with subsequent reports that Yki/YAP/TAZ promoted proliferation of stem and progenitor cells from a variety of tissues and that activation of the core cassette of the Hippo Pathway restricted their proliferation (Camargo et al., 2007). For example, YAP promoted the proliferation of muscle satellite cells (stem cells in skeletal muscle) and prevented their differentiation (Judson et al., 2012). In heart tissue, Hippo Pathway activation restricted cardiomyocyte proliferation whereas its inactivation or YAP expression led to their increased proliferation (Heallen et al., 2013, 2011; Lin et al., 2014; Mosqueira et al., 2014; Tian et al., 2015; Wang et al., 2014; Zi et al., 2014). In the liver, maintenance of differentiated hepatocytes requires Hippo Pathway inhibition of YAP; inactivation of the Hippo Pathway activated YAP and led to dedifferentiation and more progenitor-like cells (Yimlamai et al., 2014). Similarly, YAP promotes the proliferation of neural progenitor cells and repressed expression of NeuroM to prevent differentiation whereas YAP loss led to apoptosis or premature differentiation (Cao, Pfaff, & Gage, 2008).

Yki/YAP/TAZ does not promote proliferation and block differentiation in all contexts. There are emerging reports of a role for Yki/YAP/TAZ to promote differentiation along a specific program to reach a specific fate or to maintain a specific cell fate. In the kidney, YAP promoted transcription of genes involved in nephron formation and morphogenesis. Conditional knockout of YAP in the condensing (or cap) mesenchyme (a subset of which goes on to form the developing nephron) led to hypoplastic kidneys with multiple defects reflective of failed nephrogenesis rather than hyperplastic, overgrown kidneys (Reginensi et al., 2013). In the crypt cells of the mammalian intestine, rather than promoting hyperplastic overgrowth seen in other contexts, YAP activation led to loss of intestinal crypt cells whereas YAP loss resulted in their hyperplasia, expansion of progenitor cells, and ectopic crypts (Barry et al., 2013).

In mammalian development, a fertilized embryo undergoes rapid proliferation to reach the blastocyst stage. In the blastocyst stage, some cells commit to become trophoectoderm cells (TE), an extraembryonic tissue. The remaining cells contribute to the inner cell mass (ICM) that will become the developing embryo. This is considered the first cell fate decision made in mammalian development. TE and ICM fates are decided by the level of YAP/TEAD activity or activation of the core cassette, respectively. Nuclear localization and activation of YAP/TEAD transcriptional outputs promoted

TE fate, whereas high levels of activation of the Mst and Lats kinases promoted YAP phosphorylation and subsequent cytoplasmic localization and ICM fate (Nishioka et al., 2009; Wicklow et al., 2014).

Early studies characterizing mutations in the core cassette and Yki overexpression showed strong concordance (Harvey et al., 2003; Huang et al., 2005; Jia et al., 2003; Kango-Singh et al., 2002; Lai et al., 2005; Pantalacci et al., 2003; Tapon et al., 2002; Udan et al., 2003; Wu et al., 2003), and it was widely believed that Yki activation fully phenocopied inactivation of the core cassette. However, in recent years, a number of reports have demonstrated phenotypes associated with inactivation of the core cassette that are not phenocopied by Yki activation suggesting a wider role for the pathway than restricting Yki/YAP/TAZ activity. In some of these cases, the mechanisms underlying the Yki/YAP/TAZ-independent roles for the Hippo Pathway remain unresolved and likely involve Wts/Lats direct targets that have yet to be identified. Although some avenues in this emerging area of Yki/YAP/TAZ-independent targets of the Hippo Pathway are controversial, the number of reports is consistently growing from both *Drosophila* and mammalian systems. The following subsections review processes and targets regulated in parallel to Yki/YAP/TAZ. Section 6.2 reviews regulation of Diap1 whose protein levels are regulated parallel to Yki/YAP/TAZ and whose transcription is regulated by Yki/YAP/TAZ. Sections 6.3–6.6 review additional proteins and/or process regulated parallel to and likely independent of Yki/YAP/TAZ.

6.2 The Hippo Pathway Core Kinase Cassette Phosphorylates DIAP1 to Promote Its Degradation

The first target identified for the Hippo Pathway core kinase complex was the apoptosis inhibitor DIAP1 in flies. DIAP1 protein accumulated in tissue mutant for *hpo*, *sav*, or *wts*. These studies showed effects on both DIAP1 transcription (Jia et al., 2003; Udan et al., 2003; Wu et al., 2003) and DIAP1 protein stability (Harvey et al., 2003; Pantalacci et al., 2003). Activation of the Hpo kinase promoted phosphorylation and degradation of DIAP1 in vivo and in vitro and the core complex isolated by pull-down of Sav was sufficient to promote phosphorylation of DIAP1 protein directly (Harvey et al., 2003; Pantalacci et al., 2003). These studies did not address if Hpo or Wts acted as the direct kinase; however, one of the sites of DIAP1 phosphorylation (Harvey et al., 2003) overlaps with the loose consensus for NDR kinases such as Wts (Mah et al., 2005).

6.3 The Hippo Pathway Regulates the Actin Cytoskeleton

In *Drosophila*, loss of the core components in some contexts (Fernández et al., 2011) led to abnormal F-actin accumulation independent of Yki activation suggesting a role for the Hippo Pathway to restrict actin polymerization and regulate the actin cytoskeleton. Later studies revealed a role for the Hippo Pathway in the polarization of the actin cytoskeleton in the context of migrating border cells also in a Yki-independent manner. In this case, Wts phosphorylated and inhibited Enabled (Ena), an actin regulator that inhibits actin capping proteins thus limiting actin polymerization. By blocking the ability of Ena to inhibit actin capping, the Hippo Pathway thus restricts actin polymerization in these migrating cells to the outer rim (Lucas et al., 2013).

Regulation of F-actin by the Hippo Pathway also occurs in mammalian systems and is regulated at least in part by targeting of angiomotin (AMOT) family members AMOT, AMOTL1, and AMOTL2. AMOT proteins are not conserved in *Drosophila*. When not phosphorylated, AMOT proteins interact with F-actin filaments and stress fibers. AMOT phosphorylation by Lats1/2 disrupted this interaction, inhibited cell proliferation, and reduced actin stress fibers (Adler et al., 2013; Chan et al., 2013; Dai et al., 2013).

6.4 The Hippo Pathway Regulates Asymmetric Cell Division and Mitotic Spindle Orientation by Targeting Canoe/Afadin, Bazooka, and Mud

Asymmetric cell division is employed in a variety of developmental contexts and is also important when stem cells divide to produce both a daughter cell committed to differentiate and another daughter stem cell for self-renewal. Loss of function of Hippo Pathway core components led to defects in the asymmetric cell division of *Drosophila* neuroblasts (neural stem cells). This was not mediated by Yki but rather by the Wts-mediated phosphorylation of Canoe (Cno) and Bazooka (Baz; Afadin/AF-6 and Par3, respectively, in mammals). This role for the core components was not limited to neuroblasts but was also seen in muscle and heart progenitors (Keder et al., 2015).

Asymmetric cell division also requires proper orientation of the mitotic spindle. Within a tissue, the orientation and positioning of the mitotic spindle also contributes to tissue homeostasis. Loss of the core components of the Hippo Pathway resulted in defects in spindle orientation independent of Yki. Instead, the Hippo Pathway regulates spindle orientation by regulation of spindle pole protein Mushroom body defect (Mud, NuMA in mammals).

Wts phosphorylated Mud to prevent its self-association and allow it to interact the highly conserved protein Partner of Inscuteable (Pins) which controls spindle orientation (Dewey et al., 2015).

6.5 The Hippo Pathway Regulates Proliferation, the Mitotic Cyclins, and Organ Size by Targeting Rae1

Although increased Yki/YAP/TAZ transcriptional activity clearly played a role in the increased *cycE* transcription and cell cycle reentry phenotypes upon loss of the core components, studies of transcriptional targets had not reported a role for Yki/YAP/TAZ to regulate transcription of the mitotic cyclins, suggesting another pathway target was responsible for the accumulation of CycA and CycB upon loss of the core components. A recent report identified the WD repeat protein Rae1 as a pathway target downstream of Wts (Jahanshahi et al., 2016). Rae1 inhibition by the Hippo Pathway mediates the pathway's role to restrict CycA and CycB proteins and contributes to the pathway's role to restrict organ size (Jahanshahi et al., 2016). Rae1 is linked to regulation of microtubules, the mitotic spindle, and the Anaphase Promoting Complex/Cyclosome (APCC) (Blower, Nachury, Heald, & Weis, 2005; Brown et al., 1995; Cuende, Moreno, Bolaños, & Almeida, 2008; Jeganathan, Baker, & van Deursen, 2006; Jeganathan, Malureanu, & van Deursen, 2005; Kraemer, Dresbach, & Drenckhahn, 2001; Lee et al., 2009; Volpi, Bongiorni, Fabbretti, Wakimoto, & Prantera, 2013; Wong, Blobel, & Coutavas, 2006). Genetic interactions suggest Hippo Pathway regulation of the APCC may occur via Rae1 (Jahanshahi et al., 2016). A role for Rae1 in the pathway's regulation of the mitotic spindle and microtubules has not been addressed, but curiously, mammalian Rae1 interacts with Hippo Pathway target NuMA (Wong et al., 2006) which, as mentioned earlier, is involved in pathway regulation of spindle orientation (Dewey et al., 2015). Rae1 may not be a direct target of the Wts/Lats kinase (Jahanshahi et al., 2016), therefore this raises the question if Wts/Lats targets another effector kinase to target Rae1.

6.6 Yki/YAP/TAZ-Independent Hippo Pathway-Regulated Processes With Unresolved Mechanisms

Together with targets of the Hippo Pathway associated with broad responsibility for defining the pathway's role in growth, proliferation, differentiation, and apoptosis or in more precise and defined context-dependent functions, there are numerous roles for the Hippo Pathway revealed by characterization of mutant phenotypes and overexpression studies for which the mechanisms remain unresolved. In the epidermis of *Drosophila* embryos, mutations in *wts* and *hpo* led to defects in cell shape and in the planar polarized organization of adherens junction proteins and microtubule organization that appear independent of Yki-mediated transcription (Marcinkevicius & Zallen, 2013). Targets whose regulation ensures this organization have not been identified.

The dendrites of neurons typically form complex and branched structures that aid in the function of the neurons to receive and process signals. This arborization involves steps to establish and then maintain dendritic tiling. In *Drosophila* embryos, Wts is important for the maintenance of the dendritic field. Mutants for *wts* showed dendritic defects associated with a problem in maintaining the dendritic structures. At early time points, the dendrites were seen to tile normally, but were then lost progressively. Curiously, *hpo* mutants showed earlier defects associated with both tiling and maintenance. The broader role for Hpo likely reflects its phosphorylation and activation of Tricornered (Trc), an NDR kinase related to Wts. The roles of Hpo and Wts in dendritic tiling are independent of Yki, suggesting another target or targets important in this process (Emoto, Parrish, Jan, & Jan, 2006).

When developing *Drosophila* larvae are exposed to nonlethal doses of alcohol in ranges consistent with those found in fermenting fruit sources, resulting epithelial structures such as the eye and wing do not overgrow; they are either normal in size, or in some cases show organ size reduction (Ilanges, Jahanshahi, Balobin, & Pfleger, 2013; McClure, French, & Heberlein, 2011). Surprisingly, tissues with reduced Hpo or Wts showed enhanced overgrowth when larvae were reared on alcohol (Ilanges et al., 2013). In contrast, alcohol did not enhance overgrowth associated with Yki overexpression suggesting that the alcohol-mediated enhancement was Yki-independent and may involve other targets downstream of Hpo and/or Wts.

Wts plays a role in the degradation of *Drosophila* salivary glands via autophagy. Loss of Wts in the salivary glands resulted in a defect in autophagy accompanied by persistence of the salivary glands (Dutta & Baehrecke, 2008). This is not phenocopied by Yki activation; in contrast, Yki expression resulted in premature salivary gland degradation (Dutta & Baehrecke, 2008; Jahanshahi et al., 2016). The Wts regulation of autophagy in the salivary gland was independent of Yki and Sd and of the Yki transcriptional target *bantam* and is PI3K-dependent (Dutta & Baehrecke, 2008).

In the mammalian kidney, signaling through the Hippo Pathway is important in the restriction and self-renewal of nephron progenitors in the condensing mesenchyme. Loss of the upstream regulator *Fat4* leads or its ligand causes an expansion in the nephron progenitors of the condensing mesenchyme (Bagherie-Lachidan et al., 2015). Loss of YAP in this context does not suppress the increase in the condensing mesenchyme. In fact, as described earlier, loss of YAP on its own in the condensing mesenchyme actually leads to kidney hypoplasia (Reginensi et al., 2013). These findings indicate that the Hippo Pathway restricts the nephron progenitor pool by a YAP-independent mechanism. Additional mechanistic analysis revealed a dependence on *six2*, a gene important in progenitor cells and elevated in this context (Bagherie-Lachidan et al., 2015).

7. UPSTREAM SIGNALS AND REGULATORS AND MECHANISMS OF PATHWAY HOMEOSTASIS

A crucial feature in the regulation of the Hippo Pathway is the need to increase or decrease the activation of the core cassette as required by context. Given the potent ability of the core cassette to promote apoptosis when active, or for growth to go unchecked when the core cassette is inactive, mechanisms to ensure pathway homeostasis are crucial. Several upstream regulators affect the signaling of the pathway from an upstream step and are also regulated by the pathway as downstream targets. This section will briefly survey upstream factors and conditions reported to regulate components to disrupt or amplify signaling to downstream outputs and will focus on mechanisms to regulate Hippo Pathway homeostasis.

Consideration of a presumed scaffold (Sav) and two kinases (Hpo/MST and Wts/Lats) working together led to a model that these components in the Hippo Pathway might function together in a MAPK-like cassette. This encouraged substantial efforts by multiple groups on a larger scale to identify what upstream signals and upstream components acted to regulate activation of the core cassette. Initial efforts actively sought to identify an upstream ligand and its receptor that controlled the core kinase cassette. Numerous reports identified a number of upstream factors to reveal that, unlike many well-studied signaling pathways, the Hippo Pathway does not represent a linear pathway where an extracellular ligand (or family of ligands) binds a receptor (or family of receptors) at the cell surface to set off a linear signaling cascade aimed to modulate a transcriptional output. Instead, a diversity of upstream factors including transmembrane proteins, membrane-associated proteins, junctional complexes, stress, environmental conditions, cytoskeleton networks, cell–cell contact, and mechanical forces affect the activation of the core cassette or can modulate individual components of the core cassette (many of these are summarized in Table 4), and there numerous reports of crosstalk with other major signaling networks not reviewed here due to space constraints.

Highlighting the complex and nonlinear nature of this signaling network, upstream regulators have been reported to both activate the core cassette but also to bypass this form of activation to target specific nodes. For example, the scaffold protein Sav is an unstable protein that is stabilized by association with Hpo/MST (Aerne et al., 2015; Callus, Verhagen, & Vaux, 2006; Pantalacci et al., 2003; Wu & Wu, 2013). Sav stability can be influenced by increased insulin signaling which promotes activity of salt-inducible kinases (sik) whose phosphorylation of Sav increases its degradation (Wehr et al., 2013). In some cases, upstream components can bypass individual core components or can act at multiple steps of the pathway. For example, G protein-coupled receptors (GPCRs) are a large class of receptors that bind G proteins and respond to a variety of ligands. The signaling outputs depend on which ligands bind and activate which GPCRs and to which G proteins they are coupled. In the first instance of a traditional ligand/receptor model to activate the Hippo Pathway, lysophosphatidic acid (LPA) and sphingosine 1-phosphate (S1P) signaled through GPCRs coupled to G12/G13 to inhibit Lats1/2 and thereby activate YAP, and epinephrine and glucagon signaled through GPCRs coupled to Gs to activate Lats1/2 and thereby inhibit YAP (Yu et al., 2012). GPCR effects on Lats1/2 did not affect MST1/2 activation suggesting that GPCRs bypass the core cassette when they modulate Lats1/2. For additional in-depth reviews of the upstream regulators of the core cassette, please see Parsons, Grzeschik, Allott, and Richardson (2010), Grusche, Richardson, and Harvey (2010), Genevet and Tapon (2011), Yu and Guan (2013), Sun and Irvine (2016), and Meng, Moroishi, and Guan (2016).

7.1 FERM Domain Proteins Merlin and Expanded Regulate the Hippo Pathway

Activation of Wts/Lats is reported to occur at the membrane before signaling to downstream targets (Hergovich et al., 2006; Sun, Reddy, & Irvine, 2015; Yin et al., 2013). The Hpo/MST and Wts/Lats kinases associate with the plasma membrane in different domains. Wts/Lats is the downstream,

Table 4 Upstream Regulators of the Hippo Pathway

Fat/dachsous/zyxin	Cho et al. (2006), Sopko et al. (2009), Willecke, Hamaratoglu, Sansores-Garcia, Tao, and Halder (2008), Rauskolb, Pan, Reddy, Oh, and Irvine (2011), Reddy and Irvine (2013), Marcinkevicius and Zallen (2013), and Gaspar, Holder, Aerne, Janody, and Tapon (2015)
Ajuba LIM proteins	Das Thakur et al. (2010), Reddy and Irvine (2013), Sun and Irvine (2013), and Rauskolb, Sun, Sun, Pan, and Irvine (2014)
Protein kinase A	Yu et al. (2013) and Kim et al. (2013)
Junctional components and apical basal polarity regulators	Hamaratoglu et al. (2009), Genevet et al. (2009), Robinson, Huang, Hong, and Moberg (2010), Grzeschik, Parsons, Allott, Harvey, and Richardson (2010), Chen et al. (2010), Ling et al. (2010), Hafezi, Bosch, and Hariharan (2012), and Verghese, Waghmare, Kwon, Hanes, and Kango-Singh (2012)
α-Catenin	Schlegelmilch et al. (2011) and Silvis et al. (2011)
Echinoid	Yue, Tian, and Jiang (2012), Bossuyt et al. (2014), and Ding, Weynans, Bossing, Barros, and Berger (2016)
Ex, mer	Hamaratoglu et al. (2006) and Pellock et al. (2007)
Kibra	Yu et al. (2010), Genevet, Wehr, Brain, Thompson, and Tapon (2010), and Baumgartner, Poernbacher, Buser, Hafen, and Stocker (2010)
Herc4 ubiquitin ligase	Aerne, Gailite, Sims, and Tapon (2015)
CRL4/DCAF1	Li et al. (2010), Cooper et al. (2011), and Li et al. (2014)
β-TRCP	Zhao, Li, Tumaneng, et al. (2010) and Liu et al. (2010)
Itch ubiquitin ligase	Ho et al. (2011), Salah, Melino, and Aqeilan (2011), and Wang et al. (2012)
G protein-coupled receptors	Yu et al. (2012)

Continued

Table 4 Upstream Regulators of the Hippo Pathway—cont'd

Salt-inducible kinases (Sik) and nutrient sensing	Wehr et al. (2013) and Hirabayashi and Cagan (2015)
STRIPAK/PP2A	Ribeiro et al. (2010)
p53	Colombani, Polesello, Josué, and Tapon (2006)
RASSF/RASSF1A	Polesello, Huelsmann, Brown, and Tapon (2006), Guo et al. (2007), and Avruch, Praskova, Ortiz-Vega, Liu, and Zhang (2006)
Tao1	Poon, Lin, Zhang, and Harvey (2011) and Boggiano, Vanderzalm, and Fehon (2011)
Happyhour/MAP4K3, misshapen/MAP4K4	Li, Cho, Yue, Ip, and Jiang (2015) and Meng et al. (2015)
Protease-activated receptors (PARS)	Mo, Yu, Gong, Brown, and Guan (2012) and Huang et al. (2013)
Angiomotin family members	Wang, Huang, and Chen (2011), Zhao et al. (2011), Chan et al. (2011), Paramasivam, Sarkeshik, Yates, Fernandes, and McCollum (2011), Wang et al. (2012), Mana-Capelli, Paramasivam, Dutta, and McCollum (2014), Yi et al. (2013), and Lv et al. (2015)
Tankyrase and RNF146	Wang et al. (2015) and Troilo et al. (2016)
Actin cytoskeleton	Fernández et al. (2011), Sansores-Garcia et al. (2011), and Wada, Itoga, Okano, Yonemura, and Sasaki (2011)
Spectrin cytoskeleton	Wong et al. (2015), Fletcher et al. (2015), and Deng et al. (2015)
Microtubules and microtubule regulators	Bensenor, Barlan, Rice, Fehon, and Gelfand (2010), Aguilar et al. (2014), and Jahanshahi et al. (2016)

effector kinase and requires an activating phosphorylation from Hpo/Mst. Therefore, a limiting factor for the core cassette to signal to downstream effectors is the recruitment of Wts/Lats to the appropriate membrane region to associate with Hpo/MST. The FERM proteins (Band 4.1, Ezrin, Radixin, Moesin) Mer/NF2 (conserved in mammals) and Ex (possibly FRMD6 in mammals, although identification of a true ortholog is a matter of debate) were identified as upstream activators of the core cassette in *Drosophila*

(Hamaratoglu et al., 2006). Although individual mutations in *mer* and *ex* exhibit phenotypes distinct from mutations in core components *hpo*, *sav*, *mats*, and *wts* (Pellock et al., 2007), *mer*, *ex* double mutants phenocopy mutations in the core components (Hamaratoglu et al., 2006) suggesting they play both distinct and partially overlapping roles. More recent work revealed that Mer and Ex promote Wts/Lats recruitment to the membrane domain to receive the activating phosphorylation by Hpo/Mst (Sun et al., 2015; Yin et al., 2013). Curiously, in addition to phosphorylation-dependent regulation of Yki/YAP/TAZ by the core cassette described in Section 6.1, Ex (Badouel et al., 2009; Oh, Reddy, & Irvine, 2009), Hpo (Oh et al., 2009), and Wts (Oh et al., 2009) have each been shown to bind Yki directly. Direct binding results in cytoplasmic retention that also prevents Yki-mediated transcription in a phosphorylation-independent manner.

Yki promotes growth and proliferation but also promotes the transcription of *ex* and *mer* in *Drosophila* (Hamaratoglu et al., 2006). Thus, Yki acts in negative feedback by increasing the levels of upstream components of the pathway (Fig. 2). This mechanism of negative feedback is highly conserved; in mammalian systems, YAP promotes the transcription of *Mer/NF2* and *Lat2s* (Moroishi et al., 2015).

7.2 AMOT Family Members Regulate Lats1/2 and YAP/TAZ

As described earlier in Section 6.3, AMOT family members in mammals are targeted by Hippo Pathway-mediated phosphorylation and act in restricting cell proliferation parallel to Hippo Pathway regulation of YAP. These targets of the pathway feed back to regulate the core kinase Lats1/2 and to inhibit YAP/TAZ directly. AMOT protein AMOTL2 has been reported to bind and activate Lats2 directly, possibly functioning as a scaffold which binds both MST2 and Lats2 at tight junctions to promote their interaction (Paramasivam et al., 2011). AMOT isoform AMOTp130 is also reported to stabilize Lats1/2 protein. Lats1/2 can be targeted by the Nedd4 ubiquitin ligase family member Itch. By acting as a scaffold associating with both Itch and YAP, AMOT facilitates a switch of targeting of Lats1/2 by Itch to a targeting of YAP by Itch. In this manner, AMOT plays a dual role to stabilize Lats and destabilize YAP (Adler et al., 2013). In addition to promoting YAP instability, AMOT family members' direct binding of YAP promotes its localization and retention at tight junctions (Chan et al., 2011; Zhao et al., 2011). Thus, AMOT family members feed back to regulate both upstream steps of the Hippo Pathway as well as participating in direct regulation of

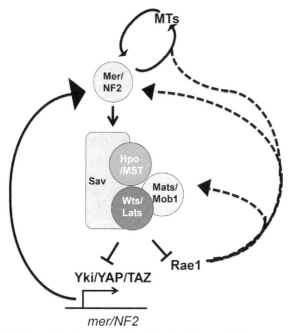

Fig. 2 Yki/YAP/TAZ and Rae1 are regulated by the Hippo Pathway core cassette and feedback to regulate the core cassette. The transcriptional regulator Yki/YAP/TAZ plays a highly conserved role to regulate the transcription of the Mer/NF2 tumor suppressor, as reviewed in Section 7.1. When the Hippo Pathway core cassette is active, Yki/YAP/TAZ is phosphorylated (as reviewed in Section 6.1) and does not promote Mer/NF2 transcription and Rae1 is phosphorylated and degraded (as reviewed in Section 6.5). When the core cassette is inactivated, Yki/YAP/TAZ becomes active and promotes Mer/NF2 transcription and Rae1 levels accumulate. Increased Mer/NF2 levels promote activation of the core cassette by localizing Wts/Lats to the membrane where it can receive its activating phosphorylation from Hpo/MST (not pictured) as reviewed in Section 7.1. Increased Rae1 levels lead to increased activation of Wts/Lats and increased levels of Hpo, Wts, and Mer by an unknown mechanism (indicated by *dotted lines*) which may be due to direct effects on individual core components and Mer or may indirect such as through microtubule regulation as reviewed in Section 7.4. In this manner, both Yki/YAP/TAZ and Rae1 act in mutually reinforcing feedback to the core cassette to negatively regulate themselves and each other. Mer/NF2 stabilizes microtubules, and microtubule polymerization and acetylation in turn promotes Mer/NF2 activity as a means of positive feedback (as reviewed in Section 7.3).

their parallel downstream output YAP to reinforce YAP targeting by the core cassette.

These studies demonstrating AMOT family member-mediated activation and stabilization of Lats1/2 and inhibition of YAP/TAZ suggest a tumor suppressor role for AMOT; however, conflicting studies suggest a cancer driver role for AMOT in some contexts. For example, AMOT has been reported to act as a cofactor for YAP to drive YAP-dependent overgrowth in the liver (Yi et al., 2013). Similarly, AMOT has been reported to promote proliferation and invasion in breast cancer (Lv et al., 2015). Additional work is needed to resolve the role of AMOT in cancer.

7.3 The Actin, Spectrin, and Microtuble Cytoskeletal Networks Regulate the Hippo Pathway

A number of cytoskeletal components have been linked to regulation of the Hippo Pathway. Of particular interest, actin is regulated by Ena/VASP downstream of Wts in flies (Lucas et al., 2013) and also by AMOT family members in mammals (Chan et al., 2013; Dai et al., 2013; Ernkvist et al., 2006; Gagné et al., 2009; Mana-Capelli et al., 2014) as described in Section 6.3. Polymerization of actin can regulate the activation status of the Hippo Pathway core cassette (Fig. 3). In *Drosophila*, F-actin accumulation caused dramatic tissue overgrowth due to inactivation of the core cassette and activation of Yki (Fernández et al., 2011; Sansores-Garcia et al., 2011). This mode of regulation is highly conserved; in mammals, YAP is also regulated by F-actin via the Hippo Pathway regulation of AMOT family members (Chan et al., 2013; Dai et al., 2013; Ernkvist et al., 2006; Gagné et al., 2009; Mana-Capelli et al., 2014).

Hippo Pathway regulation of actin and by actin signals a fundamental biological process at work. Actin cytoskeletal elements contribute

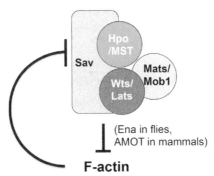

Fig. 3 Actin polymerization acts in Hippo Pathway feedback. The Hippo Pathway regulates actin dynamics as reviewed in Section 6.3, including negative regulation of actin polymerization by regulation of Ena in flies and AMOT proteins in mammals. F-actin has been shown to inhibit the activity of the Hippo Pathway core cassette as reviewed in Section 7.2 and touched on in Section 7.6. In this manner, cytoskeletal regulation is a means of feedback to ensure pathway homeostasis.

significant structural roles within a cell and work in parallel to and interface with other cytoskeletal networks including the spectrin cytoskeleton to maintain cellular architecture. These cytoskeletal networks interact with membrane proteins and junctional complexes to translate physical and mechanical cues to meaningful biological outcomes. Following studies linking actin polymerization to regulation of the Hippo Pathway, a number of groups reported the related regulation of the core cassette and Yki/YAP/TAZ by the spectrin cytoskeleton (Deng et al., 2015; Fletcher et al., 2015; Wong et al., 2015). These areas of investigation coincided with a field of research into the relationship between mechanotransduction and cell shape/cellular architecture to the Hippo signaling network indicating that the Hippo Pathway translates mechanical tension and changes in shape to biological outputs (Dupont et al., 2011; Mosqueira et al., 2014; Rauskolb et al., 2014; Wada et al., 2011). For an in-depth review of cell architecture and mechanotransduction within the Hippo Pathway field, please see Gaspar and Tapon (2014), Schroeder and Halder (2012), Hao et al. (2014), and Hariharan (2015).

Together, these reports also highlight that cytoskeletal dynamics are intimately controlled by and tied to the levels of Hippo Pathway activity. In addition to the actin and spectrin cytoskeletal networks, the microtubule cytoskeleton controls activation of the core cassette and regulation of outputs of the Hippo Pathway. Mer/NF2 interacts with and stabilizes microtubules (Muranen et al., 2007; Smole et al., 2014). Interaction with microtubules is important for the proper transport and localization of Mer/NF2 and ultimately the regulation of YAP (Aguilar et al., 2014; Bensenor et al., 2010). Thus, in addition to the complex feedback circuit between the Hippo Pathway and actin and spectrin networks, there is a positive feedback circuit between Mer/NF2 and microtubule networks (Fig. 2).

7.4 Microtubule Regulator Rae1 Feeds Back to Control Mer, Hpo, and Wts

Rae1 is downregulated by the Hippo Pathway parallel to Yki/YAP/TAZ to mediate the pathway's role in restricting mitotic cyclin levels and in regulating organ size as described in Section 6.5 (Jahanshahi et al., 2016). Surprisingly, although reducing Rae1 led to decreased organ size and increasing Rae1 increased organ size, genetic interactions revealed a role for Rae1 to inhibit Yki. Loss of Rae1 in some contexts resulted in increased Yki levels, relocalization of Yki from the cytoplasm to the nucleus, and dramatically enhanced overgrowth. Although a role for Rae1 to inhibit Yki protein directly has not been addressed, mechanistic studies revealed a role for Rae1 to promote the activation of Hpo/MST and Wts/Lats and to promote the accumulation of Mer, Hpo, and Wts proteins in a posttranscriptional manner (Jahanshahi et al., 2016). This posttranscriptional feedback circuit reinforces the transcriptional feedback described for Yki/YAP/TAZ (Fig. 2). Given that Rae1 binds and regulates microtubules (Blower et al., 2005; Brown et al., 1995; Kraemer et al., 2001), this may reflect a microtubule-mediated negative feedback mechanism adding complexity to the positive feedback circuit described above whereby Mer/NF2 stabilizes microtubules which then promote Mer localization and downregulation of YAP (Aguilar et al., 2014; Bensenor et al., 2010; Muranen et al., 2007; Smole et al., 2014) (Fig. 2).

7.5 Myc Inhibits Yki Activity

Myc transcription is promoted by Yki to promote growth as described in Section 6.1. In addition to driving growth, Myc feeds back to negatively regulate Yki (Neto-Silva et al., 2010). Loss of *myc* led to increased Yki protein levels, whereas Yki levels decreased upon *myc* overexpression. The decrease in Yki protein reflected, in part, a decrease in transcript levels. However, the regulation was not entirely transcriptional; in the absence of endogenous *yki*, the levels of a *yki* transgene under the tubulin promoter were also affected by a *myc* transgene. Thus, this negative Myc-Yki feedback circuit represents a case where Myc regulates Yki levels by both transcriptional and posttranscriptional mechanisms.

7.6 Cell Adhesion and Cell Contact Inhibition Regulate the Hippo Pathway

Immortalized nontransformed cells in culture at low density actively divide. However, once they reach confluence, the cell–cell contact inhibits subsequent proliferation, a phenomenon known as "contact inhibition" or "contact inhibition of proliferation." A number of junctional proteins and cell adhesion molecules mediate aspects of cell–cell interaction and also serve as upstream regulators of the Hippo Pathway (Table 4). Contact inhibition restricts proliferation by activating the Hippo Pathway core cassette by a number of mechanisms including signaling through cadherins, AMOT family members, and Mer/NF2 and at distinct steps in the pathway to downregulate YAP and potentially other targets (Aguilar et al., 2014; Chan et al., 2011; Curto, Cole, Lallemand, Liu, & McClatchey, 2007; Kim, Koh, Chen, & Gumbiner, 2011; Morrison et al., 2001; Okada, Lopez-Lago, &

Giancotti, 2005; Ota & Sasaki, 2008; Tikoo, Varga, Ramesh, Gusella, & Maruta, 1994; Yue et al., 2012; Zhao et al., 2011, 2007).

In contrast to cell–cell attachments to downregulate proliferation, attachment of cells to the extracellular matric stimulates growth and proliferation. Adhesion of cells to fibronectin activates focal adhesion kinase (FAK) to inhibit Hippo Pathway core component Lats1/2 to allow for YAP activation (Kim & Gumbiner, 2015).

Loss of this attachment to the extracellular matrix of cells, by extension, can lead to activation of Lats1/2 and subsequent inactivation of YAP followed by a specific type of cell death called anoikis (Zhao et al., 2012).

Cell–cell interactions and cell–substrate interactions influence the cytoskeleton and cell architecture and create mechanical tension. As discussed in Section 7.3, the cytoskeleton and cell architecture are modulated by and can modulate the activity of the Hippo Pathway core cassette; thus, cell–cell contact and cell–substrate interactions interface with homeostatic mechanisms to further define the outputs of Hippo signaling.

7.7 Oxidative Stress and MST1 Engage a Positive Feedback Circuit

Negative feedback circuits are crucial to maintain pathway homeostasis by buffering activation of the pathway within a certain range. However, some contexts call for amplification of a signal.

As noted in Sections 4 and 6, the mammalian Hippo orthologs MST1 and MST2 were originally identified as stress response kinases playing a role in apoptosis (Creasy & Chernoff, 1995a, 1995b; Kakeya et al., 1998; Taylor et al., 1996). MST1/2 can be activated by stress conditions such as the oxidative stress created by exposure to hydrogen peroxide (Kakeya et al., 1998). Active Mst1 can phosphorylate the substrate protein Peroxiredoxin-1 (PRDX1) (Rawat et al., 2013). Normally, PRDX1 reduces hydrogen peroxide to oxygen and water, but when phosphorylated, its peroxidase activity drops by approximately fourfold (Rawat et al., 2013). Therefore, the phosphorylation of PRDX1 by MST1 increases the hydrogen peroxide in a cell which can then promote and sustain the activation of MST1 as a means of positive feedback.

7.8 A Bistable Feedback Loop Between Wts and Melted in *Drosophila* Photoreceptors

In the *Drosophila* retina, postmitotic photoreceptors (R cells) can adopt different subtypes utilized in color vision depending on which rhodopsin they express. The R8 photoreceptor cell expresses rhodopsin 5 (Rh5) or rhodopsin 6 (Rh6) depending on Hippo Pathway activation or Yki activation. Wts is sufficient to promote Rh6 fate whereas the pleckstrin homology domain protein Melted (Melt) is sufficient to promote Rh5 fate. Hippo Pathway activation (and Yki inactivation) in an R8 promotes Wts-dependent repression of *melt* and thus Rh6 cell fate in most cells; Wts is then required to maintain that Rh6 fate. In contrast, Yki activity results in expression of *melt* and subsequent repression of *wts* and thus Rh5 fate in most cells (Jukam & Desplan, 2011; Jukam et al., 2013; Mikeladze-Dvali et al., 2005). In this manner, Wts and Melt mutually negatively regulate each other in a bistable loop. Yki-dependent expression of *melt* and subsequent activity of Melt to inhibit Wts acts as a positive feedback loop to maintain Yki activation and reinforce Rh5 fate.

8. DYSREGULATION OF THE HIPPO PATHWAY IN DISEASE

Given the crucial nature of the roles played by a functional Hippo Pathway, it is unsurprising that deregulation of this pathway leads to pathologic consequences. This section briefly reviews some of the implications of Hippo Pathway dysregulation in disease. For additional review of these concepts and the role of the pathway in other diseases, please see Martinez and de Iongh (2010), Emoto (2011), Plouffe, Hong, and Guan (2015), Zhou, Li, Zhao, and Guan (2015), and Wong, Meliambro, Ray, and Campbell (2016).

8.1 Eye Health and Disease

The upstream regulator Mer/NF2 is mutated in Neurofibromatosis Type 2, an autosomal dominant condition. NF2 patients have an increased incidence of cataracts and retinal abnormalities (Meyers, Gutman, Kaye, & Rothner, 1995; Parry et al., 1994, 1996; Ragge, Baser, Riccardi, & Falk, 1997), and NF2 knockout mice also develop cataracts (Giovannini et al., 2000). In Sveinsson's Chorioretinal Atrophy (SCRA), a condition characterized by choroid and retinal degeneration, Tead1 is either deleted or mutated at a residue important for its interaction with YAP/TAZ (Fossdal et al., 2004; Kitagawa, 2007). These findings suggest that the precise regulation Hippo Pathway plays an important role in maintaining eye health and preventing eye disease.

8.2 Cardiovascular Health and Heart Disease

Controlling the size of the heart is crucial for health. Cases of enlarged heart (cardiomegaly) cause a variety of symptoms and can eventually lead to congestive heart failure. In addition to controlling heart size, the Hippo Pathway role in proliferation and in progenitor cell populations affects cardiovascular health and the ability of the heart to survive injury. The human heart does not undergo significant regeneration; injury usually results in fibrotic scarring. As noted, the Hippo Pathway core cassette activation, and in particular, MST1 proapoptotic activity, promote apoptosis, and restrain the ability of cardiomyocytes to expand and regenerate cardiac tissue (Heallen et al., 2013, 2011; Lin et al., 2014; Mosqueira et al., 2014; Tian et al., 2015; Wang et al., 2014; Yan et al., 2013; You et al., 2013; Zi et al., 2014). In fact, modulation of MST1 is linked to cardiomyopathy (Chen et al., 2014). In mouse models, YAP expression improved cardiac function and regeneration following injury (Lin et al., 2014; Xin et al., 2013). These studies suggest that the Hippo Pathway restrains heart size and prevents cardiomegaly at the cost of reduced ability to repair after injury. For additional insight into the role of the Hippo Pathway in cardiovascular disease, please see Zhou et al. (2015) and Lin and Pu (2014).

8.3 Neuronal Disease

As described earlier, the Hippo Pathway regulates proliferation of *Drosophila* and mammalian neural progenitor cells (Cao et al., 2008; Keder et al., 2015) and maintenance of Rh5/Rh6 fate in *Drosophila* R8 postmitotic photoreceptor cells (Jukam & Desplan, 2011; Jukam et al., 2013; Mikeladze-Dvali et al., 2005). Numerous reports in the mammalian system extended these observations and implicated the Hippo Pathway in mammalian brain and central nervous system development including important roles in proper development of neurons, glia, and astrocytes (Ahmed et al., 2015; Cappello et al., 2013; Ciani, Patel, Allen, & Ffrench-Constant, 2003; Huang, Hu, et al., 2016; Huang, Wang, et al., 2016; Lavado, Ware, Paré, & Cao, 2014; Lin et al., 2012; Sakuma et al., 2016).

Beyond ensuring proper development, the Hippo Pathway is implicated in maintaining neuronal health. The activation of MST1/2 kinases in situations of oxidative stress puts cells experiencing oxidative stress at risk for MST1/2-induced apoptosis. In mouse models of Amyotrophic Lateral Sclerosis (ALS) mutations in Sod1 are associated with increased MST1 activity in motor neurons (Lee et al., 2013). The positive feedback between oxidative stress and MST1 (Rawat et al., 2013) could maintain the environment of oxidative stress and enhance apoptosis of affected neurons.

As discussed, the Hippo Pathway not only promotes apoptosis but also plays a role in autophagy. Autophagy can be protective of cells suggesting that activation of the Hippo Pathway core cassette might protect cells from apoptosis in some contexts rather than promote their apoptosis. In fact, the role of Fat-induced Hippo signaling is neuroprotective by promoting autophagy (Calamita & Fanto, 2011; Napoletano et al., 2011). Polyglutamine (polyQ) proteins can promote neurodegeneration. Atrophin, a polyQ protein, promotes toxicity by downregulating Fat thereby blocking autophagy.

For additional review of the roles of the Hippo Pathway in neuronal diseases and neurodegeneration, see Emoto (2011).

8.4 Cancer and Chemotherapeutic Resistance

The roles of the Hippo Pathway to restrict growth and proliferation and promote apoptosis are key tumor suppressive functions, and the mutant phenotypes of cell cycle reentry, increased proliferation, tissue overgrowth, and cell death resistance are hallmarks of cancer cells. Maintaining a well-differentiated state rather than adopting stem/progenitor state is also a key tumor suppressive mechanism. For these reasons, a majority of the translational efforts of the field have focused on a role for the pathway in tumor suppression. The Hippo Pathway is now considered a bona fide tumor suppressor pathway, and the most characterized target Yki/YAP/TAZ a potent oncogene.

Neurofibromatosis type 2 (NF2) is an autosomal dominant familial tumor syndrome associated with mutation in Mer/NF2. NF2 is characterized by slow growing tumors of the nervous system, primarily schwannomas but also other tumor types including meningiomas, gliomas, ependymomas, and others (Gutmann, Giordano, Fishback, & Guha, 1997; Lau et al., 2008; Narod et al., 1992; Papi et al., 1995; Striedinger et al., 2008; Twist et al., 1994; Wellenreuther et al., 1995; Wolff et al., 1992; Xu, Stamenkovic, & Yu, 2010). Mer/NF2 mutations are frequently found in sporadic schwannomas, in schwannomas arising from Schwannomatosis (Kaufman et al., 2003), and in mesotheliomas (Bianchi et al., 1995; Sekido et al., 1995).

Genome-wide sequencing efforts, transcriptome analyses, and proteomic studies have yielded extensive databases from which to assess the expression levels, genomic amplification, or mutations affecting individual genes in a wide range of tumors (Bamford et al., 2004; Boon et al., 2002; Cerami et al., 2012; Gao et al., 2013; Uhlén et al., 2005). These and parallel efforts have revealed that mutations in the core components of the Hippo Pathway are found infrequently in tumors but their expression is often downregulated. In fact, downregulation of MST1/2 and Lats1/2 by silencing mechanisms is frequent in soft tissue sarcomas (Seidel et al., 2007) and also reported in other cancers such as breast cancer where it correlates with a more aggressive phenotype (Takahashi et al., 2005). YAP amplification is also common in a variety of tumors (Steinhardt et al., 2008). All together, inactivation of the core components and amplification of YAP are implicated in a range of cancers including breast cancer, gliomas, sarcomas, liver cancer, lung cancer, colorectal cancer, and many others. A role in cancer is further supported by tumor phenotypes in mouse liver models where the pathway has been extensively studied (Benhamouche et al., 2010; Lee et al., 2010; Lu et al., 2010).

Many research efforts focus on the dysregulation of the Hippo Pathway in driving tumorigenesis. A new avenue has emerged within these cancer efforts focused on cancer progression and resistance to cancer treatments. Many patients undergoing treatment show initial response to their cancer therapies, but after a period of time, their tumor cells can become resistant to the therapies. Resistant cancers typically have poor prognosis to further treatment. Alarmingly, the activation of YAP is emerging as mechanism of chemotherapeutic resistance in a growing number of contexts (Huo et al., 2013; Lin & Bivona, 2015; Lin et al., 2015; Mao et al., 2014; Marti et al., 2015; Xia, Zhang, Yu, Chang, & Fan, 2014; Yoshikawa et al., 2015).

For additional review of the Hippo Pathway in cancer, please see Lin et al. (2015), Zhao and Yang (2015), Harvey and Tapon (2007), Zeng and Hong (2008), Yu, Zhao, and Guan (2015), Pan (2010), and Zhao, Li, Lei, and Guan (2010).

9. CONCLUSIONS AND OPEN QUESTIONS

Our understanding of the Hippo signaling network has grown from its infancy and perhaps reached its adolescence. The field and concerted efforts by many groups across diverse organisms and model systems have already amassed a wealth of information, but a number of important open questions with relevance to development and disease remain.

A key question at the heart of the field is: how is the diversity of upstream outputs translated by the core cassette into diverse biological outputs as required by context? Discussions recent years have debated which upstream factors and downstream targets should be considered global components of the Hippo Pathway. This debate itself may in part answer this question—the expression, appropriate localization, or other regulatory aspects of some upstream factors and downstream targets may be restricted spatially, temporally, or in response to specific external cues to add complexity to the network. DIAP1 is regulated both transcriptionally through Yki (Huang et al., 2005; Jia et al., 2003; Udan et al., 2003; Wu et al., 2003) and post-translationally by the core complex (Harvey et al., 2003; Pantalacci et al., 2003); additional targets may undergo such dual regulation to ensure tight regulation and to fine-tune the levels of specific targets. It is striking that several processes serve as both regulators and targets of the pathway. Multiple feedback circuits including both positive and negative feedback confer additional mechanisms of complexity and control.

Discovery that the Hippo Pathway plays a role in organ size control and organ homeostasis was a major breakthrough in understanding an organ size checkpoint; however, the underlying molecular mechanisms are not fully understood. How is organ size set and then maintained? The homeostatic mechanisms that govern Hippo Pathway output and their multiple interfaces with cell architecture and tension may help cells interpret spatial positioning to integrate multiple cues locally and more broadly.

Most oncogenes and tumor suppressor signaling networks are characterized by frequent activating mutations or amplification in tumors in the case of oncogenes or by loss-of-function mutations or deletions in tumors in the case of tumor suppressors. In contrast, the tumor suppressor components of the Hippo Pathway are rarely mutated in tumors by comparison but are typically downregulated in expression, for example by silencing. This may explain why this signaling network eluded recognition and discovery for so many years but raises the question: why are the tumor suppressor components of the pathway rarely mutated in cancer? Could epigenetic strategies restore expression of silenced tumor suppressor components as a therapeutic strategy?

An ongoing area of investigation and debate is the ability of Yki/YAP/TAZ to promote growth, stemness, and/or progenitor proliferation in some contexts whereas it promotes differentiation, specific cell fates, or apoptosis in others. Similarly, whereas YAP is a potent oncogene in most contexts, there are reports of tumor suppressive functions for YAP in colorectal and breast cancers (Barry et al., 2013; Yuan et al., 2008) possibly due to its role in promoting differentiation or in promoting apoptosis when complexed with p73 (for additional discussion, see Bertini, Oka, Sudol,

Strano, & Blandino, 2009; Downward & Basu, 2008). What defines the output and role of YAP in growth, proliferation, or stemness, and driving or suppressing cancer?

The transcriptional role for Yki/YAP/TAZ is well documented and widely regarded as its primary function. Sequestration of Yki/YAP/TAZ in the cytoplasm or at junctions when complexed with 14-3-3, Ex, Hpo, Wts, or AMOT proteins prevents its transcriptional activity. These nonnuclear pools raise a controversial question: are there transcription-independent roles for Yki/YAP/TAZ? A recent report indicates a nontranscriptional role for YAP in cytokinesis (Bui et al., 2016).

Yki/YAP/TAZ is a crucial target of the pathway whose downregulation mediates numerous roles of the pathway. The emphasis and efforts focused on better understanding Yki/YAP/TAZ are well deserved. However, the recent emergence of several studies reporting Yki/YAP/TAZ-independent phenomena highlights a fundamental difference between inactivation of the core complex and activation of Yki/YAP/TAZ. Many of the seminal works identifying additional upstream regulators used Yki/YAP/TAZ transcriptional targets as readouts. Do upstream regulators preferentially target specific substrates and leave others unaffected? Are there additional upstream factors (missed by these efforts) that signal through the core cassette to regulate these Yki/YAP/TAZ-independent targets?

A number of therapeutic challenges remain. Can we exploit our knowledge of the role of the core complex and YAP/TAZ to increase the regenerative capacity of neurons and cardiac tissues to combat neurodegeneration and heart disease? Can we devise strategies to restrain YAP/TAZ oncogenic activity or to restore tumor suppression of the core cassette?

ACKNOWLEDGMENTS

Thank you to the dynamic and engaging Hippo Pathway research field whose work is contained in this review; my apologies for research not described given the depth and breadth of the field. Thank you to the Mount Sinai research community and special thanks to P. Wassarman, M. Mlodzik, R. Cagan, S. Aaronson, and J. Chipuk. Additional thank you to the New York *Drosophila* community and especially Andreas Jenny.

REFERENCES

Adler, J. J., Johnson, D. E., Heller, B. L., Bringman, L. R., Ranahan, W. P., Conwell, M. D., et al. (2013). Serum deprivation inhibits the transcriptional co-activator YAP and cell growth via phosphorylation of the 130-kDa isoform of Angiomotin by the LATS1/2 protein kinases. *Proceedings of the National Academy of Sciences of the United States of America*, *110*, 17368–17373.

Aerne, B. L., Gailite, I., Sims, D., & Tapon, N. (2015). Hippo stabilises its adaptor salvador by antagonising the HECT ubiquitin ligase Herc4. *PLoS One, 10*, e0131113.

Aguilar, A., Becker, L., Tedeschi, T., Heller, S., Iomini, C., & Nachury, M. V. (2014). α-Tubulin K40 acetylation is required for contact inhibition of proliferation and cell–substrate adhesion. *Molecular Biology of the Cell, 25*, 1854–1866.

Ahmed, A. F., de Bock, C. E., Lincz, L. F., Pundavela, J., Zouikr, I., Sontag, E., et al. (2015). FAT1 cadherin acts upstream of Hippo signalling through TAZ to regulate neuronal differentiation. *Cellular and Molecular Life Sciences, 72*, 4653–4669.

Ardestani, A., Paroni, F., Azizi, Z., Kaur, S., Khobragade, V., Yuan, T., et al. (2014). MST1 is a key regulator of beta cell apoptosis and dysfunction in diabetes. *Nature Medicine, 20*, 385–397.

Avruch, J., Praskova, M., Ortiz-Vega, S., Liu, M., & Zhang, X. F. (2006). Nore1 and RASSF1 regulation of cell proliferation and of the MST1/2 kinases. *Methods in Enzymology, 407*, 290–310.

Badouel, C., Gardano, L., Amin, N., Garg, A., Rosenfeld, R., Le Bihan, T., et al. (2009). The FERM-domain protein Expanded regulates Hippo pathway activity via direct interactions with the transcriptional activator Yorkie. *Developmental Cell, 16*, 411–420.

Bagherie-Lachidan, M., Reginensi, A., Pan, Q., Zaveri, H. P., Scott, D. A., Blencowe, B. J., et al. (2015). Stromal Fat4 acts non-autonomously with Dchs1/2 to restrict the nephron progenitor pool. *Development, 142*, 2564–2573.

Bamford, S., Dawson, E., Forbes, S., Clements, J., Pettett, R., Dogan, A., et al. (2004). The COSMIC (Catalogue of Somatic Mutations in Cancer) database and website. *British Journal of Cancer, 91*, 355–358.

Barry, E. R., Morikawa, T., Butler, B. L., Shrestha, K., de la Rosa, R., Yan, K. S., et al. (2013). Restriction of intestinal stem cell expansion and the regenerative response by YAP. *Nature, 493*, 106–110.

Basu, S., Totty, N. F., Irwin, M. S., Sudol, M., & Downward, J. (2003). Akt phosphorylates the Yes-associated protein, YAP, to induce interaction with 14-3-3 and attenuation of p73-mediated apoptosis. *Molecular Cell, 11*, 11–23.

Baumgartner, R., Poernbacher, I., Buser, N., Hafen, E., & Stocker, H. (2010). The WW domain protein Kibra acts upstream of Hippo in Drosophila. *Developmental Cell, 18*, 309–316.

Benhamouche, S., Curto, M., Saotome, I., Gladden, A. B., Liu, C. H., Giovannini, M., et al. (2010). Nf2/Merlin controls progenitor homeostasis and tumorigenesis in the liver. *Genes & Development, 24*, 1718–1730.

Bensenor, L. B., Barlan, K., Rice, S. E., Fehon, R. G., & Gelfand, V. I. (2010). Microtubule-mediated transport of the tumor-suppressor protein Merlin and its mutants. *Proceedings of the National Academy of Sciences of the United States of America, 107*, 7311–7316.

Bertini, E., Oka, T., Sudol, M., Strano, S., & Blandino, G. (2009). YAP: At the crossroad between transformation and tumor suppression. *Cell Cycle, 8*, 49–57.

Bettencourt-Dias, M., Giet, R., Sinka, R., Mazumdar, A., Lock, W. G., Balloux, F., et al. (2004). Genome-wide survey of protein kinases required for cell cycle progression. *Nature, 432*, 980–987.

Bianchi, A. B., Mitsunaga, S. I., Cheng, J. Q., Klein, W. M., Jhanwar, S. C., Seizinger, B., et al. (1995). High frequency of inactivating mutations in the neurofibromatosis type 2 gene (NF2) in primary malignant mesotheliomas. *Proceedings of the National Academy of Sciences of the United States of America, 92*, 10854–10858.

Blower, M. D., Nachury, M., Heald, R., & Weis, K. (2005). A Rae1-containing ribonucleoprotein complex is required for mitotic spindle assembly. *Cell, 121*, 223–234.

Boggiano, J. C., Vanderzalm, P. J., & Fehon, R. G. (2011). Tao-1 phosphorylates Hippo/MST kinases to regulate the Hippo–Salvador–Warts tumor suppressor pathway. *Developmental Cell, 21*, 888–895.

Boon, K., Osorio, E. C., Greenhut, S. F., Schaefer, C. F., Shoemaker, J., Polyak, K., et al. (2002). An anatomy of normal and malignant gene expression. *Proceedings of the National Academy of Sciences of the United States of America, 99*, 11287–11292.

Bossuyt, W., Chen, C. L., Chen, Q., Sudol, M., McNeill, H., Pan, D., et al. (2014). An evolutionary shift in the regulation of the Hippo pathway between mice and flies. *Oncogene, 33*, 1218–1228.

Brown, J. A., Bharathi, A., Ghosh, A., Whalen, W., Fitzgerald, E., & Dhar, R. (1995). A mutation in the Schizosaccharomyces pombe rae1 gene causes defects in poly(A)+ RNA export and in the cytoskeleton. *The Journal of Biological Chemistry, 270*, 7411–7419.

Bui, D. A., Lee, W., White, A. E., Harper, J. W., Schackmann, R. C., Overholtzer, M., et al. (2016). Cytokinesis involves a nontranscriptional function of the Hippo pathway effector YAP. *Science Signaling, 9*, ra23.

Cai, Q., Wang, W., Gao, Y., Yang, Y., Zhu, Z., & Fan, Q. (2009). Ce-wts-1 plays important roles in Caenorhabditis elegans development. *FEBS Letters, 583*, 3158–3164.

Calamita, P., & Fanto, M. (2011). Slimming down fat makes neuropathic Hippo: The Fat/Hippo tumor suppressor pathway protects adult neurons through regulation of autophagy. *Autophagy, 7*, 907–909.

Callus, B. A., Verhagen, A. M., & Vaux, D. L. (2006). Association of mammalian sterile twenty kinases, Mst1 and Mst2, with hSalvador via C-terminal coiled-coil domains, leads to its stabilization and phosphorylation. *The FEBS Journal, 273*, 4264–4276.

Camargo, F. D., Gokhale, S., Johnnidis, J. B., Fu, D., Bell, G. W., Jaenisch, R., et al. (2007). YAP1 increases organ size and expands undifferentiated progenitor cells. *Current Biology, 17*, 2054–2060.

Cao, X., Pfaff, S. L., & Gage, F. H. (2008). YAP regulates neural progenitor cell number via the TEA domain transcription factor. *Genes & Development, 22*, 3320–3334.

Cappello, S., Gray, M. J., Badouel, C., Lange, S., Einsiedler, M., Srour, M., et al. (2013). Mutations in genes encoding the cadherin receptor–ligand pair DCHS1 and FAT4 disrupt cerebral cortical development. *Nature Genetics, 45*, 1300–1308.

Cerami, E., Gao, J., Dogrusoz, U., Gross, B. E., Sumer, S. O., Aksoy, B. A., et al. (2012). The cBio cancer genomics portal: An open platform for exploring multidimensional cancer genomics data. *Cancer Discovery, 2*, 401–404.

Chan, S. W., Lim, C. J., Chong, Y. F., Pobbati, A. V., Huang, C., & Hong, W. (2011). Hippo pathway-independent restriction of TAZ and YAP by angiomotin. *The Journal of Biological Chemistry, 286*, 7018–7026.

Chan, S. W., Lim, C. J., Guo, F., Tan, I., Leung, T., & Hong, W. (2013). Actin-binding and cell proliferation activities of angiomotin family members are regulated by Hippo pathway-mediated phosphorylation. *The Journal of Biological Chemistry, 288*, 37296–37307.

Chen, C. L., Gajewski, K. M., Hamaratoglu, F., Bossuyt, W., Sansores-Garcia, L., Tao, C., et al. (2010). The apical-basal cell polarity determinant Crumbs regulates Hippo signaling in Drosophila. *Proceedings of the National Academy of Sciences of the United States of America, 107*, 15810–15815.

Chen, S. N., Gurha, P., Lombardi, R., Ruggiero, A., Willerson, J. T., & Marian, A. J. (2014). The Hippo pathway is activated and is a causal mechanism for adipogenesis in arrhythmogenic osqmyopathy. *Circulation Research, 114*, 454–468.

Cheung, W. L., Ajiro, K., Samejima, K., Kloc, M., Cheung, P., Mizzen, C. A., et al. (2003). Apoptotic phosphorylation of histone H2B is mediated by mammalian sterile twenty kinase. *Cell, 113*, 507–517.

Cho, E., Feng, Y., Rauskolb, C., Maitra, S., Fehon, R., & Irvine, K. D. (2006). Delineation of a Fat tumor suppressor pathway. *Nature Genetics*, *38*, 1142–1150.

Ciani, L., Patel, A., Allen, N. D., & Ffrench-Constant, C. (2003). Mice lacking the giant protocadherin mFAT1 exhibit renal slit junction abnormalities and a partially penetrant cyclopia and anophthalmia phenotype. *Molecular and Cellular Biology*, *23*, 3575–3582.

Colombani, J., Polesello, C., Josué, F., & Tapon, N. (2006). Dmp53 activates the Hippo pathway to promote cell death in response to DNA damage. *Current Biology*, *16*, 1453–1458.

Cooper, J., Li, W., You, L., Schiavon, G., Pepe-Caprio, A., Zhou, L., et al. (2011). Merlin/NF2 functions upstream of the nuclear E3 ubiquitin ligase CRL4DCAF1 to suppress oncogenic gene expression. *Science Signaling*, *4*, pt6.

Creasy, C. L., & Chernoff, J. (1995a). Cloning and characterization of a human protein kinase with homology to Ste20. *The Journal of Biological Chemistry*, *270*, 21695–21700.

Creasy, C. L., & Chernoff, J. (1995b). Cloning and characterization of a member of the MST subfamily of Ste20-like kinases. *Gene*, *167*, 303–306.

Cuende, J., Moreno, S., Bolaños, J. P., & Almeida, A. (2008). Retinoic acid downregulates Rae1 leading to APC(Cdh1) activation and neuroblastoma SH-SY5Y differentiation. *Oncogene*, *27*, 3339–3344.

Curto, M., Cole, B. K., Lallemand, D., Liu, C. H., & McClatchey, A. I. (2007). Contact-dependent inhibition of EGFR signaling by Nf2/Merlin. *The Journal of Cell Biology*, *177*, 893–903.

Dai, X., She, P., Chi, F., Feng, Y., Liu, H., Jin, D., et al. (2013). Phosphorylation of angiomotin by Lats1/2 kinases inhibits F-actin binding, cell migration, and angiogenesis. *The Journal of Biological Chemistry*, *288*, 34041–34051.

Das Thakur, M., Feng, Y., Jagannathan, R., Seppa, M. J., Skeath, J. B., & Longmore, G. D. (2010). Ajuba LIM proteins are negative regulators of the Hippo signaling pathway. *Current Biology*, *20*, 657–662.

Deng, H., Wang, W., Yu, J., Zheng, Y., Qing, Y., & Pan, D. (2015). Spectrin regulates Hippo signaling by modulating cortical actomyosin activity. *eLife*, *4*, e06567.

Dewey, E. B., Sanchez, D., & Johnston, C. A. (2015). Warts phosphorylates mud to promote pins-mediated mitotic spindle orientation in Drosophila, independent of Yorkie. *Current Biology*, *25*, 2751–2762.

Diepenbruck, M., Waldmeier, L., Ivanek, R., Berninger, P., Arnold, P., van Nimwegen, E., et al. (2014). Tead2 expression levels control the subcellular distribution of Yap and Taz, zyxin expression and epithelial–mesenchymal transition. *Journal of Cell Science*, *127*, 1523–1536.

Ding, R., Weynans, K., Bossing, T., Barros, C. S., & Berger, C. (2016). The Hippo signalling pathway maintains quiescence in Drosophila neural stem cells. *Nature Communications*, *7*, 10510.

Dong, J., Feldmann, G., Huang, J., Wu, S., Zhang, N., Comerford, S. A., et al. (2007). Elucidation of a universal size-control mechanism in Drosophila and mammals. *Cell*, *130*, 1120–1133.

Downward, J., & Basu, S. (2008). YAP and p73: A complex affair. *Molecular Cell*, *32*, 749–750.

Dupont, S., Morsut, L., Aragona, M., Enzo, E., Giulitti, S., Cordenonsi, M., et al. (2011). Role of YAP/TAZ in mechanotransduction. *Nature*, *474*, 179–183.

Dutta, S., & Baehrecke, E. H. (2008). Warts is required for PI3K-regulated growth arrest, autophagy, and autophagic cell death in Drosophila. *Current Biology*, *18*, 1466–1475.

Emoto, K. (2011). The growing role of the Hippo—NDR kinase signalling in neuronal development and disease. *The Journal of Biochemistry*, *150*, 133–141.

Emoto, K., Parrish, J. Z., Jan, L. Y., & Jan, Y. N. (2006). The tumour suppressor Hippo acts with the NDR kinases in dendritic tiling and maintenance. *Nature*, *443*, 210–213.

Ernkvist, M., Aase, K., Ukomadu, C., Wohlschlegel, J., Blackman, R., Veitonmäki, N., et al. (2006). p130-Angiomotin associates to actin and controls endothelial cell shape. *FEBS Journal, 273,* 2000–2011.

Fernández, B. G., Gaspar, P., Brás-Pereira, C., Jezowska, B., Rebelo, S. R., & Janody, F. (2011). Actin-capping protein and the Hippo pathway regulate F-actin and tissue growth in Drosophila. *Development, 138,* 2237–2246.

Fletcher, G. C., Elbediwy, A., Khanal, I., Ribeiro, P. S., Tapon, N., & Thompson, B. J. (2015). The Spectrin cytoskeleton regulates the Hippo signalling pathway. *The EMBO Journal, 34,* 940–954.

Fossdal, R., Jonasson, F., Kristjansdottir, G. T., Kong, A., Stefansson, H., Gosh, S., et al. (2004). A novel TEAD1 mutation is the causative allele in Sveinsson's chorioretinal atrophy (helicoid peripapillary chorioretinal degeneration). *Human Molecular Genetics, 13,* 975–981.

Frenz, L. M., Lee, S. E., Fesquet, D., & Johnston, L. H. (2000). The budding yeast Dbf2 protein kinase localises to the centrosome and moves to the bud neck in late mitosis. *Journal of Cell Science, 113,* 3399–33408.

Gagné, V., Moreau, J., Plourde, M., Lapointe, M., Lord, M., Gagnon, E., et al. (2009). Human angiomotin-like 1 associates with an angiomotin protein complex through its coiled-coil domain and induces the remodeling of the actin cytoskeleton. *Cell Motility and the Cytoskeleton, 66,* 754–768.

Gao, J., Aksoy, B. A., Dogrusoz, U., Dresdner, G., Gross, B., Sumer, S. O., et al. (2013). Integrative analysis of complex cancer genomics and clinical profiles using the cBioPortal. *Science Signaling, 6,* pl1.

Gaspar, P., Holder, M. V., Aerne, B. L., Janody, F., & Tapon, N. (2015). Zyxin antagonizes the FERM protein expanded to couple F-actin and Yorkie-dependent organ growth. *Current Biology, 25,* 679–689.

Gaspar, P., & Tapon, N. (2014). Sensing the local environment: Actin architecture and Hippo signalling. *Current Opinion in Cell Biology, 31,* 74–83.

Genevet, A., Polesello, C., Blight, K., Robertson, F., Collinson, L. M., Pichaud, F., et al. (2009). The Hippo pathway regulates apical-domain size independently of its growth-control function. *Journal of Cell Science, 122,* 2360–2370.

Genevet, A., & Tapon, N. (2011). The Hippo pathway and apico-basal cell polarity. *Biochemical Journal, 436,* 213–224.

Genevet, A., Wehr, M. C., Brain, R., Thompson, B. J., & Tapon, N. (2010). Kibra is a regulator of the Salvador/Warts/Hippo signaling network. *Developmental Cell, 18,* 300–308.

Giovannini, M., Robanus-Maandag, E., van der Valk, M., Niwa-Kawakita, M., Abramowski, V., Goutebroze, L., et al. (2000). Conditional biallelic Nf2 mutation in the mouse promotes manifestations of human neurofibromatosis type 2. *Genes & Development, 14,* 1617–1630.

Gregorieff, A., Liu, Y., Inanlou, M. R., Khomchuk, Y., & Wrana, J. L. (2015). Yap-dependent reprogramming of Lgr5+ stem cells drives intestinal regeneration and cancer. *Nature, 526,* 715–718.

Grusche, F. A., Richardson, H. E., & Harvey, K. F. (2010). Upstream regulation of the Hippo size control pathway. *Current Biology, 20,* R574–R582.

Grzeschik, N. A., Parsons, L. M., Allott, M. L., Harvey, K. F., & Richardson, H. E. (2010). Lgl, aPKC, and Crumbs regulate the Salvador/Warts/Hippo pathway through two distinct mechanisms. *Current Biology, 20,* 573–581.

Guo, C., Tommasi, S., Liu, L., Yee, J. K., Dammann, R., & Pfeifer, G. P. (2007). RASSF1A is part of a complex similar to the Drosophila Hippo/Salvador/Lats tumor-suppressor network. *Current Biology, 17,* 700–705.

Gutmann, D. H., Giordano, M. J., Fishback, A. S., & Guha, A. (1997). Loss of merlin expression in sporadic meningiomas, ependymomas and schwannomas. *Neurology, 49*, 267–270.
Hafezi, Y., Bosch, J. A., & Hariharan, I. K. (2012). Differences in levels of the transmembrane protein Crumbs can influence cell survival at clonal boundaries. *Developmental Biology, 368*, 358–369.
Hamaratoglu, F., Gajewski, K., Sansores-Garcia, L., Morrison, C., Tao, C., & Halder, G. (2009). The Hippo tumor-suppressor pathway regulates apical-domain size in parallel to tissue growth. *Journal of Cell Science, 122*, 2351–2359.
Hamaratoglu, F., Willecke, M., Kango-Singh, M., Nolo, R., Hyun, E., Tao, C., et al. (2006). The tumour-suppressor genes NF2/Merlin and Expanded act through Hippo signalling to regulate cell proliferation and apoptosis. *Nature Cell Biology, 8*, 27–36.
Hao, J., Zhang, Y., Wang, Y., Ye, R., Qiu, J., Zhao, Z., et al. (2014). Role of extracellular matrix and YAP/TAZ in cell fate determination. *Cellular Signalling, 26*, 186–191.
Hariharan, I. K. (2015). Organ size control: Lessons from Drosophila. *Developmental Cell, 34*, 255–265.
Harvey, K. F., Pfleger, C. M., & Hariharan, I. K. (2003). The Drosophila Mst ortholog, Hippo, restricts growth and cell proliferation and promotes apoptosis. *Cell, 114*, 457–467.
Harvey, K. F., & Tapon, N. (2007). The Salvador–Warts–Hippo pathway—An emerging tumour-suppressor network. *Nature Reviews Cancer, 7*, 182–191.
Hayashi, S., Ochi, H., Ogino, H., Kawasumi, A., Kamei, Y., Tamura, K., et al. (2014). Transcriptional regulators in the Hippo signaling pathway control organ growth in Xenopus tadpole tail regeneration. *Developmental Biology, 396*, 31–41.
Heallen, T., Morikawa, Y., Leach, J., Tao, G., Willerson, J. T., Johnson, R. L., et al. (2013). Hippo signaling impedes adult heart regeneration. *Development, 140*, 4683–4690.
Heallen, T., Zhang, M., Wang, J., Bonilla-Claudio, M., Klysik, E., Johnson, R. L., et al. (2011). Hippo pathway inhibits Wnt signaling to restrain cardiomyocyte proliferation and heart size. *Science, 332*, 458–461.
Hergovich, A., Kohler, R. S., Schmitz, D., Vichalkovski, A., Cornils, H., & Hemmings, B. A. (2009). The MST1 and hMOB1 tumor suppressors control human centrosome duplication by regulating NDR kinase phosphorylation. *Current Biology, 19*, 1692–1702.
Hergovich, A., Schmitz, D., & Hemmings, B. A. (2006). The human tumour suppressor LATS1 is activated by human MOB1 at the membrane. *Biochemical and Biophysical Research Communications, 345*, 50–58.
Hirabayashi, S., & Cagan, R. L. (2015). Salt-inducible kinases mediate nutrient-sensing to link dietary sugar and tumorigenesis in Drosophila. *eLife, 4*, e08501.
Ho, K. C., Zhou, Z., She, Y. M., Chun, A., Cyr, T. D., & Yang, X. (2011). Itch E3 ubiquitin ligase regulates large tumor suppressor 1 stability. *Proceedings of the National Academy of Sciences of the United States of America, 108*, 4870–4875.
Huang, Z., Hu, J., Pan, J., Wang, Y., Hu, G., Zhou, J., et al. (2016). YAP stabilizes SMAD1 and promotes BMP2-induced neocortical astrocytic differentiation. *Development, 143*, 2398–2409.
Huang, Z., Wang, Y., Hu, G., Zhou, J., Mei, L., & Xiong, W. C. (2016). YAP is a critical inducer of SOCS3, preventing reactive astrogliosis. *Cerebral Cortex, 26*, 2299–2310.
Huang, H. L., Wang, S., Yin, M. X., Dong, L., Wang, C., Wu, W., et al. (2013). Par-1 regulates tissue growth by influencing Hippo phosphorylation status and Hippo–Salvador Association. *PLoS Biology, 11*, e1001620.
Huang, J., Wu, S., Barrera, J., Matthews, K., & Pan, D. (2005). The Hippo signaling pathway coordinately regulates cell proliferation and apoptosis by inactivating Yorkie, the Drosophila homolog of YAP. *Cell, 122*, 421–434.

Huo, X., Zhang, Q., Liu, A. M., Tang, C., Gong, Y., Bian, J., et al. (2013). Overexpression of Yes-associated protein confers doxorubicin resistance in hepatocellullar carcinoma. *Oncology Reports, 29,* 840–846.

Iida, S., Hirota, T., Morisaki, T., Marumoto, T., Hara, T., Kuninaka, S., et al. (2004). Tumor suppressor WARTS ensures genomic integrity by regulating both mitotic progression and G1 tetraploidy checkpoint function. *Oncogene, 23,* 5266–5274.

Ikmi, A., Gaertner, B., Seidel, C., Srivastava, M., Zeitlinger, J., & Gibson, M. C. (2014). Molecular evolution of the Yap/Yorkie proto-oncogene and elucidation of its core transcriptional program. *Molecular Biology and Evolution, 31,* 1375–1390.

Ilanges, A., Jahanshahi, M., Balobin, D. M., & Pfleger, C. M. (2013). Alcohol interacts with genetic alteration of the Hippo tumor suppressor pathway to modulate tissue growth in Drosophila. *PLoS One, 8,* e78880.

Jahanshahi, M., Hsiao, K., Jenny, A., & Pfleger, C. M. (2016). The Hippo Pathway targets Rae1 to regulate mitosis and organ size and to feed back to regulate upstream components Merlin, Hippo, and Warts. *PLoS Genetics, 12,* e1006198.

Jang, J. W., Kim, M. K., Lee, Y. S., Lee, J. W., Kim, D. M., Song, S. H., et al. (2016). RAC-LATS1/2 signaling regulates YAP activity by switching between the YAP-binding partners TEAD4 and RUNX3. *Oncogene.* http://dx.doi.org/10.1038/onc.2016.266.

Jaspersen, S. L., Charles, J. F., Tinker-Kulberg, R. L., & Morgan, D. O. (1998). A late mitotic regulatory network controlling cyclin destruction in Saccharomyces cerevisiae. *Molecular Biology of the Cell, 9,* 2803–2817.

Jeganathan, K. B., Baker, D. J., & van Deursen, J. M. (2006). Securin associates with APCCdh1 in prometaphase but its destruction is delayed by Rae1 and Nup98 until the metaphase/anaphase transition. *Cell Cycle, 5,* 366–370.

Jeganathan, K. B., Malureanu, L., & van Deursen, J. M. (2005). The Rae1–Nup98 complex prevents aneuploidy by inhibiting securin degradation. *Nature, 438,* 1036–1039.

Jia, J., Zhang, W., Wang, B., Trinko, R., & Jiang, J. (2003). The Drosophila Ste20 family kinase dMST functions as a tumor suppressor by restricting cell proliferation and promoting apoptosis. *Genes & Development, 17,* 2514–2519.

Judson, R. N., Tremblay, A. M., Knopp, P., White, R. B., Urcia, R., De Bari, C., et al. (2012). The Hippo pathway member Yap plays a key role in influencing fate decisions in muscle satellite cells. *Journal of Cell Science, 125,* 6009–6019.

Jukam, D., & Desplan, C. (2011). Binary regulation of Hippo pathway by Merlin/NF2, Kibra, Lgl, and Melted specifies and maintains postmitotic neuronal fate. *Developmental Cell, 21,* 874–887.

Jukam, D., Xie, B., Rister, J., Terrell, D., Charlton-Perkins, M., Pistillo, D., et al. (2013). Opposite feedbacks in the Hippo pathway for growth control and neural fate. *Science, 342,* 1238016.

Justice, R. W., Zilian, O., Woods, D. F., Noll, M., & Bryant, P. J. (1995). The Drosophila tumor suppressor gene warts encodes a homolog of human myotonic dystrophy kinase and is required for the control of cell shape and proliferation. *Genes & Development, 9,* 534–546.

Kakeya, H., Onose, R., & Osada, H. (1998). Caspase-mediated activation of a 36-kDa myelin basic protein kinase during anticancer drug-induced apoptosis. *Cancer Research, 58,* 4888–4894.

Kango-Singh, M., Nolo, R., Tao, C., Verstreken, P., Hiesinger, P. R., Bellen, H. J., et al. (2002). Shar-pei mediates cell proliferation arrest during imaginal disc growth in Drosophila. *Development, 129,* 5719–5730.

Kaufman, D. L., Heinrich, B. S., Willett, C., Perry, A., Finseth, F., Sobel, R. A., et al. (2003). Somatic instability of the NF2 gene in schwannomatosis. *Archives of Neurology, 60,* 1317–1320.

Keder, A., Rives-Quinto, N., Aerne, B. L., Franco, M., Tapon, N., & Carmena, A. (2015). The Hippo pathway core cassette regulates asymmetric cell division. *Current Biology, 25*, 2739–2750.

Kim, N. G., & Gumbiner, B. M. (2015). Adhesion to fibronectin regulates Hippo signaling via the FAK-Src-PI3K pathway. *The Journal of Cell Biology, 210*, 503–515.

Kim, M., Kim, M., Lee, S., Kuninaka, S., Saya, H., Lee, H., et al. (2013). cAMP/PKA signalling reinforces the LATS-YAP pathway to fully suppress YAP in response to actin cytoskeletal changes. *The EMBO Journal, 32*, 1543–1555.

Kim, N. G., Koh, E., Chen, X., & Gumbiner, B. M. (2011). E-cadherin mediates contact inhibition of proliferation through Hippo signaling-pathway components. *Proceedings of the National Academy of Sciences of the United States of America, 108*, 11930–11935.

Kitagawa, M. (2007). A Sveinsson's chorioretinal atrophy-associated missense mutation in mouse Tead1 affects its interaction with the co-factors YAP and TAZ. *Biochemical and Biophysical Research Communications, 361*, 1022–1026.

Koontz, L. M., Liu-Chittenden, Y., Yin, F., Zheng, Y., Yu, J., Huang, B., et al. (2013). The Hippo effector Yorkie controls normal tissue growth by antagonizing scalloped-mediated default repression. *Developmental Cell, 25*, 388–401.

Kraemer, D., Dresbach, T., & Drenckhahn, D. (2001). Mrnp41 (Rae 1p) associates with microtubules in HeLa cells and in neurons. *European Journal of Cell Biology, 80*, 733–740.

Lai, Z. C., Wei, X., Shimizu, T., Ramos, E., Rohrbaugh, M., Nikolaidis, N., et al. (2005). Control of cell proliferation and apoptosis by mob as tumor suppressor, mats. *Cell, 120*, 675–685.

Lau, Y. K., Murray, L. B., Houshmandi, S. S., Xu, Y., Gutmann, D. H., & Yu, Q. (2008). Merlin is a potent inhibitor of glioma growth. *Cancer Research, 68*, 5733–5742.

Lavado, A., Ware, M., Paré, J., & Cao, X. (2014). The tumor suppressor Nf2 regulates corpus callosum development by inhibiting the transcriptional coactivator Yap. *Development, 141*, 4182–4193.

Lee, K. P., Lee, J. H., Kim, T. S., Kim, T. H., Park, H. D., Byun, J. S., et al. (2010). The Hippo–Salvador pathway restrains hepatic oval cell proliferation, liver size, and liver tumorigenesis. *Proceedings of the National Academy of Sciences of the United States of America, 107*, 8248–8253.

Lee, J.-Y., Lee, H.-S., Wi, S.-J., Park, K. Y., Schmit, A.-C., & Pai, H.-S. (2009). Dual functions of Nicotiana benthamiana Rae1 in interphase and mitosis. *The Plant Journal, 59*, 278–291.

Lee, J. K., Shin, J. H., Hwang, S. G., Gwag, B. J., McKee, A. C., Lee, J., et al. (2013). MST1 functions as a key modulator of neurodegeneration in a mouse model of ALS. *Proceedings of the National Academy of Sciences of the United States of America, 110*, 12066–12071.

Leevers, S. J., & McNeill, H. (2005). Controlling the size of organs and organisms. *Current Opinion in Cell Biology, 17*, 604–609.

Lehtinen, M. K., Yuan, Z., Boag, P. R., Yang, Y., Villén, J., Becker, E. B., et al. (2006). A conserved MST-FOXO signaling pathway mediates oxidative-stress responses and extends life span. *Cell, 125*, 987–1001.

Li, S., Cho, Y. S., Yue, T., Ip, Y. T., & Jiang, J. (2015). Overlapping functions of the MAP4K family kinases Hppy and Msn in Hippo signaling. *Cell Discovery, 1*, 15038.

Li, W., Cooper, J., Zhou, L., Yang, C., Erdjument-Bromage, H., Zagzag, D., et al. (2014). Merlin/NF2 loss-driven tumorigenesis linked to CRL4(DCAF1)-mediated inhibition of the Hippo pathway kinases Lats1 and 2 in the nucleus. *Cancer Cell, 26*, 48–60.

Li, W., You, L., Cooper, J., Schiavon, G., Pepe-Caprio, A., Zhou, L., et al. (2010). Merlin/NF2 suppresses tumorigenesis by inhibiting the E3 ubiquitin ligaseCRL4(DCAF1) in the nucleus. *Cell, 140*, 477–490.

Lian, I., Kim, J., Okazawa, H., Zhao, J., Zhao, B., Yu, J., et al. (2010). The role of YAP transcription coactivator in regulating stem cell self-renewal and differentiation. *Genes & Development, 24*, 1106–1118.

Lin, L., & Bivona, T. G. (2015). The Hippo effector YAP regulates the response of cancer cells to MAPK pathway inhibitors. *Molecular & Cellular Oncology, 3*, e1021441.

Lin, Y. T., Ding, J. Y., Li, M. Y., Yeh, T. S., Wang, T. W., & Yu, J. Y. (2012). YAP regulates neuronal differentiation through Sonic hedgehog signaling pathway. *Experimental Cell Research, 318*, 1877–1888.

Lin, A. Y., & Pearson, B. J. (2014). Planarian yorkie/YAP functions to integrate adult stem cell proliferation, organ homeostasis and maintenance of axial patterning. *Development, 141*, 1197–1208.

Lin, Z., & Pu, W. T. (2014). Harnessing Hippo in the heart: Hippo/Yap signaling and applications to heartregeneration and rejuvenation. *Stem Cell Research, 13*, 571–581.

Lin, L., Sabnis, A. J., Chan, E., Olivas, V., Cade, L., Pazarentzos, E., et al. (2015). The Hippo effector YAP promotes resistance to RAF- and MEK-targeted cancer therapies. *Nature Genetics, 47*, 250–256.

Lin, Z., von Gise, A., Zhou, P., Gu, F., Ma, Q., Jiang, J., et al. (2014). Cardiac-specific YAP activation improves cardiac function and survival in an experimental murine MI model. *Circulation Research, 115*, 354–363.

Ling, C., Zheng, Y., Yin, F., Yu, J., Huang, J., Hong, Y., et al. (2010). The apical transmembrane protein Crumbs functions as a tumor suppressor that regulates Hippo signaling by binding to Expanded. *Proceedings of the National Academy of Sciences of the United States of America, 107*, 10532–10537.

Liu, C. Y., Zha, Z. Y., Zhou, X., Zhang, H., Huang, W., Zhao, D., et al. (2010). The Hippo tumor pathway promotes TAZ degradation by phosphorylating a phosphodegron and recruiting the SCF{beta}-TrCP E3 ligase. *The Journal of Biological Chemistry, 285*, 37159–37169.

Lu, L., Li, Y., Kim, S. M., Bossuyt, W., Liu, P., Qiu, Q., et al. (2010). Hippo signaling is a potent in vivo growth and tumor suppressor pathway in the mammalian liver. *Proceedings of the National Academy of Sciences of the United States of America, 107*, 1437–1442.

Lucas, E. P., Khanal, I., Gaspar, P., Fletcher, G. C., Polesello, C., Tapon, N., et al. (2013). The Hippo pathway polarizes the actin cytoskeleton during collective migration of Drosophila border cells. *The Journal of Cell Biology, 201*, 875–885.

Lv, M., Lv, M., Chen, L., Qin, T., Zhang, X., Liu, P., et al. (2015). Angiomotin promotes breast cancer cell proliferation and invasion. *Oncology Reports, 33*, 1938–1946.

Maejima, Y., Isobe, M., & Sadoshima, J. (2016). Regulation of autophagy by Beclin 1 in the heart. *Journal of Molecular and Cellular Cardiology, 95*, 19–25.

Mah, A. S., Elia, A. E., Devgan, G., Ptacek, J., Schutkowski, M., Snyder, M., et al. (2005). Substrate specificity analysis of protein kinase complex Dbf2-Mob1 by peptide library and proteome array screening. *BMC Biochemistry, 6*, 22.

Mah, A. S., Jang, J., & Deshaies, R. J. (2001). Protein kinase Cdc15 activates the Dbf2–Mob1 kinase complex. *Proceedings of the National Academy of Sciences of the United States of America, 98*(13), 7325–7330.

Mana-Capelli, S., Paramasivam, M., Dutta, S., & McCollum, D. (2014). Angiomotins link F-actin architecture to Hippo pathway signaling. *Molecular Biology of the Cell, 25*, 1676–1685.

Mao, B., Hu, F., Cheng, J., Wang, P., Xu, M., Yuan, F., et al. (2014). SIRT1 regulates YAP2-mediated cell proliferation and chemoresistance in hepatocellular carcinoma. *Oncogene, 33*, 1468–1474.

Marcinkevicius, E., & Zallen, J. A. (2013). Regulation of cytoskeletal organization and junctional remodeling by the atypical cadherin Fat. *Development, 140*, 433–443.

Marti, P., Stein, C., Blumer, T., Abraham, Y., Dill, M. T., Pikiolek, M., et al. (2015). YAP promotes proliferation, chemoresistance, and angiogenesis in human cholangiocarcinoma through TEAD transcription factors. *Hepatology*, *62*, 1497–1510.

Martinez, G., & de Iongh, R. U. (2010). The lens epithelium in ocular health and disease. *The International Journal of Biochemistry & Cell Biology*, *42*, 1945–1963.

McClure, K. D., French, R. L., & Heberlein, U. (2011). A Drosophila model for fetal alcohol syndrome disorders: Role for the insulin pathway. *Disease Models & Mechanisms*, *4*, 335–346.

Meng, Z., Moroishi, T., & Guan, K. L. (2016). Mechanisms of Hippo pathway regulation. *Genes & Development*, *30*, 1–17.

Meng, Z., Moroishi, T., Mottier-Pavie, V., Plouffe, S. W., Hansen, C. G., Hong, A. W., et al. (2015). MAP4K family kinases act in parallel to MST1/2 to activate LATS1/2 in the Hippo pathway. *Nature Communications*, *6*, 8357.

Meyers, S. M., Gutman, F. A., Kaye, L. D., & Rothner, A. D. (1995). Retinal changes associated with neurofibromatosis 2. *Transactions of the American Ophthalmological Society*, *93*, 245–252.

Mikeladze-Dvali, T., Wernet, M. F., Pistillo, D., Mazzoni, E. O., Teleman, A. A., Chen, Y. W., et al. (2005). The growth regulators warts/lats and melted interact in a bistable loop to specify opposite fates in Drosophila R8 photoreceptors. *Cell*, *122*, 775–787.

Mo, J. S., Yu, F. X., Gong, R., Brown, J. H., & Guan, K. L. (2012). Regulation of the Hippo-YAP pathway by protease-activated receptors (PARs). *Genes & Development*, *26*, 2138–2143.

Moroishi, T., Park, H. W., Qin, B., Chen, Q., Meng, Z., Plouffe, S. W., et al. (2015). A YAP/TAZ-induced feedback mechanism regulates Hippo pathway homeostasis. *Genes & Development*, *29*, 1271–1284.

Morrison, H., Sherman, L. S., Legg, J., Banine, F., Isacke, C., Haipek, C. A., et al. (2001). The NF2 tumor suppressor gene product, merlin, mediates contact inhibition of growth through interactions with CD44. *Genes & Development*, *15*, 968–980.

Mosqueira, D., Pagliari, S., Uto, K., Ebara, M., Romanazzo, S., Escobedo-Lucea, C., et al. (2014). Hippo pathway effectors control cardiac progenitor cell fate by acting as dynamic sensors of substrate mechanics and nanostructure. *ACS Nano*, *8*, 2033–2047.

Muranen, T., Gronholm, M., Lampin, A., Lallemand, D., Zhao, F., Giovannini, M., et al. (2007). The tumor suppressor merlin interacts with microtubules and modulates Schwann cell microtubule cytoskeleton. *Human Molecular Genetics*, *16*, 1742–1751.

Napoletano, F., Occhi, S., Calamita, P., Volpi, V., Blanc, E., Charroux, B., et al. (2011). Polyglutamine Atrophin provokes neurodegeneration in Drosophila by repressing fat. *The EMBO Journal*, *30*, 945–958.

Narod, S. A., Parry, D. M., Parboosingh, J., Lenoir, G. M., Ruttledge, M., Fischer, G., et al. (1992). Neurofibromatosis type 2 appears to be a genetically homogeneous disease. *The American Journal of Human Genetics*, *51*, 486–496.

Nejigane, S., Takahashi, S., Haramoto, Y., Michiue, T., & Asashima, M. (2013). Hippo signaling components, Mst1 and Mst2, act as a switch between self-renewal and differentiation in Xenopus hematopoietic and endothelial progenitors. *The International Journal of Developmental Biology*, *57*, 407–414.

Neto-Silva, R. M., de Beco, S., & Johnston, L. A. (2010). Evidence for a growth-stabilizing regulatory feedback mechanism between Myc and Yorkie, the Drosophila homolog of Yap. *Developmental Cell*, *19*, 507–520.

Nishioka, N., Inoue, K., Adachi, K., Kiyonari, H., Ota, M., Ralston, A., et al. (2009). The Hippo signaling pathway components Lats and Yap pattern Tead4 activity to distinguish mouse trophectoderm from inner cell mass. *Developmental Cell*, *16*, 398–410.

Nolo, R., Morrison, C. M., Tao, C., Zhang, X., & Halder, G. (2006). The bantam microRNA is a target of the Hippo tumor-suppressor pathway. *Current Biology, 16*, 1895–1904.

Oh, H., & Irvine, K. D. (2008). In vivo regulation of Yorkie phosphorylation and localization. *Development, 135*, 1081–1088.

Oh, H., & Irvine, K. D. (2011). Cooperative regulation of growth by Yorkie and Mad through bantam. *Developmental Cell, 20*, 109–122.

Oh, H., Reddy, B. V., & Irvine, K. D. (2009). Phosphorylation-independent repression of Yorkie in Fat–Hippo signaling. *Developmental Biology, 335*, 188–197.

Oh, H., Slattery, M., Ma, L., Crofts, A., White, K. P., Mann, R. S., et al. (2013). Genome-wide association of Yorkie with chromatin and chromatin-remodeling complexes. *Cell Reports, 3*, 309–318.

Okada, T., Lopez-Lago, M., & Giancotti, F. G. (2005). Merlin/NF-2 mediates contact inhibition of growth by suppressing recruitment of Rac to the plasma membrane. *The Journal of Cell Biology, 171*, 361–371.

Ota, M., & Sasaki, H. (2008). Mammalian Tead proteins regulate cell proliferation and contact inhibition as transcriptional mediators of Hippo signaling. *Development, 135*, 4059–4069.

Pan, D. (2010). The Hippo signaling pathway in development and cancer. *Developmental Cell, 19*, 491–505.

Pantalacci, S., Tapon, N., & Léopold, P. (2003). The Salvador partner Hippo promotes apoptosis and cell-cycle exit in Drosophila. *Nature Cell Biology, 5*, 921–927.

Papi, L., De Vitis, L. R., Vitelli, F., Ammannati, F., Mennonna, P., Montali, E., et al. (1995). Somatic mutations in the neurofibromatosis type 2 gene in sporadic meningiomas. *Human Genetics, 95*, 347–351.

Paramasivam, M., Sarkeshik, A., Yates, J. R., 3rd, Fernandes, M. J., & McCollum, D. (2011). Angiomotin family proteins are novel activators of the LATS2 kinase tumor suppressor. *Molecular Biology of the Cell, 22*, 3725–3733.

Parry, D. M., Eldridge, R., Kaiser-Kupfer, M. I., Bouzas, E. A., Pikus, A., & Patronas, N. (1994). Neurofibromatosis 2 (NF2): Clinical characteristics of 63 affected individuals and clinical evidence for heterogeneity. *American Journal of Medical Genetics, 52*, 450–461.

Parry, D. M., MacCollin, M. M., Kaiser-Kupfer, M. I., Pulaski, K., Nicholson, H. S., Bolesta, M., et al. (1996). Germ-line mutations in the neurofibromatosis 2 gene: Correlations with disease severity and retinal abnormalities. *The American Journal of Human Genetics, 59*, 529–539.

Parsons, L. M., Grzeschik, N. A., Allott, M. L., & Richardson, H. E. (2010). Lgl/aPKC and Crb regulate the Salvador/Warts/Hippo pathway. *Fly, 4*, 288–293.

Pellock, B. J., Buff, E., White, K., & Hariharan, I. K. (2007). The Drosophila tumor suppressors Expanded and Merlin differentially regulate cell cycle exit, apoptosis, and Wingless signaling. *Developmental Biology, 304*, 102–115.

Peng, H. W., Slattery, M., & Mann, R. S. (2009). Transcription factor choice in the Hippo signaling pathway: Homothorax and yorkie regulation of the microRNA bantam in the progenitor domain of the Drosophila eye imaginal disc. *Genes & Development, 23*, 2307–2319.

Plouffe, S. W., Hong, A. W., & Guan, K. L. (2015). Disease implications of the Hippo/YAP pathway. *Trends in Molecular Medicine, 21*, 212–222.

Polesello, C., Huelsmann, S., Brown, N. H., & Tapon, N. (2006). The Drosophila RASSF homolog antagonizes the Hippo pathway. *Current Biology, 16*, 2459–2465.

Poon, C. L., Lin, J. I., Zhang, X., & Harvey, K. F. (2011). The sterile 20-like kinase Tao-1 controls tissue growth by regulating the Salvador–Warts–Hippo pathway. *Developmental Cell, 21*, 896–906.

Ragge, N. K., Baser, M. E., Riccardi, V. M., & Falk, R. E. (1997). The ocular presentation of neurofibromatosis 2. *Eye, 11*, 12–18.

Rauskolb, C., Pan, G., Reddy, B. V., Oh, H., & Irvine, K. D. (2011). Zyxin links fat signaling to the Hippo pathway. *PLoS Biology, 9*, e1000624.

Rauskolb, C., Sun, S., Sun, G., Pan, Y., & Irvine, K. D. (2014). Cytoskeletal tension inhibits Hippo signaling through an Ajuba–Warts complex. *Cell, 158*, 143–156.

Rawat, S. J., Creasy, C. L., Peterson, J. R., & Chernoff, J. (2013). The tumor suppressor Mst1 promotes changes in the cellular redox state by phosphorylation and inactivation of peroxiredoxin-1 protein. *The Journal of Biological Chemistry, 288*, 8762–8771.

Reddy, B. V., & Irvine, K. D. (2013). Regulation of Hippo signaling by EGFR-MAPK signaling through Ajuba family proteins. *Developmental Cell, 24*, 459–471.

Reginensi, A., Scott, R. P., Gregorieff, A., Bagherie-Lachidan, M., Chung, C., Lim, D. S., et al. (2013). Yap- and Cdc42-dependent nephrogenesis and morphogenesis during mouse kidney development. *PLoS Genetics, 9*, e1003380.

Ren, F., Zhang, L., & Jiang, J. (2010). Hippo signaling regulates Yorkie nuclear localization and activity through 14-3-3 dependent and independent mechanisms. *Developmental Biology, 337*, 303–312.

Ribeiro, P. S., Josué, F., Wepf, A., Wehr, M. C., Rinner, O., Kelly, G., et al. (2010). Combined functional genomic and proteomic approaches identify a PP2A complex as a negative regulator of Hippo signaling. *Molecular Cell, 39*, 521–534.

Robinson, B. S., Huang, J., Hong, Y., & Moberg, K. H. (2010). Crumbs regulates Salvador/Warts/Hippo signaling in Drosophila via the FERM-domain protein Expanded. *Current Biology, 20*, 582–590.

Sakuma, C., Saito, Y., Umehara, T., Kamimura, K., Maeda, N., Mosca, T. J., et al. (2016). The Strip–Hippo pathway regulates synaptic terminal formation by modulating actin organization at the Drosophila neuromuscular synapses. *Cell Reports, 16*, 2289–2297.

Salah, Z., Melino, G., & Aqeilan, R. I. (2011). Negative regulation of the Hippo pathway by E3 ubiquitin ligase ITCH is sufficient to promote tumorigenicity. *Cancer Research, 71*, 2010–2020.

Sansores-Garcia, L., Bossuyt, W., Wada, K., Yonemura, S., Tao, C., Sasaki, H., et al. (2011). Modulating F-actin organization induces organ growth by affecting the Hippo pathway. *The EMBO Journal, 30*, 2325–2335.

Schlegelmilch, K., Mohseni, M., Kirak, O., Pruszak, J., Rodriguez, J. R., Zhou, D., et al. (2011). Yap1 acts downstream of α-catenin to control epidermal proliferation. *Cell, 144*, 782–795.

Schroeder, M. C., & Halder, G. (2012). Regulation of the Hippo pathway by cell architecture and mechanical signals. *Seminars in Cell & Developmental Biology, 23*, 803–811.

Sebé-Pedrós, A., Zheng, Y., Ruiz-Trillo, I., & Pan, D. (2012). Premetazoan origin of the Hippo signaling pathway. *Cell Reports, 1*, 13–20.

Seidel, C., Schagdarsurengin, U., Blümke, K., Würl, P., Pfeifer, G. P., Hauptmann, S., et al. (2007). Frequent hypermethylation of MST1 and MST2 in soft tissue sarcoma. *Molecular Carcinogenesis, 46*, 865–871.

Sekido, Y., Pass, H. I., Bader, S., Mew, D. J., Christman, M. F., Gazdar, A. F., et al. (1995). Neurofibromatosis type 2 (NF2) gene is somatically mutated in mesothelioma but not in lung cancer. *Cancer Research, 55*, 1227–1231.

Shimizu, T., Ho, L. L., & Lai, Z. C. (2008). The mob as tumor suppressor gene is essential for early development and regulates tissue growth in Drosophila. *Genetics, 178*, 957–965.

Silvis, M. R., Kreger, B. T., Lien, W. H., Klezovitch, O., Rudakova, G. M., Camargo, F. D., et al. (2011). α-Catenin is a tumor suppressor that controls cell accumulation by regulating the localization and activity of the transcriptional coactivator Yap1. *Science Signaling, 4*, ra33.

Smole, Z., Thoma, C. R., Applegate, K. T., Duda, M., Gutbrodt, K. L., Danuser, G., et al. (2014). Tumor suppressor NF2/merlin is a microtubule stabilizer. *Cancer Research, 74,* 353–362.

Sopko, R., Silva, E., Clayton, L., Gardano, L., Barrios-Rodiles, M., Wrana, J., et al. (2009). Phosphorylation of the tumor suppressor fat is regulated by its ligand Dachsous and the kinase discs overgrown. *Current Biology, 19,* 1112–1117.

Steinhardt, A. A., Gayyed, M. F., Klein, A. P., Dong, J., Maitra, A., Pan, D., et al. (2008). Expression of Yes-associated protein in common solid tumors. *Human Pathology, 39,* 1582–1589.

Strano, S., Monti, O., Pediconi, N., Baccarini, A., Fontemaggi, G., Lapi, E., et al. (2005). The transcriptional coactivator Yes-associated protein drives p73 gene-target specificity in response to DNA Damage. *Molecular Cell, 18,* 447–459.

Strano, S., Munarriz, E., Rossi, M., Castagnoli, L., Shaul, Y., Sacchi, A., et al. (2001). Physical interaction with Yes-associated protein enhances p73 transcriptional activity. *The Journal of Biological Chemistry, 276,* 15164–15173.

Striedinger, K., VandenBerg, S. R., Baia, G. S., McDermott, M. W., Gutmann, D. H., et al. (2008). The neurofibromatosis 2 tumor suppressor gene product, merlin, regulates human meningioma cell growth by signaling through YAP. *Neoplasia, 10,* 1204–1212.

Sun, G., & Irvine, K. D. (2013). Ajuba family proteins link JNK to Hippo signaling. *Science Signaling, 6,* ra81.

Sun, S., & Irvine, K. D. (2016). Cellular organization and cytoskeletal regulation of the Hippo signaling network. *Trends in Cell Biology, 26,* 694–704.

Sun, S., Reddy, B. V. V. G., & Irvine, K. D. (2015). Localization of Hippo signalling complexes and Warts activation in vivo. *Nature Communications, 6,* 8402.

Suzuki, H., Yabuta, N., Okada, N., Torigata, K., Aylon, Y., Oren, M., et al. (2013). Lats2 phosphorylates p21/CDKN1A after UV irradiation and regulates apoptosis. *Journal of Cell Science, 126,* 4358–4368.

Takahashi, Y., Miyoshi, Y., Takahata, C., Irahara, N., Taguchi, T., Tamaki, Y., et al. (2005). Down-regulation of LATS1 and LATS2 mRNA expression by promoter hypermethylation and its association with biologically aggressive phenotype in human breast cancers. *Clinical Cancer Research, 11,* 1380–1385.

Tang, F., Gill, J., Ficht, X., Barthlott, T., Cornils, H., Schmitz-Rohmer, D., et al. (2015). The kinases NDR1/2 act downstream of the Hippo homolog MST1 to mediate both egress of thymocytes from the thymus and lymphocyte motility. *Science Signaling, 8,* ra100.

Tapon, N., Harvey, K. F., Bell, D. W., Wahrer, D. C., Schiripo, T. A., Haber, D., et al. (2002). Salvador promotes both cell cycle exit and apoptosis in Drosophila and is mutated in human cancer cell lines. *Cell, 110,* 467–478.

Taylor, L. K., Wang, H. C., & Erikson, R. L. (1996). Newly identified stress-responsive protein kinases, Krs-1 and Krs-2. *Proceedings of the National Academy of Sciences of the United States of America, 93,* 10099–10104.

Thompson, B. J., & Cohen, S. M. (2006). The Hippo pathway regulates the bantam microRNA to control cell proliferation and apoptosis in Drosophila. *Cell, 2006*(126), 767–774.

Tian, Y., Liu, Y., Wang, T., Zhou, N., Kong, J., Chen, L., et al. (2015). A microRNA–Hippo pathway that promotes cardiomyocyte proliferation and cardiac regeneration in mice. *Science Translational Medicine, 7,* 279ra38.

Tikoo, A., Varga, M., Ramesh, V., Gusella, J., & Maruta, H. (1994). An anti-Ras function of neurofibromatosis type 2 gene product (NF2/Merlin). *Journal of Biological Chemistry, 269,* 23387–23390.

Troilo, A., Benson, E. K., Esposito, D., Garibsingh, R. A., Reddy, E. P., Mungamuri, S. K., et al. (2016). Angiomotin stabilization by tankyrase inhibitors antagonizes constitutive

TEAD-dependent transcription and proliferation of human tumor cells with Hippo pathway core component mutations. *Oncotarget, 7*, 28765–28782.

Twist, E. C., Ruttledge, M. H., Rousseau, M., Sanson, M., Papi, L., Merel, P., et al. (1994). The neurofibromatosis type 2 gene is inactivated in schwannomas. *Human Molecular Genetics, 3*, 147–151.

Tyler, D. M., Li, W., Zhuo, N., Pellock, B., & Baker, N. E. (2007). Genes affecting cell competition in Drosophila. *Genetics, 175*, 643–657.

Udan, R. S., Kango-Singh, M., Nolo, R., Tao, C., & Halder, G. (2003). Hippo promotes proliferation arrest and apoptosis in the Salvador/Warts pathway. *Nature Cell Biology, 5*, 914–920.

Uhlén, M., Björling, E., Agaton, C., Szigyarto, C. A., Amini, B., Andersen, E., et al. (2005). A human protein atlas for normal and cancer tissues based on antibody proteomics. *Molecular & Cellular Proteomics, 4*, 1920–1932.

Verghese, S., Bedi, S., & Kango-Singh, M. (2012). Hippo signalling controls Dronc activity to regulate organ size in Drosophila. *Cell Death and Differentiation, 19*, 1664–1676.

Verghese, S., Waghmare, I., Kwon, H., Hanes, K., & Kango-Singh, M. (2012). Scribble acts in the Drosophila fat–Hippo pathway to regulate warts activity. *PLoS One, 7*, e47173.

Vichalkovski, A., Gresko, E., Cornils, H., Hergovich, A., Schmitz, D., & Hemmings, B. A. (2008). NDR kinase is activated by RASSF1A/MST1 in response to Fas receptor stimulation and promotes apoptosis. *Current Biology, 18*, 1889–1895.

Volpi, S., Bongiorni, S., Fabbretti, F., Wakimoto, B. T., & Prantera, G. (2013). Drosophila rae1 is required for male meiosis and spermatogenesis. *Journal of Cell Science, 126*, 3541–3551.

Wada, K., Itoga, K., Okano, T., Yonemura, S., & Sasaki, H. (2011). Hippo pathway regulation by cell morphology and stress fibers. *Development, 138*, 3907–3914.

Wang, C., An, J., Zhang, P., Xu, C., Gao, K., Wu, D., et al. (2012). The Nedd4-like ubiquitin E3 ligases target angiomotin/p130 to ubiquitin-dependent degradation. *The Biochemical Journal, 444*, 279–289.

Wang, W., Huang, J., & Chen, J. (2011). Angiomotin-like proteins associate with and negatively regulate YAP1. *Journal of Biological Chemistry, 286*, 4364–4370.

Wang, W., Li, N., Li, X., Tran, M. K., Han, X., & Chen, J. (2015). Tankyrase inhibitors target YAP by stabilizing angiomotin family proteins. *Cell Reports, 13*, 524–532.

Wang, P., Mao, B., Luo, W., Wei, B., Jiang, W., Liu, D., et al. (2014). The alteration of Hippo/YAP signaling in the development of hypertrophic cardiomyopathy. *Basic Research in Cardiology, 109*, 435.

Wehr, M. C., Holder, M. V., Gailite, I., Saunders, R. E., Maile, T. M., Ciirdaeva, E., et al. (2013). Salt-inducible kinases regulate growth through the Hippo signalling pathway in Drosophila. *Nature Cell Biology, 15*, 61–71.

Wei, X., Shimizu, T., & Lai, Z. C. (2007). Mob as tumor suppressor is activated by Hippo kinase for growth inhibition in Drosophila. *The EMBO Journal, 26*, 1772–1781.

Wellenreuther, R., Kraus, J. A., Lenartz, D., Menon, A. G., Schramm, J., Louis, D. N., et al. (1995). Analysis of the neurofibromatosis 2 gene reveals molecular variants of meningioma. *The American Journal of Pathology, 146*, 827–832.

Wen, W., Zhu, F., Zhang, J., Keum, Y. S., Zykova, T., Yao, K., et al. (2010). MST1 promotes apoptosis through phosphorylation of histone H2AX. *The Journal of Biological Chemistry, 285*, 39108–39116.

Wicklow, E., Blij, S., Frum, T., Hirate, Y., Lang, R. A., Sasaki, H., et al. (2014). HIPPO pathway members restrict SOX2 to the inner cell mass where it promotes ICM fates in the mouse blastocyst. *PLoS Genetics, 10*, e1004618.

Wilkinson, D. S., & Hansen, M. (2015). LC3 is a novel substrate for the mammalian Hippo kinases, STK3/STK4. *Autophagy, 11*, 856–857.

Wilkinson, D. S., Jariwala, J. S., Anderson, E., Mitra, K., Meisenhelder, J., Chang, J. T., et al. (2015). Phosphorylation of LC3 by the Hippo kinases STK3/STK4 is essential for autophagy. *Molecular Cell, 57*, 55–68.

Willecke, M., Hamaratoglu, F., Sansores-Garcia, L., Tao, C., & Halder, G. (2008). Boundaries of Dachsous Cadherin activity modulate the Hippo signaling pathway to induce cell proliferation. *Proceedings of the National Academy of Sciences of the United States of America, 105*, 14897–14902.

Wolff, R. K., Frazer, K. A., Jackler, R. K., Lanser, M. J., Pitts, L. H., & Cox, D. R. (1992). Analysis of chromosome 22 deletions in neurofibromatosis type 2-related tumors. *The American Journal of Human Genetics, 51*, 478–485.

Wong, R. W., Blobel, G., & Coutavas, E. (2006). Rae1 interaction with NuMA is required for bipolar spindle formation. *Proceedings of the National Academy of Sciences of the United States of America, 103*, 19783–19787.

Wong, K. K., Li, W., An, Y., Duan, Y., Li, Z., Kang, Y., et al. (2015). β-Spectrin regulates the Hippo signaling pathway and modulates the basal actin network. *Journal of Biological Chemistry, 290*, 6397–6407.

Wong, J. S., Meliambro, K., Ray, J., & Campbell, K. N. (2016). Hippo signaling in the kidney: The good and the bad. *The American Journal of Physiology—Renal Physiology, 311*, F241–F248.

Wu, S., Huang, J., Dong, J., & Pan, D. (2003). Hippo encodes a Ste-20 family protein kinase that restricts cell proliferation and promotes apoptosis in conjunction with salvador and warts. *Cell, 114*, 445–456.

Wu, S., Liu, Y., Zheng, Y., Dong, J., & Pan, D. (2008). The TEAD/TEF family protein Scalloped mediates transcriptional output of the Hippo growth-regulatory pathway. *Developmental Cell, 14*, 388–398.

Wu, D., & Wu, S. (2013). A conserved serine residue regulates the stability of Drosophila Salvador and human WW domain-containing adaptor 45 through proteasomal degradation. *Biochemical & Biophysical Research Communications, 433*, 538–541.

Xia, Y., Zhang, Y. L., Yu, C., Chang, T., & Fan, H. Y. (2014). YAP/TEAD co-activator regulated pluripotency and chemoresistance in ovarian cancer initiated cells. *PLoS One, 9*, e109575.

Xin, M., Kim, Y., Sutherland, L. B., Murakami, M., Qi, X., McAnally, J., et al. (2013). Hippo pathway effector Yap promotes cardiac regeneration. *Proceedings of the National Academy of Sciences of the United States of America, 110*, 13839–13844.

Xu, Y., Stamenkovic, I., & Yu, Q. (2010). CD44 attenuates activation of the Hippo signaling pathway and is a prime therapeutic target for glioblastoma. *Cancer Research, 70*, 2455–2464.

Xu, T., Wang, W., Zhang, S., Stewart, R. A., & Yu, W. (1995). Identifying tumor suppressors in genetic mosaics: The Drosophila lats gene encodes a putative protein kinase. *Development, 121*, 1053–1063.

Yan, G., Qin, Q., Yi, B., Chuprun, K., Sun, H., Huang, S., et al. (2013). Protein-L-isoaspartate (D-aspartate) O-methyltransferase protects cardiomyocytes against hypoxia induced apoptosis through inhibiting proapoptotic kinase Mst1. *International Journal of Cardiology, 168*, 3291–3299.

Yi, C., Shen, Z., Stemmer-Rachamimov, A., Dawany, N., Troutman, S., Showe, L. C., et al. (2013). The p130 isoform of angiomotin is required for Yap-mediated hepatic epithelial cell proliferation and tumorigenesis. *Science Signaling, 6*, ra77.

Yimlamai, D., Christodoulou, C., Galli, G. G., Yanger, K., Pepe-Mooney, B., Gurung, B., et al. (2014). Hippo pathway activity influences liver cell fate. *Cell, 157*, 1324–1338.

Yin, F., Yu, J., Zheng, Y., Chen, Q., Zhang, N., & Pan, D. (2013). Spatial organization of Hippo signaling at the plasma membrane mediated by the tumor suppressor Merlin/NF2. *Cell, 154*, 1342–1355.

Yoshikawa, K., Noguchi, K., Nakano, Y., Yamamura, M., Takaoka, K., Hashimoto-Tamaoki, T., et al. (2015). The Hippo pathway transcriptional co-activator, YAP, confers resistance to cisplatin in human oral squamous cell carcinoma. *International Journal of Oncology*, *46*, 2364–2370.

You, B., Huang, S., Qin, Q., Yi, B., Yuan, Y., Xu, Z., et al. (2013). Glyceraldehyde-3-phosphate dehydrogenase interacts with proapoptotic kinase mst1 to promote cardiomyocyte apoptosis. *PLoS One*, *8*, e58697.

You, B., Yan, G., Zhang, Z., Yan, L., Li, J., Ge, Q., et al. (2009). Phosphorylation of cardiac troponin I by mammalian sterile 20-like kinase 1. *Biochemical Journal*, *418*, 93–101.

Yu, F. X., & Guan, K. L. (2013). The Hippo pathway: Regulators and regulations. *Genes & Development*, *27*, 355–371.

Yu, F. X., Zhang, Y., Park, H. W., Jewell, J. L., Chen, Q., Deng, Y., et al. (2013). Protein kinase A activates the Hippo pathway to modulate cell proliferation and differentiation. *Genes & Development*, *27*, 1223–1232.

Yu, F. X., Zhao, B., & Guan, K. L. (2015). Hippo pathway in organ size control, tissue homeostasis, and cancer. *Cell*, *163*, 811–828.

Yu, F. X., Zhao, B., Panupinthu, N., Jewell, J. L., Lian, I., Wang, L. H., et al. (2012). Regulation of the Hippo-YAP pathway by G-protein-coupled receptor signaling. *Cell*, *150*, 780–791.

Yu, J., Zheng, Y., Dong, J., Klusza, S., Deng, W. M., & Pan, D. (2010). Kibra functions as a tumor suppressor protein that regulates Hippo signaling in conjunction with Merlin and Expanded. *Developmental Cell*, *18*, 288–299.

Yuan, M., Tomlinson, V., Lara, R., Holliday, D., Chelala, C., Harada, T., et al. (2008). Yes-associated protein (YAP) functions as a tumor suppressor in breast. *Cell Death and Differentiation*, *15*, 1752–1759.

Yue, T., Tian, A., & Jiang, J. (2012). The cell adhesion molecule echinoid functions as a tumor suppressor and upstream regulator of the Hippo signaling pathway. *Developmental Cell*, *22*, 255–267.

Zaidi, S. K., Sullivan, A. J., Medina, R., Ito, Y., van Wijnen, A. J., Stein, J. L., et al. (2004). Tyrosine phosphorylation controls Runx2-mediated subnuclear targeting of YAP to repress transcription. *The EMBO Journal*, *23*, 790–799.

Zeng, Q., & Hong, W. (2008). The emerging role of the Hippo pathway in cell contact inhibition, organ size control, and cancer development in mammals. *Cancer Cell*, *13*, 188–192.

Zhang, H., Pasolli, H. A., & Fuchs, E. (2011). Yes-associated protein (YAP) transcriptional coactivator functions in balancing growth and differentiation in skin. *Proceedings of the National Academy of Sciences of the United States of America*, *108*, 2270–2275.

Zhang, L., Ren, F., Zhang, Q., Chen, Y., Wang, B., & Jiang, J. (2008). The TEAD/TEF family of transcription factor Scalloped mediates Hippo signaling in organ size control. *Developmental Cell*, *14*, 377–387.

Zhang, C., Robinson, B. S., Xu, W., Yang, L., Yao, B., Zhao, H., et al. (2015). The ecdysone receptor coactivator Taiman links Yorkie to transcriptional control of germline stem cell factors in somatic tissue. *Developmental Cell*, *34*, 168–180.

Zhao, B., Li, L., Lei, Q., & Guan, K. L. (2010). The Hippo-YAP pathway in organ size control and tumorigenesis: An updated version. *Genes & Development*, *24*, 862–874.

Zhao, B., Li, L., Lu, Q., Wang, L. H., Liu, C. Y., Lei, Q., et al. (2011). Angiomotin is a novel Hippo pathway component that inhibits YAP oncoprotein. *Genes & Development*, *25*, 51–63.

Zhao, B., Li, L., Tumaneng, K., Wang, C. Y., & Guan, K. L. (2010). A coordinated phosphorylation by Lats and CK1 regulates YAP stability through SCF(beta-TRCP). *Genes & Development*, *24*, 72–85.

Zhao, B., Li, L., Wang, L., Wang, C. Y., Yu, J., & Guan, K. L. (2012). Cell detachment activates the Hippo pathway via cytoskeleton reorganization to induce anoikis. *Genes & Development, 26*, 54–68.

Zhao, B., Wei, X., Li, W., Udan, R. S., Yang, Q., Kim, J., et al. (2007). Inactivation of YAP oncoprotein by the Hippo pathway is involved in cell contact inhibition and tissue growth control. *Genes & Development, 21*, 2747–2761.

Zhao, Y., & Yang, X. (2015). The Hippo pathway in chemotherapeutic drug resistance. *International Journal of Cancer, 137*, 2767–2773.

Zhao, B., Ye, X., Yu, J., Li, L., Li, W., Li, S., et al. (2008). TEAD mediates YAP-dependent gene induction and growth control. *Genes & Development, 22*, 1962–1971.

Zhou, D., Conrad, C., Xia, F., Park, J. S., Payer, B., Yin, Y., et al. (2009). Mst1 and Mst2 maintain hepatocyte quiescence and suppress hepatocellular carcinoma development through inactivation of the Yap1 oncogene. *Cancer Cell, 16*, 425–438.

Zhou, Q., Li, L., Zhao, B., & Guan, K. L. (2015). The Hippo pathway in heart development, regeneration, and diseases. *Circulation Research, 116*, 1431–1447.

Zi, M., Maqsood, A., Prehar, S., Mohamed, T. M., Abou-Leisa, R., Robertson, A., et al. (2014). The mammalian Ste20-like kinase 2 (Mst2) modulates stress-induced cardiac hypertrophy. *Journal of Biological Chemistry, 289*, 24275–24288.

Ziosi, M., Baena-López, L. A., Grifoni, D., Froldi, F., Pession, A., Garoia, F., et al. (2010). dMyc functions downstream of Yorkie to promote the supercompetitive behavior of Hippo pathway mutant cells. *PLoS Genetics, 6*, e1001140.

CHAPTER SEVEN

Regulation of Embryonic and Postnatal Development by the CSF-1 Receptor

Violeta Chitu, E. Richard Stanley[1]

Albert Einstein College of Medicine, Bronx, NY, United States
[1]Corresponding author: e-mail address: richard.stanley@einstein.yu.edu

Contents

1. Introduction	230
2. The CSF-1R in Mononuclear Phagocyte Development	233
2.1 Role of the CSF-1R in Monocyte and Tissue Macrophage Development	233
2.2 Regulation of Monocyte Subsets by the CSF-1R	236
2.3 Unique Roles of the CSF-1R Ligands, CSF-1 and IL-34, in Mononuclear Phagocyte Development	237
2.4 Functional Significance of CSF-1R-Dependent Mononuclear Phagocytes	238
3. CSF-1-Regulated Macrophages in Tissue Morphogenesis and Organismal Growth	239
3.1 Regulation of Angiogenesis and Lymphangiogenesis	240
3.2 Regulation of Neurogenesis, Neuronal Wiring, and Synaptic Pruning by Microglia	242
3.3 Mammary Gland Development	243
3.4 Pancreatic Development	243
3.5 Colon Development	244
3.6 Potential Contribution of CSF-1R to Heart and Kidney Development	244
3.7 Contribution of CSF-1 to Maintenance of Lung Tissue Integrity	245
3.8 Role of CSF-1-Regulated Macrophages in Growth	245
4. Regulation of Osteoclasts and Bone Development by CSF-1	246
4.1 CSF-1R Signaling in Osteoclastogenesis	249
4.2 CSF-1R Regulation of Mature OC Migration and Bone Resorption	254
4.3 CSF-1R Signaling for OC Survival	256
5. Direct Regulation of Nonhematopoietic Cells by the CSF-1R	256
5.1 Neuronal Stem Cells	256
5.2 Paneth Cells	257
5.3 Direct Regulation of Oocytes, Preimplantation Embryos, and Trophoblastic Cells	258
6. Conclusions	258
Acknowledgments	260
References	260

Abstract

Macrophages are found in all tissues and regulate tissue morphogenesis during development through trophic and scavenger functions. The colony stimulating factor-1 (CSF-1) receptor (CSF-1R) is the major regulator of tissue macrophage development and maintenance. In combination with receptor activator of nuclear factor κB (RANK), the CSF-1R also regulates the differentiation of the bone-resorbing osteoclast and controls bone remodeling during embryonic and early postnatal development. CSF-1R-regulated macrophages play trophic and remodeling roles in development. Outside the mononuclear phagocytic system, the CSF-1R directly regulates neuronal survival and differentiation, the development of intestinal Paneth cells and of preimplantation embryos, as well as trophoblast innate immune function. Consistent with the pleiotropic roles of the receptor during development, CSF-1R deficiency in most mouse strains causes embryonic or perinatal death and the surviving mice exhibit multiple developmental and functional deficits. The CSF-1R is activated by two dimeric glycoprotein ligands, CSF-1, and interleukin-34 (IL-34). Homozygous *Csf1*-null mutations phenocopy most of the deficits of *Csf1r*-null mice. In contrast, *Il34*-null mice have no gross phenotype, except for decreased numbers of Langerhans cells and microglia, indicating that CSF-1 plays the major developmental role. Homozygous inactivating mutations of the *Csf1r* or its ligands have not been reported in man. However, heterozygous inactivating mutations in the *Csf1r* lead to a dominantly inherited adult-onset progressive dementia, highlighting the importance of CSF-1R signaling in the brain.

1. INTRODUCTION

Colony stimulating factors (CSFs) promote the proliferation and differentiation of immature bone marrow progenitor cells to form colonies of mature granulocytes and macrophages (Bradley & Metcalf, 1966; Pluznik & Sachs, 1965). The first of these to be purified was the homodimeric glycoprotein colony stimulating factor-1 (CSF-1) (also known as M-CSF), which stimulates the formation of pure macrophage colonies (Stanley, Chen, & Lin, 1978; Stanley & Heard, 1977). Using radiolabeled CSF-1, a cell surface CSF-1 receptor (CSF-1R, also known as Fms, c-Fms, CD115, FIM2, or M-CSF receptor) was identified and shown to be selectively expressed on macrophages and their progenitors (Byrne, Guilbert, & Stanley, 1981; Guilbert & Stanley, 1980, 1986). Sequencing of purified human CSF-1 led to the cDNA cloning of both human and mouse CSF-1 genes (Kawasaki et al., 1985; Ladner et al., 1987, 1988). The receptor was then purified, shown to possess tyrosine kinase activity (Yeung, Jubinsky, Sengupta, Yeung, & Stanley, 1987) and to be encoded by the c-*fms* protooncogene (Sherr et al., 1985). Subsequent sequencing of the c-*fms* cDNA revealed that it encoded a class III receptor tyrosine kinase (Coussens et al., 1986). Radioimmuno- and radioreceptor assays revealed that CSF-1 was present at biologically active concentrations in the circulation and in most tissues (Das, Stanley, Guilbert, & Forman, 1980; Stanley, 1979). When injected into the circulation, it increased blood monocyte and tissue macrophage levels (Hume, Pavli, Donahue, & Fidler, 1988), and local and circulating concentrations of CSF-1 were shown to increase dramatically in response to challenges increasing the demand for macrophages (e.g., bacterial infection) (Roth, Bartocci, & Stanley, 1997). The identification of an inactivating mutation in the *Csf1* gene (Wiktor-Jedrzejczak et al., 1990; Yoshida et al., 1990) in the *osteopetrotic* (*op*) mouse (Marks & Lane, 1976) demonstrated that CSF-1 also regulated the development of osteoclasts. Subsequently, *Csf1* deficiency was shown to cause a similar osteopetrotic phenotype in the *toothless* (*tl*) rat (Van Wesenbeeck et al., 2002). Osteoclast development was shown to depend on synergism between CSF-1 and the chemokine RANK ligand (Nakagawa et al., 1998), deficiency of which also resulted in osteopetrosis (Kong et al., 1999). In addition, the $Csf1^{op/op}$ mouse was shown to exhibit developmental defects resulting in growth (Marks & Lane, 1976; Ryan et al., 2001; Wiktor-Jedrzejczak, Ahmed, Szczylik, & Skelly, 1982), neurologic (Michaelson et al., 1996), and reproductive (Cohen, Chisholm, Arceci, Stanley, & Pollard, 1996; Cohen, Nishimura, Zhu, & Pollard, 1999; Cohen, Zhu, Nishimura, & Pollard, 2002) phenotypes. Expression of full-length CSF-1 under the control of the *Csf1* promoter rescued the phenotype of $Csf1^{op/op}$ mice, confirming that CSF-1 deficiency was the sole cause of these developmental deficits in the $Csf1^{op/op}$ mouse (Ryan et al., 2001).

Consistent with the pleiotropic effects of CSF-1 deficiency (Pollard & Stanley, 1996), a *Csf1* promoter lacZ transgene revealed its widespread occurrence in tissues. Several different cell types, including epithelial cells, endothelial cells, osteoblasts, and neurons, were shown to synthesize CSF-1 (Huynh et al., 2009; Nandi et al., 2012; Ryan et al., 2001). Biochemical studies revealed the existence of three distinct biologically active CSF-1 isoforms, a secreted glycoprotein (sgCSF-1) (Price, Choi, Rosenberg, & Stanley, 1992; Stanley & Heard, 1977), a secreted proteoglycan (spCSF-1) (Price et al., 1992; Suzu et al., 1992), and a membrane-spanning cell surface glycoprotein (csCSF-1) (Rettenmier et al., 1987; reviewed in Pixley & Stanley, 2004), each of which have overlapping, yet distinct effects when exclusively expressed in $Csf1^{op/op}$ mice (Akcora et al., 2013; Dai, Zong, Sylvestre, & Stanley, 2004; Huynh et al., 2009; Nandi, Akhter, Seifert, Dai, & Stanley, 2006).

Apart from its expression on cells of hematopoietic origin (macrophages, osteoclasts, dendritic cells, and their precursors), the CSF-1R is also expressed on nonhematopoietic cells (reviewed in Chitu, Caescu, et al., 2015; Chitu, Gokhan, Nandi, Mehler, & Stanley, 2016; Chitu & Stanley, 2006), including Paneth cells (PC) (Huynh et al., 2009), epithelial intestinal cells of the colon (Huynh et al., 2013), renal proximal tubule epithelial cells (Menke et al., 2009), neural progenitor cells (NPCs; Nandi et al., 2012), and several subpopulations of neurons (Luo et al., 2013; Nandi et al., 2012; Wang, Berezovska, & Fedoroff, 1999), oocytes, preimplantation embryos, decidual, and trophoblastic cells (Arceci, Pampfer, & Pollard, 1992; Arceci, Shanahan, Stanley, & Pollard, 1989; Pampfer, Arceci, & Pollard, 1991; Regenstreif & Rossant, 1989; Sasmono et al., 2003). In man, two separate *CSF1R* promoters drive expression in myeloid cells and cells of the trophoblast (Visvader & Verma, 1989).

The more severe phenotype of the $Csf1r^{-/-}$ mouse compared with the $Csf1^{op/op}$ mouse predicted the existence of a second ligand (Dai et al., 2002). The new ligand, interleukin-34 (IL-34), was discovered by functional screening of the secreted proteins and extracellular domains of the human proteome (Lin et al., 2008). IL-34 binds the CSF-1R with high affinity and competes with CSF-1 for binding (Lin et al., 2008; Wei et al., 2010) to the D2-D3 extracellular domains (Chen, Liu, Focia, Shim, & He, 2008; Elegheert et al., 2011; Felix et al., 2013; Ma et al., 2012). IL-34, but not CSF-1, acts via a second receptor, the chondroitin sulfate-containing proteoglycan, protein tyrosine phosphatase-zeta (Nandi et al., 2013), which interestingly, is not expressed on microglia (Lorenzetto et al., 2014) or macrophages (Nandi, S. & Stanley, E.R., personal observation) but is coexpressed with the CSF-1R on neural (Nandi et al., 2012; von Holst, Sirko, & Faissner, 2006) and hematopoietic (Himburg et al., 2012; Sarrazin et al., 2009) stem cells. IL-34, but not CSF-1, also binds another cell surface chondroitin sulfate-containing proteoglycan, syndecan-1, and this interaction enhances its ability to stimulate CSF-1R-directed macrophage migration (Segaliny, Brion, Mortier, et al., 2015). In addition, it has been suggested that IL-34 and CSF-1 homodimers interact with low affinity to form an active heterotetramer with altered receptor stimulatory activity (Segaliny, Brion, Brulin, et al., 2015). However, the significance of this heterotetramer in vivo is unclear, as the expression patterns of IL-34 and CSF-1 are mostly nonoverlapping. For example, IL-34 is highly expressed in epidermis and forebrain and low in the female reproductive tract, dermis, heart, and hindbrain, where CSF-1 is highly expressed (Chitu et al., 2016; Nandi et al., 2012; Wang et al., 2016; Wei et al., 2010).

Csf1r−/− mice on most strain backgrounds die perinatally (Dai et al., 2002; Dai, Zong, Akhter, & Stanley, 2004) and the surviving mice exhibit multiple developmental deficits (Dai et al., 2002; Dai, Zong, Akhter, et al., 2004; Nandi et al., 2012). Although slightly attenuated, these phenotypes are reproduced in $Csf1^{op/op}$ mice (Chitu, Caescu, et al., 2015). In contrast, $Il34^{-/-}$ mice are grossly normal (Greter et al., 2012; Wang et al., 2012). Homozygous inactivating mutations of the human *Csf1r*, *Csf1,* or *Il34* genes have not been reported. However, *Csf1r* haploinsufficiency in man leads to an adult-onset progressive dementia, predominantly affecting the cerebral white matter (Konno et al., 2014; Nicholson et al., 2013; Rademakers et al., 2011) highlighting the importance of CSF-1R signaling in the nervous system (Chitu et al., 2016).

This review focuses on the function of the CSF-1R in development. We discuss its major role in the regulation of mononuclear phagocyte development, osteoclastogenesis, and bone formation and its direct regulation of other cell types. The functions of the CSF-1R in the adult have recently been discussed in a comprehensive review (Chitu, Caescu, et al., 2015). The present review complements other focused reviews from this laboratory of CSF-1R signaling in myeloid cells (Pixley & Stanley, 2004; Stanley & Chitu, 2014), CSF-1 in immunity and inflammation (Chitu & Stanley, 2006), and of the CSF-1R and its ligands in brain (Chitu et al., 2016).

2. THE CSF-1R IN MONONUCLEAR PHAGOCYTE DEVELOPMENT

2.1 Role of the CSF-1R in Monocyte and Tissue Macrophage Development

In metazoans, all tissues contain resident mononuclear phagocytes which, in the adult, represent as much as 15% of the total cells (Hume, 2008). During mouse embryonic development, macrophages (MΦ) are generated in three distinct waves (Fig. 1A). A first transient wave occurs between embryonic day (E) 7.5–8.0 and gives rise to primordial, maternally derived, yolk sac (YS) CD45^{high} c-Kit$^-$ MΦ (Bertrand et al., 2005; Kierdorf et al., 2013). Consistent with their maturity, these cells express the F4/80 marker and are nonproliferative (Bertrand et al., 2005). Between E7.5 and 8.0 (Fig. 1A first, left panel), the YS contains up to 100 maternally derived MΦ, located outside the blood islands, in the vicinity of the neural folds and the caudal part of the embryo (Bertrand et al., 2005) that may provide scavenger or trophic functions. Colonization of the YS by maternal MΦ is a transient process, their numbers declining to negligible levels

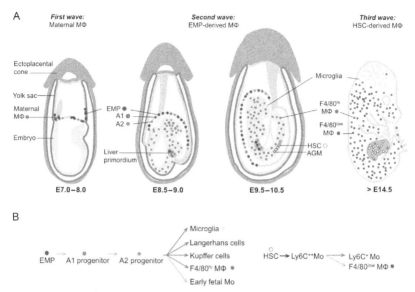

Fig. 1 Regulation of mononuclear phagocyte development by CSF-1R. (A) Ontogenically distinct mouse macrophages are generated in three successive waves. (B) Contribution of EMP and HSC to tissue macrophage populations. *Red* and *blue arrows* indicate critical and partial contribution of CSF-1R signaling to development, respectively, *gray arrows* indicate possible contribution, *black arrows*, lack of requirement.

(2–5 MΦ/YS) between E8.5 and E9 (Bertrand et al., 2005) (Fig. 1A second panel). From E7.0 to E8.0, precursors with erythromyeloid potential (EMP) emerge in the CD45$^-$ c-Kit$^+$ YS cell fraction (Bertrand et al., 2005; Kierdorf et al., 2013; Palis, Robertson, Kennedy, Wall, & Keller, 1999) (Fig. 1A first panel). E8 EMPs represent the earliest MΦ precursor from which two subsets of YS MΦ progenitors, A1 (CD45$^+$ c-Kit$^{-/lo}$ F4/80^{lo} CX$_3$CR1$^-$) and A2 (CD45$^+$ c-Kit^{hi} F4/80^{hi} CX$_3$CR1$^+$), arise by E9 (Hagemeyer et al., 2016; Kierdorf et al., 2013) (Fig. 1A, second panel; Fig. 1B). Depending on the detection system used, CSF-1R expression was reported to first occur at E8 in EMP (Gomez Perdiguero et al., 2015), or at E9 in A2 progenitors (Kierdorf et al., 2013). After the circulation is established at E8.5, these progenitors colonize the developing tissues giving rise to the second wave of MΦ (Hagemeyer et al., 2016; Kierdorf et al., 2013). EMPs colonizing the fetal liver give rise to fetal erythrocytes, granulocytes, and monocytes (Gomez Perdiguero et al., 2015). A2 progenitors colonize the brain giving rise to microglia (Hagemeyer et al., 2016; Kierdorf et al., 2013), meningeal, and perivascular MΦ (Goldmann et al., 2016), as well as other regions of the embryo proper, where they give rise to liver Kupffer cells and to the vast majority of F4/80^{high} CD11b^{low} tissue MΦ (Ginhoux & Prinz, 2015; Goldmann et al., 2016; Gomez Perdiguero et al., 2015; Hagemeyer et al., 2016; Schulz et al., 2012) (Fig. 1A, second and third panels; Fig. 1B). Fetal liver EMPs and all tissue MΦ are of YS origin up to E14.5. The first hematopoietic stem cells emerge in the aorta-gonad-mesonephros region after E9.0 (Golub & Cumano, 2013) (Fig. 1A, third panel). From E14.5 to E18.5 (Fig. 1A, fourth panel; Fig. 1B), HSCs replace the YS-derived myeloid progenitors in the liver and subsequently the EMP-derived fetal monocytes, granulocytes, and erythrocytes (Gomez Perdiguero et al., 2015). The contribution of the third wave of HSC-derived monocytes to tissue-resident MΦ varies with tissue and age. Monocytes give rise to F4/80^{low} CD11b^{high} tissue macrophages (Gomez Perdiguero et al., 2015; Schulz et al., 2012) and, except for the kidney, contribute only to a minor fraction of F4/80^{high} tissue MΦ, microglia, Kupffer, or Langerhans cells (LC), in the young adult (Gomez Perdiguero et al., 2015; Hagemeyer et al., 2016). In some tissues, the composition of MΦ populations changes over time from predominantly EMP− to predominantly monocyte derived. Intestinal EMP-derived MΦ are replaced by Ly6C^{++} monocyte-derived MΦ within the first 2 weeks of postnatal life (Bain et al., 2014), while choroid plexus and lung alveolar MΦ are progressively replaced with age (Goldmann et al., 2016; Gomez Perdiguero et al., 2015; Hagemeyer et al., 2016; Schulz et al., 2012). YS-derived MΦ express high levels of CSF-1R mRNA, while HSC-derived MΦ express low levels (Hagemeyer et al., 2016; Schulz et al., 2012). Germline ablation and inhibitor studies show that, consistent with its high expression in YS-derived MΦ, the CSF-1R is critical for the development of LC and microglia, partially required for the development of other tissue MΦ and not required for monocytopoiesis (Cecchini et al., 1994; Chitu et al., 2012; Dai et al., 2002; Ginhoux et al., 2010; Hoeffel et al., 2012; Lenzo et al., 2012; MacDonald et al., 2010; Merad et al., 2002; Schulz et al., 2012).

CSF-1R signaling also regulates the development and migration of fish (Hanington, Hitchen, Beamish, & Belosevic, 2009; Herbomel, Thisse, & Thisse, 2001; Wang et al., 2013), amphibian (Grayfer & Robert, 2013), and avian (Garceau et al., 2010) MΦ and CSF-1R-related receptors regulate plasmatocytes, the functional equivalent of MΦ in insects (reviewed in Ratheesh, Belyaeva, & Siekhaus, 2015). However, the finding that CSF-1R-expressing YS-derived-tissue MΦ were not retained in the hatched birds, while a CSF-1R$^+$ MΦ-restricted bone marrow progenitor could efficiently generate adult tissue MΦ (Garceau et al., 2015), suggests that the ontogeny of tissue MΦ may differ among vertebrates.

2.2 Regulation of Monocyte Subsets by the CSF-1R

Three subsets of monocytes—classical, intermediate, and nonclassical—with distinct functional features, have been described in mice and humans, using a recent classification approved by the Nomenclature Committee of the International Union of Immunological Societies, in which $^+$ denotes an expression level of $\sim$10-fold, and $^{++}$ $\sim$100-fold, above the level of the isotype control (Ziegler-Heitbrock et al., 2010). Classical monocytes (CD14^{++}CD16$^-$ in humans and Ly6C^{++}CD43$^+$ in mouse) and intermediate monocytes (CD14^{++}CD16$^+$ in humans and Ly6C^{++}CD43^{++} in mouse) express high levels of the chemokine receptor CCR2 and are efficiently recruited to sites of injury, where they differentiate into inflammatory MΦ. Nonclassical monocytes (CD14$^+$CD16^{++} in humans and Ly6C$^+$CD43^{++} in mouse) express low levels of CCR2 but high levels of CX3CR1. Also known as patrolling monocytes, they scan the resting vasculature to promote the safe disposal of damaged endothelial cells, acting as housekeepers of the vasculature (Auffray et al., 2007; Carlin et al., 2013). Lineage tracing studies in mice have shown that circulating Ly6C^{++} classical monocytes undergo stepwise differentiation into Ly6C$^+$ nonclassical monocytes (Yona et al., 2013). Data from transcriptional profiling studies (Ancuta et al., 2009) suggest that this developmental relationship is conserved in humans. Conversely, blockade of CSF-1R signaling does not impair the generation of Ly6C^{++} immature monocytes but leads to the selective depletion of the mature Ly6C$^+$ blood monocytes (Lenzo et al., 2012; MacDonald et al., 2010; Yona et al., 2013), suggesting that the CSF-1R promotes the maturation of Ly6C^{++} blood monocytes into Ly6C$^+$ monocytes, or the survival of the latter. Clinical trials involving the administration of recombinant CSF-1 (Weiner et al., 1994) suggest a similar role of CSF-1R signaling in the development of the nonclassical monocyte subset in man. Furthermore, patients with adult-onset leukoencephalopathy with spheroids and pigmented glia (ALSP), a neurologic disease associated with mutations in *CSF1R* that reduce cell surface expression and perturb kinase activity (reviewed in Chitu et al., 2016), exhibit a selective reduction of the CD14$^+$ CD16^{++} monocytes (Hofer et al., 2015). In mouse, the equivalent monocyte subset, Ly6C$^+$, can infiltrate the brain, differentiate into perivascular MΦ, which maintain the blood–brain barrier (Audoy-Remus et al., 2008) and have neuroprotective functions following brain or spinal cord injury (Bellavance, Gosselin, Yong, Stys, & Rivest, 2015; Shechter et al., 2013). Thus, it would be of interest to establish whether the loss of the Ly6C$^+$ subset contributes to the pathogenesis of ALSP in a mouse model (Chitu, Gokhan, et al., 2015).

2.3 Unique Roles of the CSF-1R Ligands, CSF-1 and IL-34, in Mononuclear Phagocyte Development

CSF-1 and IL-34 are differentially expressed both in development and in adults. In the developing mouse embryo, CSF-1, but not IL-34, is expressed in the YS (Johnson & Burgess, 1978; Wang et al., 2016). While CSF-1R signaling is crucial for the development of E12.5 YS primitive MΦ, absence of IL-34 does not affect their production or liver monocytopoiesis at E18.5 (Wang et al., 2016). These data suggest that CSF-1 is the primary regulator of embryonic MΦ development. Consistent with this, studies in $Csf1^{op/op}$ mice showed that CSF-1 is a major regulator of the development and maintenance of most tissue MΦ (Cecchini et al., 1994). In contrast, $Il34^{-/-}$ mice have no apparent deficits in most tissue MΦ or blood monocytes (Wang et al., 2012). IL-34 primarily regulates the development and maintenance of Langerhans cells, whereas regulation of microglia is shared with CSF-1, in a brain region-specific manner (Chitu et al., 2016; Ginhoux et al., 2010; Greter et al., 2012; Wang et al., 2012). An interesting example of nonredundant sequential regulation by CSF-1 and IL-34 has been documented in LC development (Dai et al., 2002; Wang et al., 2016). CSF-1R is essential for the development and maintenance of LC (Ginhoux et al., 2006). LC are reduced in newborn, but recover in adult $Csf1^{op/op}$ mice (Cecchini et al., 1994; Dai et al., 2002), while $Il34^{-/-}$ mice have a severe reduction in LC throughout life (Greter et al., 2012; Wang et al., 2012, 2016), suggesting that both CSF-1 and IL-34 contribute to LC development but that IL-34 alone is required for LC maintenance. Wang et al. (2016) showed that in the developing skin IL-34 expression commences at E17.5, is high in the epidermis and low in the dermis, while CSF-1 exhibits the opposite pattern. IL-34 is not necessary for the infiltration and survival of LC precursors in the immature dermis (possibly because of high local levels of CSF-1) but is necessary to promote the proliferation and survival of LC precursors in the mature epidermis. This CSF-1/IL-34 complementarity is conserved in the adult skin where regeneration of LC after UV irradiation is dependent on CSF-1, while their maintenance depends on IL-34 (Wang et al., 2016).

CSF-1 and IL-34 may also have unique roles in innate immunity. Adult $Csf1^{op/op}$ mice exhibit only a mild reduction in monocytes and mild immunological deficits (reviewed in Chitu & Stanley, 2006). A recent study (Yamane et al., 2014) reported that follicular dendritic cell-derived IL-34, but not CSF-1, triggered the CSF-1R-dependent differentiation of splenic myeloid progenitors to follicular dendritic cell-induced monocytes, a new type of monocytic cell with B cell-stimulating activity. Other studies show that following immune challenges, IL-34 and CSF-1 are differentially expressed. In *Xenopus laevis* tadpoles infected with Ranavirus Frog virus 3 (RV FV3), IL-34 and CSF-1 have divergent patterns of expression and opposite roles in antiviral immunity, with IL-34 promoting immunity and CSF-1 increasing the susceptibility to RV FV3 infection (Grayfer & Robert, 2014, 2015). In rainbow trout (*Oncorhynchus mykiss*), IL-34, but not CSF-1, is induced in primary head kidney MΦ by pathogen-associated molecular patterns, inflammatory cytokines, and parasitic kidney infection (Wang et al., 2013). Furthermore, TNF-α induces IL-34, rather than CSF-1 expression in fibroblast-like synovial cells isolated from patients with rheumatoid arthritis (Hwang et al., 2012). Together, these data indicate that IL-34 plays a nonredundant role in the control of MΦ activation during inflammation. Unlike CSF-1, IL-34 can also signal via a second receptor, protein tyrosine phosphatase-zeta (Nandi et al., 2013), which is coexpressed with the CSF-1R in some cell types (Chitu et al., 2016). Furthermore, IL-34 interaction with Syndecan modulates IL-34 activation of the CSF-1R (Segaliny, Brion, Mortier, et al., 2015). These distinguishing properties of IL-34 may explain some of its nonredundant biological activities.

2.4 Functional Significance of CSF-1R-Dependent Mononuclear Phagocytes

MΦ exhibit a remarkable functional plasticity reflected in their ability to adapt to developmental demands and inflammatory challenges (reviewed in Murray & Wynn, 2011; Pollard, 2009). As discussed earlier, the CSF-1R is dispensable for monocytopoiesis from HSC, suggesting that it primarily regulates the generation of YS-derived MΦ. Several studies suggest that MΦ ontogeny may also influence their responses to microenvironmental cues. EMP- and HSC-derived MΦ exhibit unique gene expression signatures. YS-derived MΦ express high levels of F4/80, CSF-1R and CX_3CR1 (Hagemeyer et al., 2016; Schulz et al., 2012), which are associated with monocyte/MΦ maturation, survival, and trophic functions (Caescu et al., 2015; Chitu, Caescu, et al., 2015; Hamilton, Zhao, Pavicic, & Datta,

2014; Nikolic, de Bruijn, Lutz, & Leenen, 2003; White, McNeill, Channon, & Greaves, 2014). In contrast, HSC-derived MΦ express low levels of F4/80, CSF-1R and CX_3CR1 mRNA, and high levels of genes responsible for antigen presentation and CCR2, a chemokine receptor that mediates recruitment to sites of inflammation (Hagemeyer et al., 2016; Schulz et al., 2012). These differences may reflect a higher propensity of YS-derived MΦ for trophic functions and of HSC-derived MΦ for inflammatory functions. Indeed, in models of tissue injury, CSF-1R-dependent resident MΦ were shown to adopt reparative functions (Leblond et al., 2015), while infiltration of CSF-1R-independent $Ly6C^{++}$ monocytes was a transient event with the potential to cause tissue damage (Ajami, Bennett, Krieger, McNagny, & Rossi, 2011; Croxford et al., 2015). Furthermore, CSF-1R signaling studies indicate that CSF-1 signaling promotes a trophic state in macrophages (Caescu et al., 2015; Fleetwood, Lawrence, Hamilton, & Cook, 2007). Induction of miR-21 downstream of CSF-1R phosphotyrosine 721-mediated signaling directly suppresses the expression of over 30 proinflammatory mRNAs, while increasing the expression of several trophic mRNAs (Caescu et al., 2015). The control of this proinflammatory to trophic switch is independent of signaling via phosphotyrosines 544, 559, and 807 that regulate the macrophage proliferative response (Yu et al., 2012), suggesting that distinct pathways regulate tissue macrophage development/maintenance and remodeling/trophic functions.

3. CSF-1-REGULATED MACROPHAGES IN TISSUE MORPHOGENESIS AND ORGANISMAL GROWTH

The dynamic changes in tissue CSF-1 levels and synthesis during development suggest that CSF-1 plays an important role in the late gestational and early postnatal periods (Roth & Stanley, 1995, 1996). In $Csf1^{op/op}$ and $Csf1r^{-/-}$ mice, organogenesis is relatively normal, except for disturbed skeletal development, due to the absence of osteoclasts (Dai et al., 2002; Dai, Zong, Akhter, et al., 2004; Marks & Lane, 1976) and abnormal brain development (Erblich, Zhu, Etgen, Dobrenis, & Pollard, 2011; Nandi et al., 2012). Despite this, $Csf1r^{-/-}$ mutants die perinatally (Chitu et al., 2016; Dai et al., 2002; Dai, Zong, Akhter, et al., 2004). It is not clear which organ failures are responsible for lethality, although loss of direct regulation of NPCs may contribute (see later). $Csf1^{op/op}$ mice survive for longer periods than $Csf1r^{-/-}$ mice, and the functional consequences of this mutation have been studied in more detail. $Csf1^{op/op}$ mice have reduced weight, low growth rate, extensive skeletal abnormalities, hearing, vision and olfactory deficits, abnormal intestinal organization, Paneth cell deficiency, infertility, reduced mammary gland development, altered angiogenesis and lymphangiogenesis and altered neurogenesis (Kondo et al., 2013; Ochsenbein et al., 2016; reviewed in Arnold & Betsholtz, 2013; Chitu, Caescu, et al., 2015; Chitu et al., 2016; Harvey & Gordon, 2012; Pollard, 2009; Pollard & Stanley, 1996). CSF-1 regulates tissue development through its action on both hematopoietic (macrophages and osteoclasts) and nonhematopoietic cells. Macrophages contribute to tissue and organ morphogenesis by releasing growth factors, remodeling the extracellular matrix, eliciting developmental apoptosis, and clearing apoptotic cells (reviewed in Jones & Ricardo, 2013; Pollard, 2009; Stefater, Ren, Lang, & Duffield, 2011). Mice lacking CSF-1 or the CSF-1R exhibit significant depletion of macrophages in some tissues. In other tissues, their numbers are close to normal, but their functions are impaired (Cecchini et al., 1994; Dai et al., 2002; Ginhoux et al., 2010; Roth & Stanley, 1996). Those deficits that are caused by loss of CSF-1R regulation of tissue macrophages are discussed later and summarized in Fig. 2.

3.1 Regulation of Angiogenesis and Lymphangiogenesis
3.1.1 Angiogenesis
The role of CSF-1R in developmental angiogenesis is inferred from the vascular deficits in the retina and the hindbrain of $Csf1^{op/op}$ mice (reviewed in Arnold & Betsholtz, 2013). $Csf1^{op/op}$ hindbrains have reduced complexity of the subventricular vascular plexus (Fantin et al., 2010), and $Csf1^{op/op}$ retina exhibits significantly reduced vessel branching and insufficient arterial–venous patterning (Kubota et al., 2009). The retinal vascular phenotype was also observed after postnatal treatment with anti-CSF-1R antibody or a small molecule inhibitor (Kubota et al., 2009). Depletion and restoration studies have shown that microglia play an important role in developmental angiogenesis (Checchin, Sennlaub, Levavasseur, Leduc, & Chemtob, 2006). Microglia are dramatically reduced in the hindbrain and retina of $Csf1^{op/op}$ mice (Cecchini et al., 1994; Ginhoux et al., 2010) and the reduced retinal vascular branching recovered as development progressed (Kubota et al., 2009), concomitantly with the recovery of microglia (Cecchini et al., 1994). Although other studies suggested that macrophages promote angiogenesis by producing vascular endothelial growth factor (VEGF), neither VEGF levels nor the numbers of endothelial tip cells and filopodia were altered in $Csf1^{op/op}$ retina (Kubota et al., 2009) indicating

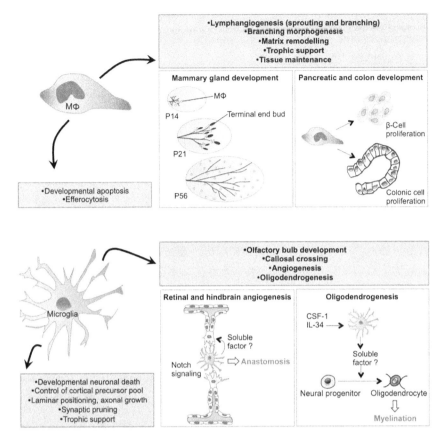

Fig. 2 Developmental functions of tissue macrophages. *Bold letters* indicate processes shown to be dysregulated in $Csf1r^{-/-}$ or $Csf1^{op/op}$ mice. The drawings show details of the cellular and molecular interactions involved.

that CSF-1 contributes to developmental vascular remodeling independently of VEGF (Kubota et al., 2009). Other macrophage-derived candidate mediators are the Wnts, which both positively and negatively regulate angiogenesis (reviewed in Newman & Hughes, 2012). However, direct cell-to-cell contacts may be necessary for directional, vascular morphogenesis in vivo (Fantin et al., 2010; Rymo et al., 2011). Fantin et al. demonstrated a close spatiotemporal relationship between macrophages and sprouting vessels, with macrophages associating at all times with endothelial tip cells and with vessel junctions and suggested that microglia promote vascular anastomosis in the developing hindbrain by physically assisting endothelial tip cell fusion through direct cell-to-cell contacts (Fantin et al., 2010) (Fig. 2). Indeed, other studies show that during retinal angiogenesis in mice, activation of Notch1 signaling in microglia by Delta-like 4 positive tip cells is important for endothelial cell anastomosis (Outtz, Tattersall, Kofler, Steinbach, & Kitajewski, 2011).

3.1.2 Lymphangiogenesis

Defects in lymphangiogenesis have been demonstrated in both $Csf1^{op/op}$ and $Csf1r^{-/-}$ mice. Gordon et al. found that the lymphatic vasculature was hyperplastic in $Csf1r^{-/-}$ embryos (Gordon et al., 2010). In contrast, Kubota et al. showed reduced lymphatic vessel branching in the postnatal trachea and ears of $Csf1^{op/op}$ mice and a slower drainage of tissue fluids in their limbs and ears (Kubota et al., 2009). Early postnatal inhibition of CSF-1 signaling in wild-type mice (P8–P15) also decreased macrophage numbers and vessel branching in the ears and trachea (Kubota et al., 2009). Focusing on postnatal lymphatic vessel development on the pleural side of the diaphragmatic muscle in $Csf1r^{-/-}$ mice, Ochsenbein et al. found that macrophage ablation in $Csf1r^{-/-}$ mice did not inhibit lymphatic vessel development in the diaphragm, but increased branch formation of lymphatic sprouts. Consistent with a role of macrophages in attenuating lymphatic vessel sprouting, reduced sprouting was observed in lymphatic endothelial cells cultured with conditioned medium from P7 diaphragmatic macrophages (Ochsenbein et al., 2016). Thus, while it is clear that CSF-1R-dependent macrophages regulate developmental lymphangiogenesis, they can play opposing roles, depending on the developmental stage and tissue involved.

3.2 Regulation of Neurogenesis, Neuronal Wiring, and Synaptic Pruning by Microglia

During brain development, microglia support neurogenesis, brain wiring and regulate synaptic refinement (recently reviewed in Casano & Peri, 2015; Frost & Schafer, 2016; Schafer & Stevens, 2015). Microglia are maintained through the combined action of the CSF-1R ligands (Ginhoux et al., 2010; Greter et al., 2012; Wang et al., 2012). Characterization of brain development in $Csf1r^{-/-}$ mice revealed multiple structural and functional abnormalities (reviewed in Chitu et al., 2016). Studies in $Nestin\text{-}cre/+;Csf1r^{fl/fl}$ mice with neural lineage-specific deletion of the $Csf1r$ showed that some of these abnormalities were caused by loss of CSF-1R regulation of NPCs. Others, including increased brain mass and lateral ventricle size, atrophy of the olfactory bulb, failure of callosal axonal crossing, and decreased subcortical oligodendrocyte differentiation, were not present in the $Nestin\text{-}cre/+;Csf1r^{fl/fl}$ mice and may be due to the dramatic reduction of microglia in the $Csf1r^{-/-}$ mice (Nandi et al., 2012). Indeed, the finding that CSF-1 enhances oligodendrocyte differentiation in cultures of NPCs containing microglia but not in microglia-free cultures, identifies microglia as mediators of CSF-1R effects on oligodendrogenesis (Fig. 2).

3.3 Mammary Gland Development

CSF-1 mRNA is expressed in the mammary ducts and intensely at the terminal end buds (TEBs) (Ryan et al., 2001) which grow out into the postnatal fat pad as the ducts themselves become branched during puberty. Macrophages, the only CSF-1R-expressing cells in mammary gland, are localized to the invading TEBs, are associated with collagen fibers particularly localized along the necks of the TEBs, and are motile within the collagen matrix (Gouon-Evans, Rothenberg, & Pollard, 2000; Ingman, Wyckoff, Gouon-Evans, Condeelis, & Pollard, 2006) (Fig. 2). In $Csf1^{op/op}$ mammary glands, macrophages are severely depleted, gland development is retarded, and the mice have reduced TEB numbers, ductal branching, and elongation, with a rounder, less elongated, TEB shape. $Csf1^{op/op}$ mice have normal estrogen levels throughout puberty, but a compromised estrogen ductal outgrowth response (Gouon-Evans et al., 2000). Abnormal mammary gland ductal development also occurs in $Csf1^{op/op}$ mice during pregnancy (Pollard & Hennighausen, 1994). Local mammary ductal expression of CSF-1 specifically restores mammary gland macrophages and corrects the branching morphogenesis defect (Van Nguyen & Pollard, 2002). How macrophages mediate their effects on mammary gland development remains to be established (Brady, Chuntova, & Schwertfeger, 2016). However, as macrophages promote collagen fibrillogenesis, rather than collagen I synthesis, it appears that the collagen laid down by fibroblasts is remodeled by the macrophages to maintain TEB structure during ductal outgrowth (Ingman et al., 2006).

3.4 Pancreatic Development

Macrophages of the developing mouse pancreas are dispersed in the mesenchyme and connective tissue around vessels, ducts, islets, and interlobular septa, particularly at sites of islet neogenesis at the duct–islet interface (Charre et al., 2002). Coincident with a wave of β-cell proliferation and islet formation, the number of macrophages and CSF-1R$^+$ monocytes near insulin$^+$ β-cells increases from E14.5 to E17.5 (Geutskens, Otonkoski, Pulkkinen, Drexhage, & Leenen, 2005). Fetal $Csf1^{op/op}$ mice are severely depleted of macrophages and have a major insulin$^+$ β-cell mass deficit, due to a combination of reduced β-cell proliferation and β-cell hypotrophy (Banaei-Bouchareb et al., 2004). By P21, insulin$^+$ cell mass and proliferation recover, but the hypotrophy persists. Conversely, CSF-1 added to fetal pancreas organ cultures increased the number of macrophages, which was paralleled by an increase in the number of insulin-producing cells (Geutskens et al., 2005). Consistent with a role of the CSF-1R during pancreatic development across species, CSF-1 mRNA expression is detected in ducts, aggregated epithelial cells, and vessel endothelial cells in the human embryonic pancreas (Banaei-Bouchareb, Peuchmaur, Czernichow, & Polak, 2006).

3.5 Colon Development

Basolateral immunohistochemical expression of CSF-1R has been reported by immunohistochemical methods on epithelial cells of both mouse and human colonic crypts (Huynh et al., 2013; Ramsay et al., 2004). Both $Csf1^{op/op}$ and $Csf1r^{-/-}$ mice were found to have abnormal colon organization, with altered enterocyte and enteroendocrine cell fates, excessive goblet cell staining, reduced cell proliferation and stem cell marker (Lgr5) expression (Huynh et al., 2013). However, neither CSF-1R ligand elicited significant in vitro proliferative responses from immortalized mouse colonic epithelial cells, indicating that the CSF-1R does not directly regulate colonocyte proliferation in vivo. As studies of $Csf1^{op/op}$ mice have revealed that macrophages are required for the amplification of mouse colonic epithelial progenitors in response to dextran sodium sulfate wounding of the epithelium (Pull, Doherty, Mills, Gordon, & Stappenbeck, 2005), the most likely mediators of this colonocyte proliferation in development are macrophages, the other major CSF-1R-expressing cell in colon.

3.6 Potential Contribution of CSF-1R to Heart and Kidney Development

CSF-1 concentrations increase in the kidney perinatally (Roth & Stanley, 1996), associated with the postnatal increase in macrophage numbers in the interstitial regions between developing tubules (Cecchini et al., 1994; Hume & Gordon, 1983; Morris, Graham, & Gordon, 1991; Rae et al., 2007). Renal proximal tubules express CSF-1 mRNA (Jang et al., 2006; Wang et al., 2015), and macrophages are severely depleted in the embryonic and postnatal kidneys of $Csf1^{op/op}$ mice (Cecchini et al., 1994; Roth,

Dominguez, & Stanley, 1998). CSF-1 addition to embryonic kidney explants increases renal growth and ureteric bud branching (Rae et al., 2007). However, kidney anatomy and function in $Csf1^{op/op}$ mice has not been examined in detail to determine potential renal deficits.

Direct evidence for an important role of macrophages in heart development comes from studies in *X. laevis*, where loss of macrophage production leads to a failure of the myocardium to form the myocardial trough that gives rise to the heart tube. Interestingly, the knockdown of lurp1, a secreted macrophage Ly6 protein of unknown function, results in a similar phenotype (Smith & Mohun, 2011). CSF-1 protein and mRNA are present in both embryonic and adult mouse heart (Roth & Stanley, 1996; Ryan et al., 2001). YS-derived cardiac tissue macrophages appear in the mouse myocardium at E9.5 (Epelman et al., 2014) and CSF-1R inhibitor studies indicate that the vast majority of heart macrophages are dependent on CSF-1R signaling (Leblond et al., 2015). It remains to be established how the CSF-1R contributes to heart development.

3.7 Contribution of CSF-1 to Maintenance of Lung Tissue Integrity

CSF-1 and CSF-1 mRNA are both present in embryonic and postnatal lung, and the development of lung macrophages is highly dependent on CSF-1 (Roth et al., 1998; Roth & Stanley, 1996). The concentration of macrophages within branch points during embryonic branching morphogenesis and, postnatally, during alveolarization (P14–P21) (Jones et al., 2013) suggests that they could regulate lung development. However, despite the fact that P20 $Csf1^{op/op}$ mice have reduced numbers of alveolar macrophages (Shibata, Zsengeller, Otake, Palaniyar, & Trapnell, 2001), no effect on lung morphogenesis has been reported. Rather CSF-1R signaling appears to be necessary for maintenance of tissue integrity. The alveolar macrophage deficiency in $Csf1^{op/op}$ mice spontaneously corrects with age but is associated with increased lung IL-3 that promotes matrix metalloproteinase expression by alveolar macrophages and destruction of the lung tissue leading to pulmonary emphysema (Shibata et al., 2001).

3.8 Role of CSF-1-Regulated Macrophages in Growth

The reduced postnatal growth rate of $Csf1^{op/op}$ and $Csf1r^{-/-}$ mice (Dai et al., 2002; Marks & Lane, 1976; Ryan et al., 2001), together with in vitro studies of the effects of CSF-1 on embryonic tissue explants (Geutskens et al., 2005;

Michaelson et al., 1996; Rae et al., 2007) and in vivo (Alikhan et al., 2011), clearly indicate that CSF-1 regulates tissue growth and development. Furthermore, the correlation between CSF-1 levels and tissue macrophage densities in vivo and in explant cultures (Alikhan et al., 2011; Cecchini et al., 1994; Roth et al., 1998) suggests that the trophic effects of CSF-1 are mediated, at least in part, by macrophages. The reduced growth rate of $Csf1^{op/op}$ mice is rescued by the $Csf1$ promoter-first intron-driven expression of full-length CSF-1 that gives rise to both secreted isoforms and also, through incomplete proteolytic cleavage, to cell surface CSF-1 (Ryan et al., 2001). Subsequent studies showed that the secreted glycoprotein or proteoglycan isoforms of CSF-1 failed (Nandi et al., 2006), while the cell surface isoform alone was sufficient (Dai, Zong, Sylvestre, et al., 2004) to rescue growth, highlighting the importance of local regulation by CSF-1 in vivo. Several studies suggest that CSF-1 stimulates tissue growth by promoting the production of insulin-like growth factor-1 (IGF-1) by macrophages (Arkins, Rebeiz, Biragyn, Reese, & Kelley, 1993; Arkins, Rebeiz, Brunke-Reese, Minshall, & Kelley, 1995; Gow, Sester, & Hume, 2010). $Csf1^{op/op}$ and $Igf1^{-/-}$ mice share growth and developmental phenotypes (Liu, Yakar, & LeRoith, 2000), and circulating IGF-1 and CSF-1 levels are positively correlated (Gow et al., 2010). However, macrophage-specific $Igf1$ inactivation in lean mice does not affect growth (Chang, Kim, Xu, & Ferrante, 2016), suggesting that regulation of IGF-1 production by macrophages is not a major contributor to organismal growth. Considering the limitation on size placed by the reduced skeleton (Thompson, 1942) and the complete correction of the $Csf1^{op/op}$ growth deficit by osteoblast-specific promoter-driven expression of CSF-1 (Abboud, Woodruff, Liu, Shen, & Ghosh-Choudhury, 2002) or by $Csf1r$ promoter-driven expression of cell surface CSF-1, it is possible that the low growth rate is secondary to defects in bone formation.

4. REGULATION OF OSTEOCLASTS AND BONE DEVELOPMENT BY CSF-1

CSF-1R signaling plays a central role in osteoclast and bone development (Dai et al., 2002; Norgard, Marks, Reinholt, & Andersson, 2003; Pollard & Stanley, 1996; Wiktor-Jedrzejczak et al., 1990; Yoshida et al., 1990). CSF-1 plays a nonredundant function in osteoclastogenesis that cannot be substituted by GM-CSF, FLT3 ligand, or VEGF (Hodge,

Kirkland, & Nicholson, 2004). Furthermore, despite the functional redundancy of IL-34 and CSF-1 in osteoclastogenesis in vitro (Wei et al., 2010), $Il\text{-}34^{-/-}$ mice have no osteopetrotic phenotype (Greter et al., 2012), reflecting the differential expression of CSF-1 and IL-34 in vivo (Wei et al., 2010).

One of the most striking developmental defects in rodent $Csf1r$ or $Csf1$ deficiencies is the presence of bone abnormalities, including delayed tooth eruption, auditory ossicle abnormalities, narrowing of the bone marrow cavity, increased metaphyseal bone density, reduced cortical thickness, and increased porosity (Fig. 3). These defects are associated with impaired bone remodeling due to a severe osteoclast deficiency that secondarily impairs osteoblast function as well as osteocyte survival and function (Abboud et al., 2002; Aharinejad et al., 1999; Dai et al., 2002; Dai, Zong, Akhter, et al., 2004; Harris et al., 2012; Ryan et al., 2001; Wiktor-Jedrzejczak et al., 1990; Yoshida et al., 1990) (Fig. 3). In postnatal $Csf1^{op/op}$ and $Csf1r^{-/-}$ mice, osteoclast numbers progressively increase, leading to the resorption of excess trabecular bone and partial restoration of the bone marrow cavity (Begg & Bertoncello, 1993; Dai et al., 2002). The partial restoration in $Csf1^{op/op}$ mice can be blocked by suppressing VEGF (Niida et al., 1999, 2005), or FLT3 (Lean, Fuller, & Chambers, 2001), signaling. However, since in contrast to FLT3, VEGF does not induce osteoclastic differentiation, nor RANK mRNA expression in vitro (Lean et al., 2001), it is likely that the actions of VEGF are indirect. Long-term CSF-1R inhibitor treatment in adult wild-type mice increases bone density (Chitu et al., 2012), suggesting that these compensatory mechanisms are limited to a narrow developmental time window.

Administration of recombinant CSF-1 partially rescued the bone phenotype of $Csf1^{op/op}$ mice (Cecchini et al., 1994; Pollard & Stanley, 1996; Sundquist, Cecchini, & Marks, 1995), whereas $Csf1$ promoter and first intron-driven transgenic expression of full-length CSF-1 (Ryan et al., 2001), or osteoblast-specific promoter-driven expression of CSF-1 (Abboud et al., 2002) completely rescued, emphasizing the importance of local regulation by the CSF-1R. In addition, an important role of the chondroitin sulfate chain of proteoglycan CSF-1 is indicated by the complete rescue of osteoclast development and skeletal abnormalities of $Csf1^{op/op}$ mice by $Csf1$ promoter and first intron-driven transgenic expression of the secreted proteoglycan isoform (Nandi et al., 2006), but not by the secreted glycoprotein or cell surface CSF-1 isoforms (Dai, Zong, Sylvestre, et al., 2004; Nandi et al., 2006).

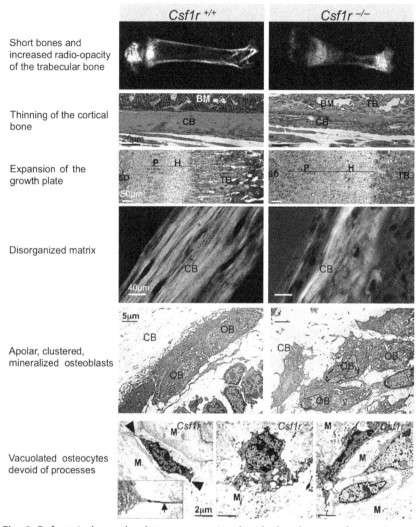

Fig. 3 Defects in bone development associated with the absence of osteoclasts in $Csf1r^{-/-}$ mice. *BM*, bone marrow; *CB*, cortical bone; *H*, hypertrophic chondrocyte region; *M*, matrix; *OB*, osteoblasts; *P*, proliferating chondrocyte region; *SO*, secondary ossification center; *TB*, trabecular bone. The *arrowheads* in the lower left panel point to canaliculi in the matrix. The *arrow* in the inset indicates an osteocyte process. *Panels republished with permission of John Wiley and Sons Inc., from Dai, X.-M., Zong, X.-H., Akhter, M. P., & Stanley, E. R. (2004). Osteoclast deficiency results in disorganized matrix, reduced mineralization, and abnormal osteoblast behavior in developing bone.* Journal of Bone and Mineral Research, *19(9), 1441–1451; permission conveyed through Copyright Clearance Center, Inc.*

4.1 CSF-1R Signaling in Osteoclastogenesis

In combination with RANKL, CSF-1 drives OC differentiation (Arai et al., 1999; Nakagawa et al., 1998), sustains the survival of mature OC (Bouyer et al., 2007; Fuller et al., 1993), and stimulates their spreading (Palacio & Felix, 2001; Sakai et al., 2006) and migration (Fuller et al., 1993). In vitro, the effects of CSF-1 during OC differentiation are biphasic, with low concentrations (in the normal physiological range) enhancing OC formation and high concentrations promoting macrophage differentiation (Hodge, Kirkland, & Nicholson, 2007; Perkins & Kling, 1995). Blockade of CSF-1 during the proliferative phase of OC formation dramatically inhibits osteoclastogenesis but has very little effect after the onset of OC fusion (Biskobing, Fan, & Rubin, 1995; Hodge et al., 2007), suggesting that the major role of CSF-1 is to drive the expansion of OC precursors and their maturation to a fusion-competent state but that CSF-1R signaling is not required for the process of fusion (Fig. 4). The cross talk between CSF-1R and RANK signaling pathways during OC committment, differentiation, survival and function is illustrated in Fig. 5 and discussed below.

4.1.1 OC Precursor Proliferation

Although the precise identity of OC precursors has not been established, it is known that following exposure to CSF-1 and RANKL, both pluripotent and commited myeloid progenitors as well as circulating monocytes can activate the expression of OC-specific genes and fuse into multinucleated OC-like cells. The efficiency of OC formation is proportional to the proliferative potential and the degree of pluripotency of the progenitors

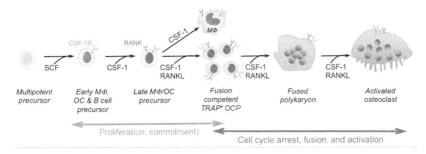

Fig. 4 Regulation of OC differentiation by CSF-1 and RANKL.

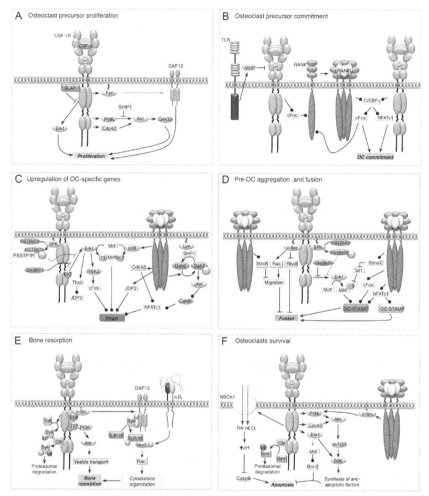

Fig. 5 CSF-1R signaling for osteoclast precursor proliferation (A) and commitment (B), the upregulation of osteoclast-specific genes (C), preosteoclast fusion (D), bone resorption by mature osteoclasts (E), and osteoclast survival (F) is shown.

(Arai et al., 1999; Hodge, Kirkland, Aitken, et al., 2004). Studies in bone marrow cell cultures suggest that CSF-1 regulates OC precursor proliferation through several mechanisms including activation of Cdc42/PI3K/Akt (Ito et al., 2010), Erk1 (He et al., 2011; Kim, Zou, et al., 2010), and DNAX-activating protein of molecular mass 12 kDa (DAP12) pathways (Despars et al., 2013; Humphrey et al., 2004) (Fig. 5A).

Cdc42 is required for CSF-1 activation of the Akt/GSK3β pathway, but not of ERKs. Cdc42 promotes CSF-1-driven proliferation in OC precursors (Ito et al., 2010). In vivo, OC-specific ablation of Cdc42 using cathepsin K-Cre leads to increased bone density and protection from ovariectomy-induced bone loss, while a Cdc42 gain-of-function mouse model exhibits a reciprocal phenotype (Ito et al., 2010). $Ship^{-/-}$ mice have severe osteoporosis (Takeshita et al., 2002), and their bone marrow-derived macrophages exhibit increased osteoclastogenic responses due to increased CSF-1-driven proliferation of OC precursors that is associated with increased PI3K/Akt activation, but unaltered MAPK pathway activation (Zhou et al., 2006). These data suggest that the PI3K/Akt pathway plays a central role in OC precursor proliferation.

Genetic ablation of Erk1, but not Erk2, reduced bone marrow bipotent OC/macrophage progenitor cell numbers, tartrate-resistant acid phosphatase (TRAP) expression, and multinucleation and diminished bone resorption leading to an increase in bone density (He et al., 2011). SLAP is an adaptor protein that associates with the CSF-1R in lipid rafts and inhibits CSF-1-induced Erk1/2 activation and proliferation without affecting RANKL signaling. In vitro, $Slap^{-/-}$ mouse bone marrow cultures exhibit increased osteoclastogenic responses but also reduced survival of mature OC following the withdrawal of CSF-1 and RANKL, features that may explain why $Slap^{-/-}$ mice maintain normal OC numbers (Kim, Zou, et al., 2010).

DAP12 is a transmembrane adaptor that mediates the proliferative signals of the CSF-1R (Otero et al., 2009). DAP12-deficient mice exhibit mild osteopetrosis (Humphrey et al., 2004), while mice overexpressing DAP12 have osteopenia and increased numbers of OC (Despars et al., 2013). This is associated with greater abundance of OC precursors in the spleen. DAP12-overexpressing splenocytes hyperproliferate in the presence of CSF-1 and RANKL but do not exhibit hyperresponsiveness to increasing concentrations of RANKL (Despars et al., 2013), indicating that RANK signaling in osteoclast precursors is not regulated by DAP12.

$Fyn^{-/-}$ mice exhibit normal basal osteoclastogenesis and bone density. However, their responses to RANKL are attenuated, indicating that although not essential in vivo, Fyn plays a regulatory role. In vitro, Fyn deficiency reduced CSF-1-stimulated proliferation of OC precursors, delayed their differentiation, diminished osteoclastogenesis, and increased the susceptibilty of mature OC to apoptosis, without affecting the activation of Erk1/2 or Akt (Kim, Warren, et al., 2010).

4.1.2 OC Precursor Commitment

Signals emerging from the RANK–RANKL complex are crucial for the differentiation of bipotent macrophage/OC precursors to OC (Arai et al., 1999). Examination of early stages of osteoclastogenesis revealed that the CSF-1R induces RANK expression in multipotent myeloid precursors (Arai et al., 1999) in a c-Fos-dependent manner (Arai et al., 2012) (Fig. 5B). This process is inhibited by activated Toll-like receptors, which promote matrix metalloproteinase-mediated CSF-1R shedding, thus diverting OC precursors to a monocytic fate (Ji et al., 2009). The combined actions of CSF-1 and RANKL induce high C/EBPα expression in murine bone marrow cells (Chen et al., 2013). Ectopic expression of C/EBPα induced the expression of other transcription factors essential for OC development such as c-Fos and NFATc1, and reprogrammed monocytic cells to OC-like cells in the absence of RANKL, suggesting that C/EBPα is the key transcriptional regulator of OC lineage commitment (Chen et al., 2013).

Although the precise point of osteoclastic commitment has not been defined, the upregulation of bone-resorbing enzymes, including TRAP and Cathepsin K, is likely to occur early after commitment. Inhibition of JNK at the $TRAP^+$ mononuclear cell (preosteoclast, pOC) stage leads to the downregulation of CaMK and NFATc1 levels resulting in the conversion of $TRAP^+$ cells to $TRAP^-$ cells and concomitant upregulation of macrophage markers, phagocytic activity, and dendritic cell potential (Chang et al., 2008). Thus, JNK signaling plays a critical role in maintaining OC commitment (Fig. 5C).

4.1.3 Upregulation of OC-Specific Genes

CSF-1R-activated, Erk1/2-mediated phosphorylation of microphthalmia-associated transcription factor (Mitf) at Ser73 and RANKL-activated, p38-dependent phosphorylation of Mitf at Ser307 stimulate the upregulation of TRAP activity (Ito et al., 2010; Mansky, Sankar, Han, & Ostrowski, 2002; Weilbaecher et al., 2001) (Fig. 5C). Two adaptor proteins, PSTPIP2 and SH3BP1, have been reported to suppress the activation of Erk1/2 downstream of CSF-1R (Chitu et al., 2009; Ueki et al., 2007) and the expression of TRAP (Chitu et al., 2012; Kawaida et al., 2003; Ueki et al., 2007). While the mechanism through which SH3BP1 acts is unclear, PSTPIP2-mediated suppression of TRAP expression requires both its association with PEST-family phosphatases and its CSF-1R-induced tyrosine phosphorylation (Chitu et al., 2012). Another event critical for induction of TRAP expression is the Tbx3-dependent induction of Jun dimerization protein

2 (JDP2) expression by the CSF-1/Erk1/2 pathway (Yao et al., 2014). JDP2 mediates RANKL-induced upregulation of TRAP and Cathepsin K (Kawaida et al., 2003). It has been reported that the Src family kinase, Lyn, forms a complex with RANK, the tyrosine phosphatase, SHP-1, and the adaptor protein, Grb2-associated binder 2 (Gab2) that impedes Gab2 phosphorylation and downstream activation of JNK and NF-kappaB and subsequent OC development (Kim et al., 2009). While $Lyn^{-/-}$ mice have no overt bone phenotype, they exhibit increased osteoclastogenesis in response to RANKL administration (Kim et al., 2009), indicating that Lyn could mediate bone loss in conditions associated with increased RANKL levels such as postmenopausal bone loss (Mundy, 2007).

4.1.4 Pre-OC Aggregation and Fusion

Dendritic cell-specific transmembrane protein (DC-STAMP)-deficient mice cannot develop multinucleated OCs (Yagi et al., 2005), suggesting that DC-STAMP is a crucial factor for OC fusion. Signals emerging from both CSF-1R and RANK converge to upregulate the expression of DC-STAMP (Fig. 5D). CSF-1R-induced and Erk1/2-mediated phosphorylation of Mitf at serine residue 73 stimulates Mitf transcriptional activity and OC fusion (Weilbaecher et al., 2001). RANKL provides a permissive signal by upregulating the expression of the DExD/H helicase family corepressor strawberry notch homologue 2 (Sbno2) (Maruyama et al., 2013). Sbno2 binds to T cell acute lymphocytic leukemia 1 (Tal1), attenuating Tal1 inhibition of DC-STAMP expression and permitting the activation of the DC-STAMP promoter by Mitf (Maruyama et al., 2013). Independently of the Mitf pathway, RANK promotes the transcription of the fusogenic receptors DC-STAMP and OC-STAMP by upregulating the expression of c-Fos and NFATc1 (Dou et al., 2014; Miyamoto et al., 2012). In addition, regulation of the actin cytoskeleton by CSF-1R and RANK also plays an important role in promoting pOC fusion. Downstream of CSF-1R, c-Src activates Rac and inhibits Rho, two small GTPases that regulate the cytoskeleton. While Rac activity is dispensable for osteoclast development and multinucleation (Croke et al., 2011), Src-mediated inhibition of Rho activity is necessary for efficient fusion (Takito et al., 2015). Activated RANK upregulates the expression of the Rho family GTP-ase Wrch (a.k.a. RhoU) (Brazier et al., 2006) which inhibits CSF-1-induced pOC migration and integrin-mediated adhesion while simultaneously promoting their aggregation and fusion (Brazier, Pawlak, Vives, & Blangy, 2009).

4.2 CSF-1R Regulation of Mature OC Migration and Bone Resorption

During bone resorption, the OCs alternate between phases of matrix degradation and nonresorptive stages of migration. Nonresorbing OCs move on the bone surface while bone-resorbing OCs are nonmotile and form an adhesion structure known as the sealing zone that separates the ruffled border from the rest of the cell membrane (Lakkakorpi & Vaananen, 1996). The sealing zone contains F-actin-rich podosomes and the $\alpha_v\beta_3$ integrin, a receptor for vitronectin and osteopontin (OPN) (Faccio et al., 2002; Insogna, Sahni, et al., 1997; Nakamura et al., 1999). Inhibition of integrin-mediated adhesion leads to the disruption of both the sealing zone and bone resorption (Nakamura et al., 1999). Paradoxically, stimulation of either human or rodent mature OCs by CSF-1 induces a rapid spreading and chemotactic response (Faccio et al., 2002; Faccio, Novack, Zallone, Ross, & Teitelbaum, 2003; Pilkington, Sims, & Dixon, 1998), while also increasing bone resorption (Faccio et al., 2002; Hodge, Collier, Pavlos, Kirkland, & Nicholson, 2011). A possible explanation resides in fact that CSF-1R collaborates with $\alpha_v\beta_3$ integrin to promote the reorganization of the OC cytoskeleton (Faccio et al., 2003). CSF-1 treatment increases OC chemotaxis to OPN coated-surfaces (Faccio et al., 2002). Like CSF-1, OPN increases both OC migration and bone resorption. It has been suggested that, via interaction with the $\alpha_v\beta_3$ integrin on the OC membrane, the OPN bound to the bone matrix mediates the anchorage of OCs to the bone, thus favoring resorption, while subsequent dephosphorylation of OPN by OC-secreted TRAP inhibits adhesion and initiates migration (Ek-Rylander & Andersson, 2010; Reinholt, Hultenby, Oldberg, & Heinegard, 1990).

In vitro, OCs plated on glass form a peripheral podosome belt which is analogous to the sealing zone. Acute stimulation with CSF-1 induces a spreading response accompanied by the dissolution of the podosome belt and the reorganization of F-actin and $\alpha_v\beta_3$ integrin in peripheral ruffles and perinuclear punctae (Faccio et al., 2007; Insogna, Tanaka, et al., 1997; Shinohara et al., 2012). While not sufficient for the transduction of CSF-1R-mediated cytoskeletal changes (Faccio et al., 2007), activation of Src by its interaction with the phosphorylated tyrosine residue 559 (Y559) of CSF-1R plays a central role in this process (Fig. 5E). The Y559F CSF1R mutation blocks CSF-1-induced cytoskeletal reorganization (Faccio et al., 2007), and several lines of evidence suggest that c-Src is the Src family kinase (SFK) regulating the pathway mediating CSF-1R-induced actin remodelling in OC: (1) $Src^{-/-}$ OC fail to form a sealing zone and their F-actin reorganization in response to CSF-1 is abolished (Insogna, Tanaka, et al., 1997); (2) Fyn or Lyn deficiencies do not affect actin organization in OC (Kim, Warren, et al., 2010; Kim et al., 2009), and (3) Hck controls podosome formation in pOC before Src expression is upregulated, but is not essential for podosome formation or bone resorption by mature OC in which Src is highly expressed (Vérollet et al., 2013). A proposed pathway through which CSF-1R Y559-dependent activation of Src regulates OC cytoskeletal organization involves Src phosphorylation of the immunoreceptor tyrosine-based activation motifs in DAP12, which triggers the recruitment and activation the nonreceptor tyrosine kinase, Syk, via Src homology region 2 (SH2) domain-mediated interactions (Zou, Reeve, Liu, Teitelbaum, & Ross, 2008) (Fig. 5E). Activated Syk phosphorylates the SH2 domain-containing leukocyte protein-76 (SLP-76) adaptor which in turn recruits and mediates the tyrosine phosphorylation of Vav3 (Faccio et al., 2005; Vérollet et al., 2013). Vav3 activates Rac in OCs downstream of both the CSF-1R and $\alpha_v\beta_3$ integrin (Faccio et al., 2005; Reeve et al., 2009). Rac1 and 2 mediate CSF-1-induced chemotactic migration in pre-OCs and mature OCs (Faccio et al., 2007; Itokowa et al., 2011; Wang et al., 2008). Deletion of DAP12, Syk, SLP-76, Vav3, or Rac1 and 2 yields OCs that fail to organize their cytoskeleton and exhibit defects in bone resorption (Croke et al., 2011; Faccio et al., 2005; Mócsai et al., 2004; Reeve et al., 2009; Zou et al., 2008; Zou, Reeve, Zhao, Ross, & Teitelbaum, 2009). In addition, the CSF-1R also initiates a negative feedback pathway leading to phosphorylation of Syk at Tyr-317, which promotes Syk association with Cbl, the ubiquitination and degradation of Syk, and attenuation of OC function (Zou et al., 2009). The Y317F mutation abolishes the association of Syk with Cbl, resulting in CSF-1-induced hyperphosphorylation of the cytoskeleton-organizing molecules, SLP-76, Vav3, and PLCγ2, and increased resorptive capacity of the OCs (Zou et al., 2009).

Recently, the PI3K/Akt pathway has been described to regulate bone resorption downstream of CSF-1R, but not the cytoskeletal remodeling. Disruption of the CSF-1R-p85 binding site (CSF-1R Y721F) or combined targeting of the p85 α and β genes in OCs did not suppress cytoskeletal remodeling or actin ring formation (Faccio et al., 2007). However, p85 deficiency leads to an osteopetrotic phenotype caused by the absence of a ruffled border and defective transport of Cathepsin K-containing vesicles at the plasma membrane (Shinohara et al., 2012).

4.3 CSF-1R Signaling for OC Survival

CSF-1R signaling is necessary for the survival of OCs and their precursors. In OC precursors, CSF-1 exerts antiapoptotic effects through the upregulation of the antiapoptotic protein Bcl-X(L) which in turn inhibits caspase-9-mediated apoptosis (Woo, Kim, & Ko, 2002). In mature OC, CSF-1R-activated pathways that promote survival include activation of mTOR/S6 kinase, the Cbl ubiquitin ligase and the electroneutral Na/HCO$_3$ cotransporter, NBCn1 (Fig. 5F). Activation of mTOR/S6 kinase is regulated by multiple CSF-1R effectors, including PI3K/AKT, Erk, and geranylgeranylated proteins, such as Cdc42 (Glantschnig, Fisher, Wesolowski, Rodan, & Reszka, 2003; Kim, Zou, et al., 2010; Miyazaki et al., 2000). Activated Cbl ubiquitinates the proapoptotic Bcl-2 family member, Bim, leading to its proteasomal degradation (Akiyama et al., 2003). CSF-1 stimulation activates NBCn1, leading to alkalinization of intracellular pH and inhibition of caspase-8-mediated apoptosis (Bouyer et al., 2007). In addition to its role in OC differentiation, Mitf is required for the expression of the antiapoptotic protein Bcl-2 in the OC lineage. Since both Mitf- and Bcl2-deficient mice are osteopetrotic (McGill et al., 2002), it is possible that Mitf activation by CSF-1 and RANKL also contributes to OC survival.

5. DIRECT REGULATION OF NONHEMATOPOIETIC CELLS BY THE CSF-1R

The CSF-1R is expressed on several nonmononuclear phagocytic cell types, including neuronal stem cells, PC, oocytes, early embryonic cells, and trophoblastic cells that are directly regulated by its ligands (Fig. 6).

5.1 Neuronal Stem Cells

Apart from microglia, the CSF-1R is expressed on NPC and several subpopulations of neurons in the brain (reviewed in Chitu et al., 2016). CSF-1, and CSF-1 and IL-34 were shown to directly regulate NPC self-renewal, differentiation, and survival in vitro (Nandi et al., 2012). Targeted deletion of the *Csf1r* in neural progenitors of *Nestin-Cre/+;Csf1r$^{fl/fl}$* mice did not affect the development of tissue macrophages, osteoclasts, or microglia, yet recapitulated several neurological defects of the *Csf1r$^{-/-}$* mouse including a smaller

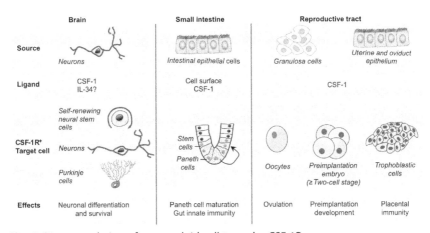

Fig. 6 Direct regulation of nonmyeloid cell types by CSF-1R.

brain size, expansion of forebrain neural progenitors and elevated forebrain apoptosis as well as the perinatal lethality. These findings raise the possibility that neurological deficits significantly contribute to the early lethality of $Csf1r^{-/-}$ mice.

5.2 Paneth Cells

PC are granulated cells located at the base of small intestinal crypts that play a critical role in the innate immune responses by releasing antimicrobial substances in response to microbial challenge. They have additional roles as regulators of detoxification, digestion, and in the regulation of stem cells as components of the stem cell niche (Clevers & Bevins, 2013; Porter, Bevins, Ghosh, & Ganz, 2002; Wehkamp & Stange, 2006). In both $Csf1^{op/op}$ and $Csf1r^{-/-}$ mice, there is abnormal organization of the small intestine, with marked reductions of epithelial cells, including PC and intestinal stem cells, and of interstitial macrophages of the villi (Akcora et al., 2013; Cecchini et al., 1994; Huynh et al., 2009). PCs express the CSF-1R and reside in close proximity to CSF-1-expressing epithelial cells within the crypt (Huynh et al., 2009). *Csf1* promoter-driven transgenic expression of the cell surface isoform of CSF-1 in $Csf1^{op/op}$ mice rescues the macrophage and PC deficiencies, as well as stem cell activity suggesting that CSF-1 regulates intestinal development, in a juxtacrine or paracrine manner (Huynh et al., 2009). In addition, intestine-specific deletion of *Csf1r* reproduced the $Csf1r^{-/-}$ phenotype, formally proving that CSF-1R cell autonomously supports PC development (Akcora et al., 2013). These studies also showed that the CSF-1R is essential for de novo generation of PC, but not for their survival and demonstrated the critical role of CSF-1R regulation of PC in the maintenance of intestinal stem cells.

5.3 Direct Regulation of Oocytes, Preimplantation Embryos, and Trophoblastic Cells

Defects in the hypothalamic–pituitary feedback system of $Csf1^{op/op}$ mice lead to abnormal pubertal development and impaired fertility (reviewed in Chitu, Caescu, et al., 2015). Apart from its effects on the hypothalamic–pituitary feedback system, direct regulation by the CSF-1R in nonmononuclear phagocytic cells plays important roles in embryonic development and in the innate immune protection of the fetus at the maternal/fetal interface. CSF-1R is expressed on oocytes, early mouse embryos, and cells of the trophoblast, and CSF-1 is expressed by granulosa cells, oviduct, and highly in the uterine epithelium of pregnant mice. $Csf1^{op/op}$ females have extended estrous cycles and poor ovulation rates, but once oocytes are fertilized, their implantation rates approach those of wild-type mice (Araki et al., 1996; Cohen, Zhu, & Pollard, 1997). $Csf1^{op/op}$ females have fewer ovarian antral follicles and markedly reduced numbers of ovarian macrophages (Araki et al., 1996). Administration of CSF-1 to rats increased both antral follicle and ovarian macrophage numbers (Nishimura et al., 1995) suggesting that the reduced ovulation could result from loss of trophic effects of the macrophages, or loss of direct regulation of oocytes. Studies with cultured preimplantation mouse embyros showed that CSF-1 accelerates the formation of the blastocyst cavity and increases the number of trophoblast cells (Bhatnagar, Papaioannou, & Biggers, 1995; Pampfer et al., 1991). Direct CSF-1R regulation in trophoblastic cells is essential for the placental immune response to *Listeria monocytogenes,* by signaling the synthesis of chemoattractants that recruit neutrophils to the site of infection (Guleria & Pollard, 2000).

6. CONCLUSIONS

This review has focused on the developmental roles played by the CSF-1R. The CSF-1R is the central regulator of osteoclasts and microglia as reflected in the bone and neurological phenotypes of the CSF-1R-deficient mouse. Tissue-resident macrophages play an important role in organismal development. All tissue macrophages express the CSF-1R but CSF-1R-signaling is not essential for the development and survival of all macrophages. Recent studies indicate that the CSF-1R may be most critical for the development of EMP-derived tissue macrophages. However, additional studies are required to formally prove this. Of the two known CSF-1R ligands, CSF-1 is most critical for organismal development. Apart from effects mediated by its action on myeloid cells, CSF-1 also directly regulates certain nonmononuclear phagocytic cells, including neural stem cells, PC and cells of the female reproductive tract. A snapshot encapsulating both the developmental and homeostatic roles of the CSF-1R across species is summarized in Fig. 7. Initial observations showing that CSF-1 mRNA is highly expressed in cells proximal to tissue macrophages, or other CSF-1R expressing cells (Bartocci, Pollard, & Stanley, 1986; Gouon-Evans et al., 2000; Huynh et al., 2009; Nandi et al., 2012; Pollard et al., 1987; Ryan et al., 2001) suggested the importance of local regulation, which was subsequently confirmed by reconstituting CSF-1 expression either

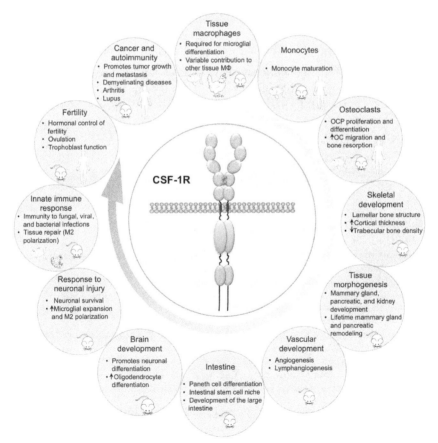

Fig. 7 Roles of CSF-1R in development and adult physiology. The organism in which the relevant studies were performed is depicted.

locally or systemically. Although activation of the CSF-1R on myeloid cells by either CSF-1, or IL-34, results in similar biochemical and biological outcomes, their in vivo patterns of expression are mostly distinct, both during development and inflammation, explaining the striking differences in their null phenotypes. The coexpression of the CSF-1R with the additional IL-34 receptor, PTPζ, on stem cells raises the possibility that the two CSF-1R ligands may elicit different biological outcomes on these cells.

What are the major unanswered questions? It would be of interest to determine whether the small differences in CSF-1R signaling between myeloid cells responding to CSF-1, IL-34, and IL-34/syndecan contribute to the specification of different tissue macrophage populations, e.g., Langerhans cells and dermal macrophages. Investigation of how the CSF-1R regulates macrophage function has been hampered by the fact that *Csf1r* deficiency affects their viability and proliferation. The recent finding that the pathways regulating macrophage survival/proliferation are distinct from those regulating their trophic/inflammatory phenotypes provides the basis for a new strategy to study the role of the CSF-1R in macrophage function. Further understanding of the signaling pathways critical for CSF-1R-regulated functions in nonmyeloid cells is also needed, especially in view of the development of CSF-1R inhibitors for long-term clinical treatments.

ACKNOWLEDGMENTS

We thank Dr. Şölen Gökhan for critically reviewing the manuscript and the editor for his helpful comments. This work was supported by NIH Grants R01 NS091519 and P01 CA100324 (to E.R.S.).

REFERENCES

Abboud, S. L., Woodruff, K., Liu, C., Shen, V., & Ghosh-Choudhury, N. (2002). Rescue of the osteopetrotic defect in op/op mice by osteoblast-specific targeting of soluble colony-stimulating factor-1. *Endocrinology*, *143*, 1942–1949.

Aharinejad, S., Grossschmidt, K., Franz, P., Streicher, J., Nourani, F., MacKay, C. A., et al. (1999). Auditory ossicle abnormalities and hearing loss in the toothless (osteopetrotic) mutation in the rat and their improvement after treatment with colony-stimulating factor-1. *Journal of Bone and Mineral Research*, *14*, 415–423.

Ajami, B., Bennett, J. L., Krieger, C., McNagny, K. M., & Rossi, F. M. (2011). Infiltrating monocytes trigger EAE progression, but do not contribute to the resident microglia pool. *Nature Neuroscience*, *14*, 1142–1149.

Akcora, D., Huynh, D., Lightowler, S., Germann, M., Robine, S., de May, J. R., et al. (2013). The CSF-1 receptor fashions the intestinal stem cell niche. *Stem Cell Research*, *10*, 203–212.

Akiyama, T., Bouillet, P., Miyazaki, T., Kadono, Y., Chikuda, H., Chung, U. I., et al. (2003). Regulation of osteoclast apoptosis by ubiquitylation of proapoptotic BH3-only Bcl-2 family member Bim. *EMBO Journal*, *22*, 6653–6664.

Alikhan, M. A., Jones, C. V., Williams, T. M., Beckhouse, A. G., Fletcher, A. L., Kett, M. M., et al. (2011). Colony-stimulating factor-1 promotes kidney growth and repair via alteration of macrophage responses. *The American Journal of Pathology, 179*, 1243–1256.

Ancuta, P., Liu, K. Y., Misra, V., Wacleche, V. S., Gosselin, A., Zhou, X., et al. (2009). Transcriptional profiling reveals developmental relationship and distinct biological functions of CD16+ and CD16− monocyte subsets. *BMC Genomics, 10*, 403.

Arai, F., Miyamoto, T., Ohneda, O., Inada, T., Sudo, T., Brasel, K., et al. (1999). Commitment and differentiation of osteoclast precursor cells by the sequential expression of c-Fms and receptor activator of nuclear factor kappaB (RANK) receptors. *The Journal of Experimental Medicine, 190*, 1741–1754.

Arai, A., Mizoguchi, T., Harada, S., Kobayashi, Y., Nakamichi, Y., Yasuda, H., et al. (2012). Fos plays an essential role in the upregulation of RANK expression in osteoclast precursors within the bone microenvironment. *Journal of Cell Science, 125*, 2910–2917.

Araki, M., Fukumatsu, Y., Katabuchi, H., Shultz, L. D., Takahashi, K., & Okamura, H. (1996). Follicular development and ovulation in macrophage colony-stimulating factor-deficient mice homozygous for the osteopetrosis (op) mutation. *Biology of Reproduction, 54*, 478–484.

Arceci, R. J., Pampfer, S., & Pollard, J. W. (1992). Expression of CSF-1/c-fms and SF/c-kit mRNA during preimplantation mouse development. *Developmental Biology, 151*, 1–8.

Arceci, R. J., Shanahan, F., Stanley, E. R., & Pollard, J. W. (1989). Temporal expression and location of colony-stimulating factor 1 (CSF-1) and its receptor in the female reproductive tract are consistent with CSF-1-regulated placental development. *Proceedings of the National Academy of Sciences of the United States of America, 86*, 8818–8822.

Arkins, S., Rebeiz, N., Biragyn, A., Reese, D. L., & Kelley, K. W. (1993). Murine macrophages express abundant insulin-like growth factor-I class I Ea and Eb transcripts. *Endocrinology, 133*, 2334–2343.

Arkins, S., Rebeiz, N., Brunke-Reese, D. L., Minshall, C., & Kelley, K. W. (1995). The colony-stimulating factors induce expression of insulin-like growth factor-I messenger ribonucleic acid during hematopoiesis. *Endocrinology, 136*, 1153–1160.

Arnold, T., & Betsholtz, C. (2013). Correction: The importance of microglia in the development of the vasculature in the central nervous system. *Vascular Cell, 5*, 12.

Audoy-Remus, J., Richard, J. F., Soulet, D., Zhou, H., Kubes, P., & Vallieres, L. (2008). Rod-Shaped monocytes patrol the brain vasculature and give rise to perivascular macrophages under the influence of proinflammatory cytokines and angiopoietin-2. *The Journal of Neuroscience, 28*, 10187–10199.

Auffray, C., Fogg, D., Garfa, M., Elain, G., Join-Lambert, O., Kayal, S., et al. (2007). Monitoring of blood vessels and tissues by a population of monocytes with patrolling behavior. *Science, 317*, 666–670.

Bain, C. C., Bravo-Blas, A., Scott, C. L., Gomez Perdiguero, E., Geissmann, F., Henri, S., et al. (2014). Constant replenishment from circulating monocytes maintains the macrophage pool in the intestine of adult mice. *Nature Immunology, 15*, 929–937.

Banaei-Bouchareb, L., Gouon-Evans, V., Samara-Boustani, D., Castellotti, M. C., Czernichow, P., Pollard, J. W., et al. (2004). Insulin cell mass is altered in Csf1op/Csf1op macrophage-deficient mice. *Journal of Leukocyte Biology, 76*, 359–367.

Banaei-Bouchareb, L., Peuchmaur, M., Czernichow, P., & Polak, M. (2006). A transient microenvironment loaded mainly with macrophages in the early developing human pancreas. *The Journal of Endocrinology, 188*, 467–480.

Bartocci, A., Pollard, J. W., & Stanley, E. R. (1986). Regulation of colony-stimulating factor 1 during pregnancy. *Journal of Experimental Medicine, 164*, 956–961.

Begg, S. K., & Bertoncello, I. (1993). The hematopoietic deficiencies in osteopetrotic (op/op) mice are not permanent, but progressively correct with age. *Experimental Hematology, 21*, 493–495.

Bellavance, M. A., Gosselin, D., Yong, V. W., Stys, P. K., & Rivest, S. (2015). Patrolling monocytes play a critical role in CX3CR1-mediated neuroprotection during excitotoxicity. *Brain Structure & Function, 220,* 1759–1776.

Bertrand, J. Y., Jalil, A., Klaine, M., Jung, S., Cumano, A., & Godin, I. (2005). Three pathways to mature macrophages in the early mouse yolk sac. *Blood, 106,* 3004–3011.

Bhatnagar, P., Papaioannou, V. E., & Biggers, J. D. (1995). CSF-1 and mouse preimplantation development in vitro. *Development, 121,* 1333–1339.

Biskobing, D. M., Fan, X., & Rubin, J. (1995). Characterization of MCSF-induced proliferation and subsequent osteoclast formation in murine marrow culture. *Journal of Bone and Mineral Research, 10,* 1025–1032.

Bouyer, P., Sakai, H., Itokawa, T., Kawano, T., Fulton, C. M., Boron, W. F., et al. (2007). Colony-stimulating factor-1 increases osteoclast intracellular pH and promotes survival via the electroneutral Na/HCO3 cotransporter NBCn1. *Endocrinology, 148,* 831–840.

Bradley, T. R., & Metcalf, D. (1966). The growth of mouse bone marrow cells in vitro. *Australian Journal of Experimental Biology and Medical Science, 44,* 287–299.

Brady, N. J., Chuntova, P., & Schwertfeger, K. L. (2016). Macrophages: Regulators of the inflammatory microenvironment during mammary gland development and breast cancer. *Mediators of Inflammation, 2016,* 4549676.

Brazier, H., Pawlak, G., Vives, V., & Blangy, A. (2009). The Rho GTPase Wrch1 regulates osteoclast precursor adhesion and migration. *The International Journal of Biochemistry & Cell Biology, 41,* 1391–1401.

Brazier, H., Stephens, S., Ory, S., Fort, P., Morrison, N., & Blangy, A. (2006). Expression profile of RhoGTPases and RhoGEFs during RANKL-stimulated osteoclastogenesis: Identification of essential genes in osteoclasts. *Journal of Bone and Mineral Research, 21,* 1387–1398.

Byrne, P. V., Guilbert, L. J., & Stanley, E. R. (1981). Distribution of cells bearing receptors for a colony-stimulating factor (CSF) in murine tissues. *The Journal of Cell Biology, 91,* 848–853.

Caescu, C. I., Guo, X., Tesfa, L., Bhagat, T. D., Verma, A., Zheng, D., et al. (2015). Colony stimulating factor-1 receptor signaling networks inhibit mouse macrophage inflammatory responses by induction of microRNA-21. *Blood, 125,* e1–e13.

Carlin, L. M., Stamatiades, E. G., Auffray, C., Hanna, R. N., Glover, L., Vizcay-Barrena, G., et al. (2013). Nr4a1-dependent Ly6C(low) monocytes monitor endothelial cells and orchestrate their disposal. *Cell, 153,* 362–375.

Casano, A. M., & Peri, F. (2015). Microglia: Multitasking specialists of the brain. *Developmental Cell, 32,* 469–477.

Cecchini, M. G., Dominguez, M. G., Mocci, S., Wetterwald, A., Felix, R., Fleisch, H., et al. (1994). Role of colony stimulating factor-1 in the establishment and regulation of tissue macrophages during postnatal development of the mouse. *Development, 120,* 1357–1372.

Chang, E. J., Ha, J., Huang, H., Kim, H. J., Woo, J. H., Lee, Y., et al. (2008). The JNK-dependent CaMK pathway restrains the reversion of committed cells during osteoclast differentiation. *Journal of Cell Science, 121,* 2555–2564.

Chang, H. R., Kim, H. J., Xu, X., & Ferrante, A. W., Jr. (2016). Macrophage and adipocyte IGF1 maintain adipose tissue homeostasis during metabolic stresses. *Obesity (Silver Spring), 24,* 172–183.

Charre, S., Rosmalen, J. G., Pelegri, C., Alves, V., Leenen, P. J., Drexhage, H. A., et al. (2002). Abnormalities in dendritic cell and macrophage accumulation in the pancreas of nonobese diabetic (NOD) mice during the early neonatal period. *Histology and Histopathology, 17,* 393–401.

Checchin, D., Sennlaub, F., Levavasseur, E., Leduc, M., & Chemtob, S. (2006). Potential role of microglia in retinal blood vessel formation. *Investigative Ophthalmology & Visual Science, 47,* 3595–3602.

Chen, X., Liu, H., Focia, P. J., Shim, A. H., & He, X. (2008). Structure of macrophage colony stimulating factor bound to FMS: Diverse signaling assemblies of class III receptor tyrosine kinases. *Proceedings of the National Academy of Sciences of the United States of America, 105*, 18267–18272.

Chen, W., Zhu, G., Hao, L., Wu, M., Ci, H., & Li, Y. P. (2013). C/EBPalpha regulates osteoclast lineage commitment. *Proceedings of the National Academy of Sciences of the United States of America, 110*, 7294–7299.

Chitu, V., Caescu, C. I., Stanley, E. R., Lennartsson, J., Ronnstrand, L., & Heldin, C.-H. (2015). PDGF receptor family. In D. L. Wheeler & Y. Yarden (Eds.), *The receptor tyrosine kinases: Family and subfamilies* (pp. 373–538). New York: Springer Science.

Chitu, V., Ferguson, P. J., de Bruijn, R., Schlueter, A. J., Ochoa, L. A., Waldschmidt, T. J., et al. (2009). Primed innate immunity leads to autoinflammatory disease in PSTPIP2-deficient cmo mice. *Blood, 114*, 2497–2505.

Chitu, V., Gokhan, S., Gulinello, M., Branch, C. A., Patil, M., Basu, R., et al. (2015). Phenotypic characterization of a Csf1r haploinsufficient mouse model of adult-onset leukodystrophy with axonal spheroids and pigmented glia (ALSP). *Neurobiology of Disease, 74*, 219–228.

Chitu, V., Gokhan, S., Nandi, S., Mehler, M. F., & Stanley, E. R. (2016). Emerging roles for CSF-1 receptor and its ligands in the nervous system. *Trends in Neurosciences, 39*, 378–393.

Chitu, V., Nacu, V., Charles, J. F., Henne, W. M., McMahon, H. T., Nandi, S., et al. (2012). PSTPIP2 deficiency in mice causes osteopenia and increased differentiation of multipotent myeloid precursors into osteoclasts. *Blood, 120*, 3126–3135.

Chitu, V., & Stanley, E. R. (2006). Colony-stimulating factor-1 in immunity and inflammation. *Current Opinion in Immunology, 18*, 39–48.

Clevers, H. C., & Bevins, C. L. (2013). Paneth cells: Maestros of the small intestinal crypts. *Annual Review of Physiology, 75*, 289–311.

Cohen, P. E., Chisholm, O., Arceci, R. J., Stanley, E. R., & Pollard, J. W. (1996). Absence of colony-stimulating factor-1 in osteopetrotic ($csfm^{op}/csfm^{op}$) mice results in male fertility defects. *Biology of Reproduction, 55*, 310–317.

Cohen, P. E., Nishimura, K., Zhu, L., & Pollard, J. W. (1999). Macrophages: Important accessory cells for reproductive function. *Journal of Leukocyte Biology, 66*, 765–772.

Cohen, P. E., Zhu, L., Nishimura, K., & Pollard, J. W. (2002). Colony-stimulating factor 1 regulation of neuroendocrine pathways that control gonadal function in mice. *Endocrinology, 143*, 1413–1422.

Cohen, P. E., Zhu, L., & Pollard, J. W. (1997). Absence of colony stimulating factor-1 in osteopetrotic (csfmop/csfmop) mice disrupts estrous cycles and ovulation. *Biology of Reproduction, 56*, 110–118.

Coussens, L., Van Beveren, C., Smith, D., Chen, E., Mitchell, R. L., Isacke, C. M., et al. (1986). Structural alteration of viral homologue of receptor proto-oncogene fms at carboxyl terminus. *Nature, 320*, 277–280.

Croke, M., Ross, F. P., Korhonen, M., Williams, D. A., Zou, W., & Teitelbaum, S. L. (2011). Rac deletion in osteoclasts causes severe osteopetrosis. *Journal of Cell Science, 124*, 3811–3821.

Croxford, A. L., Lanzinger, M., Hartmann, F. J., Schreiner, B., Mair, F., Pelczar, P., et al. (2015). The cytokine GM-CSF drives the inflammatory signature of CCR2 monocytes and licenses autoimmunity. *Immunity, 43*, 502–514.

Dai, X. M., Ryan, G. R., Hapel, A. J., Dominguez, M. G., Russell, R. G., Kapp, S., et al. (2002). Targeted disruption of the mouse colony-stimulating factor 1 receptor gene results in osteopetrosis, mononuclear phagocyte deficiency, increased primitive progenitor cell frequencies, and reproductive defects. *Blood, 99*, 111–120.

Dai, X. M., Zong, X. H., Akhter, M. P., & Stanley, E. R. (2004a). Osteoclast deficiency results in disorganized matrix, reduced mineralization, and abnormal osteoblast behavior in developing bone. *Journal of Bone and Mineral Research, 19*, 1441–1451.

Dai, X. M., Zong, X. H., Sylvestre, V., & Stanley, E. R. (2004b). Incomplete restoration of colony-stimulating factor 1 (CSF-1) function in CSF-1-deficient Csf1op/Csf1op mice by transgenic expression of cell surface CSF-1. *Blood, 103*, 1114–1123.

Das, S. K., Stanley, E. R., Guilbert, L. J., & Forman, L. W. (1980). Discrimination of a colony stimulating factor subclass by a specific receptor on a macrophage cell line. *Journal of Cellular Physiology, 104*, 359–366.

Despars, G., Pandruvada, S. N., Anginot, A., Domenget, C., Jurdic, P., & Mazzorana, M. (2013). DAP12 overexpression induces osteopenia and impaired early hematopoiesis. *PloS One*, 8e65297.

Dou, C., Zhang, C., Kang, F., Yang, X., Jiang, H., Bai, Y., et al. (2014). MiR-7b directly targets DC-STAMP causing suppression of NFATc1 and c-Fos signaling during osteoclast fusion and differentiation. *Biochimica et Biophysica Acta, 1839*, 1084–1096.

Ek-Rylander, B., & Andersson, G. (2010). Osteoclast migration on phosphorylated osteopontin is regulated by endogenous tartrate-resistant acid phosphatase. *Experimental Cell Research, 316*, 443–451.

Elegheert, J., Desfosses, A., Shkumatov, A. V., Wu, X., Bracke, N., Verstraete, K., et al. (2011). Extracellular complexes of the hematopoietic human and mouse CSF-1 receptor are driven by common assembly principles. *Structure, 19*, 1762–1772.

Epelman, S., Lavine, K. J., Beaudin, A. E., Sojka, D. K., Carrero, J. A., Calderon, B., et al. (2014). Embryonic and adult-derived resident cardiac macrophages are maintained through distinct mechanisms at steady state and during inflammation. *Immunity, 40*, 91–104.

Erblich, B., Zhu, L., Etgen, A. M., Dobrenis, K., & Pollard, J. W. (2011). Absence of colony stimulation factor-1 receptor results in loss of microglia, disrupted brain development and olfactory deficits. *PloS One*, 6e26317.

Faccio, R., Grano, M., Colucci, S., Villa, A., Giannelli, G., Quaranta, V., et al. (2002). Localization and possible role of two different alpha v beta 3 integrin conformations in resting and resorbing osteoclasts. *Journal of Cell Science, 115*, 2919–2929.

Faccio, R., Novack, D. V., Zallone, A., Ross, F. P., & Teitelbaum, S. L. (2003). Dynamic changes in the osteoclast cytoskeleton in response to growth factors and cell attachment are controlled by beta3 integrin. *The Journal of Cell Biology, 162*, 499–509.

Faccio, R., Takeshita, S., Colaianni, G., Chappel, J., Zallone, A., Teitelbaum, S. L., et al. (2007). M-CSF regulates the cytoskeleton via recruitment of a multimeric signaling complex to c-Fms Tyr-559/697/721. *The Journal of Biological Chemistry, 282*, 18991–18999.

Faccio, R., Teitelbaum, S. L., Fujikawa, K., Chappel, J., Zallone, A., Tybulewicz, V. L., et al. (2005). Vav3 regulates osteoclast function and bone mass. *Nature Medicine, 11*, 284–290.

Fantin, A., Vieira, J. M., Gestri, G., Denti, L., Schwarz, Q., Prykhozhij, S., et al. (2010). Tissue macrophages act as cellular chaperones for vascular anastomosis downstream of VEGF-mediated endothelial tip cell induction. *Blood, 116*, 829–840.

Felix, J., Elegheert, J., Gutsche, I., Shkumatov, A. V., Wen, Y. R., Bracke, N., et al. (2013). Human IL-34 and CSF-1 establish structurally similar extracellular assemblies with their common hematopoietic receptor. *Structure, 21*, 528–539.

Fleetwood, A. J., Lawrence, T., Hamilton, J. A., & Cook, A. D. (2007). Granulocyte-macrophage colony-stimulating factor (CSF) and macrophage CSF-dependent macrophage phenotypes display differences in cytokine profiles and transcription factor activities: Implications for CSF blockade in inflammation. *Journal of Immunology, 178*, 5245–5252.

Frost, J. L., & Schafer, D. P. (2016). Microglia: Architects of the developing nervous system. *Trends in Cell Biology, 26,* 587–597.
Fuller, K., Owens, J. M., Jagger, C. J., Wilson, A., Moss, R., & Chambers, T. J. (1993). Macrophage colony-stimulating factor stimulates survival and chemotactic behavior in isolated osteoclasts. *Journal of Experimental Medicine, 178,* 1733–1744.
Garceau, V., Balic, A., Garcia-Morales, C., Sauter, K. A., McGrew, M. J., Smith, J., et al. (2015). The development and maintenance of the mononuclear phagocyte system of the chick is controlled by signals from the macrophage colony-stimulating factor receptor. *BMC Biology, 13,* 12.
Garceau, V., Smith, J., Paton, I. R., Davey, M., Fares, M. A., Sester, D. P., et al. (2010). Pivotal advance: Avian colony-stimulating factor 1 (CSF-1), interleukin-34 (IL-34), and CSF-1 receptor genes and gene products. *Journal of Leukocyte Biology, 87,* 753–764.
Geutskens, S. B., Otonkoski, T., Pulkkinen, M. A., Drexhage, H. A., & Leenen, P. J. (2005). Macrophages in the murine pancreas and their involvement in fetal endocrine development in vitro. *Journal of Leukocyte Biology, 78,* 845–852.
Ginhoux, F., Greter, M., Leboeuf, M., Nandi, S., See, P., Gokhan, S., et al. (2010). Fate mapping analysis reveals that adult microglia derive from primitive macrophages. *Science, 330,* 841–845.
Ginhoux, F., & Prinz, M. (2015). Origin of microglia: Current concepts and past controversies. *Cold Spring Harbor Perspectives in Biology, 7,* a020537.
Ginhoux, F., Tacke, F., Angeli, V., Bogunovic, M., Loubeau, M., Dai, X. M., et al. (2006). Langerhans cells arise from monocytes in vivo. *Nature Immunology, 7,* 265–273.
Glantschnig, H., Fisher, J. E., Wesolowski, G., Rodan, G. A., & Reszka, A. A. (2003). M-CSF, TNFalpha and RANK ligand promote osteoclast survival by signaling through mTOR/S6 kinase. *Cell Death and Differentiation, 10,* 1165–1177.
Goldmann, T., Wieghofer, P., Jordao, M. J., Prutek, F., Hagemeyer, N., Frenzel, K., et al. (2016). Origin, fate and dynamics of macrophages at central nervous system interfaces. *Nature Immunology, 17,* 797–805.
Golub, R., & Cumano, A. (2013). Embryonic hematopoiesis. *Blood Cells, Molecules & Diseases, 51,* 226–231.
Gomez Perdiguero, E., Klapproth, K., Schulz, C., Busch, K., Azzoni, E., Crozet, L., et al. (2015). Tissue-resident macrophages originate from yolk-sac-derived erythro-myeloid progenitors. *Nature, 518,* 547–551.
Gordon, E. J., Rao, S., Pollard, J. W., Nutt, S. L., Lang, R. A., & Harvey, N. L. (2010). Macrophages define dermal lymphatic vessel calibre during development by regulating lymphatic endothelial cell proliferation. *Development, 137,* 3899–3910.
Gouon-Evans, V., Rothenberg, M. E., & Pollard, J. W. (2000). Postnatal mammary gland development requires macrophages and eosinophils. *Development, 127,* 2269–2282.
Gow, D. J., Sester, D. P., & Hume, D. A. (2010). CSF-1, IGF-1, and the control of postnatal growth and development. *Journal of Leukocyte Biology, 88,* 475–481.
Grayfer, L., & Robert, J. (2013). Colony-stimulating factor-1-responsive macrophage precursors reside in the Amphibian (Xenopus laevis) bone marrow rather than the hematopoietic subcapsular liver. *Journal of Innate Immunity, 5,* 531–542.
Grayfer, L., & Robert, J. (2014). Divergent antiviral roles of amphibian (Xenopus laevis) macrophages elicited by colony-stimulating factor-1 and interleukin-34. *Journal of Leukocyte Biology, 96,* 1143–1153.
Grayfer, L., & Robert, J. (2015). Distinct functional roles of amphibian (Xenopus laevis) colony-stimulating factor-1- and interleukin-34-derived macrophages. *Journal of Leukocyte Biology, 98,* 641–649.
Greter, M., Lelios, I., Pelczar, P., Hoeffel, G., Price, J., Leboeuf, M., et al. (2012). Stroma-derived interleukin-34 controls the development and maintenance of langerhans cells and the maintenance of microglia. *Immunity, 37,* 1050–1060.

Guilbert, L. J., & Stanley, E. R. (1980). Specific interaction of murine colony-stimulating factor with mononuclear phagocytic cells. *The Journal of Cell Biology*, 85, 153–159.

Guilbert, L. J., & Stanley, E. R. (1986). The interaction of ^{125}I-colony stimulating factor-1 with bone marrow-derived macrophages. *The Journal of Biological Chemistry*, 261, 4024–4032.

Guleria, I., & Pollard, J. W. (2000). The trophoblast is a component of the innate immune system during pregnancy. *Nature Medicine*, 6, 589–593.

Hagemeyer, N., Kierdorf, K., Frenzel, K., Xue, J., Ringelhan, M., Abdullah, Z., et al. (2016). Transcriptome-based profiling of yolk sac-derived macrophages reveals a role for Irf8 in macrophage maturation. *The EMBO Journal*, 35, 1730–1744.

Hamilton, T. A., Zhao, C., Pavicic, P. G., Jr., & Datta, S. (2014). Myeloid colony-stimulating factors as regulators of macrophage polarization. *Frontiers in Immunology*, 5, 554.

Hanington, P. C., Hitchen, S. J., Beamish, L. A., & Belosevic, M. (2009). Macrophage colony stimulating factor (CSF-1) is a central growth factor of goldfish macrophages. *Fish & Shellfish Immunology*, 26, 1–9.

Harris, S. E., MacDougall, M., Horn, D., Woodruff, K., Zimmer, S. N., Rebel, V. I., et al. (2012). Meox2Cre-mediated disruption of CSF-1 leads to osteopetrosis and osteocyte defects. *Bone*, 50, 42–53.

Harvey, N. L., & Gordon, E. J. (2012). Deciphering the roles of macrophages in developmental and inflammation stimulated lymphangiogenesis. *Vascular Cell*, 4, 15.

He, Y., Staser, K., Rhodes, S. D., Liu, Y., Wu, X., Park, S. J., et al. (2011). Erk1 positively regulates osteoclast differentiation and bone resorptive activity. *PloS One*, 6e24780.

Herbomel, P., Thisse, B., & Thisse, C. (2001). Zebrafish early macrophages colonize cephalic mesenchyme and developing brain, retina, and epidermis through a M-CSF receptor-dependent invasive process. *Developmental Biology*, 238, 274–288.

Himburg, H. A., Harris, J. R., Ito, T., Daher, P., Russell, J. L., Quarmyne, M., et al. (2012). Pleiotrophin regulates the retention and self-renewal of hematopoietic stem cells in the bone marrow vascular niche. *Cell Reports*, 2, 964–975.

Hodge, J. M., Collier, F. M., Pavlos, N. J., Kirkland, M. A., & Nicholson, G. C. (2011). M-CSF potently augments RANKL-induced resorption activation in mature human osteoclasts. *PloS One*, 6e21462.

Hodge, J. M., Kirkland, M. A., Aitken, C. J., Waugh, C. M., Myers, D. E., Lopez, C. M., et al. (2004a). Osteoclastic potential of human CFU-GM: Biphasic effect of GM-CSF. *Journal of Bone and Mineral Research*, 19, 190–199.

Hodge, J. M., Kirkland, M. A., & Nicholson, G. C. (2004b). GM-CSF cannot substitute for M-CSF in human osteoclastogenesis. *Biochemical and Biophysical Research Communications*, 321, 7–12.

Hodge, J. M., Kirkland, M. A., & Nicholson, G. C. (2007). Multiple roles of M-CSF in human osteoclastogenesis. *Journal of Cellular Biochemistry*, 102, 759–768.

Hoeffel, G., Wang, Y., Greter, M., See, P., Teo, P., Malleret, B., et al. (2012). Adult Langerhans cells derive predominantly from embryonic fetal liver monocytes with a minor contribution of yolk sac-derived macrophages. *The Journal of Experimental Medicine*, 209, 1167–1181.

Hofer, T. P., Zawada, A. M., Frankenberger, M., Skokann, K., Satzl, A. A., Gesierich, W., et al. (2015). slan-defined subsets of CD16-positive monocytes: Impact of granulomatous inflammation and M-CSF receptor mutation. *Blood*, 126, 2601–2610.

Hume, D. A. (2008). Macrophages as APC and the dendritic cell myth. *Journal of Immunology*, 181, 5829–5835.

Hume, D. A., & Gordon, S. (1983). Mononuclear phagocyte system of the mouse defined by immunohistochemical localization of antigen F4/80. Identification of resident macrophages in renal medullary and cortical interstitium and the juxtaglomerular complex. *The Journal of Experimental Medicine*, 157, 1704–1709.

Hume, D. A., Pavli, P., Donahue, R. E., & Fidler, I. J. (1988). The effect of human recombinant macrophage colony-stimulating factor (CSF-1) on the murine mononuclear phagocyte system in vivo. *Journal of Immunology*, *141*, 3405–3409.

Humphrey, M. B., Ogasawara, K., Yao, W., Spusta, S. C., Daws, M. R., Lane, N. E., et al. (2004). The signaling adapter protein DAP12 regulates multinucleation during osteoclast development. *Journal of Bone and Mineral Research*, *19*, 224–234.

Huynh, D., Akcora, D., Malaterre, J., Chan, C. K., Dai, X. M., Bertoncello, I., et al. (2013). CSF-1 receptor-dependent colon development, homeostasis and inflammatory stress response. *PloS One*, *8*e56951.

Huynh, D., Dai, X. M., Nandi, S., Lightowler, S., Trivett, M., Chan, C. K., et al. (2009). Colony stimulating factor-1 dependence of paneth cell development in the mouse small intestine. *Gastroenterology*, *137*, 136–144. 144.e1-3.

Hwang, S. J., Choi, B., Kang, S. S., Chang, J. H., Kim, Y. G., Chung, Y. H., et al. (2012). Interleukin-34 produced by human fibroblast-like synovial cells in rheumatoid arthritis supports osteoclastogenesis. *Arthritis Research & Therapy*, *14*, R14.

Ingman, W. V., Wyckoff, J., Gouon-Evans, V., Condeelis, J., & Pollard, J. W. (2006). Macrophages promote collagen fibrillogenesis around terminal end buds of the developing mammary gland. *Developmental Dynamics*, *235*, 3222–3229.

Insogna, K. L., Sahni, M., Grey, A. B., Tanaka, S., Horne, W. C., Neff, L., et al. (1997). Colony-stimulating factor-1 induces cytoskeletal reorganization and c-src-dependent tyrosine phosphorylation of selected cellular proteins in rodent osteoclasts. *The Journal of Clinical Investigation*, *100*, 2476–2485.

Insogna, K., Tanaka, S., Neff, L., Horne, W., Levy, J., & Baron, R. (1997). Role of c-Src in cellular events associated with colony-stimulating factor-1-induced spreading in osteoclasts. *Molecular Reproduction and Development*, *46*, 104–108.

Ito, Y., Teitelbaum, S. L., Zou, W., Zheng, Y., Johnson, J. F., Chappel, J., et al. (2010). Cdc42 regulates bone modeling and remodeling in mice by modulating RANKL/M-CSF signaling and osteoclast polarization. *The Journal of Clinical Investigation*, *120*, 1981–1993.

Itokowa, T., Zhu, M. L., Troiano, N., Bian, J., Kawano, T., & Insogna, K. (2011). Osteoclasts lacking Rac2 have defective chemotaxis and resorptive activity. *Calcified Tissue International*, *88*, 75–86.

Jang, M. H., Herber, D. M., Jiang, X., Nandi, S., Dai, X. M., Zeller, G., et al. (2006). Distinct in vivo roles of colony-stimulating factor-1 isoforms in renal inflammation. *Journal of Immunology*, *177*, 4055–4063.

Ji, J. D., Park-Min, K. H., Shen, Z., Fajardo, R. J., Goldring, S. R., McHugh, K. P., et al. (2009). Inhibition of RANK expression and osteoclastogenesis by TLRs and IFN-gamma in human osteoclast precursors. *Journal of Immunology*, *183*, 7223–7233.

Johnson, G. R., & Burgess, A. W. (1978). Molecular and biological properties of a macrophage colony-stimulating factor from mouse yolk sacs. *The Journal of Cell Biology*, *77*, 35–47.

Jones, C. V., & Ricardo, S. D. (2013). Macrophages and CSF-1: Implications for development and beyond. *Organogenesis*, *9*, 249–260.

Jones, C. V., Williams, T. M., Walker, K. A., Dickinson, H., Sakkal, S., Rumballe, B. A., et al. (2013). M2 macrophage polarisation is associated with alveolar formation during postnatal lung development. *Respiratory Research*, *14*, 41.

Kawaida, R., Ohtsuka, T., Okutsu, J., Takahashi, T., Kadono, Y., Oda, H., et al. (2003). Jun dimerization protein 2 (JDP2), a member of the AP-1 family of transcription factor, mediates osteoclast differentiation induced by RANKL. *The Journal of Experimental Medicine*, *197*, 1029–1035.

Kawasaki, E. S., Ladner, M. B., Wang, A. M., Van Arsdell, J., Warren, M. K., Coyne, M. Y., et al. (1985). Molecular cloning of a complementary DNA encoding human macrophage-specific colony-stimulating factor (CSF-1). *Science*, *230*, 291–296.

Kierdorf, K., Erny, D., Goldmann, T., Sander, V., Schulz, C., Perdiguero, E. G., et al. (2013). Microglia emerge from erythromyeloid precursors via Pu.1- and Irf8-dependent pathways. *Nature Neuroscience, 16,* 273–280.

Kim, H. J., Warren, J. T., Kim, S. Y., Chappel, J. C., DeSelm, C. J., Ross, F. P., et al. (2010). Fyn promotes proliferation, differentiation, survival and function of osteoclast lineage cells. *Journal of Cellular Biochemistry, 111,* 1107–1113.

Kim, H. J., Zhang, K., Zhang, L., Ross, F. P., Teitelbaum, S. L., & Faccio, R. (2009). The Src family kinase, Lyn, suppresses osteoclastogenesis in vitro and in vivo. *Proceedings of the National Academy of Sciences of the United States of America, 106,* 2325–2330.

Kim, H. J., Zou, W., Ito, Y., Kim, S. Y., Chappel, J., Ross, F. P., et al. (2010). Src-like adaptor protein regulates osteoclast generation and survival. *Journal of Cellular Biochemistry, 110,* 201–209.

Kondo, Y., Ramaker, J. M., Radcliff, A. B., Baldassari, S., Mayer, J. A., Ver Hoeve, J. N., et al. (2013). Spontaneous optic nerve compression in the osteopetrotic (op/op) mouse: A novel model of myelination failure. *The Journal of Neuroscience, 33,* 3514–3525.

Kong, Y. Y., Yoshida, H., Sarosi, I., Tan, H. L., Timms, E., Capparelli, C., et al. (1999). OPGL is a key regulator of osteoclastogenesis, lymphocyte development and lymph-node organogenesis. *Nature, 397,* 315–323.

Konno, T., Tada, M., Tada, M., Koyama, A., Nozaki, H., Harigaya, Y., et al. (2014). Haploinsufficiency of CSF-1R and clinicopathologic characterization in patients with HDLS. *Neurology, 82,* 139–148.

Kubota, Y., Takubo, K., Shimizu, T., Ohno, H., Kishi, K., Shibuya, M., et al. (2009). M-CSF inhibition selectively targets pathological angiogenesis and lymphangiogenesis. *The Journal of Experimental Medicine, 206,* 1089–1102.

Ladner, M. B., Martin, G. A., Noble, J. A., Nikoloff, D. M., Tal, R., Kawasaki, E. S., et al. (1987). Human CSF-1: Gene structure and alternative splicing of mRNA precursors. *The EMBO Journal, 6,* 2693–2698.

Ladner, M. B., Martin, G. A., Noble, J. A., Wittman, V. P., Warren, M. K., McGrogan, M., et al. (1988). cDNA cloning and expression of murine macrophage colony-stimulating factor from L929 cells. *Proceedings of the National Academy of Sciences of the United States of America, 85,* 6706–6710.

Lakkakorpi, P. T., & Vaananen, H. K. (1996). Cytoskeletal changes in osteoclasts during the resorption cycle. *Microscopy Research and Technique, 33,* 171–181.

Lean, J. M., Fuller, K., & Chambers, T. J. (2001). FLT3 ligand can substitute for macrophage colony-stimulating factor in support of osteoclast differentiation and function. *Blood, 98,* 2707–2713.

Leblond, A. L., Klinkert, K., Martin, K., Turner, E. C., Kumar, A. H., Browne, T., et al. (2015). Systemic and cardiac depletion of M2 macrophage through CSF-1R signaling inhibition alters cardiac function post myocardial infarction. *PloS One, 10*e0137515.

Lenzo, J. C., Turner, A. L., Cook, A. D., Vlahos, R., Anderson, G. P., Reynolds, E. C., et al. (2012). Control of macrophage lineage populations by CSF-1 receptor and GM-CSF in homeostasis and inflammation. *Immunology and Cell Biology, 90,* 429–440.

Lin, H., Lee, E., Hestir, K., Leo, C., Huang, M., Bosch, E., et al. (2008). Discovery of a cytokine and its receptor by functional screening of the extracellular proteome. *Science, 320,* 807–811.

Liu, J. L., Yakar, S., & LeRoith, D. (2000). Conditional knockout of mouse insulin-like growth factor-1 gene using the Cre/loxP system. *Proceedings of the Society for Experimental Biology and Medicine, 223,* 344–351.

Lorenzetto, E., Moratti, E., Vezzalini, M., Harroch, S., Sorio, C., & Buffelli, M. (2014). Distribution of different isoforms of receptor protein tyrosine phosphatase gamma (Ptprg-RPTP gamma) in adult mouse brain: Upregulation during neuroinflammation. *Brain Structure & Function, 219,* 875–890.

Luo, J., Elwood, F., Britschgi, M., Villeda, S., Zhang, H., Ding, Z., et al. (2013). Colony-stimulating factor 1 receptor (CSF1R) signaling in injured neurons facilitates protection and survival. *The Journal of Experimental Medicine, 210,* 157–172.

Ma, X., Lin, W. Y., Chen, Y., Stawicki, S., Mukhyala, K., Wu, Y., et al. (2012). Structural basis for the dual recognition of helical cytokines IL-34 and CSF-1 by CSF-1R. *Structure, 20,* 676–687.

MacDonald, K. P., Palmer, J. S., Cronau, S., Seppanen, E., Olver, S., Raffelt, N. C., et al. (2010). An antibody against the colony-stimulating factor 1 receptor depletes the resident subset of monocytes and tissue- and tumor-associated macrophages but does not inhibit inflammation. *Blood, 116,* 3955–3963.

Mansky, K. C., Sankar, U., Han, J., & Ostrowski, M. C. (2002). Microphthalmia transcription factor is a target of the p38 MAPK pathway in response to receptor activator of NF-kappa B ligand signaling. *The Journal of Biological Chemistry, 277,* 11077–11083.

Marks, S. C., Jr., & Lane, P. W. (1976). Osteopetrosis, a new recessive skeletal mutation on chromosome 12 of the mouse. *The Journal of Heredity, 67,* 11–18.

Maruyama, K., Uematsu, S., Kondo, T., Takeuchi, O., Martino, M. M., Kawasaki, T., et al. (2013). Strawberry notch homologue 2 regulates osteoclast fusion by enhancing the expression of DC-STAMP. *The Journal of Experimental Medicine, 210,* 1947–1960.

McGill, G. G., Horstmann, M., Widlund, H. R., Du, J., Motyckova, G., Nishimura, E. K., et al. (2002). Bcl2 regulation by the melanocyte master regulator Mitf modulates lineage survival and melanoma cell viability. *Cell, 109,* 707–718.

Menke, J., Iwata, Y., Rabacal, W. A., Basu, R., Yeung, Y. G., Humphreys, B. D., et al. (2009). CSF-1 signals directly to renal tubular epithelial cells to mediate repair in mice. *The Journal of Clinical Investigation, 119,* 2330–2342.

Merad, M., Manz, M. G., Karsunky, H., Wagers, A., Peters, W., Charo, I., et al. (2002). Langerhans cells renew in the skin throughout life under steady-state conditions. *Nature Immunology, 3,* 1135–1141.

Michaelson, M. D., Bieri, P. L., Mehler, M. F., Xu, H., Arezzo, J. C., Pollard, J. W., et al. (1996). CSF-1 deficiency in mice results in abnormal brain development. *Development, 122,* 2661–2672.

Miyamoto, H., Suzuki, T., Miyauchi, Y., Iwasaki, R., Kobayashi, T., Sato, Y., et al. (2012). Osteoclast stimulatory transmembrane protein and dendritic cell-specific transmembrane protein cooperatively modulate cell–cell fusion to form osteoclasts and foreign body giant cells. *Journal of Bone and Mineral Research, 27,* 1289–1297.

Miyazaki, T., Katagiri, H., Kanegae, Y., Takayanagi, H., Sawada, Y., Yamamoto, A., et al. (2000). Reciprocal role of ERK and NF-kappaB pathways in survival and activation of osteoclasts. *The Journal of Cell Biology, 148,* 333–342.

Mócsai, A., Humphrey, M. B., Van Ziffle, J. A., Hu, Y., Burghardt, A., Spusta, S. C., et al. (2004). The immunomodulatory adapter proteins DAP12 and Fc receptor gamma-chain (FcRgamma) regulate development of functional osteoclasts through the Syk tyrosine kinase. *Proceedings of the National Academy of Sciences of the United States of America, 101,* 6158–6163.

Morris, L., Graham, C. F., & Gordon, S. (1991). Macrophages in haemopoietic and other tissues of the developing mouse detected by the monoclonal antibody F4/80. *Development, 112,* 517–526.

Mundy, G. R. (2007). Osteoporosis and inflammation. *Nutrition Reviews, 65,* S147–S151.

Murray, P. J., & Wynn, T. A. (2011). Protective and pathogenic functions of macrophage subsets. *Nature Reviews Immunology, 11,* 723–737.

Nakagawa, N., Kinosaki, M., Yamaguchi, K., Shima, N., Yasuda, H., Yano, K., et al. (1998). RANK is the essential signaling receptor for osteoclast differentiation factor in osteoclastogenesis. *Biochemical and Biophysical Research Communications, 253,* 395–400.

Nakamura, I., Pilkington, M. F., Lakkakorpi, P. T., Lipfert, L., Sims, S. M., Dixon, S. J., et al. (1999). Role of alpha(v)beta(3) integrin in osteoclast migration and formation of the sealing zone. *Journal of Cell Science, 112*, 3985–3993.
Nandi, S., Akhter, M. P., Seifert, M. F., Dai, X. M., & Stanley, E. R. (2006). Developmental and functional significance of the CSF-1 proteoglycan chondroitin sulfate chain. *Blood, 107*, 786–795.
Nandi, S., Cioce, M., Yeung, Y. G., Nieves, E., Tesfa, L., Lin, H., et al. (2013). Receptor-type protein-tyrosine phosphatase zeta is a functional receptor for interleukin-34. *The Journal of Biological Chemistry, 288*, 21972–21986.
Nandi, S., Gokhan, S., Dai, X. M., Wei, S., Enikolopov, G., Lin, H., et al. (2012). The CSF-1 receptor ligands IL-34 and CSF-1 exhibit distinct developmental brain expression patterns and regulate neural progenitor cell maintenance and maturation. *Developmental Biology, 367*, 100–113.
Newman, A. C., & Hughes, C. C. (2012). Macrophages and angiogenesis: A role for Wnt signaling. *Vascular Cell, 4*, 13.
Nicholson, A. M., Baker, M. C., Finch, N. A., Rutherford, N. J., Wider, C., Graff-Radford, N. R., et al. (2013). CSF1R mutations link POLD and HDLS as a single disease entity. *Neurology, 80*, 1033–1040.
Niida, S., Kaku, M., Amano, H., Yoshida, H., Kataoka, H., Nishikawa, S., et al. (1999). Vascular endothelial growth factor can substitute for macrophage colony-stimulating factor in the support of osteoclastic bone resorption. *The Journal of Experimental Medicine, 190*, 293–298.
Niida, S., Kondo, T., Hiratsuka, S., Hayashi, S., Amizuka, N., Noda, T., et al. (2005). VEGF receptor 1 signaling is essential for osteoclast development and bone marrow formation in colony-stimulating factor 1-deficient mice. *Proceedings of the National Academy of Sciences of the United States of America, 102*, 14016–14021.
Nikolic, T., de Bruijn, M. F., Lutz, M. B., & Leenen, P. J. (2003). Developmental stages of myeloid dendritic cells in mouse bone marrow. *International Immunology, 15*, 515–524.
Nishimura, K., Tanaka, N., Ohshige, A., Fukumatsu, Y., Matsuura, K., & Okamura, H. (1995). Effects of macrophage colony-stimulating factor on folliculogenesis in gonadotrophin-primed immature rats. *Journal of Reproduction and Fertility, 104*, 325–330.
Norgard, M., Marks, S. C., Jr., Reinholt, F. P., & Andersson, G. (2003). The effects of colony-stimulating factor-1 (CSF-1) on the development of osteoclasts and their expression of tartrate-resistant acid phosphatase (TRAP) in toothless (tl-osteopetrotic) rats. *Critical Reviews in Eukaryotic Gene Expression, 13*, 117–132.
Ochsenbein, A. M., Karaman, S., Proulx, S. T., Goldmann, R., Chittazhathu, J., Dasargyri, A., et al. (2016). Regulation of lymphangiogenesis in the diaphragm by macrophages and VEGFR-3 signaling. *Angiogenesis, 19*, 513–524.
Otero, K., Turnbull, I. R., Poliani, P. L., Vermi, W., Cerutti, E., Aoshi, T., et al. (2009). Macrophage colony-stimulating factor induces the proliferation and survival of macrophages via a pathway involving DAP12 and beta-catenin. *Nature Immunology, 10*, 734–743.
Outtz, H. H., Tattersall, I. W., Kofler, N. M., Steinbach, N., & Kitajewski, J. (2011). Notch1 controls macrophage recruitment and Notch signaling is activated at sites of endothelial cell anastomosis during retinal angiogenesis in mice. *Blood, 118*, 3436–3439.
Palacio, S., & Felix, R. (2001). The role of phosphoinositide 3-kinase in spreading osteoclasts induced by colony-stimulating factor-1. *European Journal of Endocrinology, 144*, 431–440.
Palis, J., Robertson, S., Kennedy, M., Wall, C., & Keller, G. (1999). Development of erythroid and myeloid progenitors in the yolk sac and embryo proper of the mouse. *Development, 126*, 5073–5084.

Pampfer, S., Arceci, R. J., & Pollard, J. W. (1991). Role of colony stimulating factor-1 (CSF-1) and other lympho-hematopoietic growth factors in mouse pre-implantation development. *Bioessays, 13*, 535–540.

Perkins, S. L., & Kling, S. J. (1995). Local concentrations of macrophage colony-stimulating factor mediate osteoclastic differentiation. *The American Journal of Physiology, 269*, E1024–E1030.

Pilkington, M. F., Sims, S. M., & Dixon, S. J. (1998). Wortmannin inhibits spreading and chemotaxis of rat osteoclasts in vitro. *Journal of Bone and Mineral Research, 13*, 688–694.

Pixley, F. J., & Stanley, E. R. (2004). CSF-1 regulation of the wandering macrophage: Complexity in action. *Trends in Cell Biology, 14*, 628–638.

Pluznik, D. H., & Sachs, L. (1965). The cloning of normal "mast" cells in tissue culture. *Journal of Cellular and Comparative Physiology, 66*, 319–324.

Pollard, J. W. (2009). Trophic macrophages in development and disease. *Nature Reviews. Immunology, 9*, 259–270.

Pollard, J. W., Bartocci, A., Arceci, R., Orlofsky, A., Ladner, M. B., & Stanley, E. R. (1987). Apparent role of the macrophage growth factor, CSF-1, in placental development. *Nature, 330*, 484–486.

Pollard, J. W., & Hennighausen, L. (1994). Colony-stimulating factor-1 is required for mammary-gland development during pregnancy. *Proceedings of the National Academy of Sciences of the United States of America, 91*, 9312–9316.

Pollard, J. W., & Stanley, E. R. (1996). Pleiotropic roles for CSF-1 in development defined by the mouse mutation osteopetrotic. *Advances in Developmental Biochemistry, 4*, 153–193.

Porter, E. M., Bevins, C. L., Ghosh, D., & Ganz, T. (2002). The multifaceted Paneth cell. *Cellular and Molecular Life Sciences, 59*, 156–170.

Price, L. K. H., Choi, H. U., Rosenberg, L., & Stanley, E. R. (1992). The predominant form of secreted colony stimulating factor-1 is a proteoglycan. *The Journal of Biological Chemistry, 267*, 2190–2199.

Pull, S. L., Doherty, J. M., Mills, J. C., Gordon, J. I., & Stappenbeck, T. S. (2005). Activated macrophages are an adaptive element of the colonic epithelial progenitor niche necessary for regenerative responses to injury. *Proceedings of the National Academy of Sciences of the United States of America, 102*, 99–104.

Rademakers, R., Baker, M., Nicholson, A. M., Rutherford, N. J., Finch, N., Soto-Ortolaza, A., et al. (2011). Mutations in the colony stimulating factor 1 receptor (CSF1R) gene cause hereditary diffuse leukoencephalopathy with spheroids. *Nature Genetics, 44*, 200–205.

Rae, F., Woods, K., Sasmono, T., Campanale, N., Taylor, D., Ovchinnikov, D. A., et al. (2007). Characterisation and trophic functions of murine embryonic macrophages based upon the use of a Csf1r-EGFP transgene reporter. *Developmental Biology, 308*, 232–246.

Ramsay, R. G., Micallef, S. J., Williams, B., Lightowler, S., Vincan, E., Heath, J. K., et al. (2004). Colony-stimulating factor-1 promotes clonogenic growth of normal murine colonic crypt epithelial cells in vitro. *Journal of Interferon & Cytokine Research, 24*, 416–427.

Ratheesh, A., Belyaeva, V., & Siekhaus, D. E. (2015). Drosophila immune cell migration and adhesion during embryonic development and larval immune responses. *Current Opinion in Cell Biology, 36*, 71–79.

Reeve, J. L., Zou, W., Liu, Y., Maltzman, J. S., Ross, F. P., & Teitelbaum, S. L. (2009). SLP-76 couples Syk to the osteoclast cytoskeleton. *Journal of Immunology, 183*, 1804–1812.

Regenstreif, L. J., & Rossant, J. (1989). Expression of the c-fms proto-oncogene and of the cytokine, CSF-1, during mouse embryogenesis. *Developmental Biology, 133*, 284–294.

Reinholt, F. P., Hultenby, K., Oldberg, A., & Heinegard, D. (1990). Osteopontin—A possible anchor of osteoclasts to bone. *Proceedings of the National Academy of Sciences of the United States of America, 87*, 4473–4475.

Rettenmier, C. W., Roussel, M. F., Ashmun, R. A., Ralph, P., Price, K., & Sherr, C. J. (1987). Synthesis of membrane-bound colony-stimulating factor 1 (CSF-1) and downmodulation of CSF-1 receptors in NIH 3T3 cells transformed by cotransfection of the human CSF-1 and c-fms (CSF-1 receptor) genes. *Molecular and Cellular Biology, 7,* 2378–2387.

Roth, P., Bartocci, A., & Stanley, E. R. (1997). Lipopolysaccharide induces synthesis of mouse colony-stimulating factor-1 in vivo. *Journal of Immunology, 158,* 3874–3880.

Roth, P., Dominguez, M. G., & Stanley, E. R. (1998). The effects of colony-stimulating factor-1 on the distribution of mononuclear phagocytes in the developing osteopetrotic mouse. *Blood, 91,* 3773–3783.

Roth, P., & Stanley, E. R. (1995). Colony-stimulating factor-1 expression in the human fetus and newborn. *Journal of Leukocyte Biology, 58,* 432–437.

Roth, P., & Stanley, E. R. (1996). Colony stimulating factor-1 expression is developmentally regulated in the mouse. *Journal of Leukocyte Biology, 59,* 817–823.

Ryan, G. R., Dai, X. M., Dominguez, M. G., Tong, W., Chuan, F., Chisholm, O., et al. (2001). Rescue of the colony-stimulating factor 1 (CSF-1)-nullizygous mouse (Csf1 (op)/Csf1(op)) phenotype with a CSF-1 transgene and identification of sites of local CSF-1 synthesis. *Blood, 98,* 74–84.

Rymo, S. F., Gerhardt, H., Wolfhagen Sand, F., Lang, R., Uv, A., & Betsholtz, C. (2011). A two-way communication between microglial cells and angiogenic sprouts regulates angiogenesis in aortic ring cultures. *PloS One, 6,* e15846.

Sakai, H., Chen, Y., Itokawa, T., Yu, K. P., Zhu, M. L., & Insogna, K. (2006). Activated c-Fms recruits Vav and Rac during CSF-1-induced cytoskeletal remodeling and spreading in osteoclasts. *Bone, 39,* 1290–1301.

Sarrazin, S., Mossadegh-Keller, N., Fukao, T., Aziz, A., Mourcin, F., Vanhille, L., et al. (2009). MafB restricts M-CSF-dependent myeloid commitment divisions of hematopoietic stem cells. *Cell, 138,* 300–313.

Sasmono, R. T., Oceandy, D., Pollard, J. W., Tong, W., Pavli, P., Wainwright, B. J., et al. (2003). A macrophage colony-stimulating factor receptor-green fluorescent protein transgene is expressed throughout the mononuclear phagocyte system of the mouse. *Blood, 101,* 1155–1163.

Schafer, D. P., & Stevens, B. (2015). Microglia function in central nervous system development and plasticity. *Cold Spring Harbor Perspectives in Biology, 7,* a020545.

Schulz, C., Gomez Perdiguero, E., Chorro, L., Szabo-Rogers, H., Cagnard, N., Kierdorf, K., et al. (2012). A lineage of myeloid cells independent of Myb and hematopoietic stem cells. *Science, 336,* 86–90.

Segaliny, A. I., Brion, R., Brulin, B., Maillasson, M., Charrier, C., Teletchea, S., et al. (2015a). IL-34 and M-CSF form a novel heteromeric cytokine and regulate the M-CSF receptor activation and localization. *Cytokine, 76,* 170–181.

Segaliny, A., Brion, R., Mortier, E., Maillasson, M., Cherel, M., Jacques, Y., et al. (2015b). Syndecan-1 regulates the biological activities of interleukin-34. *Biochimica et Biophysica Acta, 1853,* 1010–1021.

Shechter, R., Miller, O., Yovel, G., Rosenzweig, N., London, A., Ruckh, J., et al. (2013). Recruitment of beneficial M2 macrophages to injured spinal cord is orchestrated by remote brain choroid plexus. *Immunity, 38,* 555–569.

Sherr, C. J., Rettenmier, C. W., Sacca, R., Roussel, M. F., Look, A. T., & Stanley, E. R. (1985). The c-fms proto-oncogene product is related to the receptor for the mononuclear phagocyte growth factor, CSF-1. *Cell, 41,* 665–676.

Shibata, Y., Zsengeller, Z., Otake, K., Palaniyar, N., & Trapnell, B. C. (2001). Alveolar macrophage deficiency in osteopetrotic mice deficient in macrophage colony-stimulating factor is spontaneously corrected with age and associated with matrix metalloproteinase expression and emphysema. *Blood, 98,* 2845–2852.

Shinohara, M., Nakamura, M., Masuda, H., Hirose, J., Kadono, Y., Iwasawa, M., et al. (2012). Class IA phosphatidylinositol 3-kinase regulates osteoclastic bone resorption through Akt-mediated vesicle transport. *Journal of Bone and Mineral Research*, 27, 2464–2475.

Smith, S. J., & Mohun, T. J. (2011). Early cardiac morphogenesis defects caused by loss of embryonic macrophage function in Xenopus. *Mechanisms of Development*, 128, 303–315.

Stanley, E. R. (1979). Colony-stimulating factor (CSF) radioimmunoassay: Detection of a CSF subclass stimulating macrophage production. *Proceedings of the National Academy of Sciences of the United States of America*, 76, 2969–2973.

Stanley, E. R., Chen, D. M., & Lin, H.-S. (1978). Induction of macrophage production and proliferation by a purified colony stimulating factor. *Nature*, 274, 168–170.

Stanley, E. R., & Chitu, V. (2014). CSF-1 receptor signaling in myeloid cells. *Cold Spring Harbor Perspectives in Biology*, 6, 1–21.

Stanley, E. R., & Heard, P. M. (1977). Factors regulating macrophage production and growth. Purification and some properties of the colony stimulating factor from medium conditioned by mouse L cells. *The Journal of Biological Chemistry*, 252, 4305–4312.

Stefater, J. A., 3rd, Ren, S., Lang, R. A., & Duffield, J. S. (2011). Metchnikoff's policemen: Macrophages in development, homeostasis and regeneration. *Trends in Molecular Medicine*, 17, 743–752.

Sundquist, K. T., Cecchini, M. G., & Marks, S. C., Jr. (1995). Colony-stimulating factor-1 injections improve but do not cure skeletal sclerosis in osteopetrotic (op) mice. *Bone*, 16, 39.

Suzu, S., Ohtsuki, T., Yanai, N., Takatsu, Z., Kawashima, T., Takaku, F., et al. (1992). Identification of a high molecular weight macrophage colony-stimulating factor as a glycosaminoglycan-containing species. *The Journal of Biological Chemistry*, 267, 4345–4348.

Takeshita, S., Namba, N., Zhao, J. J., Jiang, Y., Genant, H. K., Silva, M. J., et al. (2002). SHIP-deficient mice are severely osteoporotic due to increased numbers of hyper-resorptive osteoclasts. *Nature Medicine*, 8, 943–949.

Takito, J., Otsuka, H., Yanagisawa, N., Arai, H., Shiga, M., Inoue, M., et al. (2015). Regulation of osteoclast multinucleation by the actin cytoskeleton signaling network. *Journal of Cellular Physiology*, 230, 395–405.

Thompson, D. W. (1942). *On growth and form*. Cambridge, UK: University Press.

Ueki, Y., Lin, C. Y., Senoo, M., Ebihara, T., Agata, N., Onji, M., et al. (2007). Increased myeloid cell responses to M-CSF and RANKL cause bone loss and inflammation in SH3BP2 "cherubism" mice. *Cell*, 128, 71–83.

Van Nguyen, A., & Pollard, J. W. (2002). Colony stimulating factor-1 is required to recruit macrophages into the mammary gland to facilitate mammary ductal outgrowth. *Developmental Biology*, 247, 11–25.

Van Wesenbeeck, L., Odgren, P. R., MacKay, C. A., D'Angelo, M., Safadi, F. F., Popoff, S. N., et al. (2002). The osteopetrotic mutation toothless (tl) is a loss-of-function frameshift mutation in the rat Csf1 gene: Evidence of a crucial role for CSF-1 in osteoclastogenesis and endochondral ossification. *Proceedings of the National Academy of Sciences of the United States of America*, 99, 14303–14308.

Vérollet, C., Gallois, A., Dacquin, R., Lastrucci, C., Pandruvada, S. N., Ortega, N., et al. (2013). Hck contributes to bone homeostasis by controlling the recruitment of osteoclast precursors. *The FASEB Journal*, 27, 3608–3618.

Visvader, J., & Verma, I. M. (1989). Differential transcription of exon 1 of the human c-fms gene in placental trophoblasts and monocytes. *Molecular and Cellular Biology*, 9, 1336–1341.

von Holst, A., Sirko, S., & Faissner, A. (2006). The unique 473HD-Chondroitinsulfate epitope is expressed by radial glia and involved in neural precursor cell proliferation. *The Journal of Neuroscience*, 26, 4082–4094.

Wang, Y., Berezovska, O., & Fedoroff, S. (1999). Expression of colony stimulating factor-1 receptor (CSF-1R) by CNS neurons in mice. *Journal of Neuroscience Research, 57*, 616–632.
Wang, Y., Bugatti, M., Ulland, T. K., Vermi, W., Gilfillan, S., & Colonna, M. (2016). Nonredundant roles of keratinocyte-derived IL-34 and neutrophil-derived CSF1 in Langerhans cell renewal in the steady state and during inflammation. *European Journal of Immunology, 46*, 552–559.
Wang, Y., Chang, J., Yao, B., Niu, A., Kelly, E., Breeggemann, M. C., et al. (2015). Proximal tubule-derived colony stimulating factor-1 mediates polarization of renal macrophages and dendritic cells, and recovery in acute kidney injury. *Kidney International, 88*, 1274–1282.
Wang, T., Kono, T., Monte, M. M., Kuse, H., Costa, M. M., Korenaga, H., et al. (2013). Identification of IL-34 in teleost fish: Differential expression of rainbow trout IL-34, MCSF1 and MCSF2, ligands of the MCSF receptor. *Molecular Immunology, 53*, 39–409.
Wang, Y., Lebowitz, D., Sun, C., Thang, H., Grynpas, M. D., & Glogauer, M. (2008). Identifying the relative contributions of Rac1 and Rac2 to osteoclastogenesis. *Journal of Bone and Mineral Research, 23*, 260–270.
Wang, Y., Szretter, K. J., Vermi, W., Gilfillan, S., Rossini, C., Cella, M., et al. (2012). IL-34 is a tissue-restricted ligand of CSF1R required for the development of Langerhans cells and microglia. *Nature Immunology, 13*, 753–760.
Wehkamp, J., & Stange, E. F. (2006). Paneth cells and the innate immune response. *Current Opinion in Gastroenterology, 22*, 644–650.
Wei, S., Nandi, S., Chitu, V., Yeung, Y. G., Yu, W., Huang, M., et al. (2010). Functional overlap but differential expression of CSF-1 and IL-34 in their CSF-1 receptor-mediated regulation of myeloid cells. *Journal of Leukocyte Biology, 88*, 495–505.
Weilbaecher, K. N., Motyckova, G., Huber, W. E., Takemoto, C. M., Hemesath, T. J., Xu, Y., et al. (2001). Linkage of M-CSF signaling to Mitf, TFE3, and the osteoclast defect in Mitf(mi/mi) mice. *Molecular Cell, 8*, 749–758.
Weiner, L. M., Li, W., Holmes, M., Catalano, R. B., Dovnarsky, M., Padavic, K., et al. (1994). Phase I trial of recombinant macrophage colony-stimulating factor and recombinant gamma-interferon: Toxicity, monocytosis, and clinical effects. *Cancer Research, 54*, 4084–4090.
White, G. E., McNeill, E., Channon, K. M., & Greaves, D. R. (2014). Fractalkine promotes human monocyte survival via a reduction in oxidative stress. *Arteriosclerosis, Thrombosis, and Vascular Biology, 34*, 2554–2562.
Wiktor-Jedrzejczak, W., Ahmed, A., Szczylik, C., & Skelly, R. R. (1982). Hematological characterization of congenital osteopetrosis in op/op mouse. *The Journal of Experimental Medicine, 156*, 1516–1527.
Wiktor-Jedrzejczak, W., Bartocci, A., Ferrante, A. W., Jr., Ahmed-Ansari, A., Sell, K. W., Pollard, J. W., et al. (1990). Total absence of colony-stimulating factor 1 in the macrophage-deficient osteopetrotic (op/op) mouse. *Proceedings of the National Academy of Sciences of the United States of America, 87*, 4828–4832.
Woo, K. M., Kim, H. M., & Ko, J. S. (2002). Macrophage colony-stimulating factor promotes the survival of osteoclast precursors by up-regulating Bcl-X(L). *Experimental & Molecular Medicine, 34*, 340–346.
Yagi, M., Miyamoto, T., Sawatani, Y., Iwamoto, K., Hosogane, N., Fujita, N., et al. (2005). DC-STAMP is essential for cell-cell fusion in osteoclasts and foreign body giant cells. *The Journal of Experimental Medicine, 202*, 345–351.
Yamane, F., Nishikawa, Y., Matsui, K., Asakura, M., Iwasaki, E., Watanabe, K., et al. (2014). CSF-1 receptor-mediated differentiation of a new type of monocytic cell with B cell-stimulating activity: Its selective dependence on IL-34. *Journal of Leukocyte Biology, 95*, 19–31.

Yao, C., Yao, G. Q., Sun, B. H., Zhang, C., Tommasini, S. M., & Insogna, K. (2014). The transcription factor T-box3 regulates colony stimulating factor 1-dependent Jun dimerization protein 2 expression and plays an important role in osteoclastogenesis. *The Journal of Biological Chemistry, 289*, 6775–6790.

Yeung, Y.-G., Jubinsky, P. T., Sengupta, A., Yeung, D. C.-Y., & Stanley, E. R. (1987). Purification of the colony-stimulating factor 1 receptor and demonstration of its tyrosine kinase activity. *Proceedings of the National Academy of Sciences of the United States of America, 84*, 1268–1271.

Yona, S., Kim, K. W., Wolf, Y., Mildner, A., Varol, D., Breker, M., et al. (2013). Fate mapping reveals origins and dynamics of monocytes and tissue macrophages under homeostasis. *Immunity, 38*, 79–91.

Yoshida, H., Hayashi, S., Kunisada, T., Ogawa, M., Nishikawa, S., Okamura, H., et al. (1990). The murine mutation osteopetrosis is in the coding region of the macrophage colony stimulating factor gene. *Nature, 345*, 442–444.

Yu, W., Chen, J., Xiong, Y., Pixley, F. J., Yeung, Y. G., & Stanley, E. R. (2012). Macrophage proliferation is regulated through CSF-1 receptor tyrosines 544, 559, and 807. *The Journal of Biological Chemistry, 287*, 13694–13704.

Zhou, P., Kitaura, H., Teitelbaum, S. L., Krystal, G., Ross, F. P., & Takeshita, S. (2006). SHIP1 negatively regulates proliferation of osteoclast precursors via Akt-dependent alterations in D-type cyclins and p27. *Journal of Immunology, 177*, 8777–8784.

Ziegler-Heitbrock, L., Ancuta, P., Crowe, S., Dalod, M., Grau, V., Hart, D. N., et al. (2010). Nomenclature of monocytes and dendritic cells in blood. *Blood, 116*, e74–e80.

Zou, W., Reeve, J. L., Liu, Y., Teitelbaum, S. L., & Ross, F. P. (2008). DAP12 couples c-Fms activation to the osteoclast cytoskeleton by recruitment of Syk. *Molecular Cell, 31*, 422–431.

Zou, W., Reeve, J. L., Zhao, H., Ross, F. P., & Teitelbaum, S. L. (2009). Syk tyrosine 317 negatively regulates osteoclast function via the ubiquitin-protein isopeptide ligase activity of Cbl. *The Journal of Biological Chemistry, 284*, 18833–18839.

CHAPTER EIGHT

Glycogen Synthase Kinase 3: A Kinase for All Pathways?

Prital Patel, James R. Woodgett[1]
Lunenfeld-Tanenbaum Research Institute, Sinai Health System & University of Toronto, Toronto, ON, Canada
[1]Corresponding author: e-mail address: woodgett@lunenfeld.ca

Contents

1. The Early Years — 278
2. GSK-3 Isoforms, Orthologues, and Expression — 278
3. GSK-3 Structure — 280
4. Regulation — 282
5. Signaling Infidelity — 284
   5.1 Insulin (and Receptor Tyrosine Kinases Acting Through Phosphatidylinositol 3′ Kinase Activation) — 284
   5.2 Wnt Signaling — 286
   5.3 Notch Signaling — 288
   5.4 Hedgehog Signaling — 289
   5.5 Transforming Growth Factor β Signaling — 290
6. So How Is Specificity Achieved? — 292
7. Lessons From GSK-3 KO Mice — 292
   7.1 Roles of GSK-3 in Mammary Development and Tumorigenesis — 293
   7.2 GSK-3 in Brain Development — 293
   7.3 GSK-3 in Skeletal Muscle — 294
8. Therapeutic Perspectives — 294
   8.1 Alzheimer's Disease — 296
9. Conclusions and Perspectives — 297
Acknowledgments — 297
References — 297

Abstract

Glycogen synthase kinase-3 (GSK-3) is an unusual protein-serine kinase in that it is primarily regulated by inhibition and lies downstream of multiple cell signaling pathways. This raises a variety of questions in terms of its physiological role(s), how signaling specificity is maintained and why so many eggs have been placed into one basket. There are actually two baskets, as there are two isoforms, GSK-3α and β, that are highly related and largely redundant. Their many substrates range from regulators of cellular metabolism to molecules that control growth and differentiation. In this chapter, we review the characteristics of GSK-3, update progress in understanding the kinase, and try to answer some of the questions raised by its unusual properties. Indeed, the kinase may trigger transformation in our thinking of how cellular signals are organized and controlled.

1. THE EARLY YEARS

The existence of glycogen synthase kinase-3 (GSK-3) was first uncovered through studies of glycogen metabolism. Insulin was known to promote glycogen deposition in muscle and liver, and the rate-limiting enzyme in the pathway to formation of glycogen was determined to be glycogen synthase (Larner et al., 1968). Moreover, this enzyme, like phosphorylase, which plays the opposite role in degrading glycogen to release glucose units, was shown to be regulated by phosphorylation. Insulin triggered the dephosphorylation of glycogen synthase. This raised the question of whether insulin inhibited a protein kinase(s) or activated a protein phosphatase—a conundrum that took over 15 years to sort out (see Section 5.1). In the 1980s, the field became further complicated by finding that glycogen synthase could be phosphorylated by at least five different protein kinases in vitro (and dephosphorylated by several phosphatases). Based on their elution profiles from phosphocellulose resin, these kinases were sequentially named GSK-1 to -5 (Cohen et al., 1982). GSK-1 was identified as cyclic AMP-dependent protein kinase, GSK-2 as phosphorylase kinase, and GSK-5 as Casein Kinase-II (Embi, Rylatt, & Cohen, 1980). However, none of these three kinases was found to target the sites that responded to insulin.

Site-specific phosphopeptide analysis of which of the phosphorylated serines and threonines on glycogen synthase were changed upon exposure of tissues to insulin revealed four residues that were dephosphorylated. These were the sites that GSK-3 targeted in vitro (Parker, Caudwell, & Cohen, 1983). Thus, one of the first physiologically relevant substrates of GSK-3 was discovered and the name has stuck ever since. Since this initial discovery, nearly 100 proteins have been added to the list of putative GSK-3 targets (Sutherland, 2011) with many more predicted (Linding et al., 2007).

2. GSK-3 ISOFORMS, ORTHOLOGUES, AND EXPRESSION

Cloning of GSK-3 first revealed it to comprise two distinct genes (termed α and β) that are highly homologous (Woodgett, 1990). The two isoforms are each comprised of 11 exons and are 98% similar within their catalytic domain. However, they differ significantly in their N- and C-terminal domains. GSK-3α has a molecular weight of 51 kDa, whereas GSK-3β is smaller at 47 kDa. Unique to the α isoform is an N-terminal domain consisting of 63 residues that is glycine rich (71% glycine). The C-terminal regions of GSK-3α and β also differ, with the last 80 residues sharing only 34% similarity. In addition to these individual isoforms, a splice variant of GSK-3β that contains a 13-amino acid insert within its kinase domain is found in the brain (Mukai, Ishiguro, Sano, & Fujita, 2002). The cloning of the mammalian GSK-3 genes was contemporaneous with isolation of the *Drosophila* orthologue, termed Zeste-White3 or Shaggy (Bourouis et al., 1990; Siegfried, Perkins, Capaci, & Perrimon, 1990). Indeed, the fruitfly genetic analysis revealed the first hints that GSK-3 had far broader functions than sugar metabolism (not that that isn't important!).

The multifaceted roles of GSK-3 in development quickly surfaced from analysis of genetically tractable eukaryotes such as *Dictyostelium* and *Drosophila*. *Dictyostelium* expresses a single homologue of GSK-3 that bears 71% identity to the catalytic domain of human GSK-3β. During development, the integration of growth/differentiation signals with low-nutrient conditions results in exit from cell cycle and induction of differentiation in *Dictyostelium*. Two main cell fates, the spore and stalk fates, arise in the differentiation process. In GSK-3 null mutants, less than 5% of cells differentiate into spores, and the majority of the cells differentiated into stalk cells compared with 80% spore formation in wild-type *Dictyostelium* (Harwood, Plyte, Woodgett, Strutt, & Kay, 1995).

Studies on the GSK-3 orthologue in *Drosophila* (Zeste-white3/Shaggy) demonstrated involvement in regulating segment polarity and wing organization. Segment polarity genes are required for patterning within each embryonic segment and mutations in genes involved with segment polarity lead to the duplication or deletion of zones within segments. Homozygous mutants of the GSK-3 orthologue lacked differentiated structures that appear in the most anterior portion of segments, known as denticles (Siegfried et al., 1990). In addition to segment polarity, studies on dorsal–ventral axis formation in *Drosophila* as well as epistasis experiments placed Shaggy as a key mediator within the Wingless (wg) developmental pathway. In vertebrates, global organizers are groups of cells that cause a reorganization of spatial pattern when transplanted into ectopic locations. Wnt expressing cells were found to reorganize the dorsal–ventral patterning in the *Drosophila* leg. Similarly, mutants of Shaggy were shown to possess ventral organizer-like activity. These mutants acted autonomously to specify ventral cell fate as well as nonautonomously to reorganize dorsal–ventral patterning in the *Drosophila* leg (Diaz-Benjumea & Cohen, 1994).

3. GSK-3 STRUCTURE

As its name implies, GSK-3α and GSK-3β fall into the CMGC (containing CDK, MAPK, GSK-3, CLK families) group of protein kinases. The two isoforms are encoded on two separate chromosomes (GSK3α: mouse chromosome 7, human chromosome 19 and GSK-3β: mouse chromosome 16, human chromosome 3). GSK-3 is an ubiquitously expressed protein with both gene products expressed in all mammalian tissues (Lau, Miller, Anderton, & Shaw, 1999; Yao, Shaw, Wong, & Wan, 2002). GSK-3 homologues are conserved throughout eukaryotic evolution from yeast to plants and mammals (Ali, Hoeflich, & Woodgett, 2001). Invertebrates and at least some birds tend to only express one isoform of GSK-3 and this is most closely related to GSK-3β (Alon et al., 2011; Bianchi, Plyte, Kreis, & Woodgett, 1993; Plyte, Hughes, Nikolakaki, Pulverer, & Woodgett, 1992).

Similar to typical serine/threonine kinases, GSK-3 has a small N-terminal domain (residues 25–134 in GSK-3β) and a larger C-terminal domain (residues 135–380). The N-terminal region of GSK-3β comprises seven antiparallel β-strands and an α-helix (Fig. 1). The activation domain starts with the DFG motif and ends with the APE motif. GSK-3's structure most closely resembles mitogen-activated protein kinases (MAPKs). MAPKs require phosphorylation of two residues within their T-loop for activation. These residues include a tyrosine and a threonine. Similar to MAPKs, GSK-3 is phosphorylated (constitutively) in its active state at tyrosine 216 (GSK-3β) and tyrosine 279 (GSK-3α) (see Fig. 1, red circle) (Hughes, Nikolakaki, Plyte, Totty, & Woodgett, 1993). However, unlike MAPKs, GSK-3 lacks a phosphorylated threonine in its T-loop. Tyrosine phosphorylation of GSK-3 is most likely a cotranslational autophosphorylation event, much like DYRK (Lochhead et al., 2006) and, as such, GSK-3 tyrosine phosphorylation, although required for activity, is not regulated as it is for MAPKs.

Prior to determination of its crystal structure, GSK-3 was shown to preferentially bind substrates that were prephosphorylated four residues C-terminal to its phosphorylation site (Fiol, Mahrenholz, Wang, Roeske, & Roach, 1987). Substrates lacking this priming phosphorylation are not recognized by GSK-3. In retrospect, this is where a quirk of nature

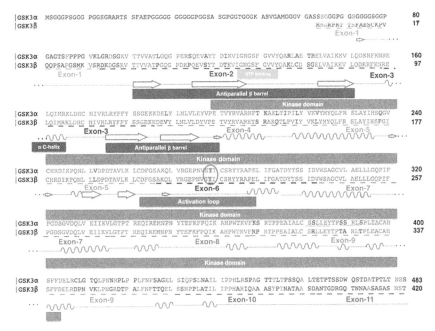

Fig. 1 GSK-3α and GSK-3β sequence alignment, secondary structures, and functional domains. Sequence alignment of human GSK-3α (gi|49574532) and β (gi|49168506). *Arrows* represent β-sheets, and *waves* represent α-helices. GSK-3α and β show high sequence similarity in their activation loops and kinase domains. GSK-3α has a unique glycine-rich N-terminal domain. Sequences in *green* represent matches, whereas *purple* sequences represent differences between the two kinases. FASTA sequences of the two kinases were aligned using the multalin online tool (http://multalin.toulouse.inra.fr/multalin/). Secondary structures were annotated from PDB structure 1i09. ATP-binding site and kinase domain residue identities were derived from UnitProt (UniProtKB—P49841). All secondary structures and domains are representative of those derived from the crystal structure of GSK-3β. Phosphorylated tyrosines circled in *red*.

was fortunate as the GSK-3 priming site on glycogen synthase is targeted by Casein Kinase-II (Fiol et al., 1987), is largely stoichiometrically phosphorylated, and is a poor target for most phosphatases. Hence, purified glycogen synthase was well prepared for in vitro phosphorylation by GSK-3. It is also important to note that allosteric regulation of glycogen synthase by glucose 6-phosphate likely plays a more important role than phosphorylation (Bouskila et al., 2010).

Where on GSK-3 the phosphorylated, primed substrates bound was not identified until 2001, when the structure of GSK-3 was solved (Bax et al., 2001; Dajani et al., 2001; ter Haar et al., 2001). Three basic residues arginines 96 and 180, and lysine 205 within GSK-3β form a binding pocket for the phosphate group of the primed substrate. Binding of a prephosphorylated/primed substrate to the T-loop of GSK-3 may compensate for the lack of a phosphothreonine moiety that is required for activation in kinases with similar structures to GSK-3.

4. REGULATION

As mentioned earlier, unlike most kinases, GSK-3 is active under basal conditions and requires extracellular signaling for its inactivation. Regulation of pathways in which GSK-3 is involved may occur through four mechanisms (Fig. 2):
(a) inactivating phosphorylation by other protein kinases on serine-21 of GSK-3α or serine-9 of GSK-3β (Fig. 2D),
(b) changes in subcellular compartmentalization of GSK-3α or GSK-3β,
(c) changes in tyrosine phosphorylation of GSK-3α at Tyr-279/GSK-3β at Tyr-216 (theoretical),
(d) through inactivation of kinases that act to prime GSK-3 substrate proteins (Fig. 2B).

Insulin, growth factors, or certain amino acids can trigger inactivation of GSK-3 through phosphorylation at serine 21 (GSK-3α) or serine 9 (GSK-3β) through the action of kinases such as AKT/PKB, p90rsk, and p70rsk (Cross, Alessi, Cohen, Andjelkovich, & Hemmings, 1995; Eldar-Finkelman, Seger, Vandenheede, & Krebs, 1995). When phosphorylated, the amino terminal domain located serine residues of GSK-3α and β bind to the substrate-binding pocket and hence act as pseudosubstrates, impeding actual substrate binding (Fig. 2D) (Dajani et al., 2001). Priming kinases can vary for each substrate. For example, the priming kinase for several sites on the microtubule-associated Tau protein is CDK5 (Li, Hawkes, Qureshi, Kar, & Paudel, 2006).

In contrast to serine phosphorylation of GSK-3, tyrosine phosphorylation is activating. Mutation of GSK-3β's T-loop tyrosine residue (Y216) to an nonphosphorylatable phenylalanine reside results in a 5- to 10-fold decrease in activity, and phosphorylation of GSK-3β at Y216 increases its activity by fivefold (Dajani et al., 2003; Hughes et al., 1993). Phosphorylation of GSK-3α and β at their respective tyrosine residues has been shown to occur, at least in vitro, through kinases such as FYN2 and PYK2 (Hartigan, Xiong, & Johnson, 2001; Lesort, Jope, & Johnson, 1999), and some evidence of increased phosphorylation and activity of

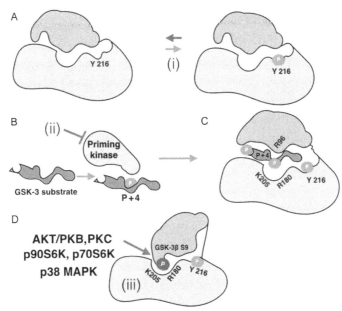

Fig. 2 Mechanisms of GSK-3 action, activation, and inactivation. (A) GSK-3β is phosphorylated constitutively at tyrosine 216. This phosphorylation may play a role in increasing accessibility to the substrate-binding groove. (B) The majority of GSK-3's substrates require a priming phosphorylation four (P+4) residues C-terminal to the GSK-3 phosphorylation site. (C) Binding of the primed residue into the binding pocket allows for the proper positioning of substrate for phosphorylation by GSK-3 on the N residue. (D) Phosphorylation on serine 9 of GSK-3β leads to the kinase folding into its own priming site binding pocket, acting like a pseudosubstrate and thereby preventing substrate entry into the GSK-3 substrate-binding groove. Arrows in red denote mechanisms by which GSK-3 can be inhibited: (i) inhibition of phosphorylation on tyrosine 216 may prevent efficient access of substrates to the GSK-3 substrate-binding site, (ii) inhibition of the priming kinase can also prevent phosphorylation of GSK-3 substrates, and (iii) kinases such as AKT/PKB can phosphorylate GSK-3 at serine 9 on GSK-3β and serine 21 on GSK-3α. GSK-3 recognizes many of its substrates through a primed phosphorylation motif of S/TXXXpS/T. Priming phosphorylation entails the phosphorylation of a serine/threonine residues (P+4) from the phosphorylation site of GSK-3. The priming kinases vary for each substrate and account for the ability of GSK-3 to regulate many pleiotropic pathways simultaneously.

GSK-3 in response to growth factor withdrawal in neuronal cells has been demonstrated to occur in both wild-type and catalytically inactive mutants of GSK-3. More recent data suggest, however, that wild-type GSK-3, but not its catalytically inactive form, is capable of autophosphorylating the tyrosine residue (Cole, Frame, & Cohen, 2004; Lochhead et al., 2006).

5. SIGNALING INFIDELITY

The combination of biochemical analysis and results from genetically tractable models provided key insights into how GSK-3 is regulated and what physiological processes it might be involved in. But these studies also raised many questions. How might a protein play a role in glycogen metabolism and developmental pathways such as Wnt? The situation became much more complex with the implication of GSK-3 playing roles in other signaling pathways:

5.1 Insulin (and Receptor Tyrosine Kinases Acting Through Phosphatidylinositol 3' Kinase Activation)

As described earlier, early hints of GSK-3's involvement in the insulin signaling pathway came from studies in rabbit skeletal muscle. Of the five distinct kinases that were shown to phosphorylate glycogen synthase, the sites phosphorylated by GSK-3 (3a, b, c, 4) were shown to be dephosphorylated in response to insulin treatment (Parker et al., 1983).

Ligand-activated insulin receptor (IR) directly phosphorylates insulin receptor substrates (IRS-1, 2, 3) on multiple tyrosine residues (Fig. 3). IRS proteins have a pleckstrin-homology domain and a phosphotyrosine-binding module, the latter of which accounts for their high affinity to the IR postinsulin stimulation (White, Maron, & Kahn, 1985). The phosphotyrosines on the receptor-associated proteins bind the Src Homology (SH2) domains on the regulatory subunit (p85) of phosphoinositide-3-kinase (PI3K) which increases the catalytic activity of the PI3K's p110 subunit and brings PI3K into close proximity to its substrate phosphatidylinositol (4,5) bisphosphate (PtdIns(4,5)P2). PtdIns(4,5)P2 becomes phosphorylated at the D3 position of the inositol ring to generate PtdIns(3,4,5)P. PtdIns(3,4,5)P recruits the serine kinases PDK-1, PKB/AKT, and PKC to the plasma membrane via their cognate pleckstrin homology (PH) domains, where PDK-1 phosphorylates and activates (in concert with mTORC2) PKB/AKT. The now active PKB/AKT in turn phosphorylates and inactivates GSK-3α and β. The reduced GSK-3 activity allows for phosphatases to act on glycogen synthase to reduce its inhibitory phosphorylation, resulting in promotion of glycogen accumulation.

In animal models of diabetes, pharmacological inhibition of GSK-3 has resulted in improvement of insulin signaling and glucose levels (Eldar-Finkelman, Schreyer, Shinohara, LeBoeuf, & Krebs, 1999; Meijer,

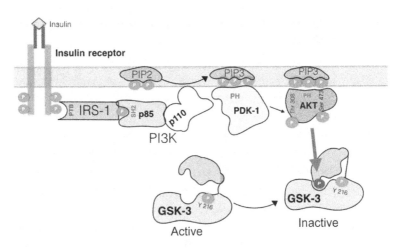

Fig. 3 Insulin signaling. Insulin binding to its receptor leads to a conformational change allowing for cross phosphorylation of its tyrosine residues. These residues create a docking site for the insulin receptor substrate-1 (IRS-1). PI3K's regulatory p85 subunit then docks onto the SH2 domains created by phosphorylation of IRS-1, leading to the activation of its p110 catalytic subunit which then phosphorylates phosphoinositide (4,5) phosphate generating phosphoinositide (3,4,5) phosphate. This then creates a pleckstrin homology (PH)-binding motif that leads to the membrane localization of PDK-1 and AKT/PKB. PDK-1 phosphorylates and activates PKBAKT at threonine 308. A subsequent phosphorylation event leads to a more activated AKT. Phosphorylation of GSK-3 by activated AKT at Ser 9 on GSK-3β or Ser-21 on GSK-3α inhibits the kinase, allowing for phosphatases to act on downstream targets such as glycogen synthase, thereby converting glycogen synthase into an activated state. *Black arrows* and *green circles* represent activating phosphorylations, whereas the *red arrow* and *red circle* represents inactivating phosphorylation.

Flajolet, & Greengard, 2004). Additionally, acute treatment of human skeletal muscle cells with GSK-3 inhibitors leads to an increase in glycogen synthase activation, increased glycogen deposition and increased levels of IRS-1 (Nikoulina et al., 2002). Generation of genetic models has furthered the validation of utilizing GSK-3 inhibitors in the treatment of T2DM. Genetic manipulation of GSK-3β rendering it inactive to inhibition via insulin signaling by conversion of serine 9 into alanine (GSK-3βS9A) reduced muscle glycogen deposition (Pearce et al., 2004). GSK-3 phosphorylates residues of kinesin motor proteins which mediate the translocation of the GLUT4 glucose transporters, resulting in their inhibition (Morfini, Szebenyi, Elluru, Ratner, & Brady, 2002). Phosphorylation of serine/threonine residues of IRS-1 (as opposed to tyrosine) reduces the ability of

IR to phosphorylate IRS on stimulatory tyrosines and promotes turnover of the IRS protein (Greene & Garofalo, 2002), and increased serine phosphorylation of IRS proteins has been implicated in insulin resistance (Hotamisligil et al., 1996). GSK-3 phosphorylates serine 332 of IRS-1 with serine 336 acting as the priming site and mutation of serine 322 promotes stimulatory tyrosine phosphorylation of IRS-1 (Liberman & Eldar-Finkelman, 2005).

5.2 Wnt Signaling

Wnt ligands control processes in the cell that affect cell proliferation, cell polarity, and cell fate determination (Logan & Nusse, 2004). Wnts are evolutionarily conserved glycoproteins that couple to various coreceptors activating diverse cellular responses. Wnt signaling is classified into either the canonical (β-catenin-dependent) pathway or the noncanonical (β-catenin-independent) pathway. The diversity in responses and cross talk stemming from 19 Wnt coding genes and 15 different Wnt receptors and coreceptors make this bipartite classification inexact. Under basal conditions, the cytoplasmic levels of a key component of the canonical pathway, β-catenin, kept low via phosphorylation events that tag the protein for ubiquitin-mediated proteolysis (Fig. 4). In addition to studies in *Drosophila*, evidence from *Xenopus laevis* has also pointed to the involvement of GSK-3 in this developmental pathway.

In *Xenopus* embryos, dominant-negative mutants of GSK-3 induce axis duplication—a pathology known to be associated with abrogated Wnt/β-catenin signaling activation (Dominguez, Itoh, & Sokol, 1995; He, Saint-Jeannet, Woodgett, Varmus, & Dawid, 1995). Subsequent studies demonstrated that a small fraction of GSK-3 (<10% of the cellular total) associates with a "destruction complex" consisting of several proteins including Axin, β-catenin, and the tumor suppressor, adenomatous polyposis coli (APC). GSK-3 phosphorylates β-catenin on threonine 41 and serines 37 and 33 after priming phosphorylation of β-catenin by casein-kinase 1α (CK1α) at serine 45. Phosphorylation of β-catenin promotes interaction with APC that enhances its recognition by the SKP1-Cullin1-F-box E3 ligase and β-TrCP complex, triggering its ubiquitination, followed by degradation via the 26S proteasome.

Wnt binding to its serpentine receptor, Frizzled, leads to association with low-density lipoprotein receptor 5/6 (LRP5/6). Phosphorylation of LRP5/6 on SSS(T/S)S motifs, by GSK-3 and CK-1, creates high-affinity

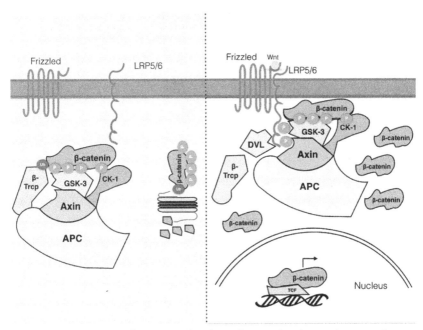

Fig. 4 Wnt signaling. Under unstimulated conditions, the destruction complex consisting of APC, Axin, GSK-3, CK-1, and β-TrCP coordinate the degradation of β-catenin. Wnt binding to its receptor, Frizzled, leads to colocalization with the LRP5/6 receptor. LRP5/6 then gets phosphorylated, creating high-affinity docking sites that lead to the subsequent translocation of the destruction complex to the membrane with the exception of β-TrCP. Newly transcribed β-catenin can translocate to the nucleus and regulate gene expression by binding to TCF/LEF transcription factors. In mammals, R-spondin ligands potentiate Wnt signaling by binding LGR4/5 receptors. Binding leads to the clearance of E3 ubiquitin ligases (RNF43 and ZNRF3) that degrade Wnt receptors.

binding sites for Axin and recruits the entire complex (with the exception of β-TrCP) to the plasma membrane. Newly transcribed β-catenin avoids tagging by the relocated complex, accumulates in the cytoplasm, and translocates to the nucleus, where it binds to TCF/LEF DNA-binding factors and regulates a program of gene expression. An additional level of regulation in the Wnt pathway in mammals, at least, is carried out by the R-spondin–LGR4/5–ZNF4/RNF43 complex. R-spondin interaction with LGR4/5 receptors potentiates Wnt signaling. R-spondin binding to LGR4/5 leads to the clearance of ZNRF3 and RNF43, two transmembrane E3 ubiquitin ligases that target Wnt receptors Frizzled and LRP6 for degradation (Planas-Paz et al., 2016).

5.3 Notch Signaling

Conserved across metazoa, Notch signaling has important roles in tissue patterning, cell fate, and cell polarity. Notch is a ligand-activated single pass transmembrane receptor that acts as a membrane-tethered transcription factor. Cleavage of the notch intracellular domain (NICD) requires activity of two enzymes, TACE and γ-secretase and is initiated by the binding of Notch ligands such as Delta-like, Jagged, and Serrate. Freshly cleaved NICD translocates to the nucleus and binds to the CSL family of transcription factors (CBF1, Suppressor of Hairless, Lag-1), to transactivate genes that include *Hes1* and *Hey1* (Fig. 5).

The role of GSK-3 in the Notch pathway is less clear than in the Wnt pathway. GSK-3 has been shown to phosphorylate NICD; however, the cellular readouts in terms of activation and inhibition are conflicting (Han, Ju, & Shin, 2012). GSK-3 has been reported to decrease Notch1 protein levels and decreases transcriptional activity of both Notch 1 and 2 (Espinosa, Ingles-Esteve, Aguilera, & Bigas, 2003; Jin, Kim, Oh, Ki, &

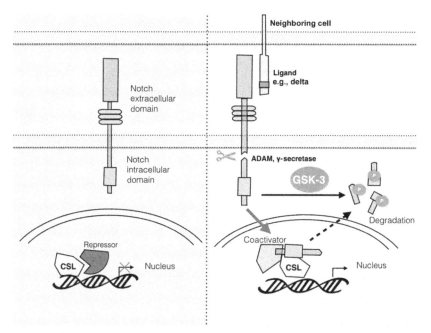

Fig. 5 Notch signaling. Notch effector CSL is bound to a repressor under unstimulated conditions. Upon binding of the Notch ligand, Delta, and the recruitment of proteases ADAM and γ-secretase, the notch intracellular domain (NICD) is released and translocates to the nucleus. Binding of the NICD to CSL (CBF1, Suppressor of Hairless, Lag-1) along with other coactivators leads to the transcription of genes such as *Hes* and *Hey*.

Kim, 2009). Contrary results indicate that GSK-3β-mediated phosphorylation may stabilize and activate Notch1 through inhibition of proteasomal degradation of Notch1 (Foltz, Santiago, Berechid, & Nye, 2002).

5.4 Hedgehog Signaling

Like Wnt and Notch, the Hedgehog signaling pathway also plays an important role in the development of metazoa. Components of Hedgehog signaling were initially identified in *Drosophila* and are highly conserved between flies and vertebrates. The binding of Hedgehog ligand (mammalian forms include: Indian Hedgehog, Desert Hedgehog, and Sonic Hedgehog) to its receptor, Patched, is promoted by two transmembrane proteins, CDO and BOC (Ihog/boi in *Drosophila*). In its unbound state, Patched represses Smoothened, a G-protein-coupled receptor-like protein (Fig. 6). Under

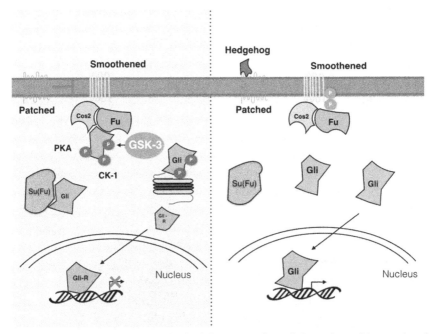

Fig. 6 Hedgehog signaling. Patched inhibits Smoothened. Smoothened is associated with SuFu and Cos2, and in this complex, three kinases, GSK-3, CK-1, and PKA, phosphorylate Gli2/3. This phosphorylation targets Gli2/3 by the proteasome, generating a truncated form that can translocate into the nucleus and repress gene transcription. Upon binding of Hedgehog to Patched, the inhibition on Smoothened is released, leading to the disassembly of SuFu and Cos2, and release of unphosphorylated Gli2/3. Full-length Gli2/3 can then translocate into the nucleus and activate gene expression of target genes.

basal conditions, processing of members of the Gli transcription factor family occurs through the binding of these proteins to Kif7/Cos and SuFu. Full-length Gli proteins act as transcriptional activators; however, sequential phosphorylation within their carboxy-terminal domain by PKA, GSK-3, and CK1 targets them for processing via the F-box protein βTrCP (Price & Calderon, 2002; Tempe, Casas, Karas, Blanchet-Tournier, & Concordet, 2006). Unlike their full-length counterparts, these modified transcription factors act to repress gene transcription.

Ligand binding releases inhibition of Smoothened by Patched. Smoothened, in its active state, promotes its dissociation from Kif7/Cos and SuFu, leading to the release of full-length Gli2/3 leading to the translocation of these proteins into the nucleus and transcriptional activation of genes that includes *Ptch1*, which attenuates the signal, or *Gli1*, which amplifies the signal. Target genes of Hedgehog signaling also include cell cycle regulators such as *Myc*, *Cyclin D*, and *E* (Fig. 6).

5.5 Transforming Growth Factor β Signaling

Transforming growth factor β (TGFβ) is a polypeptide ligand that drives many processes that regulate development, proliferation, differentiation, morphogenesis, and regeneration. The cellular outcomes of TGFβ signaling are highly context-dependent and targets range from a few in embryonic cells to the order of hundreds in differentiated cells. The negative regulators of TGFβ signaling pathways including SMAD7 and SKIL are conserved across cellular contexts. The SMAD signaling pathway is a central component of major TGFβ responses. Two ligand subfamilies define TGFβ signaling: the TGFβ-Activin-Nodal subfamily and the BMP family. Ligands bind to serine/threonine tetramers consisting of two Type I (signal-propagating) and two Type II (activating) receptors. Ligand binding permits the phosphorylation of the Type I receptor in a glycine- and serine-rich domain, leading to the docking of Smad transcription factors to the receptor. TGFβ signaling occurs through Smad2/3, whereas BMP signaling occurs through Smad1, 5, and 8 (Fig. 7). Docking results in the phosphorylation of Smads at two serine residues at the end of their carboxyl terminal region, leading to dissociation of Smads, subsequent binding to Smad4 (common to both pathways), and translocation of the entire Smad complex into the nucleus. Smads can bind to DNA directly using their MH1 domains, with the exception of Smad2, which binds through DNA-binding partners. Signaling, via Smads, is terminated by transcriptional targets of Smad: Smad6 and 7. Interaction of Smad6 and 7 with receptors leads to the recruitment of Smurf and related

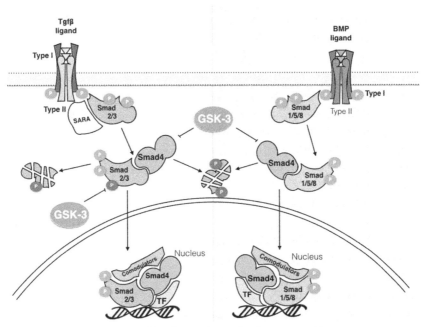

Fig. 7 Tgf-β signaling. Ligand binding induces the formation of heterotetramers consisting of two Type I and two Type II receptors. The Type II receptors then phosphorylate a glycine–serine-rich region in the Type I receptor. Signaling through the TGFβ pathway occurs with the help of Smad anchor for receptor activation (SARA) through Smad2/3, whereas signaling through the BMP pathway occurs via activation of Smad1/5/8. The Type I receptor kinase then phosphorylates the respective Smads, and this leads to their dissociation from the receptor complex and association with the common Smad4. Subsequent translocation to the nucleus and binding with additional comodulators leads to the activation of gene transcription.

ubiquitin ligases of the Nedd4 family and induces receptor degradation. Several additional layers of regulation occur within the TGFβ pathway. GSK-3 has been shown to phosphorylate Smad3 at threonine 66 in its Axin-associated complex, leading to Smad3 degradation (Guo et al., 2008). GSK-3 is also involved in Smad4 regulation. Specifically, activation of the MAPK pathway via fibroblast growth factor primes Smad4 for phosphorylation via GSK-3 at three residues, leading to creation of a phosphodegron that is subsequently ubiquitinated and degraded by the E3-ligase, βTrCP. Furthermore, Wnt signaling leads to inhibition of GSK-3, preventing Smad4 degradation and, in conjunction with FGF signaling, leads to prolonged TGFβ signaling (Demagny, Araki, & De Robertis, 2014).

6. SO HOW IS SPECIFICITY ACHIEVED?

Clearly, with so many distinct signaling pathways having at least part of their mechanism of action directed through GSK-3, there must be means by which the integrity and selectivity of their signal is preserved? Important clues have emerged from study of differential binding and anchoring proteins such as the AKAPs that sequester a small portion of a signaling protein within a subcellular domain. This occurs with the Axin scaffolding proteins in Wnt signaling. Since Axin is expressed at less than a 10th level of GSK-3 and has high affinity, it is always saturated with the kinase yet monopolizes only a small fraction of it. This is why Wnt pathway activation is largely unaffected until more than 80% of the GSK-3 is genetically removed (Doble, Patel, Wood, Kockeritz, & Woodgett, 2007). Similarly, Akt/PKB-dependent phosphorylation of the N-terminal domain of GSK-3, while affecting phosphatidylinositol $3'$ kinase-mediated targets, has little to no effect on β-catenin (Voskas, Ling, & Woodgett, 2010). It is therefore a mistake to view all of the GSK-3 molecules in a cell as equivalent. Small subsets are coupled to distinct signaling conduits. They may exchange between compartments and/or complexes, but within the dynamic timeframes of signaling, may as well be completely distinct from one another. Unfortunately for us, our tools of small molecule inhibitors, siRNA, CRISPR, and/or knockout (KO) by homologous recombination, are totally incapable of targeting such subcellular compartmentalization. We need new tools and probes.

7. LESSONS FROM GSK-3 KO MICE

Despite the high degree of homology between the GSK-3α and β isoforms, the global GSK-3 KO models revealed nonredundancies in signaling pathways. GSK-3α global KO mice are viable and demonstrate enhanced glucose and insulin sensitivity, enhanced glycogen deposition in liver, and increased IRS-1 expression (MacAulay et al., 2007). The phenotype of GSK-3α global KO mice was later determined, however, to be specific to KOs on an ICR background since when GSK-3α global KO mice were backcrossed onto the Black6 strain background they showed glucose and insulin sensitivities comparable to wild-type mice (Patel, Macaulay, & Woodgett, 2011).

Unlike GSK-3α null animals, GSK-3β global KO mice are inviable. By embryonic day (E)13.5–14.5, GSK-3β KO embryos appear pale.

Examination of fetal livers at E13.5 revealed multifocal hemorrhagic degeneration in the liver (Hoeflich et al., 2000). The mechanism behind liver apoptosis was determined to be through the absence of an antiapoptotic NF-κB-mediated response to TNF-α. Maintaining mice in a sterile environment avoids the exposure of embryos to infection-induced TNF-α responses. Indeed, when GSK-3β KO mice are housed in sterile facilities, they survive up until birth but die immediately thereafter due to heart patterning defects. These defects include a double outlet in the right ventricle, ventricular septal defects, and cardiomyocyte hyperproliferation that leads to hypertrophic myopathy (Kerkela et al., 2008).

The distinctive phenotypes between GSK-3α and β prompted study of the individual roles of these kinases. To date, none of the chemical inhibitors so far identified that act on GSK-3 distinguish between the two isoforms—although via perusal of PubMed abstracts ~90% of GSK-3 inhibitor studies suggest the authors believed the inhibitors they employed were specific for GSK-3. The subsequent generation of GSK-3 conditional KO models allowed for tissue-specific deletion of isoforms.

7.1 Roles of GSK-3 in Mammary Development and Tumorigenesis

Crossing of global GSK-3α KO mice with doubly floxed GSK-3β alleles and mice with Cre under control of the whey acidic protein promoter resulted in loss of all GSK-3 alleles specifically in the mammary gland. These mice developed adenosquamous carcinomas at 6 months of age, a tumor phenotype associated with the activation of β-catenin. Additional deletion of β-catenin in this model resulted in a delay in tumor development and a change in tumor morphology. The resultant cancer cells exhibited activation of Hedgehog and Notch signaling along with increased expression of plakoglobin/γ-catenin (Dembowy, Adissu, Liu, Zacksenhaus, & Woodgett, 2015). These experiments revealed that multiple tumor-promoting pathways are induced upon suppression of GSK-3, with canonical Wnt being most predominant.

7.2 GSK-3 in Brain Development

Similar to the mammary gland conditional KO model, GSK-3 deletion in the brain was carried out by crossing GSK-3α global null mice and GSK-3β doubly floxed mice with Nestin-Cre. Complete ablation of GSK-3 in the developing brain resulted in a late embryonic lethal phenotype. Massive hyperproliferation of neural progenitors was observed, while generation of postmitotic neurons was suppressed. There was loss of radial polarity and neuronal migration. Hedgehog, Notch, and Wnt signaling targets were elevated in GSK-3 KO brains (Kim et al., 2009).

7.3 GSK-3 in Skeletal Muscle

Conditional KO of GSK-3β or GSK-3α in muscle was generated by crossing floxed lines with mice expressing Cre-recombinase under the control of the myosin light chain kinase promoter. GSK-3β muscle-specific KOs demonstrated improved performance following glucose challenge, enhanced insulin-stimulated glycogen synthase regulation, and glycogen deposition (Patel et al., 2008). Deletion of GSK-3α, however, did not result in improved glucose tolerance or insulin sensitivity revealing tissue-specific isoform phenotypes of GSK-3 in muscle.

8. THERAPEUTIC PERSPECTIVES

A number of diverse chemical inhibitors of GSK-3 have been developed including cations, synthetic small-molecule competitive inhibitors, ATP-competitive and noncompetitive inhibitors, and substrate-competitive inhibitors. The majority of drugs that have been designed compete with ATP for access to GSK-3's ATP-binding site. ATP-binding sites have a high degree of conservation across kinase families, and therefore, selectivity of chemical inhibitors that target the ATP-binding site is relatively low. Drugs that bind outside of the ATP-binding regions or to the prephosphorylated/primed substrate-binding site of GSK-3 may offer more selective inhibition and therefore may be more suited to clinical application.

Lithium was the first GSK-3 inhibitor to be discovered (Klein & Melton, 1996) and shown to be effective in intact cells (Stambolic, Ruel, & Woodgett, 1996). Its mechanisms of action are multifold: it competes with Mg^{2+} for binding, through enhanced serine phosphorylation of GSK-3 and autoregulation (De Sarno, Li, & Jope, 2002; Kirshenboim, Plotkin, Shlomo, Kaidanovich-Beilin, & Eldar-Finkelman, 2004; Klein & Melton, 1996; Zhang, Phiel, Spece, Gurvich, & Klein, 2003). Among cations, zinc, beryllium, and sodium tungstate are other potent GSK-3 inhibitors, with the latter acting through indirect mechanisms. A search for drugs to inhibit cyclin-dependent kinases led to the chance discovery of potent GSK-3 inhibitors owing to the high degree of homology (86%) in their ATP-binding domains. These inhibitors fall under the class of ATP-competitive inhibitors such as 6-BIO and Meridianins. Synthetic inhibitors that compete for ATP binding have also been generated and include the aminopyrimidines, arylindolemaleimides, thiazoles, and paullones.

ATP noncompetitive inhibitors have been isolated from marine organisms and include manzamines and furanosesquiterpenes. Organically synthesized compounds under the thiadiazolidinediones and halomethylketone class have also been characterized. HMK-32, a halomethylketone, is the first irreversible inhibitor of GSK-3 that functions via formation of permanent sulfur–carbon bonds. Peptide inhibitors that mimic GSK-3 substrate motifs are also attractive inhibitors of GSK-3, for example, the peptide L803-mts was derived from GSK-3's substrate, Heat Shock Factor-1 (Plotkin, Kaidanovich, Talior, & Eldar-Finkelman, 2003). This inhibitor has recently been modified to allow improved cell entry and has demonstrated efficacy in several disease models (Licht-Murava et al., 2016).

As mentioned earlier, there are, as yet, no isoform-selective inhibitors for GSK-3. There is a broad collection of chemically diverse small molecules that do potently target GSK-3 and several have been developed—primarily with the aim to reverse insulin-resistance in type-II diabetes or to slow progression of neurodegeneration.

There is good evidence that patients with Type II diabetes show elevated GSK-3 levels and activity in skeletal muscle, suggesting that therapeutic inhibition of this kinase may be beneficial. Several lines of in vitro and in vivo work have yielded promising results for GSK-3 inhibition in the treatment of diabetes. The weak GSK-3 inhibitor, lithium, was demonstrated to activate glycogen synthase and to stimulate the synthesis of glucose in primary rat adipocytes (Cheng, Creacy, & Larner, 1983) and to stimulate synthesis of glycogen in muscle. More potent inhibitors, SB 216763 and SB 415286, showed increased rates of glycogen synthesis in HeLa cells (Coghlan et al., 2000). The Chiron-derived GSK-3 inhibitors, when used in vitro, activate glycogen synthase and promote glycogen synthesis in muscle cell lines. In vivo results using these inhibitors improved performance in oral and intraperitoneal glucose tolerance tests in normal and diabetic animals. Furthermore, increased translocation of the insulin-inducible glucose transporter, GLUT4, was also observed (Cline et al., 2002; Henriksen et al., 2003; Nikoulina et al., 2002; Ring et al., 2003). Micromolar concentrations of the substrate competitive inhibitor L803-mts-activated adipocyte glucose transport and activation of glycogen synthase. Moreover, improvements in glucose tolerance in normal and diabetic animals were observed with this inhibitor (Cohen & Goedert, 2004; Plotkin et al., 2003). Development of GSK-3 as a therapeutic target in diabetes is, however, largely on hold due to concerns of the potential for growth promoting and therefore carcinogenic effects.

8.1 Alzheimer's Disease

Alzheimer's disease (AD) is characterized by the formation of neurofibrillary tangles consisting of hyperphosphorylated Tau and amyloid plaques consisting predominately of amyloid-β (Aβ) fibrils. GSK-3 has been implicated in both pathological hallmarks of AD. Nonphosphorylated Tau is associated with microtubules, whereas phosphorylation of Tau on its microtubule-binding repeats decreases its ability to bind to microtubules and leads to Tau aggregation when the protein is hyperphosphorylated. Whether the hyperphosphorylation of Tau is due to increased levels of kinase activity and/or decreased levels of phosphatase activity is unclear. In vitro, GSK-3 has been shown to phosphorylate over 20 sites on Tau in physiological and pathological states (Hanger, Anderton, & Noble, 2009). Processing of amyloid precursor proteins requires two separate cleavages, the first of which is carried out by β-secretase, BACE, and the second by γ-secretase. The final product of cleavage results in Aβ, which is then released (Hardy & Higgins, 1992). γ-Secretase is found in a complex with Presenilin-1, and familial mutations in Presenilin-1 have been associated with early onset AD. GSK-3 has been shown to coimmunoprecipitate with Presenilin-1 in human brain samples and is also involved in Presenilin phosphorylation at Thr 743 (Aplin, Gibb, Jacobsen, Gallo, & Anderton, 1996; Takashima et al., 1998).

Transgenic mouse models overexpressing the APP intracellular domain, as well as models carrying AD-associated, familial mutations of APP, have demonstrated beneficial effects of GSK-3 inhibition through lithium and valproic acid (a mood stabilizer that inhibits GSK-3 as well as having other effects) (Ryan & Pimplikar, 2005; Su et al., 2004). Lithium has been used in the treatment of mood disorders for over 50 years. However, clinical trials exploring its utility in the treatment of AD have, so far, yielded conflicting results (Macdonald et al., 2008). The GSK-3 inhibitor NP-12, a member of the non-ATP competitive TDZD class of inhibitors, was the first to reach Phase II clinical trials for the treatment of AD. NP-12 passed Phase IIa trials based on patient tolerance, however, due to the small study size, conclusive data on the benefit to patients were not established (del Ser et al., 2013; Hampel et al., 2009).

There are also possible therapeutic indications in certain forms of autism and myotonic dystrophy (Jones et al., 2012).

9. CONCLUSIONS AND PERSPECTIVES

GSK-3 stands out among the 500+ eukaryotic protein kinases for several reasons including its unusual modes of regulation, its intercalation into a myriad of signaling pathways and the many proteins it regulates (we have only scratched the surface here). Tracing back its history of characterization, it has led scientists on a bewildering ride, slotting into new roles as our understanding of biology increased. It has also been strongly implicated in a series of disorders, particularly relating to brain development, mental health, and neurodegeneration (again, we have scratched the surface; for more, see Eldar-Finkelman & Martinez, 2011). Despite the unquestionable progress, there remain many questions. Why has nature turned to the kinase for so many tasks—after all, there are only 15 or so distinct signaling pathways, and GSK-3 either plays a direct role or an modulating role in most of them. Surely, this has greater meaning in terms of cellular coordination—perhaps contributing to biological robustness. The functions of the two isoforms also leave much to be desired. It may simply be that they are redundant in most functions and the extra copy reflects a fail-safe in case of loss of one. But there is also compelling evidence for some distinct roles—perhaps for cell-specific processes that we have not yet designed probes to detect. What is perhaps most clear is that understanding of the role of GSK-3 has helped illuminate areas of protein kinase biology that have challenged our thinking—moving us away from linear pathways and toward a far more integrated and interactive network of regulatory systems.

ACKNOWLEDGMENTS

J.R.W. is supported by a CIHR Foundation Grant. P.P. thanks the Canadian Liver Foundation for studentship support.

REFERENCES

Ali, A., Hoeflich, K. P., & Woodgett, J. R. (2001). Glycogen synthase kinase-3: Properties, functions, and regulation. *Chemical Reviews, 101*, 2527–2540.

Alon, L. T., Pietrokovski, S., Barkan, S., Avrahami, L., Kaidanovich-Beilin, O., Woodgett, J. R., et al. (2011). Selective loss of glycogen synthase kinase-3alpha in birds reveals distinct roles for GSK-3 isozymes in tau phosphorylation. *FEBS Letters, 585*, 1158–1162.

Aplin, A. E., Gibb, G. M., Jacobsen, J. S., Gallo, J. M., & Anderton, B. H. (1996). In vitro phosphorylation of the cytoplasmic domain of the amyloid precursor protein by glycogen synthase kinase-3beta. *Journal of Neurochemistry, 67*, 699–707.

Bax, B., Carter, P. S., Lewis, C., Guy, A. R., Bridges, A., Tanner, R., et al. (2001). The structure of phosphorylated GSK-3beta complexed with a peptide, FRATtide, that inhibits beta-catenin phosphorylation. *Structure, 9*, 1143–1152.

Bianchi, M. W., Plyte, S. E., Kreis, M., & Woodgett, J. R. (1993). A Saccharomyces cerevisiae protein-serine kinase related to mammalian glycogen synthase kinase-3 and the Drosophila melanogaster gene shaggy product. *Gene, 134*, 51–56.

Bourouis, M., Moore, P., Ruel, L., Grau, Y., Heitzler, P., & Simpson, P. (1990). An early embryonic product of the gene shaggy encodes a serine/threonine protein kinase related to the CDC28/cdc2+ subfamily. *The EMBO Journal, 9*, 2877–2884.

Bouskila, M., Hunter, R. W., Ibrahim, A. F., Delattre, L., Peggie, M., van Diepen, J. A., et al. (2010). Allosteric regulation of glycogen synthase controls glycogen synthesis in muscle. *Cell Metabolism, 12*, 456–466.

Cheng, K., Creacy, S., & Larner, J. (1983). 'Insulin-like' effects of lithium ion on isolated rat adipocytes. II. Specific activation of glycogen synthase. *Molecular and Cellular Biochemistry, 56*, 183–189.

Cline, G. W., Johnson, K., Regittnig, W., Perret, P., Tozzo, E., Xiao, L., et al. (2002). Effects of a novel glycogen synthase kinase-3 inhibitor on insulin-stimulated glucose metabolism in Zucker diabetic fatty (fa/fa) rats. *Diabetes, 51*, 2903–2910.

Coghlan, M. P., Culbert, A. A., Cross, D. A., Corcoran, S. L., Yates, J. W., Pearce, N. J., et al. (2000). Selective small molecule inhibitors of glycogen synthase kinase-3 modulate glycogen metabolism and gene transcription. *Chemistry & Biology, 7*, 793–803.

Cohen, P., & Goedert, M. (2004). GSK3 inhibitors: Development and therapeutic potential. *Nature Reviews. Drug Discovery, 3*, 479–487.

Cohen, P., Yellowlees, D., Aitken, A., Donella-Deana, A., Hemmings, B. A., & Parker, P. J. (1982). Separation and characterisation of glycogen synthase kinase 3, glycogen synthase kinase 4 and glycogen synthase kinase 5 from rabbit skeletal muscle. *European Journal of Biochemistry, 124*, 21–35.

Cole, A., Frame, S., & Cohen, P. (2004). Further evidence that the tyrosine phosphorylation of glycogen synthase kinase-3 (GSK3) in mammalian cells is an autophosphorylation event. *The Biochemical Journal, 377*, 249–255.

Cross, D. A., Alessi, D. R., Cohen, P., Andjelkovich, M., & Hemmings, B. A. (1995). Inhibition of glycogen synthase kinase-3 by insulin mediated by protein kinase B. *Nature, 378*, 785–789.

Dajani, R., Fraser, E., Roe, S. M., Yeo, M., Good, V. M., Thompson, V., et al. (2003). Structural basis for recruitment of glycogen synthase kinase 3beta to the axin-APC scaffold complex. *The EMBO Journal, 22*(3), 494–501.

Dajani, R., Fraser, E., Roe, S. M., Young, N., Good, V., Dale, T. C., et al. (2001). Crystal structure of glycogen synthase kinase 3 beta: Structural basis for phosphate-primed substrate specificity and autoinhibition. *Cell, 105*(6), 721–732.

De Sarno, P., Li, X., & Jope, R. S. (2002). Regulation of Akt and glycogen synthase kinase-3 beta phosphorylation by sodium valproate and lithium. *Neuropharmacology, 43*(7), 1158–1164.

del Ser, T., Steinwachs, K. C., Gertz, H. J., Andres, M. V., Gomez-Carrillo, B., Medina, M., et al. (2013). Treatment of Alzheimer's disease with the GSK-3 inhibitor tideglusib: A pilot study. *Journal of Alzheimer's Disease, 33*(1), 205–215.

Demagny, H., Araki, T., & De Robertis, E. M. (2014). The tumor suppressor Smad4/DPC4 is regulated by phosphorylations that integrate FGF, Wnt, and TGF-beta signaling. *Cell Reports, 9*(2), 688–700.

Dembowy, J., Adissu, H. A., Liu, J. C., Zacksenhaus, E., & Woodgett, J. R. (2015). Effect of glycogen synthase kinase-3 inactivation on mouse mammary gland development and oncogenesis. *Oncogene, 34*(27), 3514–3526.

Diaz-Benjumea, F. J., & Cohen, S. M. (1994). wingless acts through the shaggy/zeste-white 3 kinase to direct dorsal-ventral axis formation in the Drosophila leg. *Development, 120*(6), 1661–1670.

Doble, B. W., Patel, S., Wood, G. A., Kockeritz, L. K., & Woodgett, J. R. (2007). Functional redundancy of GSK-3alpha and GSK-3beta in Wnt/beta-catenin signaling shown by using an allelic series of embryonic stem cell lines. *Developmental Cell, 12*(6), 957–971.

Dominguez, I., Itoh, K., & Sokol, S. Y. (1995). Role of glycogen synthase kinase 3 beta as a negative regulator of dorsoventral axis formation in Xenopus embryos. *Proceedings of the National Academy of Sciences of the United States of America, 92*(18), 8498–8502.

Eldar-Finkelman, H., & Martinez, A. (2011). GSK-3 inhibitors: Preclinical and clinical focus on CNS. *Frontiers in Molecular Neuroscience, 4*, 32.

Eldar-Finkelman, H., Schreyer, S. A., Shinohara, M. M., LeBoeuf, R. C., & Krebs, E. G. (1999). Increased glycogen synthase kinase-3 activity in diabetes- and obesity-prone C57BL/6J mice. *Diabetes, 48*(8), 1662–1666.

Eldar-Finkelman, H., Seger, R., Vandenheede, J. R., & Krebs, E. G. (1995). Inactivation of glycogen synthase kinase-3 by epidermal growth factor is mediated by mitogen-activated protein kinase/p90 ribosomal protein S6 kinase signaling pathway in NIH/3T3 cells. *The Journal of Biological Chemistry, 270*(3), 987–990.

Embi, N., Rylatt, D. B., & Cohen, P. (1980). Glycogen synthase kinase-3 from rabbit skeletal muscle. Separation from cyclic-AMP-dependent protein kinase and phosphorylase kinase. *European Journal of Biochemistry, 107*(2), 519–527.

Espinosa, L., Ingles-Esteve, J., Aguilera, C., & Bigas, A. (2003). Phosphorylation by glycogen synthase kinase-3 beta down-regulates Notch activity, a link for Notch and Wnt pathways. *The Journal of Biological Chemistry, 278*(34), 32227–32235.

Fiol, C. J., Mahrenholz, A. M., Wang, Y., Roeske, R. W., & Roach, P. J. (1987). Formation of protein kinase recognition sites by covalent modification of the substrate. Molecular mechanism for the synergistic action of casein kinase II and glycogen synthase kinase 3. *The Journal of Biological Chemistry, 262*(29), 14042–14048.

Foltz, D. R., Santiago, M. C., Berechid, B. E., & Nye, J. S. (2002). Glycogen synthase kinase-3beta modulates notch signaling and stability. *Current Biology: CB, 12*(12), 1006–1011.

Greene, M. W., & Garofalo, R. S. (2002). Positive and negative regulatory role of insulin receptor substrate 1 and 2 (IRS-1 and IRS-2) serine/threonine phosphorylation. *Biochemistry, 41*(22), 7082–7091.

Guo, X., Ramirez, A., Waddell, D. S., Li, Z., Liu, X., & Wang, X. F. (2008). Axin and GSK3-control Smad3 protein stability and modulate TGF-signaling. *Genes & Development, 22*(1), 106–120.

Hampel, H., Ewers, M., Burger, K., Annas, P., Mortberg, A., Bogstedt, A., et al. (2009). Lithium trial in Alzheimer's disease: A randomized, single-blind, placebo-controlled, multicenter 10-week study. *The Journal of Clinical Psychiatry, 70*(6), 922–931.

Han, X., Ju, J. H., & Shin, I. (2012). Glycogen synthase kinase 3-beta phosphorylates novel S/T-P-S/T domains in Notch1 intracellular domain and induces its nuclear localization. *Biochemical and Biophysical Research Communications, 423*(2), 282–288.

Hanger, D. P., Anderton, B. H., & Noble, W. (2009). Tau phosphorylation: The therapeutic challenge for neurodegenerative disease. *Trends in Molecular Medicine, 15*(3), 112–119.

Hardy, J. A., & Higgins, G. A. (1992). Alzheimer's disease: The amyloid cascade hypothesis. *Science, 256*(5054), 184–185.

Hartigan, J. A., Xiong, W. C., & Johnson, G. V. (2001). Glycogen synthase kinase 3beta is tyrosine phosphorylated by PYK2. *Biochemical and Biophysical Research Communications, 284*(2), 485–489.

Harwood, A. J., Plyte, S. E., Woodgett, J., Strutt, H., & Kay, R. R. (1995). Glycogen synthase kinase 3 regulates cell fate in Dictyostelium. *Cell, 80*(1), 139–148.

He, X., Saint-Jeannet, J. P., Woodgett, J. R., Varmus, H. E., & Dawid, I. B. (1995). Glycogen synthase kinase-3 and dorsoventral patterning in Xenopus embryos. *Nature*, *374*(6523), 617–622.
Henriksen, E. J., Kinnick, T. R., Teachey, M. K., O'Keefe, M. P., Ring, D., Johnson, K. W., et al. (2003). Modulation of muscle insulin resistance by selective inhibition of GSK-3 in Zucker diabetic fatty rats. *American Journal of Physiology. Endocrinology and Metabolism*, *284*(5), E892–E900.
Hoeflich, K. P., Luo, J., Rubie, E. A., Tsao, M. S., Jin, O., & Woodgett, J. R. (2000). Requirement for glycogen synthase kinase-3beta in cell survival and NF-kappaB activation. *Nature*, *406*(6791), 86–90.
Hotamisligil, G. S., Peraldi, P., Budavari, A., Ellis, R., White, M. F., & Spiegelman, B. M. (1996). IRS-1-mediated inhibition of insulin receptor tyrosine kinase activity in TNF-alpha- and obesity-induced insulin resistance. *Science*, *271*(5249), 665–668.
Hughes, K., Nikolakaki, E., Plyte, S. E., Totty, N. F., & Woodgett, J. R. (1993). Modulation of the glycogen synthase kinase-3 family by tyrosine phosphorylation. *The EMBO Journal*, *12*(2), 803–808.
Jin, Y. H., Kim, H., Oh, M., Ki, H., & Kim, K. (2009). Regulation of Notch1/NICD and Hes1 expressions by GSK-3alpha/beta. *Molecules and Cells*, *27*(1), 15–19.
Jones, K., Wei, C., Iakova, P., Bugiardini, E., Schneider-Gold, C., Meola, G., et al. (2012). GSK3β mediates muscle pathology in myotonic dystrophy. *The Journal of Clinical Investigation*, *122*(12), 4461–4472.
Kerkela, R., Kockeritz, L., Macaulay, K., Zhou, J., Doble, B. W., Beahm, C., et al. (2008). Deletion of GSK-3beta in mice leads to hypertrophic cardiomyopathy secondary to cardiomyoblast hyperproliferation. *The Journal of Clinical Investigation*, *118*(11), 3609–3618.
Kim, W. Y., Wang, X., Wu, Y., Doble, B. W., Patel, S., Woodgett, J. R., et al. (2009). GSK-3 is a master regulator of neural progenitor homeostasis. *Nature Neuroscience*, *12*(11), 1390–1397.
Kirshenboim, N., Plotkin, B., Shlomo, S. B., Kaidanovich-Beilin, O., & Eldar-Finkelman, H. (2004). Lithium-mediated phosphorylation of glycogen synthase kinase-3beta involves PI3 kinase-dependent activation of protein kinase C-alpha. *Journal of Molecular Neuroscience*, *24*(2), 237–245.
Klein, P. S., & Melton, D. A. (1996). A molecular mechanism for the effect of lithium on development. *Proceedings of the National Academy of Sciences of the United States of America*, *93*(16), 8455–8459.
Larner, J., Villar-Palasi, C., Goldberg, N. D., Bishop, J. S., Huijing, F., Wenger, J. I., et al. (1968). Hormonal and non-hormonal control of glycogen synthesis-control of transferase phosphatase and transferase I kinase. *Advances in Enzyme Regulation*, *6*, 409–423.
Lau, K. F., Miller, C. C., Anderton, B. H., & Shaw, P. C. (1999). Expression analysis of glycogen synthase kinase-3 in human tissues. *The Journal of Peptide Research*, *54*(1), 85–91.
Lesort, M., Jope, R. S., & Johnson, G. V. (1999). Insulin transiently increases tau phosphorylation: Involvement of glycogen synthase kinase-3beta and Fyn tyrosine kinase. *Journal of Neurochemistry*, *72*(2), 576–584.
Li, T., Hawkes, C., Qureshi, H. Y., Kar, S., & Paudel, H. K. (2006). Cyclin-dependent protein kinase 5 primes microtubule-associated protein tau site-specifically for glycogen synthase kinase 3beta. *Biochemistry*, *45*(10), 3134–3145.
Liberman, Z., & Eldar-Finkelman, H. (2005). Serine 332 phosphorylation of insulin receptor substrate-1 by glycogen synthase kinase-3 attenuates insulin signaling. *The Journal of Biological Chemistry*, *280*(6), 4422–4428.
Licht-Murava, A., Paz, R., Vaks, L., Avrahami, L., Plotkin, B., Eisenstein, M., et al. (2016). A unique type of GSK-3 inhibitor brings new opportunities to the clinic. *Science Signaling*, *9*, ra110.

Linding, R., Jensen, L. J., Ostheimer, G. J., van Vugt, M. A. T. M., Jørgensen, C., Miron, I. M., et al. (2007). Systematic discovery of in vivo phosphorylation networks. *Cell, 129*(7), 1415–1426.
Lochhead, P. A., Kinstrie, R., Sibbet, G., Rawjee, T., Morrice, N., & Cleghon, V. (2006). A chaperone-dependent GSK3beta transitional intermediate mediates activation-loop autophosphorylation. *Molecular Cell, 24*(4), 627–633.
Logan, C. Y., & Nusse, R. (2004). The Wnt signaling pathway in development and disease. *Annual Review of Cell and Developmental Biology, 20*, 781–810.
MacAulay, K., Doble, B. W., Patel, S., Hansotia, T., Sinclair, E. M., Drucker, D. J., et al. (2007). Glycogen synthase kinase 3alpha-specific regulation of murine hepatic glycogen metabolism. *Cell Metabolism, 6*(4), 329–337.
Macdonald, A., Briggs, K., Poppe, M., Higgins, A., Velayudhan, L., & Lovestone, S. (2008). A feasibility and tolerability study of lithium in Alzheimer's disease. *International Journal of Geriatric Psychiatry, 23*(7), 704–711.
Meijer, L., Flajolet, M., & Greengard, P. (2004). Pharmacological inhibitors of glycogen synthase kinase 3. *Trends in Pharmacological Sciences, 25*(9), 471–480.
Morfini, G., Szebenyi, G., Elluru, R., Ratner, N., & Brady, S. T. (2002). Glycogen synthase kinase 3 phosphorylates kinesin light chains and negatively regulates kinesin-based motility. *The EMBO Journal, 21*(3), 281–293.
Mukai, F., Ishiguro, K., Sano, Y., & Fujita, S. C. (2002). Alternative splicing isoform of tau protein kinase I/glycogen synthase kinase 3beta. *Journal of Neurochemistry, 81*(5), 1073–1083.
Nikoulina, S. E., Ciaraldi, T. P., Mudaliar, S., Carter, L., Johnson, K., & Henry, R. R. (2002). Inhibition of glycogen synthase kinase 3 improves insulin action and glucose metabolism in human skeletal muscle. *Diabetes, 51*(7), 2190–2198.
Parker, P. J., Caudwell, F. B., & Cohen, P. (1983). Glycogen synthase from rabbit skeletal muscle; effect of insulin on the state of phosphorylation of the seven phosphoserine residues in vivo. *European Journal of Biochemistry, 130*(1), 227–234.
Patel, S., Doble, B. W., MacAulay, K., Sinclair, E. M., Drucker, D. J., & Woodgett, J. R. (2008). Tissue-specific role of glycogen synthase kinase 3beta in glucose homeostasis and insulin action. *Molecular and Cellular Biology, 28*(20), 6314–6328.
Patel, S., Macaulay, K., & Woodgett, J. R. (2011). Tissue-specific analysis of glycogen synthase kinase-3alpha (GSK-3alpha) in glucose metabolism: Effect of strain variation. *PLoS One, 6*(1), e15845.
Pearce, N. J., Arch, J. R., Clapham, J. C., Coghlan, M. P., Corcoran, S. L., Lister, C. A., et al. (2004). Development of glucose intolerance in male transgenic mice overexpressing human glycogen synthase kinase-3beta on a muscle-specific promoter. *Metabolism, 53*(10), 1322–1330.
Planas-Paz, L., Orsini, V., Boulter, L., Calabrese, D., Pikiolek, M., Nigsch, F., et al. (2016). The RSPO-LGR4/5-ZNRF3/RNF43 module controls liver zonation and size. *Nature Cell Biology, 18*(5), 467–479.
Plotkin, B., Kaidanovich, O., Talior, I., & Eldar-Finkelman, H. (2003). Insulin mimetic action of synthetic phosphorylated peptide inhibitors of glycogen synthase kinase-3. *The Journal of Pharmacology and Experimental Therapeutics, 305*(3), 974–980.
Plyte, S. E., Hughes, K., Nikolakaki, E., Pulverer, B. J., & Woodgett, J. R. (1992). Glycogen synthase kinase-3: Functions in oncogenesis and development. *Biochimica Biophysica Acta, 1114*(2–3), 147–162.
Price, M. A., & Calderon, D. (2002). Proteolysis of the hedgehog signaling effector cubitus interruptus requires phosphorylation by glycogen synthase kinase 3 and casein kinase 1. *Cell, 108*(6), 823–835.
Ring, D. B., Johnson, K. W., Henriksen, E. J., Nuss, J. M., Goff, D., Kinnick, T. R., et al. (2003). Selective glycogen synthase kinase 3 inhibitors potentiate insulin activation of glucose transport and utilization in vitro and in vivo. *Diabetes, 52*(3), 588–595.

Ryan, K. A., & Pimplikar, S. W. (2005). Activation of GSK-3 and phosphorylation of CRMP2 in transgenic mice expressing APP intracellular domain. *The Journal of Cell Biology, 171*(2), 327–335.

Siegfried, E., Perkins, L. A., Capaci, T. M., & Perrimon, N. (1990). Putative protein kinase product of the Drosophila segment-polarity gene zeste-white3. *Nature, 345*(6278), 825–829.

Stambolic, V., Ruel, L., & Woodgett, J. R. (1996). Lithium inhibits glycogen synthase kinase-3 activity and mimics wingless signalling in intact cells. *Current Biology: CB, 6*(12), 1664–1668.

Su, Y., Ryder, J., Li, B., Wu, X., Fox, N., Solenberg, P., et al. (2004). Lithium, a common drug for bipolar disorder treatment, regulates amyloid-beta precursor protein processing. *Biochemistry, 43*(22), 6899–6908.

Sutherland, C. (2011). What are the bona fide GSK3 substrates? *International Journal of Alzheimer's Disease, 2011*, 505607.

Takashima, A., Murayama, M., Murayama, O., Kohno, T., Honda, T., Yasutake, K., et al. (1998). Presenilin 1 associates with glycogen synthase kinase-3beta and its substrate tau. *Proceedings of the National Academy of Sciences of the United States of America, 95*(16), 9637–9641.

Tempe, D., Casas, M., Karas, S., Blanchet-Tournier, M. F., & Concordet, J. P. (2006). Multisite protein kinase A and glycogen synthase kinase 3beta phosphorylation leads to Gli3 ubiquitination by SCFbetaTrCP. *Molecular and Cellular Biology, 26*(11), 4316–4326.

ter Haar, E., Coll, J. T., Austen, D. A., Hsiao, H. M., Swenson, L., & Jain, J. (2001). Structure of GSK3beta reveals a primed phosphorylation mechanism. *Nature Structural Biology, 8*(7), 593–596.

Voskas, D., Ling, L. S., & Woodgett, J. R. (2010). Does GSK-3 provide a shortcut for PI3K activation of Wnt signaling? *F1000 Biology Reports, 2*, 82.

White, M. F., Maron, R., & Kahn, C. R. (1985). Insulin rapidly stimulates tyrosine phosphorylation of a Mr-185,000 protein in intact cells. *Nature, 318*(6042), 183–186.

Woodgett, J. R. (1990). Molecular cloning and expression of glycogen synthase kinase-3/factor A. *The EMBO Journal, 9*(8), 2431–2438.

Yao, H. B., Shaw, P. C., Wong, C. C., & Wan, D. C. (2002). Expression of glycogen synthase kinase-3 isoforms in mouse tissues and their transcription in the brain. *Journal of Chemical Neuroanatomy, 23*(4), 291–297.

Zhang, F., Phiel, C. J., Spece, L., Gurvich, N., & Klein, P. S. (2003). Inhibitory phosphorylation of glycogen synthase kinase-3 (GSK-3) in response to lithium. Evidence for autoregulation of GSK-3. *The Journal of Biological Chemistry, 278*, 33067–33077.

CHAPTER NINE

CK1 in Developmental Signaling: Hedgehog and Wnt

Jin Jiang[1]

University of Texas Southwestern Medical Center at Dallas, Dallas, TX, United States
[1]Corresponding author: e-mail address: jin.jiang@utsouthwestern.edu

Contents

1. Introduction — 304
   1.1 Overview of the CK1 Family Kinases — 304
   1.2 Overview of Canonical Hh and Wnt Pathways — 305
2. CK1 in the Regulation of Hh Pathway — 308
   2.1 Regulation of Ci/Gli Processing by CK1 — 308
   2.2 Regulation of Smo Activation by CK1 in *Drosophila* — 311
   2.3 A Conserved Role of CK1 in Mammalian Smo Activation — 314
   2.4 CK1 Positively Regulates Hh Signaling Downstream of Smo — 316
3. CK1 in Wnt Signaling — 317
   3.1 Regulation of β-Catenin Destruction: A Negative Role of CK1 in Wnt Signaling — 318
   3.2 Regulation of LRP6 Phosphorylation: A Positive Role of CK1 in Wnt Signaling — 319
4. Regulation of CK1 in Hh and Wnt Signaling — 320
5. Conclusion — 322
Acknowledgments — 323
References — 323

Abstract

The casein kinase 1 (CK1) family of serine (Ser)/threonine (Thr) protein kinases participates in a myriad of cellular processes including developmental signaling. Hedgehog (Hh) and Wnt pathways are two major and evolutionarily conserved signaling pathways that control embryonic development and adult tissue homeostasis. Deregulation of these pathways leads to many human disorders including birth defects and cancer. Here, I review the role of CK1 in the regulation of Hh and Wnt signal transduction cascades from the membrane reception systems to the transcriptional effectors. In both Hh and Wnt pathways, multiple CK1 family members regulate signal transduction at several levels of the pathways and play either positive or negative roles depending on the signaling status, individual CK1 isoforms involved, and the specific substrates they phosphorylate. A common mechanism underlying the control of CK1-mediated phosphorylation of Hh and Wnt pathway components is the regulation of CK1/substrate interaction within large protein complexes. I will highlight this feature in the context of Hh signaling and draw interesting parallels between the Hh and Wnt pathways.

1. INTRODUCTION
1.1 Overview of the CK1 Family Kinases

CK1 protein kinases belong to a large family of evolutionarily conserved monomeric Ser/Thr kinases in eukaryotes (Cheong & Virshup, 2011). The CK1 family has been implicated in the regulation of diverse cellular processes, including membrane receptor trafficking, cytoskeleton maintenance, cell division, DNA damage response and repair, and nuclear translocation (Gross & Anderson, 1998; Knippschild et al., 2005). CK1 also participates in complex developmental and physiological processes such as Hh and Wnt signaling and circadian rhythms by phosphorylating key regulatory components in these processes (Chen & Jiang, 2013; MacDonald, Tamai, & He, 2009).

CK1 family members are found in eukaryotic organisms ranging from yeast to human with mammals possessing seven family members: α, β, γ1, γ2, γ3, δ, and ε. *Drosophila* has eight CK1 family members, whereas *C. elegans* has up to 87 members (Fig. 1) (Plowman, Sudarsanam, Bingham, Whyte, & Hunter, 1999; Zhang, Jia, et al., 2006). The CK1 family members contain highly conserved kinase domains but differ in their N- and C-terminal domains in terms of the length and amino acid sequence. Unlike other CK1 family members that are found in the cytosol, CK1γ is membrane associated due to palmitoylation of its C-terminus (Davidson et al., 2005). Due to their sharing of a highly conserved kinase domain, the CK1 family kinases phosphorylate similar target sites that conform the following consensus: $D/E/(p)S/T(X)_{1-3}\underline{S/T}$, with underlined S/T as the phosphoacceptor site, (p)S/T as phosphorylated residue, and X representing any amino acid (Knippschild et al., 2005). Although the spacing between the phosphoacceptor site and the upstream acid residue or phosphorylated S/T can be one to three amino acids, an optimal space is usually two amino acids and the phosphorylation efficiency can also be influenced by the surrounding sequence. For example, a cluster of acidic residues at the N-terminus of the consensus site confer increased phosphorylation efficiency by CK1. Many CK1 target sites contain S/T instead of an acid residue at -2, -3, or -4 position, thus require a prior phosphorylation by a priming kinase. In general, CK1 family kinases are constitutively active; however, the C-terminal extensions of CK1δ and CK1ε are autophosphorylated, which inhibits the activity of their kinase domains, although in vivo phosphatases keep them constitutively active in many cases (Rivers, Gietzen, Vielhaber, & Virshup, 1998).

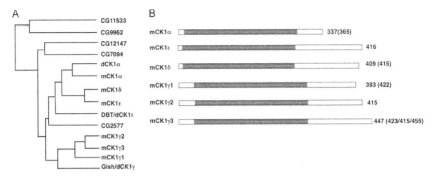

Fig. 1 CK1 family of kinases. (A) Family tree of CK1 isoforms from *Drosophila* and mammals. In addition to CK1α, CK1ε, and CK1γ, CG12147, CG7094, and CG9962 have also been implicated in Wnt signaling in *Drosophila* (Serysheva et al., 2013; Zhang, Jia, et al., 2006). (B) Schematic drawings of mammalian CK1 family members with kinase domains depicted in *gray*. Numbers in parentheses indicate the lengths of CK1 isoforms generated from alternative splicing.

1.2 Overview of Canonical Hh and Wnt Pathways

Hh and Wnt pathways are two major signaling pathways that control embryonic development and adult tissue homeostasis (Jiang & Hui, 2008; MacDonald et al., 2009). Malfunction of these signaling pathways have been attributed to numerous human diseases including birth defects and cancer. The canonical Hh and Wnt pathways share many pathway components, employ the same logic for pathway regulation, and thus are often considered as "sister" pathways.

Hh acts through a largely conserved pathway to regulate the balance between activator and repressor forms of the Gli family of zinc finger transcription factors (Gli^A and Gli^R; Fig. 2A). While *Drosophila* only has one Hh and one Gli protein, Cubitus interruptus (Ci), mammals have three Hh family members Shh, Ihh, and Dhh and three Gli proteins: Gli1, Gli2, and Gli3 (Jiang & Hui, 2008). In mice, Gli^R function is mainly derived from Gli3 through proteolytic processing whereas Gli^A activity is primarily contributed by Gli2. Gli1 is a transcriptional target of Hh signaling and acts in a positive feedback to reinforce Gli^A activity.

The reception system for Hh signals consists of a 12 transmembrane protein Patched (Ptc) that binds directly to Hh and a seven transmembrane GPCR family protein Smoothened (Smo) that transduces the signal into the cytoplasm (Jiang & Hui, 2008). In the absence of Hh, the constitutive activity of Ptc blocks Smo activation, allowing the proteolytic processing of full-length Gli/Ci to generate C-terminally truncated Gli^R/Ci^R that represses a subset of Hh target genes. Binding of Hh to Ptc alleviates

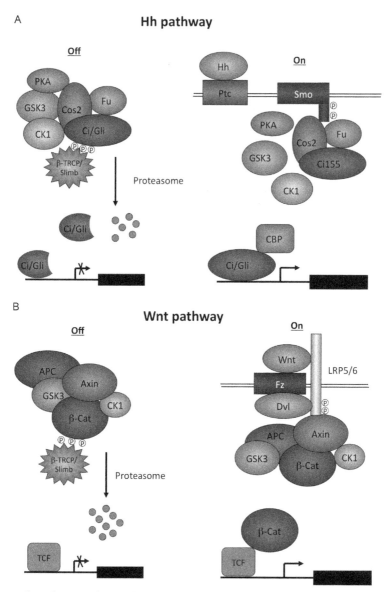

Fig. 2 Hh and Wnt pathways. (A) *Drosophila* Hh pathway. In the "signaling off" state, Ci and its kinases PKA, GSK3, and CK1 form a large protein complex scaffolded by Cos2 and Fu, leading to its phosphorylation and proteolysis to generate a truncated repressor form that inhibits the expression of Hh target genes. In the "signaling on" state, Hh binds Ptc and releases its inhibition of Smo, leading to Smo phosphorylation and activation. Smo interacts with Cos2/Fu and promotes Fu phosphorylation and activation but inhibits PKA/CK1-mediated Ci phosphorylation and processing. Activated Fu converts full-length Ci into an active form that stimulates the expression of Hh target genes.

inhibition of Smo, and activated Smo signals intracellularly to block Gli^R/Ci^R production and promotes Gli^A/Ci^A activation. The fundamentals of *Drosophila* and mammalian Hh signal transduction pathways are very similar, although major differences can be found in several regulatory steps including the utilization of primary cilia for Hh signal transduction in mammals but not in *Drosophila* (Huangfu & Anderson, 2006).

The primary cilium is a microtubule-based membrane protrusion and antenna-like cellular structure (Goetz & Anderson, 2010). Genetic screens in mice identified multiple intraflagellar transport proteins critical for appropriate Hh signaling (Garcia-Garcia et al., 2005; Huangfu & Anderson, 2005). The major pathway components including Ptc, Smo, and Gli proteins are present in the primary cilium in a manner regulated by Hh signaling. Hh promotes ciliary exit of Ptc but ciliary accumulation of Smo (Corbit et al., 2005; Rohatgi, Milenkovic, & Scott, 2007). Hh also stimulates the accumulation of Gli proteins at the cilia tip, a step likely reflecting Gli^A formation. Disruption of primary cilia impedes the formation of both Gli^R and Gli^A (Haycraft et al., 2005; Huangfu & Anderson, 2005; Liu, Wang, & Niswander, 2005). Thus, the primary cilium may function as a signaling center to orchestrate the molecular events leading to Gli processing in the absence of Hh as well as Gli activation in response to Hh although the exact biochemical mechanisms remain poorly understood.

In the canonical Wnt pathway, Wnt acts through a receptor complex consisting of Frizzle (FZ) family of seven transmembrane proteins related to Smo and lipoprotein receptor-related protein LRP5/6, which signal intracellularly to stabilize the transcription effector β-catenin (Fig. 2B) (MacDonald et al., 2009). In the absence of Wnt, β-catenin is degraded by the action of a destruction complex consisting of Axin, APC, and several kinases including CK1 and GSK3. Binding of Wnt to FZ and LRP5/6 leads to phosphorylation and activation of LRP5/6, which binds to and inhibits the Axin/APC destruction complex, leading to stabilization and nuclear translocation of β-catenin. In the nucleus, β-catenin binds the TCF family of transcription factor to regulate Wnt pathway target genes.

(B) Wnt/β-catenin pathway. In the "signaling off" state, β-catenin and its kinases GSK3 and CK1α form a destruction complex scaffolded by Axin and APC, leading to its phosphorylation and degradation. In the "signaling on" state, Wnt binds Fz and LRP5/6, leading to phosphorylation of LRP5/6 by GSK3 and CK1. LRP5/6 phosphorylation on the PPPSP motifs recruits the destruction complex through Axin, leading to a blockage of β-catenin phosphorylation. As a consequence, β-catenin accumulates, translocate to the nucleus, and binds TCF to turn on Wnt responsive genes.

Hence, in both Hh and Wnt pathways, the absence of pathway ligands allows proteolytic degradation of the pathway transcription factor/effector by large "destruction complexes," whereas ligand stimulation stabilize the pathway transcription factor/effector by inhibiting the destruction complexes. In addition, Hh and Wnt pathways share many common pathway components including GSK3, CK1, and E3 ubiquitin ligase SCF-Slimb/βTRCP.

2. CK1 IN THE REGULATION OF Hh PATHWAY

CK1 was initially identified as a negative regulator of Hh signaling at the level of transcription factor Ci by genetic studies in *Drosophila* and genome-wide RNAi screen in *Drosophila* cultured cells (Jia et al., 2005; Lum et al., 2003; Price & Kalderon, 2002). Overexpression of a dominant negative form of CK1ε resulted in an ectopic expression of an Hh target gene *decapentaplegic* (*dpp*) in wing imaginal discs and duplication of adult wings (Jia et al., 2005), phenocopying ectopic Hh expression (Basler & Struhl, 1994). However, further genetic study revealed that CK1 also plays a positive role in the Hh pathway at the level of Smo (Jia, Tong, Wang, Luo, & Jiang, 2004). A kinome-wide RNAi screen identified CK1α as a positive regulator of Hh signaling in mammalian cultured cells (Evangelista et al., 2008), and a subsequent study demonstrated that CK1α regulates Hh signaling by phosphorylating and activating Smo (Chen et al., 2011). In *Drosophila*, CK1 also exerts its positive role by phosphorylating Ci to stabilize the Ci^A and by phosphorylating fused (Fu) to promote its activation (Shi et al., 2014; Zhou & Kalderon, 2011). Hence CK1 plays a dual role in Hh signaling and acts at multiple levels.

2.1 Regulation of Ci/Gli Processing by CK1

An initial hint that CK1 plays a role in Hh signaling came from the study on Ci regulation in *Drosophila*. Ci existed in a full-length form (Ci^F) and a C-terminally truncated repressor form (Ci76 or Ci^R) that is derived from Ci^F by proteolytic processing in the absence of Hh signal (Aza-Blanc, Ramirez-Weber, Laget, Schwartz, & Kornberg, 1997; Methot & Basler, 1999). Ci processing requires protein kinase A (PKA) and the F-box protein Slimb, the *Drosophila* Ortholog of mammalian β-TRCP that functions as a substrate recognition component of the CUL1-based SCF E3 ubiquitin ligase complex (Jiang & Struhl, 1995, 1998; Li, Ohlmeyer, Lane, & Kalderon, 1995; Price & Kalderon, 1999; Wang, Wang, & Jiang, 1999). Genetic inactivation of either PKA or Slimb in wing imaginal discs resulted in accumulation of Ci^F at the expense of Ci^R and ectopic activation of *dpp*, leading to wing duplication, a phenotype similar to that caused by ectopic Hh expression (Jiang & Struhl, 1995, 1998; Li et al., 1995; Pan & Rubin, 1995). An important feature of the SCF family of E3s is that they only recognize substrates after the substrates are phosphorylated, thus providing a link between kinase-mediated phosphorylation and ubiquitin/proteasome-mediated proteolysis (Spencer, Jiang, & Chen, 1999; Winston et al., 1999). Indeed, PKA can directly phosphorylate five Ser/Thr residues in the C-terminal half of Ci, and mutating any one of the three PKA sites (sites 1–3) abolishes Ci processing (Fig. 3A) (Price & Kalderon, 1999; Wang et al., 1999). However, phosphorylation of Ci by PKA alone does not confer recognition by Slimb, implying that additional phosphorylation events are required for Ci processing. Further genetic studies identified CK1 and Shaggy (Sgg), the *Drosophila* GSK3, as two kinases that act in conjunction with PKA to promote Ci processing (Jia et al., 2002, 2005; Price & Kalderon, 2002). Inactivation of either CK1α or CK1ε/DBT alone did not significantly affect Ci processing but their combined inactivation resulted in a blockage of Ci processing in wing discs (Jia et al., 2005). PKA, GSK3, and CK1 phosphorylate Ci sequentially at three S/T clusters, with PKA serving as the priming kinase for GSK3 and CK1 (Fig. 3A), and these phosphorylation events create docking sites for SCF^{Slimb} (Jia et al., 2005; Smelkinson & Kalderon, 2006). Mutating individual phosphorylation clusters diminished the production of Ci^R in vivo, suggesting that these phosphorylation events act in concert to promote Ci processing (Jia et al., 2002, 2005). A more careful analysis revealed that an extended phosphorylation cluster primed by PKA sites 1 and 2 (SpTpYYGSp) closely resembles the Slimb/β-TRCP-binding site consensus $DSpGX_{2-4}Sp$ and provides the primary contact site for Slimb (Fig. 3A) (Smelkinson, Zhou, & Kalderon, 2007). In addition, multiple phosphorylation events may recruit two copies of SCF^{Slimb} complex that bind Ci simultaneously (Smelkinson et al., 2007). Thus, efficient Slimb binding and Ci processing require coordinated phosphorylation at multiple sites by these three kinases, which may render the regulation of Ci processing very sensitive to Hh. Indeed, low levels of Hh signaling appear to be sufficient to block Ci processing (Strigini & Cohen, 1997). Similar phosphorylation events mediated by PKA, GSK3, and CK1 regulate proteolysis of Gli2 and Gli3 by recruiting β-TRCP (Bhatia et al., 2006; Pan, Bai, Joyner, & Wang, 2006; Tempe, Casas, Karaz, Blanchet-Tournier, & Concordet, 2006; Wang, Fallon, & Beachy, 2000; Wang & Li, 2006; Wen et al., 2010). In contrast to Gli3 where

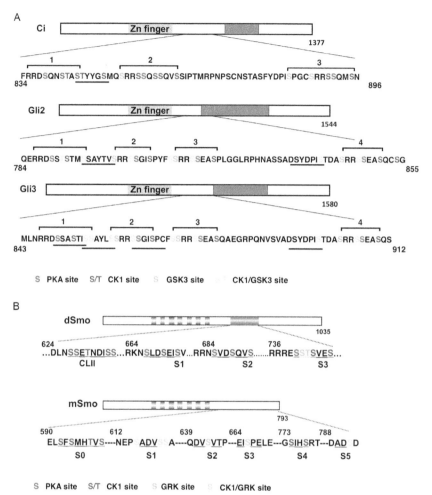

Fig. 3 CK1 phosphorylates both Ci/Gli and Smo in the Hh pathway. (A) Diagrams of Ci, mouse Gli2, and human Gli3 showing the PKA/GSK3/CK1 phosphorylation clusters in Ci/Gli. Putative Slimb/β-TRCP-binding sites in Ci/Gli are underlined. *Gray* and *blue boxes* denote a Zn-finger DNA-binding domain and a transactivation domain, respectively. Phosphorylation sites for the indicated kinases are color coded. (B) Diagrams showing the Smo phosphorylation sites in *Drosophila* (*top*) and mammalian Smo (*bottom*). *Gray boxes* indicate transmembrane helixes. The *red box* in *Drosophila* Smo C-tail denotes the SAID domain. The phosphorylation clusters are *underlined* and phosphorylation sites for the indicated kinases are color coded.

β-TRCP-mediated proteolysis leads to partial degradation and therefore the production of Gli^R, Gli2 proteolysis often leads to complete degradation of the protein, consistent with the genetic studies suggesting that Gli3 is the major contributor of Gli^R (Hui & Angers, 2011).

In the absence of Hh, Ci forms a complex with several Hh pathway components including the kinesin-like protein Costal2 (Cos2) and the S/T kinase fused (Fu) (Robbins et al., 1997; Sisson, Ho, Suyama, & Scott, 1997). In addition to restricting Ci nuclear translocation, the Cos2/Fu complex also interacts with PKA, GSK3, and CK1 to facilitate Ci phosphorylation and processing (Fig. 2A) (Zhang et al., 2005). In the presence of Hh, the ability of Cos2/Fu complex to recruit PKA, GSK3, and CK1 is compromised (Shi, Li, Jia, & Jiang, 2011; Zhang et al., 2005), leading to diminished Ci phosphorylation by these kinases; as a consequence, Ci processing and the production of Ci^R are blocked. Exactly how Hh inhibits Cos2-Fu-Ci-kinase complex formation is still not understood. Ci also forms a complex with suppressor of fused (Sufu), another inhibitory component of the Hh pathway, which impedes Ci nuclear translocation and inhibits Ci^A activity in the nucleus (Methot & Basler, 2000; Ohlmeyer & Kalderon, 1998; Wang, Amanai, Wang, & Jiang, 2000).

In mammals, Gli proteins also interact with Sufu and Kif7, the mammalian homolog of Cos2. In contrast to *Drosophila* where Cos2 plays a major role in inhibiting Ci in the absence of Hh, Sufu plays a major role whereas Kif7 plays a minor role in restricting Hh pathway activity in mammals (Cheung et al., 2009; Cooper et al., 2005; Endoh-Yamagami et al., 2009; Kise, Morinaka, Teglund, & Miki, 2009; Liem, He, Ocbina, & Anderson, 2009; Svard et al., 2006). Both Sufu and Kif7 are required for Gli3 processing (Chen et al., 2009; Cheung et al., 2009; Endoh-Yamagami et al., 2009; Humke, Dorn, Milenkovic, Scott, & Rohatgi, 2010; Kise et al., 2009; Liem et al., 2009; Wang, Pan, & Wang, 2010), raising a possibility that they may also regulate Gli3 phosphorylation. Indeed, Sufu can simultaneously bind GSK3 and Gli3 and thus recruit GSK3 to phosphorylate Gli3 (Kise et al., 2009). Shh signaling may inhibit Gli phosphorylation by dissociating Sufu–Gli–kinase complex (Humke et al., 2010; Kise et al., 2009; Tukachinsky, Lopez, & Salic, 2010). It remains to be determined whether Sufu/Kif7 recruit PKA and CK1 to promote Gli3 phosphorylation and processing.

2.2 Regulation of Smo Activation by CK1 in *Drosophila*

Smo belongs to the GPCR family of transmembrane proteins and is an obligatory signal transducer of the canonical Hh signaling pathway in both

Drosophila and vertebrates (Jiang & Hui, 2008). Activation mutations in *Smo* have been found in basal cell carcinoma (BCC) and medulloblastoma (Jiang & Hui, 2008; Xie et al., 1998). Hence, Smo has emerged as a prominent target for cancer therapeutics (Rubin & de Sauvage, 2006). Indeed, the FDA has proved the use of Vismodegib, a potent synthetic oral Smo inhibitor, for the treatment of advanced BCC.

In *Drosophila*, Hh induces cell surface accumulation and phosphorylation of Smo (Denef, Neubuser, Perez, & Cohen, 2000). Identification of Smo kinases came from unexpected findings that both PKA and CK1 play dual roles in Hh signaling in *Drosophila* embryos and imaginal discs (Jia et al., 2004; Ohlmeyer & Kalderon, 1997). Gain of PKA function promotes Smo accumulation and Hh pathway activation, whereas loss of PKA function blocks Hh-induced Smo accumulation as well as high levels of Hh signaling (Jia et al., 2004). Similarly, inactivation of CK1α/ε by RNAi blocks Hh-induced Smo accumulation and high levels of Hh signaling activity (Jia et al., 2004). Biochemical studies demonstrated that PKA and CK1 sequentially phosphorylate three clusters of S/T residues in the Smo carboxyl-terminal cytoplasmic tail (C-tail), with PKA serving as the priming kinase for CK1 phosphorylation (Fig. 3B) (Apionishev, Katanayeva, Marks, Kalderon, & Tomlinson, 2005; Jia et al., 2004; Zhang, Williams, Guo, Lum, & Beachy, 2004). Phospho-deficient Smo variants with either PKA or CK1 sites mutated to nonphosphorable Ala exhibit reduced cell surface expression and diminished Hh signaling activity whereas phospho-mimetic Smo variants with both PKA or CK1 sites converted to acidic residues exhibit increased cell surface level and constitutive activity (Apionishev et al., 2005; Jia et al., 2004; Zhang et al., 2004). Interestingly, increasing the number of phospho-mimetic clusters in Smo results in a progressive elevation of Smo activity, suggesting that graded Hh signals may induce different levels of Smo activity through differential phosphorylation (Jia et al., 2004). Indeed, using a phospho-specific antibody to monitor Smo phosphorylation, a later study showed that Hh induces Smo phosphorylation in a dose-dependent manner (Fan, Liu, & Jia, 2012).

Mechanistically, PKA/CK1-mediated phosphorylation promotes Smo activity via at least two paralleled mechanisms: regulating its subcellular localization and conformation. Two recent studies revealed that Hh-induced phosphorylation promotes Smo cell surface expression by inhibiting ubiquitination-mediated endocytosis and degradation of Smo (Li et al., 2012; Xia, Jia, Fan, Liu, & Jia, 2012). In the absence of Hh, Smo is both mono- and polyubiquitinated at multiple Lys residues in the carboxyl-terminal C-tail, leading to Smo endocytosis and degradation by both lysosome- and proteasome-dependent mechanisms. In addition, Smo endocytosis is also promoted by the binding of Kurtz (Krz), the *Drosophila* nonvisual arrestin (Li et al., 2012; Molnar et al., 2011). Hh inhibits Smo ubiquitination and attenuates Smo/Krz interaction, thereby stabilizing Smo on the cell surface (Li et al., 2012; Xia et al., 2012). How Hh-induced phosphorylation of Smo inhibits its ubiquitination has not been resolved. One possibility is that PKA/CK1-mediated phosphorylation of Smo C-tail may inhibit the binding of one or more Smo E3 ubiquitin ligases. The identification of Smo E3s may allow a direct test of this model.

In addition to regulating Smo trafficking, PKA/CK1-mediated phosphorylation also controls the conformation of Smo C-tail (Zhao, Tong, & Jiang, 2007). In the absence of Hh, Smo adopts a closed conformation in which the Smo C-terminus folds back to form salt bridge with multiple basic clusters (mostly Arg) in the SAID domain (Smo autoinhibition domain), which is located in the middle region of the Smo C-tail (Zhao et al., 2007). Hh-induced phosphorylation by PKA/CK1 brings negative charges to neutralize the positive charges of the nearby Arg motifs in the SAID domain, which disrupts the intramolecular electrostatic interaction that maintains the closed conformation and promotes an open conformation and clustering of Smo C-tails. The pairing of positive and negative regulatory elements offers a more precise regulation of Smo activity in response to graded Hh signals as increasing phosphorylation may gradually neutralize the negative effect of multi-Arg clusters, leading to a progressive change in Smo cell surface accumulation, conformation, and activity (Zhao et al., 2007). Indeed, decreasing the number of functional Arg motifs has the same effect of increasing the number of phospho-mimetic mutations, both leading to a progressive increase of Smo activity (Jia et al., 2004; Zhao et al., 2007).

Phospho-mimetic mutations in the three PKA/CK1 phosphorylation clusters render Smo constitutively active but fail to confer full pathway activity (Jia et al., 2004), suggesting that Smo activation may involve additional mechanisms. A mass spec analysis identified many phosphorylation sites in Smo C-tail other than the three PKA/CK1 phosphorylation clusters (Zhang et al., 2004). Subsequent studies identified both casein kinase 2 (CK2) and G protein coupled receptor kinase 2 (Gprk2 or GRK2) as additional Smo kinases (Chen et al., 2010; Jia et al., 2010). Interestingly, GRK2 promotes high-level Hh signaling through both kinase-dependent and kinase-independent mechanisms (Chen et al., 2010; Cheng, Maier, Neubueser, & Hipfner, 2010). GRK2 phosphorylates Smo C-terminal tail at Ser741/Thr742, which is facilitated by PKA/CK1-mediated phosphorylation at adjacent Ser residues (Chen et al., 2010). In addition, GRK2 forms a dimer and binds Smo to stabilize its open conformation and promote the dimerization of Smo C-tail (Chen et al., 2010). A recent study identified Gish/CK1γ as another Smo kinase that fine-tunes Hh pathway activity as loss of Gish affects the expression of high threshold Hh target genes (Li et al., 2016). Gish is a membrane-associated kinase due to its C-terminal lipid modification and membrane association appears to be critical for its function the Hh pathway. Mechanistically, Gish interacts with Smo and phosphorylates a membrane-proximal S/T cluster (CLII) on the Smo C-tail. Interestingly, interaction between Gish/Smo and subsequent phosphorylation of CLII are stimulated by Hh and promoted by PKA-mediated phosphorylation of the distal region on the Smo C-tail (Li et al., 2016). Hence the Smo kinases fall into two categories: PKA and CK1α/ε are the primary kinases whose phosphorylation of Smo at multiple sites is critical for Smo activation, whereas CK2, Gprk2, and Gish/CK1γ are the secondary kinases whose phosphorylation of Smo is promoted by the primary phosphorylation to further boost Hh pathway activity.

2.3 A Conserved Role of CK1 in Mammalian Smo Activation

Mammalian Smo (mSmo) and *Drosophila* Smo (dSmo) diverge significantly in their primary sequences. For example, mSmo does not contain three PKA/CK1 phosphorylation clusters found in dSmo (Fig. 3B). In addition, mSmo traffics into primary cilia to transduce the Hh signal, whereas the primary cilium is dispensable for *Drosophila* Hh signal transduction (Corbit et al., 2005; Rohatgi et al., 2007). These observations have led to the proposal that mSmo and dSmo are regulated by fundamentally distinct mechanisms (Huangfu & Anderson, 2006; Varjosalo, Li, & Taipale, 2006). However, subsequent studies revealed striking similarities in the activation mechanism between dSmo and mSmo: both dSmo and mSmo are regulated by multisite phosphorylation in a dose-dependent manner and phosphorylation regulates both their subcellular localization and conformation (Chen et al., 2011; Zhao et al., 2007). Furthermore, both CK1α and CK1γ are involved in the phosphorylation and activation of mSmo (Chen et al., 2011; Li et al., 2016).

A kinome RNAi screen identified CK1α as a positive regulator of mammalian Hh pathway in cultured cells (Evangelista et al., 2008). Several studies also suggested that GRK2 (ortholog of *Drosophila* Gprk2) positively regulate mammalian Hh pathways and GRK2 promotes mSmo phosphorylation (Chen et al., 2004; Evangelista et al., 2008; Meloni et al., 2006). Further study demonstrated that CK1α and GRK2 bind mSmo in response to Hh stimulation and phosphorylate mSmo C-tail at six Ser/Thr clusters (S0–S5; Fig. 3B) (Chen et al., 2011). Studies using both cultured mammalian cells and chick neural tubes suggested that multiple CK1α/GRK2 phosphorylation sites regulate mSmo activity in a dose-dependent manner, with the two membrane-proximal clusters (S0 and S1) playing a major role (Chen et al., 2011). As is the case for dSmo (Fan et al., 2012; Jia et al., 2004), mSmo phosphorylation is regulated in a dose-dependent manner with increasing amounts of Hh inducing a progressive increase of mSmo phosphorylation, suggesting that Hh gradient is translated into mSmo phosphorylation and activity gradient (Chen et al., 2011).

CK1α/GRK2-mediated phosphorylation promotes mSmo ciliary localization as phosphorylation deficient form of mSmo failed to accumulate to the primary cilium in response to Hh stimulation or oncogenic mutations whereas phospho-mimetic mutations resulted in constitutive mSmo ciliary accumulation (Chen et al., 2011). How phosphorylation promotes mSmo ciliary localization remains an unresolved issue but it correlates with enhanced binding of β-arrestin, which is thought to link mSmo to the kinesin-II motor responsible for anterograde ciliary transport (Chen et al., 2011; Kovacs et al., 2008). Besides regulating mSmo ciliary localization, Hh induces a conformational switch that results in dimerization/oligomerization of mSmo C-tail (Zhao et al., 2007). Hh-induced conformational switch is also governed by CK1α/GRK2-mediated multisite phosphorylation of mSmo C-tail in a similar manner to dSmo (Chen et al., 2011). Interestingly, cyclopamine traps ciliary localized mSmo in an unphosphorylated form that adopts an inactive conformation whereas ciliary localized mSmo in response to Hh or Smo agonists is phosphorylated and thus adopts an active conformation (Chen et al., 2011). Hence, Smo phosphorylation is a more faithful readout for pathway activation than Smo ciliary localization and can serve as a biomarker for cancers caused by deregulated Hh pathway activation.

Hh enhances the recruitment of both CK1α and GRK2 to mSmo, which may explain the signal-stimulated phosphorylation (Chen et al., 2011). In the absence of Hh, the kinase-binding pockets appear to be masked when mSmo C-tail adopts a closed conformation. Shh stimulates CK1α-binding to a juxtamembrane site, likely by inducing a conformational change in the transmembrane helices, to initiate phosphorylation and conformational change of mSmo C-tail. This further increases the binding of CK1α/GRK2 to the mSmo C-tail, forming a feed-forward mechanism to increase mSmo phosphorylation (Chen et al., 2011). Interestingly, CK1α accumulates to primary cilia in response to Hh stimulation, which may explain, at least in part, why phosphorylation of mSmo is more effective in primary cilia (Chen et al., 2011).

A recent study revealed that exogenously expressed CK1γ is localized to the primary cilia of NIH3T3 cells in a manner depending on its membrane association (Li et al., 2016). Using a phospho-specific antibody against one of the major CK1 sites (S1) on mSmo to monitor mSmo phosphorylation, Li et al. found that transfection of wild type, but not a cytosolic form (CK1γ−ΔC) of CK1γ stimulated S1 phosphorylation in a manner enhanced by Shh treatment whereas transfection of dominant negative forms of CK1γ attenuated Hh-induced S1 phosphorylation (Li et al., 2016). In addition, overexpression of wild-type CK1γ, but not CK1γ−ΔC, increased the expression of *Gli-luc* reporter gene, while the dominant negative forms attenuated Hh-induced *Gli-luc* expression (Li et al., 2016). The ability of CK1γ to promote mSmo phosphorylation and activity appears to depend on primary cilia as blockage of ciliogenesis abolished the effect of CK1γ on mSmo phosphorylation and activation (Li et al., 2016). Taken together, these observations suggest that CK1γ plays a conserved role in the Hh pathway by directly phosphorylating Smo and boosting its activity after it accumulates on the cell surface in *Drosophila* or on the primary cilia in mammals.

2.4 CK1 Positively Regulates Hh Signaling Downstream of Smo

Besides phosphorylating Smo to promote Hh pathway activation, CK1 plays additional positive roles downstream of Smo because inactivation of CK1 inhibits Hh pathway activity elicited by expressing a phospho-mimetic and constitutively active form of Smo in which both the PKA and CK1 sites are converted into acidic residues (Shi et al., 2014; Zhou & Kalderon, 2011). In *Drosophila*, the Fu kinase acts downstream of Smo to convert Ci^F into an activator form (Ci^A) in response to high levels of Hh signal (Ohlmeyer & Kalderon, 1998). Fu forms a stoichiometric complex with Cos2 and is recruited to the dimerized intracellular tail of activated Smo, leading to its own dimerization, phosphorylation, and activation (Shi et al., 2011; Zhang et al., 2011; Zhou & Kalderon, 2011). Phosphorylation of Fu occurs at multiple Ser/Thr residues in the activation loop of its kinase domain as well as in its C-terminal regulatory domain (Shi et al., 2011; Zhou & Kalderon, 2011). Phosphorylation of the Fu activation loop and C-terminal regulatory domain is mediated by Fu trans-autophosphorylation, which may prime their further phosphorylation by CK1 although direct evidence for phosphorylation of Fu by CK1 is still lacking (Shi et al., 2011; Zhou & Kalderon, 2011). Phosphorylation of Fu activation loop appears to promote Fu activation in a dose-dependent manner: increasing levels of Hh induced increasing levels of Fu phosphorylation, which in turn resulted in increasing levels of Fu activity (Shi et al., 2011). Mutating a CK1 consensus site in the activation loop severely compromised Fu activity, whereas mutating a cluster of CK1 consensus sites in the regulatory domain also attenuated Fu activity (Zhou & Kalderon, 2011). Taken together, these studies suggest that CK1 may positively regulate Hh pathway by phosphorylating both the kinase and regulatory domains of Fu.

Ci^A is a labile form of Ci that is degraded by a Cul3-based E3 ubiquitin ligase containing the BTB family protein HIB (also called Rdx) (Kent, Bush, & Hooper, 2006; Zhang, Zhang, et al., 2006; Zhang et al., 2009). Interestingly, *hib* expression is induced by Hh signaling in both *Drosophila* embryos and imaginal discs, thus forming a negative feedback loop to attenuate Hh pathway activity (Kent et al., 2006; Zhang, Zhang, et al., 2006). SPOP, which is the vertebrate homolog of HIB, may play an analogous role in fine-tuning Hh signaling by degrading Gli2/3 proteins (Chen et al., 2009; Wang et al., 2010; Wen et al., 2010). The degradation of CiA/GliA by HIB/SPOP is likely to be under tight control as excessive degradation can lead to premature loss of Hh pathway activity. Indeed, a recent study showed that Hh stimulates Ci phosphorylation by CK1 at multiple Ser/Thr-rich degrons to inhibit its recognition by HIB (Shi et al., 2014). In Hh-receiving wing disc cells, reduction of CK1 activity accelerated HIB-mediated degradation of Ci^A, leading to premature loss of pathway activity (Shi et al., 2014). Furthermore, this study showed that Gli^A is regulated by CK1 in a similar fashion and that CK1 acts downstream of Sufu to promote Shh signaling. Hence, depending on availability of the Hh ligand, CK1 can either inhibit or activate Ci by phosphorylating distinct sites.

3. CK1 IN Wnt SIGNALING

Similar to its role in the Hh pathway, CK1 also plays both positive and negative roles in the Wnt pathway by phosphorylating different pathway components. The first indication that CK1 family kinase regulates Wnt signaling is that injection of CK1ε mRNA into the ventral side of Xenopus embryos led to dorsalization and axial duplication similar to injection of Wnts (Peters, McKay, McKay, & Graff, 1999; Sakanaka, Leong, Xu, Harrison, & Williams, 1999). It was further shown that CK1ε forms a complex with Disheveled (Dvl) and Axin and positively regulates Wnt signaling by phosphorylating Dvl on multiple sites (McKay, Peters, & Graff, 2001; Peters et al., 1999; Sakanaka et al., 1999). However, the physiological relevance of Dvl phosphorylation by CK1 has not been clearly established (Penton, Wodarz, & Nusse, 2002; Strutt, Price, & Strutt, 2006; Yanfeng et al., 2011). Other CK1 family members including CK1α and CK1γ have also been implicated in Wnt signaling (see later). Although CK1ε and CK1γ mainly play a positive role in Wnt signaling through their influence on LRP6 and Dvl, whereas CK1α mainly plays a negative role through its role in the regulation of β-catenin degradation, a genetic study in *Drosophila* also reveals a negative role for CK1ε and a positive role for CK1α in a genetic sensitized background (Zhang, Jia, et al., 2006).

3.1 Regulation of β-Catenin Destruction: A Negative Role of CK1 in Wnt Signaling

Regulation of the cytoplasmic β-catenin protein abundance is a hallmark of the canonical Wnt signal transduction pathway. In the absence of Wnt, β-catenin is phosphorylated by CK1α and GSK3, which targets it for ubiquitin/proteasome-mediated degradation. β-catenin contains a degron, $DS_{33}GIHS_{37}GAVT_{41}QAPS_{45}$, in its N-terminal region (Fig. 4A), and mutations of the S/T residues stabilized β-catenin in human cancers (Polakis, 2000). Two studies showed that phosphorylation of S_{45} by CK1α primes GSK3 to phosphorylation T_{41}, S_{37}, and S_{33} consecutively (Amit et al., 2002; Liu et al., 2002), which creates a docking site for the SCF family of E3 ubiquitin ligase containing the F-box protein Slimb/β-TRCP (Jiang & Struhl, 1998; Winston et al., 1999), leading to its ubiquitination and degradation of by the proteasome.

Both CK1α and GSK3 are present in the destruction complex containing the scaffold proteins Axin and APC (Amit et al., 2002). The Axin/APC also binds β-catenin, thus bringing the substrate and kinases in close proximity to facilitate β-catenin phosphorylation (Liu et al., 2002). Interestingly, APC is also phosphorylated by CK1 (in this case, CK1ε and CK1δ) as well as GSK3 at several sites within the 20 amino acid repeats that mediate its binding to β-catenin, and these phosphorylation events greatly enhance its binding to β-catenin (Ha, Tonozuka, Stamos, Choi, & Weis, 2004; Xing et al., 2004),

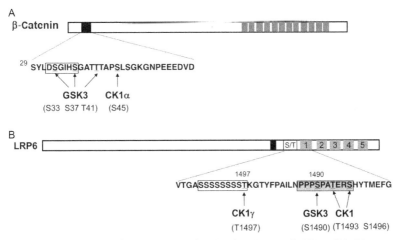

Fig. 4 Phosphorylation of β-catenin and LRP6 by CK1 and GSK3. (A) Diagram of β-catenin with the CK1 and GSK3 phosphorylation cluster shown underneath. β-TRCP-binding site is boxed. *Gray boxes* denote the Armadillo (Arm) repeats and the *black box* indicates the destruction box. (B) Diagram of LRP6 with the membrane-proximal S/T-rich sequence and the first PPPSP motif shown underneath. *Filled, open,* and *gray boxes* indicate the transmembrane domain, S/T-rich domain, and multiple PPPSP motifs.

and its ability to downregulate β-catenin (Rubinfeld, Tice, & Polakis, 2001). Phosphorylation of APC by CK1ε is greatly enhanced by the presence of an Axin fragment containing the β-catenin and CK1-binding sites (Ha et al., 2004; Rubinfeld et al., 2001). Thus, Axin functions as a scaffold in the destruction complex to facilitate phosphorylation of both β-catenin and APC.

Axin is also phosphorylated by CK1, but the biological significance of this phosphorylation event is unclear (Gao, Seeling, Hill, Yochum, & Virshup, 2002). A recent study reveals that APC promotes efficient Axin multimerization through multivalent interactions and that phosphorylation of the R2/B region of APC by GSK3/CK1 induces a conformational change in APC to release APC's Arm-repeats from Axin, allowing the release of phosphorylated β-catenin to the SCF$^{β-TRCP}$ E3 ligase, leading to ubiquitination and degradation of β-catenin (Pronobis, Rusan, & Peifer, 2015).

3.2 Regulation of LRP6 Phosphorylation: A Positive Role of CK1 in Wnt Signaling

The Wnt receptor complex contains the Fz family seven transmembrane proteins and the coreceptor LRP5/6. In the presence of Wnt, the destruction complex is recruited to activated LRP5/6 through its binding to Axin, leading to inhibition of β-catenin phosphorylation, β-catenin stabilization in the cytoplasm, and Wnt pathway activation (Mao et al., 2001; Tamai et al., 2004). LRP6 and its *Drosophila* homolog arrow contain five PPPSP motifs that mediate Axin binding when phosphorylated in response to Wnt (Fig. 4B) (Tamai et al., 2004; Zeng et al., 2005). Phosphorylation of PPPSP is mediated by GSK3, followed by CK1-mediated phosphorylation at one or more adjacent sites C-terminal to the PPPSP motifs. Multiple CK1 isoforms including CK1α, CK1ε/δ, and CK1γ have been implicated in the phosphorylation of the sites adjacent to the PPPSP motifs (Davidson et al., 2005; Zeng et al., 2005). Phosphorylation of LRP6 by both GSK3 and CK1 is required for Axin binding and phosphorylated PPPSP motifs inhibit GSK3-mediated phosphorylation of β-catenin within the destruction complex, leading to β-catenin stabilization and Wnt pathway activation (Piao et al., 2008; Wu, Huang, Garcia Abreu, & He, 2009; Zeng et al., 2005). In addition, CK1γ is involved in the phosphorylation of T1497, which is adjacent to a S/T cluster (Fig. 4) (Davidson et al., 2005). Similar to its binding to Smo, CK1γ binds LRP6 depending on its membrane association through its C-terminal palmitoylation (Davidson et al., 2005).

Wnt-induced LRP6 phosphorylation requires the Fz receptor and its downstream partner Dvl that promotes the formation of a large multimeric protein complex called LRP6 signalosome (Bilic et al., 2007; Zeng et al., 2008). Further studies reveal that Wnt3a induces the formation of PI(4 5)P2 to promote LRP6 phosphorylation and that Amer1/WTX couples the Wnt3a-induced PI(4 5)P2 to LRP6 phosphorylation (Pan et al., 2008; Tanneberger et al., 2011). Several other proteins including the transmembrane protein 198 and p120-catenin have also been implicated in the regulation of LRP6 phosphorylation (Casagolda et al., 2010; Liang et al., 2011). Furthermore, Wnt-induced LRP6 phosphorylation occurs in an acidic vesicular compartment depending on the action of prorenin receptor (PRR) and vacuolar ATPase (Cruciat et al., 2010).

Like in the Hh pathway, CK1 also exerts positive influence on Wnt signaling at multiple levels. In addition to regulating the phosphorylation of LRP6 and Dvl, CK1 also phosphorylates TCF3 to promote its interaction with β-catenin (Lee, Salic, & Kirschner, 2001).

4. REGULATION OF CK1 IN Hh AND Wnt SIGNALING

The involvement of CK1 in both Hh and Wnt signal transduction pathways and the findings that CK1 regulates each pathway at multiple levels to exert both positive and negative influence have raised import questions of how phosphorylation of CK1 substrates is regulated and how Hh and Wnt signaling achieve pathway specificity despite being regulated by a similar set of kinases. A prominent feature of Hh and Wnt signaling is that the kinases and their substrates are present in large protein complexes organized by pathway-specific scaffolding proteins (Fig. 2). In the Hh pathway, Ci and its kinases PKA, CK1, and GSK3 are present in protein complexes organized by Cos2 and Fu (Chen & Jiang, 2013; Zhang et al., 2005). Similarly in the Wnt pathway, β-catenin and its kinases CK1 and GSK3 form a complex scaffolded by Axin and APC (MacDonald et al., 2009). Therefore, to achieve pathway specificity, Hh and Wnt regulate different pools of kinases by eliciting interaction between their receptor complexes and the pathway-specific scaffold proteins, i.e., Smo/Cos2 interaction in the Hh pathway and LRP6/Axin interaction in the Wnt pathway (Jia, Tong, & Jiang, 2003; Jiang & Hui, 2008; MacDonald et al., 2009; Tamai et al., 2004). In other words, Hh and Wnt signaling are regulated by specific pool of kinases that are compartmentalized with their pathway effectors.

Another layer of CK1 regulation in the Hh and Wnt pathways is to employ different CK1 isoforms to phosphorylate distinct pathway components or even distinct sites on the same substrates. In this regard, it has been shown recently that the membrane-associated CK1 isoform CK1γ, but not the cytosolic isoform CK1α or CK1ε, is responsible for phosphorylating a membrane-proximal cluster of S/T residues on the Smo C-tail to promote high-level Hh pathway activity (Li et al., 2016). Similarly, CK1γ has been shown to phosphorylate a membrane-proximal site (T1497) on the LRP6 intracellular domain to modulate the activity of Wnt signaling (Davidson et al., 2005), and in *Drosophila*, CK1γ is the most potent CK1 isoform that phosphorylates Arrow, the *Drosophila* homolog of LRP6, likely due to its membrane association (Zhang, Jia, et al., 2006). On the other hand, CK1α is the major CK1 isoform responsible for β-catenin phosphorylation and degradation. However, both CK1α and CK1ε are involved in the regulation of Ci phosphorylation and degradation in the Hh way (Jia et al., 2005). Interestingly, Hh switches the CK1α/ε substrate from Ci to Smo, thus converting CK1α and CK1ε from negative to positive regulators of the pathway (Jia et al., 2004; Zhang et al., 2005). This conversion is achieved, at least in part, by regulating kinase/substrate interaction because Hh signaling inhibits the formation of Cos2-Ci-kinase complex but promotes the formation of Cos2–Smo–kinase complexes (Li, Ma, Wang, & Jiang, 2014; Zhang et al., 2005).

Although it is generally thought that CK1 activity is not regulated but rather its substrate accessibility is regulated in the Hh and Wnt pathways, a recent study reveals a novel mechanism of CK1 regulation by DDX3, which belongs to a family of ATP-dependent DEAD-box RNA helicases (Cruciat et al., 2013). DDX3 was identified in a genome-wide siRNA screen for novel Wnt regulators in cultured mammalian cells (Cruciat et al., 2013). DDX3 directly binds to CK1ε in a manner stimulated by Wnt, and DDX3 stimulates CK1ε kinase activity and promotes phosphorylation of Dvl2 although the physiological relevance of Dvl2 phosphorylation has not been directly tested (Cruciat et al., 2013). The enzymatic activities of DDX3, i.e., ATP hydrolysis and RNA unwinding, appear to be dispensable for binding to CK1ε, suggesting that DDX3 activates CK1ε through an allosteric mechanism (Cruciat et al., 2013). Interestingly, mutations in DDX3 were found in the Wnt-subgroup of medulloblastoma, illustrating its physiological role in Wnt/β-catenin signaling in humans (Jones et al., 2012; Pugh et al., 2012). It remains to be determined whether DDX3 regulates other CK1 isoforms in vivo and whether other DDX family helicases are involved in CK1 regulation.

5. CONCLUSION

Multiple CK1 family members regulate both Hh and Wnt signaling, and in each pathway, CK1 exerts both positive and negative influences depending on the signaling status, the specific CK1 isoforms involved, and which pathway components CK1 phosphorylates. Although many relevant CK1 substrates have been identified and the specific phosphorylation sites on individual pathway components defined, not all the biological relevant sites have been determined. For example, it has shown that CK1 activates the Hh pathway at the level of Smo as well as downstream of Smo by phosphorylating Fu and Ci (Shi et al., 2014; Zhou & Kalderon, 2011); however, the precise role and relevant CK1 sites on Fu remain to be determined. In addition, the precise function of individual CK1 isoforms in the Hh and Wnt pathways remains to be clarified, especially in the case of LRP6 phosphorylation where multiple CK1 isoforms have been implicated (Davidson et al., 2005; Zeng et al., 2005). In the past, loss-of-function study mainly employed dominant negative forms of individual isoforms, which may invoke cross-regulation among different CK1 isoforms and thus may not be exclusively "isoform specific." In addition, redundancy among different

CK1 family members might have underscored the role of individual CK1 isoforms in certain signaling processes, which is particularly problematic for CK1γ because of the presence of three CK1γ family members in mammals. Recent advance in gene editing technology, especially the CRISPR/Cas9 technology, that offers an efficient way to knock out multiple genes in the same cells (Doudna & Charpentier, 2014), will undoubtedly facilitate the loss-of-function study of individual CK1 family members in different cellular and developmental contexts and help elucidating their roles in human diseases.

ACKNOWLEDGMENTS
This work is supported by grants from the National Institutes of Health (GM118063) and Welch foundation (I-1603). J.J. is a Eugene McDermott Endowed Scholar in Biomedical Science at the University of Texas Southwestern Medical Center.

REFERENCES
Amit, S., Hatzubai, A., Birman, Y., Andersen, J. S., Ben-Shushan, E., Mann, M., et al. (2002). Axin-mediated CKI phosphorylation of beta-catenin at Ser 45: A molecular switch for the Wnt pathway. *Genes & Development, 16*, 1066–1076.
Apionishev, S., Katanayeva, N. M., Marks, S. A., Kalderon, D., & Tomlinson, A. (2005). Drosophila smoothened phosphorylation sites essential for Hedgehog signal transduction. *Nature Cell Biology, 7*, 86–92.
Aza-Blanc, P., Ramirez-Weber, F., Laget, M., Schwartz, C., & Kornberg, T. (1997). Proteolysis that is inhibited by Hedgehog targets Cubitus interruptus protein to the nucleus and converts it to a repressor. *Cell, 89*, 1043–1053.
Basler, K., & Struhl, G. (1994). Compartment boundaries and the control of Drosophila limb pattern by hedgehog protein. *Nature, 368*, 208–214.
Bhatia, N., Thiyagarajan, S., Elcheva, I., Saleem, M., Dlugosz, A., Mukhtar, H., et al. (2006). Gli2 is targeted for ubiquitination and degradation by beta-TrCP ubiquitin ligase. *The Journal of Biological Chemistry, 281*, 19320–19326.
Bilic, J., Huang, Y. L., Davidson, G., Zimmermann, T., Cruciat, C. M., Bienz, M., et al. (2007). Wnt induces LRP6 signalosomes and promotes dishevelled-dependent LRP6 phosphorylation. *Science, 316*, 1619–1622.
Casagolda, D., Del Valle-Perez, B., Valls, G., Lugilde, E., Vinyoles, M., Casado-Vela, J., et al. (2010). A p120-catenin-CK1epsilon complex regulates Wnt signaling. *Journal of Cell Science, 123*, 2621–2631.
Chen, Y., & Jiang, J. (2013). Decoding the phosphorylation code in Hedgehog signal transduction. *Cell Research, 23*, 186–200.
Chen, Y., Li, S., Tong, C., Zhao, Y., Wang, B., Liu, Y., et al. (2010). G protein-coupled receptor kinase 2 promotes high-level Hedgehog signaling by regulating the active state of Smo through kinase-dependent and kinase-independent mechanisms in Drosophila. *Genes & Development, 24*, 2054–2067.
Chen, W., Ren, X. R., Nelson, C. D., Barak, L. S., Chen, J. K., Beachy, P. A., et al. (2004). Activity-dependent internalization of smoothened mediated by beta-arrestin 2 and GRK2. *Science, 306*, 2257–2260.

Chen, Y., Sasai, N., Ma, G., Yue, T., Jia, J., Briscoe, J., et al. (2011). Sonic Hedgehog dependent phosphorylation by CK1alpha and GRK2 is required for ciliary accumulation and activation of smoothened. *PLoS Biology, 9*, e1001083.

Chen, M. H., Wilson, C. W., Li, Y. J., Law, K. K., Lu, C. S., Gacayan, R., et al. (2009). Cilium-independent regulation of Gli protein function by sufu in Hedgehog signaling is evolutionarily conserved. *Genes & Development, 23*, 1910–1928.

Cheng, S., Maier, D., Neubueser, D., & Hipfner, D. R. (2010). Regulation of smoothened by Drosophila G-protein-coupled receptor kinases. *Developmental Biology, 337*, 99–109.

Cheong, J. K., & Virshup, D. M. (2011). Casein kinase 1: Complexity in the family. *The International Journal of Biochemistry & Cell Biology, 43*, 465–469.

Cheung, H. O., Zhang, X., Ribeiro, A., Mo, R., Makino, S., Puviindran, V., et al. (2009). The kinesin protein Kif7 is a critical regulator of Gli transcription factors in mammalian hedgehog signaling. *Science Signaling, 2*, ra29.

Cooper, A. F., Yu, K. P., Brueckner, M., Brailey, L. L., Johnson, L., McGrath, J. M., et al. (2005). Cardiac and CNS defects in a mouse with targeted disruption of suppressor of fused. *Development, 132*, 4407–4417.

Corbit, K. C., Aanstad, P., Singla, V., Norman, A. R., Stainier, D. Y., & Reiter, J. F. (2005). Vertebrate smoothened functions at the primary cilium. *Nature, 437*, 1018–1021.

Cruciat, C. M., Dolde, C., de Groot, R. E., Ohkawara, B., Reinhard, C., Korswagen, H. C., et al. (2013). RNA helicase DDX3 is a regulatory subunit of casein kinase 1 in Wnt-beta-catenin signaling. *Science, 339*, 1436–1441.

Cruciat, C. M., Ohkawara, B., Acebron, S. P., Karaulanov, E., Reinhard, C., Ingelfinger, D., et al. (2010). Requirement of prorenin receptor and vacuolar H^{+}-ATPase-mediated acidification for Wnt signaling. *Science, 327*, 459–463.

Davidson, G., Wu, W., Shen, J., Bilic, J., Fenger, U., Stannek, P., et al. (2005). Casein kinase 1 gamma couples Wnt receptor activation to cytoplasmic signal transduction. *Nature, 438*, 867–872.

Denef, N., Neubuser, D., Perez, L., & Cohen, S. M. (2000). Hedgehog induces opposite changes in turnover and subcellular localization of patched and smoothened. *Cell, 102*, 521–531.

Doudna, J. A., & Charpentier, E. (2014). Genome editing. The new frontier of genome engineering with CRISPR-Cas9. *Science, 346*, 1258096.

Endoh-Yamagami, S., Evangelista, M., Wilson, D., Wen, X., Theunissen, J. W., Phamluong, K., et al. (2009). The mammalian Cos2 homolog Kif7 plays an essential role in modulating Hh signal transduction during development. *Current Biology, 19*, 1320–1326.

Evangelista, M., Lim, T. Y., Lee, J., Parker, L., Ashique, A., Peterson, A. S., et al. (2008). Kinome siRNA screen identifies regulators of ciliogenesis and hedgehog signal transduction. *Science Signaling, 1*, ra7.

Fan, J., Liu, Y., & Jia, J. (2012). Hh-induced smoothened conformational switch is mediated by differential phosphorylation at its C-terminal tail in a dose- and position-dependent manner. *Developmental Biology, 366*, 172–184.

Gao, Z. H., Seeling, J. M., Hill, V., Yochum, A., & Virshup, D. M. (2002). Casein kinase I phosphorylates and destabilizes the beta-catenin degradation complex. *Proceedings of the National Academy of Sciences of the United States of America, 99*, 1182–1187.

Garcia-Garcia, M. J., Eggenschwiler, J. T., Caspary, T., Alcorn, H. L., Wyler, M. R., Huangfu, D., et al. (2005). Analysis of mouse embryonic patterning and morphogenesis by forward genetics. *Proceedings of the National Academy of Sciences of the United States of America, 102*, 5913–5919.

Goetz, S. C., & Anderson, K. V. (2010). The primary cilium: A signalling centre during vertebrate development. *Nature Reviews. Genetics, 11*, 331–344.

Gross, S. D., & Anderson, R. A. (1998). Casein kinase I: Spatial organization and positioning of a multifunctional protein kinase family. *Cellular Signalling, 10*, 699–711.

Ha, N. C., Tonozuka, T., Stamos, J. L., Choi, H. J., & Weis, W. I. (2004). Mechanism of phosphorylation-dependent binding of APC to beta-catenin and its role in beta-catenin degradation. *Molecular Cell, 15*, 511–521.

Haycraft, C. J., Banizs, B., Aydin-Son, Y., Zhang, Q., Michaud, E. J., & Yoder, B. K. (2005). Gli2 and gli3 localize to cilia and require the intraflagellar transport protein polaris for processing and function. *PLoS Genetics, 1*e, 53.

Huangfu, D., & Anderson, K. V. (2005). Cilia and Hedgehog responsiveness in the mouse. *Proceedings of the National Academy of Sciences of the United States of America, 102*, 11325–11330.

Huangfu, D., & Anderson, K. V. (2006). Signaling from Smo to Ci/Gli: Conservation and divergence of Hedgehog pathways from Drosophila to vertebrates. *Development, 133*, 3–14.

Hui, C. C., & Angers, S. (2011). Gli proteins in development and disease. *Annual Review of Cell and Developmental Biology, 27*, 513–537.

Humke, E. W., Dorn, K. V., Milenkovic, L., Scott, M. P., & Rohatgi, R. (2010). The output of Hedgehog signaling is controlled by the dynamic association between suppressor of fused and the Gli proteins. *Genes & Development, 24*, 670–682.

Jia, J., Amanai, K., Wang, G., Tang, J., Wang, B., & Jiang, J. (2002). Shaggy/GSK3 antagonizes Hedgehog signalling by regulating Cubitus interruptus. *Nature, 416*, 548–552.

Jia, H., Liu, Y., Xia, R., Tong, C., Yue, T., Jiang, J., et al. (2010). Casein kinase 2 promotes Hedgehog signaling by regulating both smoothened and Cubitus interruptus. *The Journal of Biological Chemistry, 285*, 37218–37226.

Jia, J., Tong, C., & Jiang, J. (2003). Smoothened transduces Hedgehog signal by physically interacting with Costal2/Fused complex through its C-terminal tail. *Genes & Development, 17*, 2709–2720.

Jia, J., Tong, C., Wang, B., Luo, L., & Jiang, J. (2004). Hedgehog signalling activity of smoothened requires phosphorylation by protein kinase A and casein kinase I. *Nature, 432*, 1045–1050.

Jia, J., Zhang, L., Zhang, Q., Tong, C., Wang, B., Hou, F., et al. (2005). Phosphorylation by double-time/CKIepsilon and CKIalpha targets cubitus interruptus for Slimb/beta-TRCP-mediated proteolytic processing. *Developmental Cell, 9*, 819–830.

Jiang, J., & Hui, C. C. (2008). Hedgehog signaling in development and cancer. *Developmental Cell, 15*, 801–812.

Jiang, J., & Struhl, G. (1995). Protein kinase A and hedgehog signaling in Drosophila limb development. *Cell, 80*, 563–572.

Jiang, J., & Struhl, G. (1998). Regulation of the Hedgehog and wingless signalling pathways by the F- box/WD40-repeat protein Slimb. *Nature, 391*, 493–496.

Jones, D. T., Jager, N., Kool, M., Zichner, T., Hutter, B., Sultan, M., et al. (2012). Dissecting the genomic complexity underlying medulloblastoma. *Nature, 488*, 100–105.

Kent, D., Bush, E. W., & Hooper, J. E. (2006). Roadkill attenuates Hedgehog responses through degradation of Cubitus interruptus. *Development, 133*, 2001–2010.

Kise, Y., Morinaka, A., Teglund, S., & Miki, H. (2009). Sufu recruits GSK3beta for efficient processing of Gli3. *Biochemical and Biophysical Research Communications, 387*, 569–574.

Knippschild, U., Gocht, A., Wolff, S., Huber, N., Lohler, J., & Stoter, M. (2005). The casein kinase 1 family: Participation in multiple cellular processes in eukaryotes. *Cellular Signalling, 17*, 675–689.

Kovacs, J. J., Whalen, E. J., Liu, R., Xiao, K., Kim, J., Chen, M., et al. (2008). Beta-arrestin-mediated localization of smoothened to the primary cilium. *Science, 320*, 1777–1781.

Lee, E., Salic, A., & Kirschner, M. W. (2001). Physiological regulation of [beta]-catenin stability by Tcf3 and CK1epsilon. *The Journal of Cell Biology, 154*, 983–993.

Li, S., Chen, Y., Shi, Q., Yue, T., Wang, B., & Jiang, J. (2012). Hedgehog-regulated ubiquitination controls smoothened trafficking and cell surface expression in Drosophila. *PLoS Biology, 10,* e1001239.

Li, S., Li, S., Han, Y., Tong, C., Wang, B., Chen, Y., et al. (2016). Regulation of smoothened phosphorylation and high-level Hedgehog signaling activity by a plasma membrane associated kinase. *PLoS Biology, 14,* e1002481.

Li, S., Ma, G., Wang, B., & Jiang, J. (2014). Hedgehog induces formation of PKA-smoothened complexes to promote smoothened phosphorylation and pathway activation. *Science Signaling, 7,* ra62.

Li, W., Ohlmeyer, J. T., Lane, M. E., & Kalderon, D. (1995). Function of protein kinase A in hedghehog signal transduction and Drosophila imaginal disc development. *Cell, 80,* 553–562.

Liang, J., Fu, Y., Cruciat, C. M., Jia, S., Wang, Y., Tong, Z., et al. (2011). Transmembrane protein 198 promotes LRP6 phosphorylation and Wnt signaling activation. *Molecular and Cellular Biology, 31,* 2577–2590.

Liem, K. F., Jr., He, M., Ocbina, P. J., & Anderson, K. V. (2009). Mouse Kif7/Costal2 is a cilia-associated protein that regulates Sonic hedgehog signaling. *Proceedings of the National Academy of Sciences of the United States of America, 106,* 13377–13382.

Liu, C., Li, Y., Semenov, M., Han, C., Baeg, G. H., Tan, Y., et al. (2002). Control of beta-catenin phosphorylation/degradation by a dual-kinase mechanism. *Cell, 108,* 837–847.

Liu, A., Wang, B., & Niswander, L. A. (2005). Mouse intraflagellar transport proteins regulate both the activator and repressor functions of Gli transcription factors. *Development, 132,* 3103–3111.

Lum, L., Yao, S., Mozer, B., Rovescalli, A., Von Kessler, D., Nirenberg, M., et al. (2003). Identification of Hedgehog pathway components by RNAi in Drosophila cultured cells. *Science, 299,* 2039–2045.

MacDonald, B. T., Tamai, K., & He, X. (2009). Wnt/beta-catenin signaling: Components, mechanisms, and diseases. *Developmental Cell, 17,* 9–26.

Mao, J., Wang, J., Liu, B., Pan, W., Farr, G. H., 3rd, Flynn, C., et al. (2001). Low-density lipoprotein receptor-related protein-5 binds to axin and regulates the canonical Wnt signaling pathway. *Molecular Cell, 7,* 801–809.

McKay, R. M., Peters, J. M., & Graff, J. M. (2001). The casein kinase I family in Wnt signaling. *Developmental Biology, 235,* 388–396.

Meloni, A. R., Fralish, G. B., Kelly, P., Salahpour, A., Chen, J. K., Wechsler-Reya, R. J., et al. (2006). Smoothened signal transduction is promoted by G protein-coupled receptor kinase 2. *Molecular and Cellular Biology, 26,* 7550–7560.

Methot, N., & Basler, K. (1999). Hedgehog controls limb development by regulating the activities of distinct transcriptional activator and repressor forms of Cubitus interruptus. *Cell, 96,* 819–831.

Methot, N., & Basler, K. (2000). Suppressor of fused opposes hedgehog signal transduction by impeding nuclear accumulation of the activator form of Cubitus interruptus. *Development, 127,* 4001–4010.

Molnar, C., Ruiz-Gomez, A., Martin, M., Rojo-Berciano, S., Mayor, F., & de Celis, J. F. (2011). Role of the Drosophila non-visual ss-arrestin kurtz in hedgehog signalling. *PLoS Genetics, 7,* e1001335.

Ohlmeyer, J. T., & Kalderon, D. (1997). Dual pathways for induction of wingless expression by protein kinase A and Hedgehog in Drosophila embryos. *Genes & Development, 11,* 2250–2258.

Ohlmeyer, J. T., & Kalderon, D. (1998). Hedgehog stimulates maturation of Cubitus interruptus into a labile transcriptional activator. *Nature, 396,* 749–753.

Pan, Y., Bai, C. B., Joyner, A. L., & Wang, B. (2006). Sonic hedgehog signaling regulates Gli2 transcriptional activity by suppressing its processing and degradation. *Molecular and Cellular Biology, 26,* 3365–3377.

Pan, W., Choi, S. C., Wang, H., Qin, Y., Volpicelli-Daley, L., Swan, L., et al. (2008). Wnt3a-mediated formation of phosphatidylinositol 4,5-bisphosphate regulates LRP6 phosphorylation. *Science, 321*, 1350–1353.

Pan, D., & Rubin, G. M. (1995). cAMP-dependent protein kinase and *hedgehog* act antagonistically in regulating *decapentaplegic* transcription in Drosophila imaginal discs. *Cell, 80*, 543–552.

Penton, A., Wodarz, A., & Nusse, R. (2002). A mutational analysis of dishevelled in Drosophila defines novel domains in the dishevelled protein as well as novel suppressing alleles of axin. *Genetics, 161*, 747–762.

Peters, J. M., McKay, R. M., McKay, J. P., & Graff, J. M. (1999). Casein kinase I transduces Wnt signals. *Nature, 401*, 345–350.

Piao, S., Lee, S. H., Kim, H., Yum, S., Stamos, J. L., Xu, Y., et al. (2008). Direct inhibition of GSK3beta by the phosphorylated cytoplasmic domain of LRP6 in Wnt/beta-catenin signaling. *PloS One, 3*, e4046.

Plowman, G. D., Sudarsanam, S., Bingham, J., Whyte, D., & Hunter, T. (1999). The protein kinases of Caenorhabditis elegans: A model for signal transduction in multicellular organisms. *Proceedings of the National Academy of Sciences of the United States of America, 96*, 13603–13610.

Polakis, P. (2000). Wnt signaling and cancer. *Genes & Development, 14*, 1837–1851.

Price, M. A., & Kalderon, D. (1999). Proteolysis of cubitus interruptus in Drosophila requires phosphorylation by protein kinase A. *Development, 126*, 4331–4339.

Price, M. A., & Kalderon, D. (2002). Proteolysis of the Hedgehog signaling effector Cubitus interruptus requires phosphorylation by glycogen synthase kinase 3 and casein kinase 1. *Cell, 108*, 823–835.

Pronobis, M. I., Rusan, N. M., & Peifer, M. (2015). A novel GSK3-regulated APC: Axin interaction regulates Wnt signaling by driving a catalytic cycle of efficient betacatenin destruction. *ELife, 4*, e08022.

Pugh, T. J., Weeraratne, S. D., Archer, T. C., Pomeranz Krummel, D. A., Auclair, D., Bochicchio, J., et al. (2012). Medulloblastoma exome sequencing uncovers subtype-specific somatic mutations. *Nature, 488*, 106–110.

Rivers, A., Gietzen, K. F., Vielhaber, E., & Virshup, D. M. (1998). Regulation of casein kinase I epsilon and casein kinase I delta by an in vivo futile phosphorylation cycle. *The Journal of Biological Chemistry, 273*, 15980–15984.

Robbins, D. J., Nybakken, K. E., Kobayashi, R., Sisson, J. C., Bishop, J. M., & Therond, P. P. (1997). Hedgehog elicits signal transduction by means of a large complex containing the kinesin-related protein costal2. *Cell, 90*, 225–234.

Rohatgi, R., Milenkovic, L., & Scott, M. P. (2007). Patched1 regulates hedgehog signaling at the primary cilium. *Science, 317*, 372–376.

Rubin, L. L., & de Sauvage, F. J. (2006). Targeting the Hedgehog pathway in cancer. *Nature Reviews. Drug Discovery, 5*, 1026–1033.

Rubinfeld, B., Tice, D. A., & Polakis, P. (2001). Axin-dependent phosphorylation of the adenomatous polyposis coli protein mediated by casein kinase 1epsilon. *The Journal of Biological Chemistry, 276*, 39037–39045.

Sakanaka, C., Leong, P., Xu, L., Harrison, S. D., & Williams, L. T. (1999). Casein kinase iepsilon in the wnt pathway: Regulation of beta-catenin function. *Proceedings of the National Academy of Sciences of the United States of America, 96*, 12548–12552.

Serysheva, E., Berhane, H., Grumolato, L., Demir, K., Balmer, S., Bodak, M., et al. (2013). Wnk kinases are positive regulators of canonical Wnt/beta-catenin signalling. *EMBO Reports, 14*, 718–725.

Shi, Q., Li, S., Jia, J., & Jiang, J. (2011). The hedgehog-induced smoothened conformational switch assembles a signaling complex that activates fused by promoting its dimerization and phosphorylation. *Development, 138*, 4219–4231.

Shi, Q., Li, S., Li, S., Jiang, A., Chen, Y., & Jiang, J. (2014). Hedgehog-induced phosphorylation by CK1 sustains the activity of Ci/Gli activator. *Proceedings of the National Academy of Sciences of the United States of America, 111*, E5651–E5660.

Sisson, J. C., Ho, K. S., Suyama, K., & Scott, M. P. (1997). Costal2, a novel kinesin-related protein in the Hedgehog signaling pathway. *Cell, 90*, 235–245.

Smelkinson, M. G., & Kalderon, D. (2006). Processing of the Drosophila hedgehog signaling effector Ci-155 to the repressor Ci-75 is mediated by direct binding to the SCF component slimb. *Current Biology, 16*, 110–116.

Smelkinson, M. G., Zhou, Q., & Kalderon, D. (2007). Regulation of Ci-SCFSlimb binding, Ci proteolysis, and hedgehog pathway activity by Ci phosphorylation. *Developmental Cell, 13*, 481–495.

Spencer, E., Jiang, J., & Chen, Z. J. (1999). Signal-induced ubiquitination of IkappaBalpha by the F-box protein Slimb/beta-TrCP. *Genes & Development, 13*, 284–294.

Strigini, M., & Cohen, S. M. (1997). A hedgehog activity gradient contributes to AP axial patterning of the Drosophila wing. *Development, 124*, 4697–4705.

Strutt, H., Price, M. A., & Strutt, D. (2006). Planar polarity is positively regulated by casein kinase iepsilon in Drosophila. *Current Biology, 16*, 1329–1336.

Svard, J., Heby-Henricson, K., Persson-Lek, M., Rozell, B., Lauth, M., Bergstrom, A., et al. (2006). Genetic elimination of suppressor of fused reveals an essential repressor function in the mammalian hedgehog signaling pathway. *Developmental Cell, 10*, 187–197.

Tamai, K., Zeng, X., Liu, C., Zhang, X., Harada, Y., Chang, Z., et al. (2004). A mechanism for Wnt coreceptor activation. *Molecular Cell, 13*, 149–156.

Tanneberger, K., Pfister, A. S., Brauburger, K., Schneikert, J., Hadjihannas, M. V., Kriz, V., et al. (2011). Amer1/WTX couples Wnt-induced formation of PtdIns(4,5)P2 to LRP6 phosphorylation. *The EMBO Journal, 30*, 1433–1443.

Tempe, D., Casas, M., Karaz, S., Blanchet-Tournier, M. F., & Concordet, J. P. (2006). Multisite protein kinase A and glycogen synthase kinase 3beta phosphorylation leads to Gli3 ubiquitination by SCFbetaTrCP. *Molecular and Cellular Biology, 26*, 4316–4326.

Tukachinsky, H., Lopez, L. V., & Salic, A. (2010). A mechanism for vertebrate Hedgehog signaling: Recruitment to cilia and dissociation of SuFu-Gli protein complexes. *The Journal of Cell Biology, 191*, 415–428.

Varjosalo, M., Li, S. P., & Taipale, J. (2006). Divergence of hedgehog signal transduction mechanism between Drosophila and mammals. *Developmental Cell, 10*, 177–186.

Wang, G., Amanai, K., Wang, B., & Jiang, J. (2000). Interactions with Costal2 and suppressor of fused regulate nuclear translocation and activity of cubitus interruptus. *Genes & Development, 14*, 2893–2905.

Wang, B., Fallon, J. F., & Beachy, P. A. (2000). Hedgehog-regulated processing of Gli3 produces an anterior/posterior repressor gradient in the developing vertebrate limb. *Cell, 100*, 423–434.

Wang, B., & Li, Y. (2006). Evidence for the direct involvement of {beta}TrCP in Gli3 protein processing. *Proceedings of the National Academy of Sciences of the United States of America, 103*, 33–38.

Wang, C., Pan, Y., & Wang, B. (2010). Suppressor of fused and Spop regulate the stability, processing and function of Gli2 and Gli3 full-length activators but not their repressors. *Development, 137*, 2001–2009.

Wang, G., Wang, B., & Jiang, J. (1999). Protein kinase A antagonizes Hedgehog signaling by regulating both the activator and repressor forms of Cubitus interruptus. *Genes & Development, 13*, 2828–2837.

Wen, X., Lai, C. K., Evangelista, M., Hongo, J. A., de Sauvage, F. J., & Scales, S. J. (2010). Kinetics of hedgehog-dependent full-length Gli3 accumulation in primary cilia and subsequent degradation. *Molecular and Cellular Biology, 30*, 1910–1922.

Winston, J. T., Strack, P., Beer-Romero, P., Chu, C. Y., Elledge, S. J., & Harper, J. W. (1999). The SCFbeta-TRCP-ubiquitin ligase complex associates specifically with phosphorylated destruction motifs in IkappaBalpha and beta-catenin and stimulates IkappaBalpha ubiquitination in vitro. *Genes & Development, 13*, 270–283.

Wu, G., Huang, H., Garcia Abreu, J., & He, X. (2009). Inhibition of GSK3 phosphorylation of beta-catenin via phosphorylated PPPSPXS motifs of Wnt coreceptor LRP6. *PloS One, 4*, e4926.

Xia, R., Jia, H., Fan, J., Liu, Y., & Jia, J. (2012). USP8 promotes smoothened signaling by preventing its ubiquitination and changing its subcellular localization. *PLoS Biology, 10*, e1001238.

Xie, J., Murone, M., Luoh, S.-M., Ryan, A., Gu, Q., Zhang, C., et al. (1998). Activating *Smoothened* mutations in sporadic basal-cell carcinoma. *Nature, 391*, 90–92.

Xing, Y., Clements, W. K., Le Trong, I., Hinds, T. R., Stenkamp, R., Kimelman, D., et al. (2004). Crystal structure of a beta-catenin/APC complex reveals a critical role for APC phosphorylation in APC function. *Molecular Cell, 15*, 523–533.

Yanfeng, W. A., Berhane, H., Mola, M., Singh, J., Jenny, A., & Mlodzik, M. (2011). Functional dissection of phosphorylation of disheveled in Drosophila. *Developmental Biology, 360*, 132–142.

Zeng, X., Huang, H., Tamai, K., Zhang, X., Harada, Y., Yokota, C., et al. (2008). Initiation of Wnt signaling: Control of Wnt coreceptor Lrp6 phosphorylation/activation via frizzled, dishevelled and axin functions. *Development, 135*, 367–375.

Zeng, X., Tamai, K., Doble, B., Li, S., Huang, H., Habas, R., et al. (2005). A dual-kinase mechanism for Wnt co-receptor phosphorylation and activation. *Nature, 438*, 873–877.

Zhang, L., Jia, J., Wang, B., Amanai, K., Wharton, K. A., Jr., & Jiang, J. (2006). Regulation of wingless signaling by the CKI family in Drosophila limb development. *Developmental Biology, 299*, 221–237.

Zhang, Y., Mao, F., Lu, Y., Wu, W., Zhang, L., & Zhao, Y. (2011). Transduction of the hedgehog signal through the dimerization of fused and the nuclear translocation of Cubitus interruptus. *Cell Research, 21*, 1436–1451.

Zhang, Q., Shi, Q., Chen, Y., Yue, T., Li, S., Wang, B., et al. (2009). Multiple Ser/Thr-rich degrons mediate the degradation of Ci/Gli by the Cul3-HIB/SPOP E3 ubiquitin ligase. *Proceedings of the National Academy of Sciences of the United States of America, 106*, 21191–21196.

Zhang, C., Williams, E. H., Guo, Y., Lum, L., & Beachy, P. A. (2004). Extensive phosphorylation of smoothened in hedgehog pathway activation. *Proceedings of the National Academy of Sciences of the United States of America, 101*, 17900–17907.

Zhang, Q., Zhang, L., Wang, B., Ou, C. Y., Chien, C. T., & Jiang, J. (2006). A hedgehog-induced BTB protein modulates hedgehog signaling by degrading Ci/Gli transcription factor. *Developmental Cell, 10*, 719–729.

Zhang, W., Zhao, Y., Tong, C., Wang, G., Wang, B., Jia, J., et al. (2005). Hedgehog-regulated costal2-kinase complexes control phosphorylation and proteolytic processing of cubitus interruptus. *Developmental Cell, 8*, 267–278.

Zhao, Y., Tong, C., & Jiang, J. (2007). Hedgehog regulates smoothened activity by inducing a conformational switch. *Nature, 450*, 252–258.

Zhou, Q., & Kalderon, D. (2011). Hedgehog activates fused through phosphorylation to elicit a full spectrum of pathway responses. *Developmental Cell, 20*, 802–814.

CHAPTER TEN

Ligand Receptor-Mediated Regulation of Growth in Plants

Miyoshi Haruta, Michael R. Sussman[1]
University of Wisconsin-Madison, Madison, WI, United States
[1]Corresponding author e-mail address: msussman@wisc.edu

Contents

1. Introduction — 332
2. Novel and Unique Plant Signaling Pathways — 333
   - 2.1 Cytokinin — 333
   - 2.2 Auxin — 334
   - 2.3 Brassinosteroid — 336
   - 2.4 Abscisic Acid — 337
   - 2.5 Ethylene — 338
   - 2.6 Gibberellin — 338
   - 2.7 Jasmonic Acid — 339
   - 2.8 Strigolactone — 340
3. Plasma Membrane as the Site of Peptide Ligand Sensing by Receptors — 341
4. FERONIA and Its Gene Family in Plant Growth and Development — 342
5. Organ Size and Growth Are Determined by Stimulation and Inhibition by Signaling Pathways in Plants and Animals — 345
6. RALF Family and Function in Plant Growth and Development — 347
7. RALF-Like Peptides Are Not Only Produced by Plants But Also by Pathogenic Microbes — 348
8. Other Endogenous Hormone-Like Peptides and Their Receptor Pairs for Plant Growth, Development, and Physiology — 349
9. Conclusion and Future Perspectives — 353
References — 354

Abstract

Growth and development of multicellular organisms are coordinately regulated by various signaling pathways involving the communication of inter- and intracellular components. To form the appropriate body patterns, cellular growth and development are modulated by either stimulating or inhibiting these pathways. Hormones and second messengers help to mediate the initiation and/or interaction of the various signaling pathways in all complex multicellular eukaryotes. In plants, hormones include small organic molecules, as well as larger peptides and small proteins, which, as in animals, act as ligands and interact with receptor proteins to trigger rapid biochemical changes and induce the intracellular transcriptional and long-term physiological responses. During the past two decades, the availability of genetic and genomic resources in the model plant species, *Arabidopsis thaliana*, has greatly helped in the discovery of plant hormone receptors and the components of signal transduction pathways and mechanisms used by these immobile but highly complex organisms. Recently, it has been shown that two of the most important plant hormones, auxin and abscisic acid (ABA), act through signaling pathways that have not yet been recognized in animals. For example, auxins stimulate cell elongation by bringing negatively acting transcriptional repressor proteins to the proteasome to be degraded, thus unleashing the gene expression program required for increasing cell size. The "dormancy" inducing hormone, ABA, binds to soluble receptor proteins and inhibits a specific class of protein phosphatases (PP2C), which activates phosphorylation signaling leading to transcriptional changes needed for the desiccation of the seeds prior to entering dormancy. While these two hormone receptors have no known animal counterparts, there are also many similarities between animal and plant signaling pathways. For example, in plants, the largest single gene family in the genome is the protein kinase family (approximately 5% of the protein coding genes), although the specific function for only a few dozen of these kinases is clearly established. Recent comparative genomics studies have revealed that parasitic nematodes and pathogenic microbes produce plant peptide hormone mimics that target specific plant plasma membrane receptor-like protein kinases, thus usurping endogenous signaling pathways for their own pathogenic purposes. With biochemical, genetic, and physiological analyses of the regulation of hormone receptor signal pathways, we are thus just now beginning to understand how plants optimize the development of their body shape and cope with constantly changing environmental conditions.

1. INTRODUCTION

Unlike animals, plants live their entire life at the place where they are first anchored. Thus, they have evolved mechanisms to adapt to their environments by creating unique body patterns and/or continually adjusting their growth rate. To do so, plants utilize ligand–receptor systems that perceive environmental changes and initiate various types of immediate responses. The first chemical compound that was proposed to be a growth stimulatory hormone found in plants many decades ago was an indole derivative, auxin (indole-3-acetic acid, IAA). IAA was first isolated from human urine then subsequently identified from plant extracts (references in Enders & Strader, 2015). Auxin induces changes in body pattern development in response to sunlight to maximize capturing the light energy and optimize growth and reproduction. After the subsequent discovery of other key small molecule plant hormones such as cytokinins (an adenine derivative), ethylene (a gas molecule), gibberellin (diterpenoid), and abscisic acid (ABA) (synthesized from carotenoid pathway), plants were traditionally, for over several decades, believed to regulate cell physiology by synthesizing and transporting only small organic hormone-like molecules. These plant-specific signaling compounds were found to initiate signal transduction by interacting with their receptor proteins localized in the cytoplasm, nucleus, or endomembranes, instead of the more commonly known plasma membrane-bound receptor kinases found at the cell surface of animal cells. It is only last 15 years or so that plant scientists realize that like animals, plants use a wide variety of peptides and small protein growth factors as extracellular signals that interact with cell-surface protein kinases, to regulate growth. The notion that plants use peptide-based signals as important regulatory pathways that alter protein phosphorylation is consistent with the fact that receptor-like kinases (RLKs) are encoded by a very large gene family in all plants. In the best-characterized model plant, *Arabidopsis thaliana*, there are over 1000 genes encoding protein kinases, out of a total of over 25,000 protein coding genes in the genome. About half of the protein kinases in plants are membrane-bound receptor-like protein kinases which contain a single transmembrane domain with an extracellularly located ectodomain where signaling molecules bind. Among those RLKs, biological functions have been defined for only a few dozen based largely on the phenotypes observed with genetic knockout mutants. Since there are only a handful of pairs of ligand–receptors with known roles in the regulation of plant growth and development, there is a rich opportunity for future research to reveal many previously unknown pathways. For this goal, our lab has been developing and applying modern genomic technologies, including mass spectrometric-based tools that quantify changes in protein phosphorylation on a proteome-wide scale.

2. NOVEL AND UNIQUE PLANT SIGNALING PATHWAYS
2.1 Cytokinin

In contrast to animals, plant cell division is restricted to a small group of cells at the very tip of the root and shoot, called a meristem. At the meristem, cells are maintained undifferentiated and primarily function to produce daughter cells. Once new cells are produced, they expand in size, which contributes to organ growth. This cell division is stimulated by cytokinin, a plant hormone that is a modified adenine derivative. Kinetin is a naturally occurring cytokinin that was first isolated and characterized at the University of

Wisconsin–Madison by Miller and Skoog (Miller, Skoog, Okumura, Von Saltza, & Strong, 1956). It is important to note that this plant growth substance is *completely* different from the better-known class of proteins called "cytokines" that are involved in many aspects of animal physiology including immune system, growth, and maturation.

Cytokinin acts through a family of receptors called histidine protein kinases, one of which was first isolated in plants via a genetic screen that identified cytokinin-insensitive plants, leading to the identification of the gene as encoding a histidine protein kinase, *cre1* (cytokinin re̲sponse 1) (Inoue et al., 2001). Histidine protein kinases are unique in plants, fungi, and bacteria and have not yet been found to play a role in animals. Cytokinin signals are thought to be perceived in the sensor domain of the receptor and transmitted to a kinase domain, in which the conserved histidine residue is phosphorylated. The phosphate group is then transferred to the aspartate residue of response regulator proteins, as first described in the osmosensor *EnvZ* of bacteria (Mahonen et al., 2006; West & Stock, 2001). Affinity measurements of a naturally occurring cytokinin, zeatin, as well as a synthetic analog, N^6-benzyladenine, to the receptors indicate that AHK3 appears to have the highest affinity (1.3 nM equilibrium dissociation constant) among histidine kinase receptor members (Romanov, Lomin, & Schmulling, 2006). Although originally thought to be localized at the cell surface, it is now accepted that cytokinin perception by the receptor is occurring in the endoplasmic reticulum and the membrane-bound receptor then relays its phosphorylation signal to the nucleus and activates or inhibits responsive genes leading to the regulation of cell division (Lomin, Yonekura-Sakakibara, Romanov, & Sakakibara, 2011; Wulfetange et al., 2011).

2.2 Auxin

After new cells are produced at the meristem, they start to expand and sometimes differentiate, which promotes organ growth and development. Many growth changes during plant body development are mediated by asymmetrical cell expansion rates across the organs. Early descriptions of experiments performed by Charles Darwin led him to postulate that aerial parts of plants bent towards light, a phenomenon known as phototropism (Darwin & Darwin, 1880). Along with geotropism and other asymmetrical growth responses, tropism is thought to be mediated by unequal rates of auxin transport down the different sides of the organ, in this case, promoting cell elongation in the shaded side of a shoot and inhibiting cell elongation in the irradiated side of the shoot. This results in the shoot bending towards the direction of the light source. Plant sensitivity to auxin is context dependent; root growth is inhibited at 100 nM IAA, a naturally occurring auxin, and shoot growth is promoted at the same concentration when exogenously applied.

Because cell expansion is fundamental to plant organ growth and development, auxin is involved in many aspects of plant life, including flower development, lateral root initiation, and embryogenesis. Forward genetic screens of Arabidopsis mutants to isolate seedlings resistant to an auxin transport inhibitor drug identified an F-box motif protein, transport inhibitor response 1 (TIR1) (Ruegger et al., 1998). Its sequence similarity to a component of E3 ubiquitin ligase complex subsequently revealed that auxin and the receptor complex are subject to movement into the proteasome and thus, targeted for degradation (Gray, Kepinski, Rouse, Leyser, & Estelle, 2001). Auxin-TIR1, a ligand–receptor pair forms a protein complex with transcriptional repressors, Aux/IAA protein family and a cofactor, inositol hexakisphosphate (InsP6), and this association causes a degradation of the repressor proteins, resulting in increased expression of auxin responsive genes. The *tir1* receptor mutant is also insensitive to auxin or its synthetic analogs such as 2,4-D (2,4-dichlorophenoxyacetic acid), because the receptor mutant cells are not able to respond a higher level of auxin accumulated in the cells when its transport out of the bottom of each cell and down the stem is inhibited.

Arabidopsis mutants which lack four members of TIR1 auxin receptor genes show severely impaired growth and insensitivity to exogenous applied auxin (Dharmasiri et al., 2005). Consistent with this, the induction of an auxin responsive marker is completely absent in the roots of a triple receptor mutant (Scheitz, Luthen, & Schenck, 2013). However, the shoot of the quadruple mutant is still capable of responding to auxin-induced growth stimulation (Schenck, Christian, Jones, & Luthen, 2010). Since there are six members of TIR F-box receptor family and 29 Aux/IAA coreceptor proteins in the Arabidopsis genome, combinatorial interaction of F-box receptor and coreceptor proteins likely account for differential sensitivities of roots and shoots to exogenously applied auxin (Calderon Villalobos et al., 2012). A membrane-bound receptor kinase system is also implicated in auxin perception (Xu et al., 2014), but the structural and regulatory basis for this function has yet to be clearly established.

2.3 Brassinosteroid

The cell elongation rate clearly influences overall plant height since cell division is restricted to a small portion of the shoots and roots. Some Arabidopsis mutants that show severe dwarfism due to reduced stem length were found that were caused by the reduced production of brassinosteroids including brassinolide (Choe et al., 1999; Noguchi et al., 1999). This compound is a steroid that was initially isolated from *Brassica napus* pollen and is capable of inducing tissue enlargement (Grove et al., 1979). A collection of Arabidopsis mutant alleles responsible to brassinosteroid-insensitive phenotype were isolated by genetic screens (Clouse, Langford, & McMorris, 1996; Li & Chory, 1997). BRI1 (Brassinosteroid-insensitive 1) encodes a leucine-rich repeat receptor-like protein kinase that was shown to interact with brassinosteroid (Wang, Seto, Fujioka, Yoshida, & Chory, 2001). BRI1 uses a coreceptor BAK1 (BRI1-associated kinase 1) which involves transphosphorylation of the two receptor kinases at the plasma membrane (Nam & Li, 2002; Russinova et al., 2004; Sun et al., 2013; Wang et al., 2008). However, brassinosteroid-induced phosphorylation of BRI1 reported to date is occurring in the time scale of 60–120 min after treatment with the ligand, which appears to be much slower than the typical ligand-regulated phosphorylation responses reported with other receptor kinases such as EGFR whose phosphorylation by EGF treatment increases within 1 min and starts decreasing after 5–10 min (Curran, Zhang, Ma, Sarkaria, & White, 2015; Oh, Wang, Clouse, & Huber, 2012; Olsen et al., 2006; Tang et al., 2008; Wang et al., 2005).

Consistent with genetic evidence showing that many critical amino-acid residues for BRI1 function are localized within the extracellular island domain, structural analyses demonstrated that brassinolide interacts with the binding pocket found in this island domain (Hothorn et al., 2011; She et al., 2011). This BRI1 receptor kinase signaling for plant steroid signal is perceived much different from animal steroid hormone receptor system, which takes place within the nucleus and directly regulates gene expression (Hall, Couse, & Korach, 2001; Shao & Brown, 2004). Despite the clear importance of the BRI1 receptor derived from genetic ablation, there is no clear evidence that brassinosteroids are differentially transported or metabolized during any aspect of plant life (Symons, Ross, Jager, & Reid, 2008). In this characteristic, brassinosteroids may be different from other plant hormones, and perhaps acting more like a cofactor than a hormone, which perhaps is a reflection of the fact that the plasma membrane is the most sterol rich membrane in the cell. Moreover, unlike other membrane-bound receptors which respond to their ligands, BRI localization, turnover, and ubiquitination are shown to be independent of brassinolide (Geldner, Hyman, Wang, Schumacher, & Chory, 2007; Martins et al., 2015).

2.4 Abscisic Acid

A critical role for protein phosphorylation in regulating plant growth is clear in the case of the ABA receptor system. ABA, first identified from abscising (a form of cell localized cell death) cotton tissue and sycamore trees, is known to be involved in a broad range of stress responses, water usage regulation, and seed desiccation/dormancy processes (Finkelstein, 2013). Although there were several earlier reports of genes encoding proteins that had characteristics of an ABA receptor, none of these proteins proved to be the actual receptor(s). The great difficulty in isolating putative ABA receptors via genetic screens hinted that ABA receptors might be members of a large redundant gene family that individually mask the phenotype, thus failing to identify receptor mutants from screens of single gene mutations. The use of a chemical genetic screen, involving small chemicals termed pyrobactins acting as a tissue-specific analog of ABA, enabled the isolation of the first bona fide ABA receptor mutants. A family of soluble ABA receptor proteins, the pyrobactins, PYRs, binds to ABA in a stereospecific manner and the bound form of PYRs can physically interact with and inhibit a specific group of protein phosphatases, PP2C. This inhibition causes the increase of phosphorylation in the activation domain of specific protein kinases, the SnRKs (Ma et al., 2009; Park et al., 2009). SnRK activation triggers phosphorylation of other target proteins including ABA-responsive transcription factors (Umezawa et al., 2013). A higher order of *pyr* receptor mutants, quadruple- or sextuple mutants, were produced and revealed relevant phenotypes (Gonzalez-Guzman et al., 2012). Consistent with ABA roles in plants as a stress hormone, *pyr* receptor mutants exhibit defects in the regulation of stomatal pore closing in response to drought stress. The use of a modified version of the PYR receptors also rendered plants able to withstand drought stress, implicating a powerful new means of modulating plant physiology via editing hormone receptor function (Park et al., 2015).

2.5 Ethylene

During the seedling growth phase at which a root and shoot first emerge and extend from a seed, growth and development are carefully coordinated. A seed deposited in soil experiences darkness and thus undergoes etiolation: elongated hypocotyl (the first stem of a seedling), shorter root, small and pale leaves, and folded cotyledonary leaves (the first leaves of a plant, which were stored in the embryo). As a plant develops, the cotyledonary leaves open, an apical hook unfolds, hypocotyl elongation ceases, and the root elongates. During this transition, a gas known as ethylene plays important roles. Treatment of germinating seedlings with ethylene causes the well known "triple response," i.e., a curved apical hook, short and thick hypocotyl, and reduced root elongation. Genetic screens identified a mutant which does not show the triple response and the gene encoding this mutation was identified as an ethylene receptor, ETR1, a protein distantly related to the receptor protein for cytokinin (Bleecker, Estelle, Somerville, & Kende, 1988; Chang, Kwok, Bleecker, & Meyerowitz, 1993) (see the above). Heterologous expression of the ETR1 receptor protein in yeast confirmed that ethylene directly binds to ETR1 (Schaller & Bleecker, 1995).

Binding of ethylene to ETR1 at the endoplasmic reticulum causes inactivation of both the receptor and its downstream protein, CTR1 (constitutive triple response), a Raf-like serine threonine kinase. These events then signal to EIN2 (ethylene insensitive 2), a transmembrane protein. The C-terminal end of EIN2 is then cleaved and translocated from the ER to the nucleus to induce activation of the EIN3 family of ethylene responsive transcription factors (Qiao et al., 2012).

2.6 Gibberellin

Stem elongation is regulated by complex signaling pathways involving various hormones, including auxin, brassinolide, and gibberellin. A diterpenoid compound, gibberellin, was first discovered from a fungus due to its effect on rice causing elongated stem. This compound was later shown to be an endogenously synthesized hormone that regulates stem cell elongation. Gibberellin acts through its nucleus-localized receptor, GID1 (Gibberellin-Insensitive Dwarf1), and like auxin, this binding induces degradation of transcriptional repressor proteins, DELLA which have a characteristic conserved motif Asp-Glu-Leu-Leu-Ala, at the amino terminus (Ueguchi-Tanaka et al., 2005). Mutations in gene sequences involved in gibberellin signaling or biosynthesis cause the production of dwarf plants, due to a reduction in stem cell length. In these mutants with shorter stems, energy from photosynthesis is diverted to grain production instead of to stem growth and this has contributed to increased crop yield responsible for the Green Revolution, for which the Nobel Peace Prize was awarded in 1970, and subsequent research demonstrated that the mutations were localized in gibberellin synthesis and receptor signaling components.

2.7 Jasmonic Acid

Previously, plant growth and development were thought to be regulated by five classical hormones, auxin, cytokinin, gibberellin, ethylene, and ABA (see the above). More recent studies now are acknowledging other small molecules, including jasmonic acid and strigolactones (see later) as either growth regulatory substances or plant hormones involved in cell-to-cell communication. Jasmonic acid, a volatile chemical derived from linolenic acid, is involved in senescence (biological aging), defense responses, and growth inhibition. The jasmonic acid biosynthesis via cyclic reaction from this fatty acid is often compared with the mammalian compound, prostaglandin, a hormone-like compound, which acts during inflammation and vasodilation (widening blood vessels), and is biosynthesized from arachidonic acid. Despite similarity in chemical structure, clearly jasmonic acid and prostaglandin share no similarities in their downstream biological response.

Jasmonic acid synthesis is induced by mechanostress and induces the expression of genes involved in defense responses. Methyl jasmonate, a volatile derivative, also diffuses into the air and induces defense responses in neighboring plants (Farmer & Ryan, 1990). Coronatine is a synthetic analog of jasmonic acid used to study jasmonic acid function and signaling in plants. A *coronatine insensitive 1* (*coi1*) mutant that was also resistant to jasmonate was shown to have a mutation in a gene encoding an F-box protein which shows a sequence similarity to TIR1, an auxin receptor (Sheard et al., 2010; Xie, Feys, James, Nieto-Rostro, & Turner, 1998). Three-dimensional analysis of the jasmonic acid receptor, COI1, revealed that COI1 can bind to a conjugated form of this hormone, jasmonate-isoleucine, a coreceptor JAZ protein, transcriptional repressors, and a cofactor, inositol pentakisphosphate (Sheard et al., 2010).

2.8 Strigolactone

Overall plant body shape at maturity is influenced by shoot branching patterns. Branch development is controlled by genetic and environmental factors, which integrate hormone transport and action, leading to the emergence and elongation of lateral branches. For example, maize (corn) and its ancestor, teosinte, are dramatically different in shoot branching patterns, presumably due to the genetic selection derived from domestication by humans over thousand years. Teosinte branched1 (TB1), encoding a transcription factor, suppresses the bud growth (Doebley, Stec, & Hubbard, 1997). An important role for TB1 as a key genetic component of branching pattern was supported by the identification of an Arabidopsis gene related to TB1. Thus, the teosinte branched 1-like 1 gene was identified and its knockout mutant exhibits a hyperbranching phenotype (Finlayson, 2007).

Characterization of bushy mutants of Arabidopsis revealed that they contain reduced concentrations of strigolactone, indicating that this may be a new type of plant hormone. Application of this compound to plants inhibits shoot branching (Umehara et al., 2008). A strigolactone-insensitive plant, d14 (Dwarf14), isolated from rice produces increased side shoots (tillers) (Arite et al., 2009). D14 of rice and its orthologs, AtD14 of Arabidopsis and *deceased apical dominance 2* (*dad2*) of petunia are strigolactone receptors, which are members of the α/β-hydrolase enzyme superfamily and distantly related to a gibberellin receptor, GID (Hamiaux et al., 2012). Interaction of strigolactone with the receptor causes hydrolysis of the ligand, which promotes the ligand–receptor complex interacting with MAX2, an F-box protein, leading to proteasome-mediated degradation of target proteins (Wang et al., 2013). The Arabidopsis mutant, *max2*, shows increased shoot branching and the gene encoding this protein is involved in the signaling of seed germination promoted by another lactone compound found in smoke, karrikin (Nelson et al., 2011; Stirnberg, van De Sande, & Leyser, 2002). Although it is well known that some plant seeds require smoke for germination and appear to be involved in the changes in floral species found within a forest after fires, there is no evidence for this compound derived from forest fire smoke to be naturally synthesized in planta to date.

In conclusion, the half dozen or so small molecules described earlier are critical growth regulators during plant growth and development. Until $\sim$20 years ago, it was believed that plants use only small molecules for regulating growth and development. However, the presence of a large family of orphan plasma membrane receptor-like kinase genes in the plant genome (but few or no G protein-coupled receptors), suggest that plant cells could be perceiving many undiscovered ligands at the plasma membrane via these kinases, perhaps to help coordinate growth and development during the many discrete steps of differentiation and function that occur at various parts and times in plant life.

3. PLASMA MEMBRANE AS THE SITE OF PEPTIDE LIGAND SENSING BY RECEPTORS

In animals, the plasma membrane sodium pump (Na^+/K^+-ATPase) provides the membrane potential and sodium gradient utilized by ion channels and cotransporters. In plants (and fungi), a plasma membrane proton pump plays a similar role in creating the primary electrochemical gradient that drives all of the channels and cotransporters. In this case, the gradient created by the pump is a much higher membrane potential (minus 250 mV resting potential) and a chemical gradient of protons. In fact, sodium is an unessential element in plants and plays no critical role in any specific plant process. Despite this important difference in fundamental energetics of the plasma membrane, changes in the cytoplasmic concentration of calcium are universal signaling events in both animals and plants. Mechanostimuli, neurotransmitters, and peptide growth factors all induce a cytoplasmic calcium transient as a second messenger, and this is often associated with phospholipid alterations and rapid changes in protein phosphorylation initiated at the plasma membrane. Although plants do not have neurons, they do exhibit rapid signal propagation via calcium waves at the organismal level in response to environmental stimuli such as abiotic stresses (Choi, Toyota, Kim, Hilleary, & Gilroy, 2014). Furthermore, the presence of many small open reading frame (ORF) genes with predicted secretion signal sequences in the genome clearly supports the idea that hormone-like peptides may play diverse physiological roles to activate intracellular signaling.

In order to biochemically identify such peptide ligands, we set up a high-throughput assay for endogenous peptides that induce a cytoplasmic calcium transient in Arabidopsis. Using a high-resolution reverse phase HPLC column, we fractionated all of the small proteins and peptides derived from a soluble extract and purified and sequenced a few micrograms of the peptide responsible for one of the major peaks of activity. The sequence of this peptide revealed that it was a homolog of RALF (rapid alkalinization factor) a

5000-Da peptide previously identified on the basis of its ability to cause alkalinization of the cellular medium upon bathing plant cells and predicted to interact with a receptor complex composed of 120 and 25 kDa proteins (Pearce, Moura, Stratmann, & Ryan, 2001b; Scheer, Pearce, & Ryan, 2005). In addition to its ability to induce cell wall alkalinization and cytoplasmic calcium transients, this peptide also suppresses cell elongation in roots when applied to plants. In order to obtain new information on the signaling pathway induced by RALF1, we performed a mass spectrometry-based quantitative analysis of the plasma membrane phosphoproteome in Arabidopsis seedlings. We observed that within 5 min of RALF treatment of Arabidopsis seedlings phosphorylation of a serine residue (S899) in the carboxy terminal regulatory tail of the Arabidopsis plasma membrane H^+-ATPase 2 (AHA2) was increased over fourfold without altering overall AHA2 protein level. In the same dataset of RALF-induced changes in the plasma membrane phosphoproteome, we identified and quantified ~600 phosphopeptides and the largest change (10- to 20-fold) in phosphopeptide abundance was observed in the FERONIA receptor kinase within two phosphopeptides located at the regulatory carboxy terminus. Based on this observation and by analogy to mammalian epidermal growth factor or insulin-induced increases in the phosphorylation of cognate tyrosine receptor kinases, we hypothesized that RALF acts through FERONIA to downregulate root growth. We tested this hypothesis using reverse genetics to study the effect of the absence of FERONIA on RALF sensitivity and also, via direct biochemical studies of RALF binding to FERONIA (Haruta, Sabat, Stecker, Minkoff, & Sussman, 2014). A recent study examining RALF-induced rapid FERONIA phosphorylation and ABA-induced FERONIA phosphorylation as a long-term response (i.e., 1–4 h incubation) suggests that FERONIA is also involved in abiotic stress responses (Chen et al., 2016).

4. FERONIA AND ITS GENE FAMILY IN PLANT GROWTH AND DEVELOPMENT

The FERONIA receptor kinase was first identified with a genetic mutant, *feronia* (or Sirene) in which the pollen tube does not stop its elongating growth when it reaches the ovule of female gamete, thus causing pollen tube overgrowth, reduced fertility, and semisterile phenotype (Escobar-Restrepo et al., 2007; Huck, Moore, Federer, & Grossniklaus, 2003; Rotman,

Gourgues, Guitton, Faure, & Berger, 2008; Rotman et al., 2003). The gene was named after the fertility Goddess, FERONIA from an Etruscan legend of ancient Italy. FERONIA is located at the surface of the synergid cells supporting the egg cells within the ovule and was thought to recognize an unknown signal molecule, presumably localized at the pollen tip. Once FERONIA interacts with signals from the pollen tube, the synergid cells undergo programmed cell death and the pollen tube bursts to release the sperm nuclei, resulting in gamete fusion. The extracellular domain of FERONIA is distantly related to malectin, a protein found in the endoplasmic reticulum of mammalian cells. Malectin is involved in protein glycosylation and interacts with a diglucose moiety (Schallus, Feher, Sternberg, Rybin, & Muhle-Goll, 2010; Schallus et al., 2008). Because of a small amount of sequence similarity of the ectodomain of FERONIA to mammalian malectin, it was speculated that FERONIA ectodomain may interact with sugar or other carbohydrate-like molecules present in the cell wall (Boisson-Dernier, Kessler, & Grossniklaus, 2011).

Although FERONIA was first discovered as a female determinant of pollen-egg recognition, the FERONIA gene is widely expressed in all vegetative tissues and developmental stages except in pollen, suggesting that its role is not limited to ovule fertilization (Lindner, Muller, Boisson-Dernier, & Grossniklaus, 2012). A role of FERONIA as a plasma membrane component involved in initiating cellular signaling in other cell types was also evident from experiments showing its interaction with guanine nucleotide exchange factors (GEFs), ROP-GEF, during root hair development (Duan, Kita, Li, Cheung, & Wu, 2010). The involvement of FERONIA in cell growth and/or polarized cell growth was observed in *feronia* mutants which display collapsed, burst, or arrested root hairs. The root hair is a single cell extended from the epidermal cell in roots and thus is often considered to share common "tip growth" regulatory mechanisms, including calcium ion, reactive oxygen species, GTP exchange reaction, with pollen tube cell elongation. Interestingly, a glycosylphosphatidylinositol (GPI) anchored protein, LORELEI-LIKE GPI ANCHORED protein 1 (LLG1), whose homologous gene LORELEI functions in ovule fertilization in a similar manner to FERONIA, is also reported to be required for root hair development. The knockout plant containing a null mutation of *llg1* showed defective root hair development resembling to *feronia* mutant. A subsequent biochemical experiment showed that LLG1 is a component of RALF1-FERONIA signaling and was demonstrated to interact with RALF1 and its receptor protein kinase, FERONIA (Li et al., 2015).

FERONIA belongs to a family of 16 receptor-like kinases known as CrRLKs (*Catharanthus roseus* receptor-like kinase), since the first gene encoding a member of this receptor kinase family was discovered from cell cultures of *C. roseus* (Madagascar periwinkle) (Schulze-Muth, Irmler, Schroder, & Schroder, 1996). A cDNA encoding this protein was expressed as a truncated version containing a protein kinase domain and it was demonstrated that the gene encodes an active protein kinase which autophosphorylates at serine and threonine residues within the protein. Another member of the CrRLK family, THESEUS, was reported in a study of suppressor screens of a cellulose synthase (*procuste1*, *prc1*) mutant, which showed defective hypocotyl elongation under dark conditions (Hematy et al., 2007). Recessive mutations in *theseus* can suppress a shorter hypocotyl phenotype seen with a cellulose synthase, *prc1* mutant, but this mutation does not affect the hypocotyl of wild-type plants. Thus, it was proposed that THESEUS is a sensor for cell wall integrity, although the molecular or biochemical definition for "cell wall integrity" remains unclear. One may also speculate that the effect of *theseus* mutation is visible only when the constraint by a rigid cell wall is reduced in the cellulose synthase mutant. If THESEUS is a negative regulator of cell expansion as is the case of FERONIA, all of the observations also fit a model in which THESEUS is normally restricting cell expansion at the plasma membrane and its knockout phenotype cannot be detected when the cell wall is restraining cell elongation, in the wild-type plant.

The idea that CrRLK is involved in sensing changes in cell wall composition hinted that FERONIA may be also playing a role during mechanosensing signal transduction. Plants naturally experience mechanical stresses from the environments or load-bearing of their own tissue weights as well as by touch from insects and fixed structures (e.g., peas wrapping their tendrils around a pole), and thus, sense changes in tensile strain by rapid changes in calcium concentration in the cytoplasm and pH in the cell wall. During this process, *feronia* mutants show impaired calcium or proton ion fluxes and reduced frequencies of penetrating into a high-density agar (Shih, Miller, Dai, Spalding, & Monshausen, 2014). In addition to the rapid changes in ion signaling, mechanical stimuli induce protein phosphorylation (Piotrowski, Liss, & Weiler, 1996). The FERONIA receptor kinase, which was shown to be an active kinase when expressed as a kinase domain in *Escherichia coli*, may be involved in mechanosensing with its kinase function since a kinase-negative mutant version of FERONIA showed a different activity compared with wild-type FERONIA. This is an intriguing result since the role of FERONIA in the ovule fertilization does not seem to rely on its kinase activity (Kessler, Lindner, Jones, & Grossniklaus, 2015); instead, a cytoplasmic protein kinase may be involved in its phosphorylation signaling downstream of the CrRLKs (Boisson-Dernier, Franck, Lituiev, & Grossniklaus, 2015). It remains to be examined whether a kinase-negative mutation of this cytoplasmic kinase affects its biological function.

Since the plasma membrane H^+-ATPase is an essential protein involved in all aspects of plant physiology (Haruta et al., 2010; Haruta & Sussman, 2012), RALF and FERONIA acting directly through changes in the catalytic activity of the pump may contribute to the pleiotropic mutant phenotypes observed with *feronia* knockout plants. Yu et al. observed that growth of the *feronia* mutant is hypersensitive to the presence of ABA (Yu et al., 2012). A genetic mutant screen of ethylene hypersensitive plants also identified *feronia* mutants (Deslauriers & Larsen, 2010). Although there is no causal link established, FERONIA has also been reported to function by directly interacting with a metabolic enzyme, glyceraldehyde-3-phosphate dehydrogenase to inhibit starch accumulation (Yang et al., 2015), as well as interacting with S-adenosylmethionine synthetase to suppresses ethylene production (Mao et al., 2015).

Two other members of the CrRLK family, ANXUR1 and ANXUR2, are collectively required for pollen tube elongation as a homozygous double knockout for the two genes caused lethality due to pollen germination and elongation defects (Boisson-Dernier et al., 2009; Miyazaki et al., 2009). The mutant pollen grains with the haplotype *anx1/anx2* can be formed, but the pollen tube ruptures before they elongate. From the mutants' phenotype, it is possible that normal function of the ANXUR receptor kinase is restricting pollen tube elongation. When restriction is removed in the double mutants, a pollen tube bursts. It remains to be tested whether FERONIA and ANXUR share the same ligands during fertilization functions.

5. ORGAN SIZE AND GROWTH ARE DETERMINED BY STIMULATION AND INHIBITION BY SIGNALING PATHWAYS IN PLANTS AND ANIMALS

In the first several days after seed imbibition starts, storage nutrients support organ expansion until a plant enters a developmental stage in which it is capable of photosynthesis. In this transition, changes in hormone metabolism occur to regulate growth and to optimally respond to the environment (Bewley, 1997; Bhalerao et al., 2002; Fait et al., 2006). Young seedlings are not photosynthetically active and their growth relies on the preserved nutrients in the seed endosperm. Thus, during the seedling stage at which plants establish photosynthetic capability, plants have sensitive sensor mechanisms to perceive light intensity and to modify growth and development. A seedling that is growing faster than necessary under a condition in which there is not enough light available, would result in deleterious consequences by using up the reserved nutrients. Plants have growth regulatory mechanisms to actively "ratchet" down the growth rate and reduce the rate of cell elongation. In the authors' previously described discovery of RALF-FERONIA, a peptide ligand and receptor kinase pair, acting as a negative regulator of cell elongation, fit into this model (Haruta et al., 2014). The primary action of this peptide–receptor signaling is to inactivate plasma membrane H^+-ATPase that provides energy to the plasma membrane, so that they can import the nutrients required for cell growth (Fig. 1).

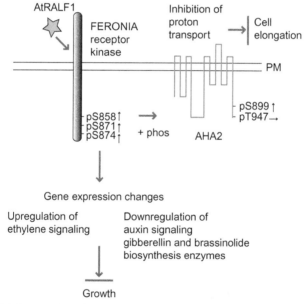

Fig. 1 Model for rapid RALF-FERONIA-regulated cell growth. RALF binding to the ectodomain of FERONIA receptor kinase induces a rapid phosphorylation at regulatory Ser residues at the carboxy terminus. RALF-induced activation of FERONIA either directly or indirectly increases the level of phosphorylation of AHA2 at Ser899. This results in inhibition of proton efflux, a rise in cell wall pH and suppression of cell growth. In the long term (e.g., 30 min), RALF modulates the expression of genes involved in cell elongation. Transcripts for genes encoding proteins involved in growth stimulatory mechanisms are reduced by RALF action.

Signaling mechanisms by which cells actively suppress growth and development are also essential in mammalian systems. Organismal growth in animals is mediated by cell proliferation, growth, and migration. Failure in appropriate regulation of such inhibitory mechanisms causes organ overgrowth or cancer development. For example, loss-of-function mutations in Hippo serine/threonine kinase of fruit fly result in organ overgrowth that disrupt eye and wing pattern development (Harvey, Pfleger, & Hariharan, 2003). Another example is kinase-mediated inhibition of cell migration. A deletion of ERK8 (extracellular signal-regulated kinase, also known as MAPK15, ERK7) causes twofold faster cell migration in Hela cells (Chia, Tham, Gill, Bard-Chapeau, & Bard, 2014). Moreover, ERK8 is known to induce degradation of the estrogen receptor and a loss of its expression is correlated with increased progression of breast cancer (Henrich et al., 2003).

6. RALF FAMILY AND FUNCTION IN PLANT GROWTH AND DEVELOPMENT

With the completion of many plant and microorganism genomes, bioinformatic analyses identified a family of RALF-like sequences throughout the plant kingdom. In the Arabidopsis genome, there are 34 RALF-like sequences. The widespread expression profiles of RALF-like sequences were also evident from some earlier mRNA sequencing analyses obtained by the expression sequence tags approach from the reproductive tissue of tomato plants (Germain, Chevalier, Caron, & Matton, 2005). A loss-of-function study of a tomato RALF (*Solanum chacoense* RALF3, ScRALF3) revealed that silencing this gene caused a reduction in seed production due to defective development of the embryo sac (Chevalier, Loubert-Hudon, & Matton, 2013). Female gamete development, referred to as megagametogenesis, involves a series of meiotic cell divisions, cell death, and mitosis to produce eight nuclei. Since our Arabidopsis RALF1/FERONIA peptide hormone receptor kinase cognate pair model predicts that RALF1 suppresses cell growth by inactivation of AHA2, the phenotype seen with ScRALF3-silenced tomato plants is consistent with this hypothesis that growth termination is a necessary process during female gametogenesis.

The involvement of RALF peptides in pollen tube elongation was examined by an in vitro assay using a tomato system. Application of SlRALF

(*Solanum lycopersicum* RALF) inhibits pollen tube elongation, but does not affect pollen grain hydration, with sequence specific and pH-dependent effects (Covey et al., 2010). Additional evidence that RALF-like peptides play roles in pollen function was observed in a transcriptome study via RNA sequencing. RALF-like 8 and 9 transcripts are among those most abundant and rank at 7th and 11th, respectively, among the 33,614 genes in pollen (Loraine, McCormick, Estrada, Patel, & Qin, 2013). RALF-like 8, which is expressed at a very low level in seedlings in the normal growth condition, was shown to be transcriptionally induced in Arabidopsis seedlings when exposed to consecutive stresses by a challenge with parasitic-nematode infestation, followed by water-deficit condition (Atkinson, Lilley, & Urwin, 2013). Overexpression of RALF-like 8 gene under a constitutive promoter resulted in shorter roots with excessive root hair production. The RALF-like 8 overexpressor is also more susceptible to drought stress. The effects of RALF peptide overexpression were also reported with RALF1 and RALF-like 23, both resulting in semi-dwarf phenotypes (Matos, Fiori, Silva-Filho, & Moura, 2008; Srivastava, Liu, Guo, Yin, & Howell, 2009). The dwarfism seen with RALF1 overexpressors was dependent on processing of the prepro-RALF precursors into the mature peptides since mutating the predicted protease recognition site of the RALF precursor abolished the dwarf phenotype.

Whereas the phenotypes observed with gene overexpression experiments require caution in interpretation due to possible indirect effects, gene knockout or silencing experiments provide more reliable clues on biological function of RALF peptide. Wu et al. reported that tobacco appears to have a smaller RALF gene family than Arabidopsis and that tobacco plants in which RALF expression is silenced showed reduced root hair development and a longer root (Wu et al., 2007).

7. RALF-LIKE PEPTIDES ARE NOT ONLY PRODUCED BY PLANTS BUT ALSO BY PATHOGENIC MICROBES

Plants in nature are continuously exposed to microbes, some of which are beneficial for normal growth and development (symbiosis) or some that are not (pathogenic infection). Both plants and symbiotic or pathogenic microbes secrete signaling molecules. Recent studies have identified genes encoding RALF-like peptides in the genome of a pathogenic fungus, *Fusarium oxysporum* (Masachis et al., 2016). The fungal-derived RALF peptide (f-RALF) was shown to act in a similar manner to endogenous plant RALF peptide and was found to be capable of inducing extracellular alkalinization in tomato suspension cells and inhibiting root elongation in tomato seedlings. More importantly, a *feronia* knockout plant which is insensitive to growth inhibition by Arabidopsis RALF1 is also insensitive to f-RALF peptide. Based on these observations, it was proposed that the fungus is suppressing plant cell growth to increase infection efficiency. In agreement with this idea, the *feronia* mutant, which does not respond to fungal RALF, is resistant to the pathogenic infection. Also consistent with this model is the observation that a deletion mutant of *F. oxysporum*, which lacks a gene encoding the f-RALF peptide infected tomato plants with a much reduced efficiency compared with the wild-type *F. oxysporum* strain.

Thynne et al. also identified RALF-like sequences in the genomes of 26 phytopathogenic fungi (Thynne et al., 2016). A RALF-like peptide, RALF-B from *F. oxysporum* f. sp. *lycopersici*, inhibits the growth of tomato seedlings. However, in this study, the authors suggested that the RALF peptide from *F. oxysporum* can influence pathogen virulence in only limited conditions, e.g., when plants were grown in media with reduced levels of nutrients.

8. OTHER ENDOGENOUS HORMONE-LIKE PEPTIDES AND THEIR RECEPTOR PAIRS FOR PLANT GROWTH, DEVELOPMENT, AND PHYSIOLOGY

Discovery of the first peptide acting as a hormone-like signal in plants dates back to 1991 when an 18-amino-acid long peptide, systemin was biochemically isolated as a factor capable of inducing a proteinase inhibitor protein from tomato leaves (Pearce, Strydom, Johnson, & Ryan, 1991). This peptide was reported to interact with a 160-kDa receptor protein at the plasma membrane and induced the downstream signals resulting in the expression of an array of defense-related proteins (Scheer & Ryan, 1999). This study opened up a paradigm of plant peptide signals, which are now perceived as a critical part of plant signaling.

Plant cells have sensitive mechanisms for responding to extracellular signal molecules. Changes in transmembrane ion fluxes and protein phosphorylation are observed very rapidly upon cellular recognition of signal compounds (Boller, 1995). Peptide-based signal molecules including systemin can elicit extracellular alkalinization of cultured cells (Felix & Boller, 1995). Thus, monitoring pH changes in the medium of cultured cells has been used as an assay to seek signal- or elicitor-like molecules which activate cellular signaling for the activation of defense responses or growth regulation at the plasma membrane (Felix, Duran, Volko, & Boller, 1999; Kunze et al., 2004). A 23-amino-acid peptide, AtPep1 (*Arabidopsis thaliana* elicitor peptide1) was isolated using this alkalinization assay and shown to induce pathogen defense in Arabidopsis through an leucine-rich repeat (LRR) receptor-like kinase PEPR1, which forms a heterodimer with the BAK1 receptor kinase (Huffaker, Pearce, & Ryan, 2006; Tang et al., 2015; Yamaguchi, Pearce, & Ryan, 2006). The AtPep1 sequence shows limited sequence conservation when compared with maize peptides, possibly suggesting that this peptide–receptor pair may be functioning in a species-specific manner (Huffaker, Dafoe, & Schmelz, 2011; Lori et al., 2015).

As is the case of mammalian peptide growth factors or hormones, posttranslational modification of peptides appears to be a common mechanism in plants. Phytosulfokine, a sulfated penta peptide isolated from asparagus, which can induce cell proliferation in cultured cells, is produced as a precursor protein and processed to a small biologically active peptide (Matsubayashi & Sakagami, 1996). A receptor for phytosulfokine, PSKR1 was identified via photoaffinity cross-linking and purification, followed by mass spectrometry sequencing. This peptide is recognized by a cell-surface receptor, LRR receptor-like kinase whose knockout caused shorter roots and overexpression caused longer roots and enhanced proliferation of cells (Matsubayashi, Ogawa, Kihara, Niwa, & Sakagami, 2006; Matsubayashi, Ogawa, Morita, & Sakagami, 2002). A crystal structural study revealed that PSK ligand binds to the PSKR1 receptor, which causes allosteric activation of the receptor and allows the ligand–receptor complex to interact with coreceptors known as somatic embryogenesis receptor-like kinases (SERKs) (Wang et al., 2015). An additional search for sulfated peptides resulted in the isolation of an 18-amino-acid tyrosine-sulfated glycopeptide, PSY1 from the media of Arabidopsis cell culture (Amano, Tsubouchi, Shinohara, Ogawa, & Matsubayashi, 2007). PSY1 stimulates root growth through a family of PSKR1, which interacts with and activates a plasma membrane H^+-ATPase (Fuglsang et al., 2014). Posttranslational modification of bioactive peptides was also reported with systemin-like peptides from the Solanaceae family. Those peptides contain hydroxylated proline residues and glycosylation composed of pentose units. Multiple peptides are produced from a single precursor preproprotein and bioactive peptides induce wound-responsive genes, which function in antiherbivory defense responses (Pearce, Moura, Stratmann, & Ryan, 2001a; Pearce & Ryan, 2003); however, receptor proteins for those systemin or systemin-like peptides remain to be clearly identified.

Some peptide–receptor pairs have been identified via genetic studies and later confirmed to directly interact using molecular biology techniques such as coimmunoprecipitation or biochemical cross-linking assay. A pair of CLAVATA3 (CLV3) ligand and CLAVATA1 (CLV1) receptor kinases is required for the maintenance of meristem cells to balance cell proliferation and differentiation, since insertion or point mutants of those genes cause increased apical bud formation (Clark, Williams, & Meyerowitz, 1997; Fletcher, Brand, Running, Simon, & Meyerowitz, 1999). The presence of an active form of CLV3 peptide was identified by in situ MALDI mass spectrometry (Kondo et al., 2006). A mass spectrum for the 12-amino-acid peptide localized at the carboxy terminus of CLV3 precursor protein was detected from Arabidopsis seedlings overexpressing the precursor protein. CLV3 peptide detected in situ contained two hydroxylated proline residues, although the hydroxylation did not influence the biological activity of the peptide. Photoaffinity labeling and a binding assay of CLV3 peptide interacting with CLV1 receptor also showed that CLV1-related receptor kinases, BAM1, 2, and 3 (Barely Any Meristem) are coreceptors for the CLV1–3 complex (DeYoung et al., 2006; Shinohara & Matsubayashi, 2015).

The Arabidopsis genome contains 26 genes encoding CLE peptides (CLAVATA3/embryo-surrounding region related, ESR) (Cock & McCormick, 2001). The CLE family shows the secretory signal sequence at the amino terminus and ~14-amino-acid conserved peptide sequence at the carboxy terminus. Application of synthetic peptides corresponding to the carboxy termini of CLV3, CLE19, or CLE40 caused shorter and thinner roots and decreases in the number of root meristematic cells (Fiers et al., 2005). Through biochemical search and purification, CLE41/44 peptide, originally named "tracheary element differentiation inhibitory factor" was found to be a factor that suppresses differentiation of mesophyll cells (cells found within leaf) into tracheary element (conducting cells in the vascular tissue) (Ito et al., 2006). CLE-like sequences are also found in the nematode genome and the precursor protein for the CLE-like peptide from potato cyst nematode *Globodera rostochiensis* (GrCLE1) was correctly processed in plants (Guo, Ni, Denver, Wang, & Clark, 2011; Wang et al., 2010). GrCLE1 inhibits root growth at the similar peptide concentration range as the endogenous peptide, CLV3. Moreover, GrCLE1 binds to Arabidopsis CLV2, BAM1, and BAM2 receptor kinase proteins, supporting a model in which the parasitic nematode uses GrCLE1 to mimic CLV3 peptide action.

Another example of a peptide–receptor pair identified initially based on a correlation of genetic mutant phenotypes is IDA (inflorescence deficient in abscission) peptide ligand, which interacts with its receptor HAESA (Latin word meaning to stick to) kinase and induces organ abscission (Butenko et al., 2003; Jinn, Stone, & Walker, 2000). In both *ida* and *haesa* mutants, floral organs including petal and stamens remain attached throughout silique (seed pod) development due to the loss of activation of abscission signaling. IDA interaction with HAESA induces the recruitment of SERKs and results in transphosphorylation of the two receptor kinases (Meng et al., 2016). Agreeing with this biochemical model, triple mutants of *serks* also exhibit defective organ abscission. Crystal structural analyses demonstrated that IDA dodeca peptide binds to HAESA with a weak affinity and SERK1 stabilizes the ligand–receptor complex (Santiago et al., 2016).

The availability of many plant genome sequences and bioinformatics tools has facilitated the identification of peptide ligands pairing with their receptor proteins. Based on the hypothesis that extracellularly secreted peptide signals are found in the genes encoding precursor proteins, small ORF genes were computationally screened. With this approach, a novel 15-amino-acid peptide, CEP1 (C-terminally encoded peptide 1) was identified (Ohyama, Ogawa, & Matsubayashi, 2008). Photoaffinity labeling assays performed with CEP1 peptide identified two LRR receptor kinases, from an Arabidopsis receptor kinase expression library produced in tobacco cells (Tabata et al., 2014). CEP-encoding genes were shown to be induced by nitrogen starvation, and together with their cognate receptor kinases, this family of peptides senses nitrogen availability in plants and regulates lateral root development. Peptide ligands, for a receptor-like kinase protein, known for regulating developmental pathways that produce or suppress guard cell production were also identified via a bioinformatic screen searching for small ORF genes. Peptides, named EPIDERMAL PATTERNING FACTOR 1 (EPF1), were found to suppress stomatal development by interacting with ERECTA receptor kinase (Hara, Kajita, Torii, Bergmann, & Kakimoto, 2007). Interestingly, stomatal development is also positively regulated by ERECTA interaction with another peptide, stomagen (Sugano et al., 2010). A current model suggests that competitive binding of stomagen and EPF1 onto the ERECTA receptor kinase determines stomatal density and patterning (Lee et al., 2015). EPF1 peptide induces heteromerization of ERECTA and SERKs in planta and this association predicts that the two receptor kinases cause transphosphorylation of the proteins (Meng et al., 2015). This biochemical result is consistent with the genetic evidence that *serk* quadruple mutant also showed increased stomatal production.

In addition to RALF and EPF1, other peptides with cysteine disulfide bonds are involved in other cell-to-cell communications. A peptide related to defensin, LURE acts as an ovule-secreted signal to attract the pollen tube. LURE peptide, first identified based on gene expression profiling, was heterogeneously produced from *E. coli* and its function of guiding the pollen tube to the ovule was demonstrated (Okuda et al., 2009). A family of RLKs that are required for sensing LURE peptide in the pollen tube was identified and some of them were shown to directly interact with the peptide (Takeuchi & Higashiyama, 2016; Wang et al., 2016). Another example for a peptide hormone with cysteine disulfide bonds is the peptide–receptor pair for pollen perception by stigma (tip of the female structure) in the Brassicaceae family including Arabidopsis. A genetic mapping study revealed that this self-incompatibility locus contained the ligand and receptor genes. The inability of self-fertilization in some species of the Brassicaceae family is mediated by a peptide ligand, S-locus cysteine-rich protein (SCR) interacting with its receptor, S-locus receptor kinase (SRK). Binding of an incompatible SCR to SRK prevents pollen grain hydration, and thus, highly polymorphic sequences of the ligands and receptors within the species contribute to a higher degree of outcrossing (Kachroo, Schopfer, Nasrallah, & Nasrallah, 2001; Schopfer, Nasrallah, & Nasrallah, 1999; Takayama et al., 2001).

9. CONCLUSION AND FUTURE PERSPECTIVES

The development and growth of plants are regulated via the action of various hormones mediating either stimulation or inhibition of cell growth. The availability of Arabidopsis genetic and genomic resources has contributed greatly to revealing how plant hormones are perceived and act. Use of novel new techniques including chemical genetics as well as functional phosphoproteomic screens have made it possible to reveal the receptor protein identities for some important hormones. As we learn more about the complexity of intersecting hormone signaling pathways, we realize the value of a systems approach that includes profiling many different biomolecules, including proteins and their posttranslational modifications. We now know that some phytohormone receptors (e.g., auxin, GA, and jasmonate) are

F-box proteins which act through the proteolysis of transcriptional repressors. The Arabidopsis genome carries ~700 F-box proteins, and moreover, more than 5% of the genome encodes the proteins involved in the ubiquitin proteasome system (Gagne, Downes, Shiu, Durski, & Vierstra, 2002; Smalle & Vierstra, 2004). It is possible that there are many more small molecule growth regulators yet to be identified. Similarly, ligands for several hundreds of membrane-bound receptor kinase proteins still remain to be identified. Finally, due to the transient nature of protein interactions and phosphorylated amino acids, the continued development of robust and high-throughput biochemical assays will be helpful in order to reveal yet to be known signaling components that lead to plant growth regulation.

REFERENCES

Amano, Y., Tsubouchi, H., Shinohara, H., Ogawa, M., & Matsubayashi, Y. (2007). Tyrosine-sulfated glycopeptide involved in cellular proliferation and expansion in Arabidopsis. *Proceedings of the National Academy of Sciences of the United States of America*, 104(46), 18333–18338.

Arite, T., Umehara, M., Ishikawa, S., Hanada, A., Maekawa, M., Yamaguchi, S., & Kyozuka, J. (2009). d14, a strigolactone-insensitive mutant of rice, shows an accelerated outgrowth of tillers. *Plant & Cell Physiology*, 50(8), 1416–1424.

Atkinson, N. J., Lilley, C. J., & Urwin, P. E. (2013). Identification of genes involved in the response of Arabidopsis to simultaneous biotic and abiotic stresses. *Plant Physiology*, 162(4), 2028–2041.

Bewley, J. D. (1997). Seed germination and dormancy. *Plant Cell*, 9(7), 1055–1066.

Bhalerao, R. P., Eklof, J., Ljung, K., Marchant, A., Bennett, M., & Sandberg, G. (2002). Shoot-derived auxin is essential for early lateral root emergence in Arabidopsis seedlings. *The Plant Journal*, 29(3), 325–332.

Bleecker, A. B., Estelle, M. A., Somerville, C., & Kende, H. (1988). Insensitivity to ethylene conferred by a dominant mutation in Arabidopsis thaliana. *Science*, 241(4869), 1086–1089.

Boisson-Dernier, A., Franck, C. M., Lituiev, D. S., & Grossniklaus, U. (2015). Receptor-like cytoplasmic kinase MARIS functions downstream of CrRLK1L-dependent signaling during tip growth. *Proceedings of the National Academy of Sciences of the United States of America*, 112(39), 12211–12216.

Boisson-Dernier, A., Kessler, S. A., & Grossniklaus, U. (2011). The walls have ears: The role of plant CrRLK1Ls in sensing and transducing extracellular signals. *Journal of Experimental Botany*, 62(5), 1581–1591.

Boisson-Dernier, A., Roy, S., Kritsas, K., Grobei, M. A., Jaciubek, M., Schroeder, J. I., & Grossniklaus, U. (2009). Disruption of the pollen-expressed FERONIA homologs ANXUR1 and ANXUR2 triggers pollen tube discharge. *Development*, 136(19), 3279–3288.

Boller, T. (1995). Chemoperception of microbial signals in plant cells. *Annual Review of Plant Physiology and Plant Molecular Biology*, 46, 189–214.

Butenko, M. A., Patterson, S. E., Grini, P. E., Stenvik, G. E., Amundsen, S. S., Mandal, A., & Aalen, R. B. (2003). Inflorescence deficient in abscission controls floral organ abscission in Arabidopsis and identifies a novel family of putative ligands in plants. *Plant Cell*, 15(10), 2296–2307.

Calderon Villalobos, L. I., Lee, S., De Oliveira, C., Ivetac, A., Brandt, W., Armitage, L., ... Estelle, M. (2012). A combinatorial TIR1/AFB-Aux/IAA co-receptor system for differential sensing of auxin. *Nature Chemical Biology*, *8*(5), 477–485.

Chang, C., Kwok, S. F., Bleecker, A. B., & Meyerowitz, E. M. (1993). Arabidopsis ethylene-response gene ETR1: Similarity of product to two-component regulators. *Science*, *262*(5133), 539–544.

Chen, J., Yu, F., Liu, Y., Du, C., Li, X., Zhu, S., ... Luan, S. (2016). FERONIA interacts with ABI2-type phosphatases to facilitate signaling cross-talk between abscisic acid and RALF peptide in Arabidopsis. *Proceedings of the National Academy of Sciences of the United States of America*, *113*(37), E5519–E5527.

Chevalier, E., Loubert-Hudon, A., & Matton, D. P. (2013). ScRALF3, a secreted RALF-like peptide involved in cell-cell communication between the sporophyte and the female gametophyte in a solanaceous species. *The Plant Journal*, *73*(6), 1019–1033.

Chia, J., Tham, K. M., Gill, D. J., Bard-Chapeau, E. A., & Bard, F. A. (2014). ERK8 is a negative regulator of O-GalNAc glycosylation and cell migration. *eLife*, *3*, e01828.

Choe, S., Dilkes, B. P., Gregory, B. D., Ross, A. S., Yuan, H., Noguchi, T., ... Feldmann, K. A. (1999). The Arabidopsis dwarf1 mutant is defective in the conversion of 24-methylenecholesterol to campesterol in brassinosteroid biosynthesis. *Plant Physiology*, *119*(3), 897–907.

Choi, W. G., Toyota, M., Kim, S. H., Hilleary, R., & Gilroy, S. (2014). Salt stress-induced Ca^{2+} waves are associated with rapid, long-distance root-to-shoot signaling in plants. *Proceedings of the National Academy of Sciences of the United States of America*, *111*(17), 6497–6502.

Clark, S. E., Williams, R. W., & Meyerowitz, E. M. (1997). The CLAVATA1 gene encodes a putative receptor kinase that controls shoot and floral meristem size in Arabidopsis. *Cell*, *89*(4), 575–585.

Clouse, S. D., Langford, M., & McMorris, T. C. (1996). A brassinosteroid-insensitive mutant in Arabidopsis thaliana exhibits multiple defects in growth and development. *Plant Physiology*, *111*(3), 671–678.

Cock, J. M., & McCormick, S. (2001). A large family of genes that share homology with CLAVATA3. *Plant Physiology*, *126*(3), 939–942.

Covey, P. A., Subbaiah, C. C., Parsons, R. L., Pearce, G., Lay, F. T., Anderson, M. A., ... Bedinger, P. A. (2010). A pollen-specific RALF from tomato that regulates pollen tube elongation. *Plant Physiology*, *153*(2), 703–715.

Curran, T. G., Zhang, Y., Ma, D. J., Sarkaria, J. N., & White, F. M. (2015). MARQUIS: A multiplex method for absolute quantification of peptides and posttranslational modifications. *Nature Communications*, *6*, 5924.

Darwin, C., & Darwin, F. (1880). *The power of movement in plants*. London: John Murray.

Deslauriers, S. D., & Larsen, P. B. (2010). FERONIA is a key modulator of brassinosteroid and ethylene responsiveness in Arabidopsis hypocotyls. *Molecular Plant*, *3*(3), 626–640.

DeYoung, B. J., Bickle, K. L., Schrage, K. J., Muskett, P., Patel, K., & Clark, S. E. (2006). The CLAVATA1-related BAM1, BAM2 and BAM3 receptor kinase-like proteins are required for meristem function in Arabidopsis. *The Plant Journal*, *45*(1), 1–16.

Dharmasiri, N., Dharmasiri, S., Weijers, D., Lechner, E., Yamada, M., Hobbie, L., ... Estelle, M. (2005). Plant development is regulated by a family of auxin receptor F box proteins. *Developmental Cell*, *9*(1), 109–119.

Doebley, J., Stec, A., & Hubbard, L. (1997). The evolution of apical dominance in maize. *Nature*, *386*(6624), 485–488.

Duan, Q., Kita, D., Li, C., Cheung, A. Y., & Wu, H. M. (2010). FERONIA receptor-like kinase regulates RHO GTPase signaling of root hair development. *Proceedings of the National Academy of Sciences of the United States of America*, *107*(41), 17821–17826.

Enders, T. A., & Strader, L. C. (2015). Auxin activity: Past, present, and future. *The American Journal of Botany, 102*(2), 180–196.
Escobar-Restrepo, J. M., Huck, N., Kessler, S., Gagliardini, V., Gheyselinck, J., Yang, W. C., & Grossniklaus, U. (2007). The FERONIA receptor-like kinase mediates male-female interactions during pollen tube reception. *Science, 317*(5838), 656–660.
Fait, A., Angelovici, R., Less, H., Ohad, I., Urbanczyk-Wochniak, E., Fernie, A. R., & Galili, G. (2006). Arabidopsis seed development and germination is associated with temporally distinct metabolic switches. *Plant Physiology, 142*(3), 839–854.
Farmer, E. E., & Ryan, C. A. (1990). Interplant communication: Airborne methyl jasmonate induces synthesis of proteinase inhibitors in plant leaves. *Proceedings of the National Academy of Sciences of the United States of America, 87*(19), 7713–7716.
Felix, G., & Boller, T. (1995). Systemin induces rapid ion fluxes and ethylene biosynthesis in Lycopersicon peruvianum cells. *Plant Journal, 7,* 381–389.
Felix, G., Duran, J. D., Volko, S., & Boller, T. (1999). Plants have a sensitive perception system for the most conserved domain of bacterial flagellin. *The Plant Journal, 18*(3), 265–276.
Fiers, M., Golemiec, E., Xu, J., van der Geest, L., Heidstra, R., Stiekema, W., & Liu, C. M. (2005). The 14-amino acid CLV3, CLE19, and CLE40 peptides trigger consumption of the root meristem in Arabidopsis through a CLAVATA2-dependent pathway. *Plant Cell, 17*(9), 2542–2553.
Finkelstein, R. (2013). Abscisic acid synthesis and response. *The Arabidopsis Book, 11,* e0166.
Finlayson, S. A. (2007). Arabidopsis Teosinte Branched1-like 1 regulates axillary bud outgrowth and is homologous to monocot Teosinte Branched1. *Plant & Cell Physiology, 48*(5), 667–677.
Fletcher, J. C., Brand, U., Running, M. P., Simon, R., & Meyerowitz, E. M. (1999). Signaling of cell fate decisions by CLAVATA3 in Arabidopsis shoot meristems. *Science, 283*(5409), 1911–1914.
Fuglsang, A. T., Kristensen, A., Cuin, T. A., Schulze, W. X., Persson, J., Thuesen, K. H., ... Palmgren, M. G. (2014). Receptor kinase-mediated control of primary active proton pumping at the plasma membrane. *The Plant Journal, 80*(6), 951–964.
Gagne, J. M., Downes, B. P., Shiu, S. H., Durski, A. M., & Vierstra, R. D. (2002). The F-box subunit of the SCF E3 complex is encoded by a diverse superfamily of genes in Arabidopsis. *Proceedings of the National Academy of Sciences of the United States of America, 99*(17), 11519–11524.
Geldner, N., Hyman, D. L., Wang, X., Schumacher, K., & Chory, J. (2007). Endosomal signaling of plant steroid receptor kinase BRI1. *Genes & Development, 21*(13), 1598–1602.
Germain, H., Chevalier, E., Caron, S., & Matton, D. P. (2005). Characterization of five RALF-like genes from Solanum chacoense provides support for a developmental role in plants. *Planta, 220*(3), 447–454.
Gonzalez-Guzman, M., Pizzio, G. A., Antoni, R., Vera-Sirera, F., Merilo, E., Bassel, G. W., ... Rodriguez, P. L. (2012). Arabidopsis PYR/PYL/RCAR receptors play a major role in quantitative regulation of stomatal aperture and transcriptional response to abscisic acid. *Plant Cell, 24*(6), 2483–2496.
Gray, W. M., Kepinski, S., Rouse, D., Leyser, O., & Estelle, M. (2001). Auxin regulates SCF(TIR1)-dependent degradation of AUX/IAA proteins. *Nature, 414*(6861), 271–276.
Grove, M., Spencer, G., Rohwedder, W., Mandava, N., Worley, J., Warthen, J., ... Cook, J. (1979). Brassinolide, a plant growth-promoting steroid isolated from Brassica napus pollen. *Nature, 281,* 216–217.

Guo, Y., Ni, J., Denver, R., Wang, X., & Clark, S. E. (2011). Mechanisms of molecular mimicry of plant CLE peptide ligands by the parasitic nematode Globodera rostochiensis. *Plant Physiology, 157*(1), 476–484.

Hall, J. M., Couse, J. F., & Korach, K. S. (2001). The multifaceted mechanisms of estradiol and estrogen receptor signaling. *The Journal of Biological Chemistry, 276*(40), 36869–36872.

Hamiaux, C., Drummond, R. S., Janssen, B. J., Ledger, S. E., Cooney, J. M., Newcomb, R. D., & Snowden, K. C. (2012). DAD2 is an alpha/beta hydrolase likely to be involved in the perception of the plant branching hormone, strigolactone. *Current Biology, 22*(21), 2032–2036.

Hara, K., Kajita, R., Torii, K. U., Bergmann, D. C., & Kakimoto, T. (2007). The secretory peptide gene EPF1 enforces the stomatal one-cell-spacing rule. *Genes & Development, 21*(14), 1720–1725.

Haruta, M., Burch, H. L., Nelson, R. B., Barrett-Wilt, G., Kline, K. G., Mohsin, S. B., ... Sussman, M. R. (2010). Molecular characterization of mutant Arabidopsis plants with reduced plasma membrane proton pump activity. *The Journal of Biological Chemistry, 285*(23), 17918–17929.

Haruta, M., Sabat, G., Stecker, K., Minkoff, B. B., & Sussman, M. R. (2014). A peptide hormone and its receptor protein kinase regulate plant cell expansion. *Science, 343*(6169), 408–411.

Haruta, M., & Sussman, M. R. (2012). The effect of a genetically reduced plasma membrane proton motive force on vegetative growth of Arabidopsis. *Plant Physiology, 158*(3), 1158–1171.

Harvey, K. F., Pfleger, C. M., & Hariharan, I. K. (2003). The Drosophila Mst ortholog, hippo, restricts growth and cell proliferation and promotes apoptosis. *Cell, 114*(4), 457–467.

Hematy, K., Sado, P. E., Van Tuinen, A., Rochange, S., Desnos, T., Balzergue, S., ... Hofte, H. (2007). A receptor-like kinase mediates the response of Arabidopsis cells to the inhibition of cellulose synthesis. *Current Biology, 17*(11), 922–931.

Henrich, L. M., Smith, J. A., Kitt, D., Errington, T. M., Nguyen, B., Traish, A. M., & Lannigan, D. A. (2003). Extracellular signal-regulated kinase 7, a regulator of hormone-dependent estrogen receptor destruction. *Molecular and Cellular Biology, 23*(17), 5979–5988.

Hothorn, M., Belkhadir, Y., Dreux, M., Dabi, T., Noel, J. P., Wilson, I. A., & Chory, J. (2011). Structural basis of steroid hormone perception by the receptor kinase BRI1. *Nature, 474*(7352), 467–471.

Huck, N., Moore, J. M., Federer, M., & Grossniklaus, U. (2003). The Arabidopsis mutant feronia disrupts the female gametophytic control of pollen tube reception. *Development, 130*(10), 2149–2159.

Huffaker, A., Dafoe, N. J., & Schmelz, E. A. (2011). ZmPep1, an ortholog of Arabidopsis elicitor peptide 1, regulates maize innate immunity and enhances disease resistance. *Plant Physiology, 155*(3), 1325–1338.

Huffaker, A., Pearce, G., & Ryan, C. A. (2006). An endogenous peptide signal in Arabidopsis activates components of the innate immune response. *Proceedings of the National Academy of Sciences of the United States of America, 103*(26), 10098–10103.

Inoue, T., Higuchi, M., Hashimoto, Y., Seki, M., Kobayashi, M., Kato, T., ... Kakimoto, T. (2001). Identification of CRE1 as a cytokinin receptor from Arabidopsis. *Nature, 409*(6823), 1060–1063.

Ito, Y., Nakanomyo, I., Motose, H., Iwamoto, K., Sawa, S., Dohmae, N., & Fukuda, H. (2006). Dodeca-CLE peptides as suppressors of plant stem cell differentiation. *Science, 313*(5788), 842–845.

Jinn, T. L., Stone, J. M., & Walker, J. C. (2000). HAESA, an Arabidopsis leucine-rich repeat receptor kinase, controls floral organ abscission. *Genes & Development, 14*(1), 108–117.
Kachroo, A., Schopfer, C. R., Nasrallah, M. E., & Nasrallah, J. B. (2001). Allele-specific receptor-ligand interactions in Brassica self-incompatibility. *Science, 293*(5536), 1824–1826.
Kessler, S. A., Lindner, H., Jones, D. S., & Grossniklaus, U. (2015). Functional analysis of related CrRLK1L receptor-like kinases in pollen tube reception. *EMBO Reports, 16*(1), 107–115.
Kondo, T., Sawa, S., Kinoshita, A., Mizuno, S., Kakimoto, T., Fukuda, H., & Sakagami, Y. (2006). A plant peptide encoded by CLV3 identified by in situ MALDI-TOF MS analysis. *Science, 313*(5788), 845–848.
Kunze, G., Zipfel, C., Robatzek, S., Niehaus, K., Boller, T., & Felix, G. (2004). The N terminus of bacterial elongation factor Tu elicits innate immunity in Arabidopsis plants. *Plant Cell, 16*(12), 3496–3507.
Lee, J. S., Hnilova, M., Maes, M., Lin, Y. C., Putarjunan, A., Han, S. K., ... Torii, K. U. (2015). Competitive binding of antagonistic peptides fine-tunes stomatal patterning. *Nature, 522*(7557), 439–443.
Li, J., & Chory, J. (1997). A putative leucine-rich repeat receptor kinase involved in brassinosteroid signal transduction. *Cell, 90*(5), 929–938.
Li, C., Yeh, F. L., Cheung, A. Y., Duan, Q., Kita, D., Liu, M. C., ... Wu, H. M. (2015). Glycosylphosphatidylinositol-anchored proteins as chaperones and co-receptors for FERONIA receptor kinase signaling in Arabidopsis. *eLife, 4*, e06587.
Lindner, H., Muller, L. M., Boisson-Dernier, A., & Grossniklaus, U. (2012). CrRLK1L receptor-like kinases: Not just another brick in the wall. *Current Opinion in Plant Biology, 15*(6), 659–669.
Lomin, S. N., Yonekura-Sakakibara, K., Romanov, G. A., & Sakakibara, H. (2011). Ligand-binding properties and subcellular localization of maize cytokinin receptors. *Journal of Experimental Botany, 62*(14), 5149–5159.
Loraine, A. E., McCormick, S., Estrada, A., Patel, K., & Qin, P. (2013). RNA-seq of Arabidopsis pollen uncovers novel transcription and alternative splicing. *Plant Physiology, 162*(2), 1092–1109.
Lori, M., van Verk, M. C., Hander, T., Schatowitz, H., Klauser, D., Flury, P., ... Bartels, S. (2015). Evolutionary divergence of the plant elicitor peptides (Peps) and their receptors: Interfamily incompatibility of perception but compatibility of downstream signalling. *Journal of Experimental Botany, 66*(17), 5315–5325.
Ma, Y., Szostkiewicz, I., Korte, A., Moes, D., Yang, Y., Christmann, A., & Grill, E. (2009). Regulators of PP2C phosphatase activity function as abscisic acid sensors. *Science, 324*(5930), 1064–1068.
Mahonen, A. P., Higuchi, M., Tormakangas, K., Miyawaki, K., Pischke, M. S., Sussman, M. R., ... Kakimoto, T. (2006). Cytokinins regulate a bidirectional phosphorelay network in Arabidopsis. *Current Biology, 16*(11), 1116–1122.
Mao, D., Yu, F., Li, J., Van de Poel, B., Tan, D., Li, J., ... Luan, S. (2015). FERONIA receptor kinase interacts with S-adenosylmethionine synthetase and suppresses S-adenosylmethionine production and ethylene biosynthesis in Arabidopsis. *Plant, Cell & Environment, 38*(12), 2566–2574.
Martins, S., Dohmann, E. M., Cayrel, A., Johnson, A., Fischer, W., Pojer, F., ... Vert, G. (2015). Internalization and vacuolar targeting of the brassinosteroid hormone receptor BRI1 are regulated by ubiquitination. *Nature Communications, 6*, 6151.
Masachis, S., Segorbe, D., Turrà, D., Leon-Ruiz, M., Fürst, U., El Ghalid, M., ... Di Pietro, A. (2016). A fungal pathogen secretes plant alkalinizing peptides to increase infection. *Nature Microbiology, 1*, 16043.

Matos, J. L., Fiori, C. S., Silva-Filho, M. C., & Moura, D. S. (2008). A conserved dibasic site is essential for correct processing of the peptide hormone AtRALF1 in Arabidopsis thaliana. *FEBS Letters, 582*(23–24), 3343–3347.

Matsubayashi, Y., Ogawa, M., Kihara, H., Niwa, M., & Sakagami, Y. (2006). Disruption and overexpression of Arabidopsis phytosulfokine receptor gene affects cellular longevity and potential for growth. *Plant Physiology, 142*(1), 45–53.

Matsubayashi, Y., Ogawa, M., Morita, A., & Sakagami, Y. (2002). An LRR receptor kinase involved in perception of a peptide plant hormone, phytosulfokine. *Science, 296*(5572), 1470–1472.

Matsubayashi, Y., & Sakagami, Y. (1996). Phytosulfokine, sulfated peptides that induce the proliferation of single mesophyll cells of Asparagus officinalis L. *Proceedings of the National Academy of Sciences of the United States of America, 93*(15), 7623–7627.

Meng, X., Chen, X., Mang, H., Liu, C., Yu, X., Gao, X., ... Shan, L. (2015). Differential function of Arabidopsis SERK family receptor-like kinases in stomatal patterning. *Current Biology, 25*(18), 2361–2372.

Meng, X., Zhou, J., Tang, J., Li, B., de Oliveira, M. V., Chai, J., ... Shan, L. (2016). Ligand-induced receptor-like kinase complex regulates floral organ abscission in Arabidopsis. *Cell Reports, 14*(6), 1330–1338.

Miller, C., Skoog, F., Okumura, F., Von Saltza, M., & Strong, F. (1956). Isolation, structure and synthesis of kinetin, a substance promoting cell division. *Journal of the American Chemical Society, 78*, 1375–1380.

Miyazaki, S., Murata, T., Sakurai-Ozato, N., Kubo, M., Demura, T., Fukuda, H., & Hasebe, M. (2009). ANXUR1 and 2, sister genes to FERONIA/SIRENE, are male factors for coordinated fertilization. *Current Biology, 19*(15), 1327–1331.

Nam, K. H., & Li, J. (2002). BRI1/BAK1, a receptor kinase pair mediating brassinosteroid signaling. *Cell, 110*(2), 203–212.

Nelson, D. C., Scaffidi, A., Dun, E. A., Waters, M. T., Flematti, G. R., Dixon, K. W., ... Smith, S. M. (2011). F-box protein MAX2 has dual roles in karrikin and strigolactone signaling in Arabidopsis thaliana. *Proceedings of the National Academy of Sciences of the United States of America, 108*(21), 8897–8902.

Noguchi, T., Fujioka, S., Takatsuto, S., Sakurai, A., Yoshida, S., Li, J., & Chory, J. (1999). Arabidopsis det2 is defective in the conversion of (24R)-24-methylcholest-4-En-3-one to (24R)-24-methyl-5alpha-cholestan-3-one in brassinosteroid biosynthesis. *Plant Physiology, 120*(3), 833–840.

Oh, M. H., Wang, X., Clouse, S. D., & Huber, S. C. (2012). Deactivation of the Arabidopsis BRASSINOSTEROID INSENSITIVE 1 (BRI1) receptor kinase by autophosphorylation within the glycine-rich loop. *Proceedings of the National Academy of Sciences of the United States of America, 109*(1), 327–332.

Ohyama, K., Ogawa, M., & Matsubayashi, Y. (2008). Identification of a biologically active, small, secreted peptide in Arabidopsis by in silico gene screening, followed by LC-MS-based structure analysis. *The Plant Journal, 55*(1), 152–160.

Okuda, S., Tsutsui, H., Shiina, K., Sprunck, S., Takeuchi, H., Yui, R., ... Higashiyama, T. (2009). Defensin-like polypeptide LUREs are pollen tube attractants secreted from synergid cells. *Nature, 458*(7236), 357–361.

Olsen, J. V., Blagoev, B., Gnad, F., Macek, B., Kumar, C., Mortensen, P., & Mann, M. (2006). Global, in vivo, and site-specific phosphorylation dynamics in signaling networks. *Cell, 127*(3), 635–648.

Park, S. Y., Fung, P., Nishimura, N., Jensen, D. R., Fujii, H., Zhao, Y., ... Cutler, S. R. (2009). Abscisic acid inhibits type 2C protein phosphatases via the PYR/PYL family of START proteins. *Science, 324*(5930), 1068–1071.

Park, S. Y., Peterson, F. C., Mosquna, A., Yao, J., Volkman, B. F., & Cutler, S. R. (2015). Agrochemical control of plant water use using engineered abscisic acid receptors. *Nature*, *520*(7548), 545–548.
Pearce, G., Moura, D. S., Stratmann, J., & Ryan, C. A. (2001a). Production of multiple plant hormones from a single polyprotein precursor. *Nature*, *411*(6839), 817–820.
Pearce, G., Moura, D. S., Stratmann, J., & Ryan, C. A., Jr. (2001b). RALF, a 5-kDa ubiquitous polypeptide in plants, arrests root growth and development. *Proceedings of the National Academy of Sciences of the United States of America*, *98*(22), 12843–12847.
Pearce, G., & Ryan, C. A. (2003). Systemic signaling in tomato plants for defense against herbivores. Isolation and characterization of three novel defense-signaling glycopeptide hormones coded in a single precursor gene. *The Journal of Biological Chemistry*, *278*(32), 30044–30050.
Pearce, G., Strydom, D., Johnson, S., & Ryan, C. A. (1991). A polypeptide from tomato leaves induces wound-inducible proteinase inhibitor proteins. *Science*, *253*(5022), 895–897.
Piotrowski, M., Liss, H., & Weiler, E. (1996). Touch-induced protein phosphorylation in mechanosensitive tendrils of Bryonia dioica Jacq. *Journal of Plant Physiology*, *147*, 539–546.
Qiao, H., Shen, Z., Huang, S. S., Schmitz, R. J., Urich, M. A., Briggs, S. P., & Ecker, J. R. (2012). Processing and subcellular trafficking of ER-tethered EIN2 control response to ethylene gas. *Science*, *338*(6105), 390–393.
Romanov, G. A., Lomin, S. N., & Schmulling, T. (2006). Biochemical characteristics and ligand-binding properties of Arabidopsis cytokinin receptor AHK3 compared to CRE1/AHK4 as revealed by a direct binding assay. *Journal of Experimental Botany*, *57*(15), 4051–4058.
Rotman, N., Gourgues, M., Guitton, A. E., Faure, J. E., & Berger, F. (2008). A dialogue between the SIRENE pathway in synergids and the fertilization independent seed pathway in the central cell controls male gamete release during double fertilization in Arabidopsis. *Molecular Plant*, *1*(4), 659–666.
Rotman, N., Rozier, F., Boavida, L., Dumas, C., Berger, F., & Faure, J. E. (2003). Female control of male gamete delivery during fertilization in Arabidopsis thaliana. *Current Biology*, *13*(5), 432–436.
Ruegger, M., Dewey, E., Gray, W. M., Hobbie, L., Turner, J., & Estelle, M. (1998). The TIR1 protein of Arabidopsis functions in auxin response and is related to human SKP2 and yeast grr1p. *Genes & Development*, *12*(2), 198–207.
Russinova, E., Borst, J. W., Kwaaitaal, M., Cano-Delgado, A., Yin, Y., Chory, J., & de Vries, S. C. (2004). Heterodimerization and endocytosis of Arabidopsis brassinosteroid receptors BRI1 and AtSERK3 (BAK1). *Plant Cell*, *16*(12), 3216–3229.
Santiago, J., Brandt, B., Wildhagen, M., Hohmann, U., Hothorn, L. A., Butenko, M. A., & Hothorn, M. (2016). Mechanistic insight into a peptide hormone signaling complex mediating floral organ abscission. *eLife*, *5*, e15075.
Schaller, G. E., & Bleecker, A. B. (1995). Ethylene-binding sites generated in yeast expressing the Arabidopsis ETR1 gene. *Science*, *270*(5243), 1809–1811.
Schallus, T., Feher, K., Sternberg, U., Rybin, V., & Muhle-Goll, C. (2010). Analysis of the specific interactions between the lectin domain of malectin and diglucosides. *Glycobiology*, *20*(8), 1010–1020.
Schallus, T., Jaeckh, C., Feher, K., Palma, A. S., Liu, Y., Simpson, J. C., ... Muhle-Goll, C. (2008). Malectin: A novel carbohydrate-binding protein of the endoplasmic reticulum and a candidate player in the early steps of protein N-glycosylation. *Molecular Biology of the Cell*, *19*(8), 3404–3414.
Scheer, J. M., Pearce, G., & Ryan, C. A. (2005). LeRALF, a plant peptide that regulates root growth and development, specifically binds to 25 and 120 kDa cell surface membrane proteins of Lycopersicon peruvianum. *Planta*, *221*(5), 667–674.

Scheer, J. M., & Ryan, C. A. (1999). A 160-kD systemin receptor on the surface of lycopersicon peruvianum suspension-cultured cells. *Plant Cell*, *11*(8), 1525–1536.

Scheitz, K., Luthen, H., & Schenck, D. (2013). Rapid auxin-induced root growth inhibition requires the TIR and AFB auxin receptors. *Planta*, *238*(6), 1171–1176.

Schenck, D., Christian, M., Jones, A., & Luthen, H. (2010). Rapid auxin-induced cell expansion and gene expression: A four-decade-old question revisited. *Plant Physiology*, *152*(3), 1183–1185.

Schopfer, C. R., Nasrallah, M. E., & Nasrallah, J. B. (1999). The male determinant of self-incompatibility in Brassica. *Science*, *286*(5445), 1697–1700.

Schulze-Muth, P., Irmler, S., Schroder, G., & Schroder, J. (1996). Novel type of receptor-like protein kinase from a higher plant (Catharanthus roseus). cDNA, gene, intramolecular autophosphorylation, and identification of a threonine important for auto- and substrate phosphorylation. *The Journal of Biological Chemistry*, *271*(43), 26684–26689.

Shao, W., & Brown, M. (2004). Advances in estrogen receptor biology: Prospects for improvements in targeted breast cancer therapy. *Breast Cancer Research*, *6*(1), 39–52.

She, J., Han, Z., Kim, T. W., Wang, J., Cheng, W., Chang, J., ... Chai, J. (2011). Structural insight into brassinosteroid perception by BRI1. *Nature*, *474*(7352), 472–476.

Sheard, L. B., Tan, X., Mao, H., Withers, J., Ben-Nissan, G., Hinds, T. R., ... Zheng, N. (2010). Jasmonate perception by inositol-phosphate-potentiated COI1-JAZ co-receptor. *Nature*, *468*(7322), 400–405.

Shih, H. W., Miller, N. D., Dai, C., Spalding, E. P., & Monshausen, G. B. (2014). The receptor-like kinase FERONIA is required for mechanical signal transduction in Arabidopsis seedlings. *Current Biology*, *24*(16), 1887–1892.

Shinohara, H., & Matsubayashi, Y. (2015). Reevaluation of the CLV3-receptor interaction in the shoot apical meristem: Dissection of the CLV3 signaling pathway from a direct ligand-binding point of view. *The Plant Journal*, *82*(2), 328–336.

Smalle, J., & Vierstra, R. D. (2004). The ubiquitin 26S proteasome proteolytic pathway. *Annual Review of Plant Biology*, *55*, 555–590.

Srivastava, R., Liu, J. X., Guo, H., Yin, Y., & Howell, S. H. (2009). Regulation and processing of a plant peptide hormone, AtRALF23, in Arabidopsis. *The Plant Journal*, *59*(6), 930–939.

Stirnberg, P., van De Sande, K., & Leyser, H. M. (2002). MAX1 and MAX2 control shoot lateral branching in Arabidopsis. *Development*, *129*(5), 1131–1141.

Sugano, S. S., Shimada, T., Imai, Y., Okawa, K., Tamai, A., Mori, M., & Hara-Nishimura, I. (2010). Stomagen positively regulates stomatal density in Arabidopsis. *Nature*, *463*(7278), 241–244.

Sun, Y., Han, Z., Tang, J., Hu, Z., Chai, C., Zhou, B., & Chai, J. (2013). Structure reveals that BAK1 as a co-receptor recognizes the BRI1-bound brassinolide. *Cell Research*, *23*(11), 1326–1329.

Symons, G. M., Ross, J. J., Jager, C. E., & Reid, J. B. (2008). Brassinosteroid transport. *Journal of Experimental Botany*, *59*(1), 17–24.

Tabata, R., Sumida, K., Yoshii, T., Ohyama, K., Shinohara, H., & Matsubayashi, Y. (2014). Perception of root-derived peptides by shoot LRR-RKs mediates systemic N-demand signaling. *Science*, *346*(6207), 343–346.

Takayama, S., Shimosato, H., Shiba, H., Funato, M., Che, F. S., Watanabe, M., ... Isogai, A. (2001). Direct ligand-receptor complex interaction controls Brassica self-incompatibility. *Nature*, *413*(6855), 534–538.

Takeuchi, H., & Higashiyama, T. (2016). Tip-localized receptors control pollen tube growth and LURE sensing in Arabidopsis. *Nature*, *531*(7593), 245–248.

Tang, W., Deng, Z., Oses-Prieto, J. A., Suzuki, N., Zhu, S., Zhang, X., ... Wang, Z. Y. (2008). Proteomics studies of brassinosteroid signal transduction using prefractionation and two-dimensional DIGE. *Molecular & Cellular Proteomics*, *7*(4), 728–738.

Tang, J., Han, Z., Sun, Y., Zhang, H., Gong, X., & Chai, J. (2015). Structural basis for recognition of an endogenous peptide by the plant receptor kinase PEPR1. *Cell Research*, *25*(1), 110–120.

Thynne, E., Saur, I. M., Simbaqueba, J., Ogilvie, H. A., Gonzalez-Cendales, Y., Mead, O., ... Solomon, P. S. (2016). Fungal phytopathogens encode functional homologues of plant rapid alkalinisation factor (RALF) peptides. *Molecular Plant Pathology*. http://dx.doi.org/10.1111/mpp.12444 [Epub ahead of print].

Ueguchi-Tanaka, M., Ashikari, M., Nakajima, M., Itoh, H., Katoh, E., Kobayashi, M., ... Matsuoka, M. (2005). GIBBERELLIN INSENSITIVE DWARF1 encodes a soluble receptor for gibberellin. *Nature*, *437*(7059), 693–698.

Umehara, M., Hanada, A., Yoshida, S., Akiyama, K., Arite, T., Takeda-Kamiya, N., ... Yamaguchi, S. (2008). Inhibition of shoot branching by new terpenoid plant hormones. *Nature*, *455*(7210), 195–200.

Umezawa, T., Sugiyama, N., Takahashi, F., Anderson, J. C., Ishihama, Y., Peck, S. C., & Shinozaki, K. (2013). Genetics and phosphoproteomics reveal a protein phosphorylation network in the abscisic acid signaling pathway in Arabidopsis thaliana. *Science Signaling*, *6*(270), rs8.

Wang, X., Goshe, M. B., Soderblom, E. J., Phinney, B. S., Kuchar, J. A., Li, J., ... Clouse, S. D. (2005). Identification and functional analysis of in vivo phosphorylation sites of the Arabidopsis BRASSINOSTEROID-INSENSITIVE1 receptor kinase. *Plant Cell*, *17*(6), 1685–1703.

Wang, X., Kota, U., He, K., Blackburn, K., Li, J., Goshe, M. B., ... Clouse, S. D. (2008). Sequential transphosphorylation of the BRI1/BAK1 receptor kinase complex impacts early events in brassinosteroid signaling. *Developmental Cell*, *15*(2), 220–235.

Wang, J., Lee, C., Replogle, A., Joshi, S., Korkin, D., Hussey, R., ... Mitchum, M. G. (2010). Dual roles for the variable domain in protein trafficking and host-specific recognition of Heterodera glycines CLE effector proteins. *The New Phytologist*, *187*(4), 1003–1017.

Wang, J., Li, H., Han, Z., Zhang, H., Wang, T., Lin, G., ... Chai, J. (2015). Allosteric receptor activation by the plant peptide hormone phytosulfokine. *Nature*, *525*(7568), 265–268.

Wang, T., Liang, L., Xue, Y., Jia, P. F., Chen, W., Zhang, M. X., ... Yang, W. C. (2016). A receptor heteromer mediates the male perception of female attractants in plants. *Nature*, *531*(7593), 241–244.

Wang, Z. Y., Seto, H., Fujioka, S., Yoshida, S., & Chory, J. (2001). BRI1 is a critical component of a plasma-membrane receptor for plant steroids. *Nature*, *410*(6826), 380–383.

Wang, Y., Sun, S., Zhu, W., Jia, K., Yang, H., & Wang, X. (2013). Strigolactone/MAX2-induced degradation of brassinosteroid transcriptional effector BES1 regulates shoot branching. *Developmental Cell*, *27*(6), 681–688.

West, A. H., & Stock, A. M. (2001). Histidine kinases and response regulator proteins in two-component signaling systems. *Trends in Biochemical Sciences*, *26*(6), 369–376.

Wu, J., Kurten, E. L., Monshausen, G., Hummel, G. M., Gilroy, S., & Baldwin, I. T. (2007). NaRALF, a peptide signal essential for the regulation of root hair tip apoplastic pH in Nicotiana attenuata, is required for root hair development and plant growth in native soils. *The Plant Journal*, *52*(5), 877–890.

Wulfetange, K., Lomin, S. N., Romanov, G. A., Stolz, A., Heyl, A., & Schmulling, T. (2011). The cytokinin receptors of Arabidopsis are located mainly to the endoplasmic reticulum. *Plant Physiology*, *156*(4), 1808–1818.

Xie, D. X., Feys, B. F., James, S., Nieto-Rostro, M., & Turner, J. G. (1998). COI1: An Arabidopsis gene required for jasmonate-regulated defense and fertility. *Science*, *280*(5366), 1091–1094.

Xu, T., Dai, N., Chen, J., Nagawa, S., Cao, M., Li, H., ... Yang, Z. (2014). Cell surface ABP1-TMK auxin-sensing complex activates ROP GTPase signaling. *Science*, *343*(6174), 1025–1028.

Yamaguchi, Y., Pearce, G., & Ryan, C. A. (2006). The cell surface leucine-rich repeat receptor for AtPep1, an endogenous peptide elicitor in Arabidopsis, is functional in transgenic tobacco cells. *Proceedings of the National Academy of Sciences of the United States of America*, *103*(26), 10104–10109.

Yang, T., Wang, L., Li, C., Liu, Y., Zhu, S., Qi, Y., ... Yu, F. (2015). Receptor protein kinase FERONIA controls leaf starch accumulation by interacting with glyceraldehyde-3-phosphate dehydrogenase. *Biochemical and Biophysical Research Communications*, *465*(1), 77–82.

Yu, F., Qian, L., Nibau, C., Duan, Q., Kita, D., Levasseur, K., ... Luan, S. (2012). FERONIA receptor kinase pathway suppresses abscisic acid signaling in Arabidopsis by activating ABI2 phosphatase. *Proceedings of the National Academy of Sciences of the United States of America*, *109*(36), 14693–14698.

CHAPTER ELEVEN

Regulation of Cell Polarity by PAR-1/MARK Kinase

Youjun Wu, Erik E. Griffin[1]

Dartmouth College, Hanover, NH, United States
[1]Corresponding author: e-mail address: erik.griffin@dartmouth.edu

Contents

1. Introduction	366
2. Structure and Regulation of PAR-1/MARK Kinases	367
2.1 Structure of PAR-1 Kinases	368
2.2 Regulation of PAR-1 Localization	369
2.3 Regulation of PAR-1 Activity	370
3. Regulation of Cell Polarity by the PAR Proteins	371
3.1 Mutual Antagonism Between the PAR Proteins	371
4. Asymmetric Division of the *C. elegans* Zygote	372
4.1 Overview of Polarization of the *C. elegans* Zygote	372
4.2 Symmetry Breaking and Polarity Establishment	373
4.3 Polarity Maintenance	376
4.4 Establishment of Cytoplasmic Asymmetries	377
4.5 Germ Plasm Segregation	378
5. Establishment of the Anterior/Posterior Axis During *Drosophila* Oogenesis	379
5.1 Overview of *Drosophila* Oogenesis	379
5.2 Oocyte Specification	380
5.3 Anterior/Posterior Axis Specification During Mid-oogenesis	381
5.4 Comparison of Anterior/Posterior Polarization in Worms and Flies	384
6. MARK Kinases and Neurogenesis	385
6.1 Neuronal Polarization	385
7. PAR-1 and Disease	386
7.1 Alzheimer's Disease	386
7.2 Cancer	387
8. Concluding Remarks	388
References	389

Abstract

PAR-1/MARK kinases are conserved serine/threonine kinases that are essential regulators of cell polarity. PAR-1/MARK kinases localize and function in opposition to the anterior PAR proteins to control the asymmetric distribution of factors in a wide variety polarized cells. In this review, we discuss the mechanisms that control the localization and activity of PAR-1/MARK kinases, including their antagonistic interactions with the anterior PAR proteins. We focus on the role PAR-1 plays in the asymmetric division of the *Caenorhabditis elegans* zygote, in the establishment of the anterior/posterior axis in the *Drosophila* oocyte and in the control of microtubule dynamics in mammalian neurons. In addition to conserved aspects of PAR-1 biology, we highlight the unique ways in which PAR-1 acts in these distinct cell types to orchestrate their polarization. Finally, we review the connections between disruptions in PAR-1/MARK function and Alzheimer's disease and cancer.

1. INTRODUCTION

PAR-1/MARK kinases are conserved serine/threonine kinases that regulate cellular organization in diverse processes including asymmetric cell division, neuronal differentiation, and epithelial organization. PAR-1 commonly functions as a member of the PAR (PARtitioning defective) network of cell polarity regulators. PAR-1 and the anterior PAR proteins (PAR-3, PAR-6, and aPKC kinase) concentrate in opposing cortical domains within polarized cells, and their asymmetric distribution underlies their ability to control cortical, cytoplasmic, and cytoskeletal asymmetries along the polarity axis (Goldstein & Macara, 2007). PAR-1 was initially discovered along with most of the other PAR proteins in a classic genetic screen for genes required for the establishment of the anterior/posterior polarity axis in the early *Caenorhabditis elegans* embryo (Kemphues, Priess, Morton, & Cheng, 1988). Shortly thereafter, PAR-1 was shown to be essential for the establishment of the anterior/posterior axis in the *Drosophila* oocyte and for apical/basal polarity in epithelial cells (Cox, Lu, Sun, Williams, & Jan, 2001; Shulman, Benton, & St Johnston, 2000). Mammalian PAR-1/MARK kinases were independently purified based on their ability to phosphorylate and thereby regulate the microtubule association of the microtubule-associated proteins TAU, MAP2, and MAP4 (Drewes, Ebneth, Preuss, Mandelkow, & Mandelkow, 1997; Drewes et al., 1995). More recently, dysregulation of PAR-1/MARK kinases has been implicated in a number of pathological settings, including in tumorigenesis and in Alzheimer's disease (Matenia & Mandelkow, 2009).

In this review, we focus on three systems in which PAR-1 function has been particularly well characterized: establishment of the anterior/posterior axis in the *C. elegans* zygote, establishment of the anterior/posterior axis in the *Drosophila* oocyte, and control of microtubule dynamics in mammalian neurons. We use these examples to illustrate both commonalities and important differences in PAR-1 biology in different cellular settings. For example, antagonistic interactions between the PAR-1 and the anterior PAR proteins are a common mechanism by which their reciprocal localization patterns are established. However, PAR-1 generates downstream cellular asymmetries through fundamentally different mechanisms in each of these examples. In the *C. elegans* zygote, PAR-1 locally controls reaction/diffusion mechanisms to rapidly generate cytoplasmic asymmetries. In the *Drosophila* oocyte, PAR-1 polarizes the microtubule cytoskeleton by locally inhibiting noncentrosomal microtubules, thus providing a foundation for long range polarized transport. In neurons, PAR-1/MARK locally controls microtubule growth and stability, which contributes to differences in the axonal and dendritic microtubule cytoskeletons.

2. STRUCTURE AND REGULATION OF PAR-1/MARK KINASES

PAR-1/MARK serine/threonine kinases are large proteins (for example, MARK1 is 88 kDa) and are members of the AMPK family of CamKII kinases (Manning, Whyte, Martinez, Hunter, & Sudarsanam, 2002). The overall architecture of PAR-1/MARK kinases is shared with most other AMPK kinases and features a kinase domain and an adjacent noncanonical ubiquitin-association (UBA) domain located near the N-terminus and a kinase-associated (KA1) membrane-binding domain near the C-terminus (Fig. 1) (Marx, Nugoor, Panneerselvam, & Mandelkow, 2010; Murphy et al., 2007; Panneerselvam, Marx, Mandelkow, & Mandelkow, 2006;

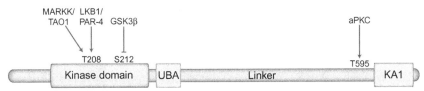

Fig. 1 Schematic of PAR-1/MARK kinase. The kinase and UBA domains are positioned near the N-terminus and a relatively long spacer domain separates the UBA domain from the C-terminal KA1 membrane-binding domain. PAR-1 kinase activity is stimulated by phosphorylation on Thr208 in the activation loop by LKB1/PAR-4 and MARKK/TAO-1 kinases. GSK3β phosphorylation on Ser212 in the activation loop inhibits PAR-1 kinase activity. aPKC phosphorylation at residue Thr595 results in the association of the 14-3-3 protein PAR-5 (not depicted) and the sequestration of PAR-1 in the cytoplasm. MARK2 residues are indicated.

Sack et al., 2016; Timm, Marx, Panneerselvam, Mandelkow, & Mandelkow, 2008). A large, variable, and relatively uncharacterized spacer domain lies between the UBA domain and the KA1 domain. Mammals encode four PAR-1/MARK kinases: MARK1 (Par1c), MARK2 (Par1b/EMK), MARK3 (Par1a/C-TAK1), and MARK4 (MARKL1/Par1d) are broadly expressed in embryonic and adult tissues and contribute to a wide range of biological processes. Both *C. elegans* and *Drosophila* encode a single PAR-1 family member that is essential for embryonic development (Guo & Kemphues, 1995; Shulman et al., 2000). The diversity of PAR-1 proteins is increased by alternative splice isoforms with differing amino or carboxy termini, which in the *Drosophila* oocyte, have distinct localization patterns and functional roles (Doerflinger, Benton, Torres, Zwart, & St Johnston, 2006). For simplicity, in this review I will refer to PAR-1/MARK kinases collectively as PAR-1 and will use specific names to refer particular family members (for example, MARK1–4 for the mammalian proteins and dPar-1 for the *Drosophila* kinase).

MARK2, MARK3, and MARK4 knockout mice are viable (the phenotypes of MARK1 deficient mice have not been reported), but have defects in energy metabolism that result in reduced body weight and adipocity, although the spectrum of metabolic defects are unique for each mutant (Bessone et al., 1999; Hurov et al., 2001; Lennerz et al., 2010; Sun et al., 2012). While the basis for these metabolic defects has not been fully elucidated, MARK4 is expressed in adipocytes where it has been shown to regulate respiration, proliferation, and apoptosis (Feng, Tian, Gan, Liu, & Sun, 2014; Liu et al., 2016). MARK2 knockout mice have a number of other phenotypes, including dwarfism, infertility, reduced learning and memory, and immune system dysfunction (Bessone et al., 1999; Hurov et al., 2001; Segu, Pascaud, Costet, Darmon, & Buhot, 2008). Additionally, MARK2 contributes to myogenesis by controlling the asymmetric division of muscle stem cells (satellite cells), and dysregulation of MARK2 was recently implicated in the loss of muscle in Duchenne muscular dystrophy (Dumont et al., 2015). MARK2/MARK3 double mutants fail to complete embryonic development indicating that in the single knockout mice, redundancy masks essential developmental functions of the MARK kinases (Lennerz et al., 2010).

2.1 Structure of PAR-1 Kinases

The PAR-1 kinase domain adopts a canonical bilobed structure that is typical of most kinases (Marx et al., 2006; Murphy et al., 2007; Panneerselvam et al.,

2006; Sack et al., 2016). Catalytic activity is stimulated by phosphorylation of the kinase domain activation loop (residue Thr208 in MARK2), which causes the activation loop to swing out of the catalytic cleft and makes the catalytic site accessible to substrates (Drewes et al., 1997; Timm et al., 2008). Phosphorylation of the PAR-1 activation loop is mediated by LKB1/PAR-4, a conserved, master regulatory kinase that phosphorylates the activation loop of all 14 members of the AMPK family of kinases (Lizcano et al., 2004). AMPK family kinases have diverse functions and it remains unclear how a single upstream kinase can specifically activate individual family members. MARK kinases can also be activated by the kinase MARKK/TAO-1, which phosphorylates the same activation loop residue as LKB1/PAR-4 (Timm et al., 2003). In addition to activating modifications, the activation loop is also a site for inhibitory phosphorylation. GSK3β kinase phosphorylates MARK2 at residue Ser212, which occludes substrate access to the catalytic site and renders the kinase inactive, even if it is also phosphorylated at Thr208 (Timm, Balusamy, et al., 2008; Timm et al., 2003).

2.2 Regulation of PAR-1 Localization

PAR-1 is typically asymmetrically distributed at the cell cortex of polarized cells. There are two elements in the C-terminus of PAR-1 that play a central role in controlling PAR-1 localization. First, the KA1 (Kinase Associated) domain is a membrane-binding domain located at the C-terminus that binds acidic phospholipids such as the plasma-membrane enriched phospholipid phosphatidylserine (Leventis & Grinstein, 2010; Moravcevic et al., 2010). Studies in *C. elegans* and mammalian cells have shown that the KA1 domain is both necessary and sufficient for cortical recruitment, consistent with the idea that recruitment to the cell cortex is mediated, at least in part, by the direct interaction between the KA1 domain and the cytoplasmic face of the plasma membrane (Goransson et al., 2006; Moravcevic et al., 2010; Motegi et al., 2011). The asymmetric distribution of PAR-1 at the cell cortex is regulated by the anterior PAR protein aPKC, which phosphorylates PAR-1 at a conserved residue near the KA1 domain (Thr595 in MARK2). Phosphorylation of PAR-1 at this residue results in the binding of the 14-3-3 protein PAR-5 to PAR-1, which sequesters PAR-1 in the cytoplasm. This mechanism prevents the association of PAR-1 with the region of the cell cortex occupied by the anterior PARs (Hurov, Watkins, & Piwnica-Worms, 2004; Motegi et al., 2011; Suzuki et al., 2004). The interactions between the PAR proteins will be discussed in more detail later.

2.3 Regulation of PAR-1 Activity

A number of mechanisms have been identified that control PAR-1 activity, including interactions with binding partners that either inhibit (for example, PAK5) or activate (for example, GAB1 or DAPK) PAR-1 kinase activity (reviewed in Matenia et al., 2005; Matenia & Mandelkow, 2009; Wu et al., 2011). Additionally, intramolecular interactions between the kinase domain, UBA domain, and KA1 domain appear to play important roles in regulating PAR-1 catalytic activity. Although it is named for its homology to the ubiquitin-binding UBA domain, the PAR-1 UBA domain possesses an extremely weak affinity for ubiquitin and does not bind ubiquitin in vivo (Marx et al., 2006; Murphy et al., 2007; Panneerselvam et al., 2006). Rather, the UBA domain binds to the back surface of the kinase domain, opposite the catalytic cleft, and plays both positive and negative roles in controlling kinase activity (reviewed in Marx et al., 2010). On one hand, the UBA domain functions as an autoinhibitory domain by holding the kinase in an inactive, "open" conformation (Marx et al., 2006; Panneerselvam et al., 2006). Similar autoinhibitory interactions between the UBA and kinase domain have been identified in other AMPK family members, although the positioning of the UBA domain on the kinase domain varies (Chen et al., 2009; Wu, Cheng, et al., 2015). On the other hand, the UBA domain is required for LKB1 phosphorylation of the activation loop, and therefore for kinase activation (Jaleel et al., 2006). Because it appears to participate both in kinase activation and inhibition, the UBA domain has been suggested to serve as a fulcrum point for the regulation of kinase activity (discussed in Marx et al., 2010).

Several lines of evidence suggest an autoinhibitory interaction between the KA1 domain and the kinase domain likely regulates PAR-1 kinase activity. Interactions between the C- and N-terminus have been detected by coimmunoprecipitation (MARK2) and by yeast two hybrid (the budding yeast homologs of PAR-1, KIN1, and KIN2) and genetic analysis in budding yeast are consistent with an autoinhibitory interaction (Elbert, Rossi, & Brennwald, 2005; Yang et al., 2012). Additionally, interaction between MARK2 and the scaffolding protein GAB1 likely stimulates MARK2 kinase activity by preventing the interaction between the N- and C-terminus (Yang et al., 2012). Further support for a potential inhibitory interaction between the C-terminus and kinase domain comes from biochemical and structural studies of SAD kinase, an AMPK kinase family member that is related to PAR-1 kinase. The SAD kinase C-terminus, including the

KA1 domain and surrounding sequences, inhibits kinase activity by folding back and strengthening the autoinhibitory interaction between the UBA domain and the kinase domain (Wu, Cheng, et al., 2015). Although the importance and mechanism of the C-terminus/KA1 domain regulation of PAR-1 kinase activity await further characterization, such a regulatory interaction would be of considerable interest because it could provide a means to couple control of PAR-1 kinase activity with control of its localization, for example, through aPKC phosphorylation or PAR-1 membrane association.

3. REGULATION OF CELL POLARITY BY THE PAR PROTEINS

PAR-1 is an essential component of a network of cell polarity regulators, the PAR proteins, which collectively function to spatially organize most polarized animal cells including epithelia, neurons, and asymmetrically dividing cells (reviewed in Goldstein & Macara, 2007; Nance & Zallen, 2011; St Johnston & Ahringer, 2010). A hallmark of the PAR proteins is that they localize to two distinct, opposing domains at the plasma membrane/cell cortex of polarized cells. One cortical domain is occupied by the anterior PAR proteins, which consist of the PDZ proteins PAR-3 (Bazooka in *Drosophila*) and PAR-6 and the kinase aPKC (PKC-3 in *C. elegans*) (Etemad-Moghadam, Guo, & Kemphues, 1995; Tabuse et al., 1998; Watts et al., 1996). The reciprocal cortical domain is occupied by PAR-1 and, in *C. elegans*, by the RING finger protein PAR-2 (Boyd, Guo, Levitan, Stinchcomb, & Kemphues, 1996; Guo & Kemphues, 1995). From these asymmetric domains, the PAR proteins control a wide range of cellular asymmetries including the polarization of the actomyosin and microtubule cytoskeletons and the partitioning of both cortical and cytoplasmic factors along the polarity axis (Goldstein & Macara, 2007; Nance & Zallen, 2011). Two additional PAR proteins, the 14-3-3 protein PAR-5 and the PAR-1-activating kinase LKB-1/PAR-4 (discussed earlier), are symmetrically distributed in the cytoplasm and at the cell cortex (Benton, Palacios, & Johnston, 2002; Morton et al., 2002; Watts, Morton, Bestman, & Kemphues, 2000).

3.1 Mutual Antagonism Between the PAR Proteins

The mechanisms that establish and maintain opposing cortical PAR domains are of great interest as they provide the foundation for the elaboration of downstream asymmetries. Studies in a number of systems have demonstrated that mutually antagonistic interactions between anterior and posterior PAR proteins provide a core means by which they concentrate in opposing cortical domains. These mechanisms have been reviewed recently (Hoege & Hyman, 2013; Motegi & Seydoux, 2013; Nance & Zallen, 2011), and will only be covered briefly here. aPKC phosphorylation of PAR-1 on a conserved residue (Thr595 in MARK2) restricts PAR-1 from concentrating in the anterior PAR cortical domain and has been observed to reduce PAR-1 kinase activity (Chen et al., 2006; Hurov et al., 2004; Motegi et al., 2011; Suzuki et al., 2004). In *C. elegans*, aPKC similarly phosphorylates the posterior PAR protein PAR-2, which prevents PAR-2 association with the anterior cortex (Hao, Boyd, & Seydoux, 2006). The phosphorylation of PAR-3 by PAR-1 restricts PAR-3 from concentrating in the PAR-1 cortical domain (Benton & St Johnston, 2003). These phosphorylation events result in binding of the 14-3-3 protein PAR-5, leading to the sequestration of the phosphorylated protein in the cytoplasm (Benton et al., 2002; Goransson et al., 2006; Hao et al., 2006; Riechmann & Ephrussi, 2004). In mammalian cells, it has been shown that the phosphorylation of PAR-3 by PAR-1 is enhanced by the scaffolding protein GAB1, which brings PAR-1 and PAR-3 together in a transient complex (Yang et al., 2012).

As discussed in the following sections, although mutual antagonism between the PAR proteins is conserved, distinct mechanisms contribute to the triggering, establishment and maintenance of PAR polarity in different polarized cells. Furthermore, PAR-1 can drive the spatial reorganization of cells through a range of different substrates and different mechanisms. For example, PAR-1 engages fundamentally different mechanisms to control the rapid polarization of *C. elegans* zygote and the relatively slow, long range polarization of the *Drosophila* oocyte. These distinct polarization mechanisms accord with the significantly different spatial and temporal scales at which asymmetries are generated in these two cells.

4. ASYMMETRIC DIVISION OF THE *C. ELEGANS* ZYGOTE
4.1 Overview of Polarization of the *C. elegans* Zygote

The first function of the PAR proteins in the *C. elegans* embryo is to orchestrate the polarization and asymmetric division of the zygote. The PAR proteins are maternally deposited in the embryo and are initially symmetrically distributed. Shortly following fertilization and the subsequent completion of meiosis, the PAR proteins redistribute to form opposing anterior (the anterior PARs) and posterior (PAR-1 and PAR-2) cortical domains. From these domains, the PAR proteins establish a number of asymmetries along the anterior/posterior axis. As will be discussed later, PAR-1 functions downstream of the other PAR proteins to direct the redistribution of somatic and germline fate determinants to the anterior and posterior cytoplasm, respectively (Fig. 2). As a result, these determinants are asymmetrically inherited by the two daughter cells, specifying the anterior daughter cell as a somatic blastomere and the posterior daughter cell as a germline blastomere (reviewed in Rose & Kemphues, 1998). In addition, the PAR proteins regulate asymmetric distribution of cortical microtubule pulling forces such that the position of the mitotic spindle is shifted to the posterior, thereby generating a smaller posterior cell. Control of spindle positioning by the PAR proteins is an essential process underlying asymmetric cell division and has been reviewed recently (Kotak & Gonczy, 2013; Lu & Johnston, 2013).

4.2 Symmetry Breaking and Polarity Establishment

Prior to the completion of meiosis, PAR-1 and PAR-2 are restricted to the cytoplasm by the anterior PARs, which are uniformly distributed throughout the cell cortex (Boyd et al., 1996; Guo & Kemphues, 1995). Upon the completion of meiosis, this symmetry is broken by the maturation of the sperm-donated centrosome, which is located near the posterior cortex in association with the sperm pronucleus (Goldstein & Hird, 1996). The centrosome triggers two symmetry-breaking mechanisms. In one mechanism, an unknown cue from the centrosome inhibits contraction of the cortical actomyosin meshwork at the posterior cortex, thus initiating flow of the actomyosin cortex toward the anterior (Motegi & Sugimoto, 2006; Munro, Nance, & Priess, 2004). These flows sweep the anterior PARs from the posterior to the anterior, thus enabling PAR-1 and PAR-2 to associate with the posterior cortex (Cowan & Hyman, 2004; Goehring, Trong, et al., 2011; Munro et al., 2004). Because actomyosin flows do not depend on PAR-1 or PAR-2, the polarized anterior PAR domain is established in *par-1* and *par-2* embryos (Boyd et al., 1996; Cuenca, Schetter, Aceto, Kemphues, & Seydoux, 2003; Etemad-Moghadam et al., 1995). The second symmetry-breaking mechanism depends on PAR-2, which binds to microtubules in a manner that protects PAR-2 from phosphorylation by aPKC (Motegi et al., 2011). Thus, microtubules that extend from the centrosome to the cortex enable PAR-2 to load onto the nearby posterior cortex despite

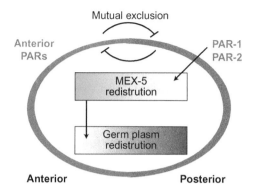
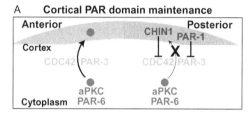
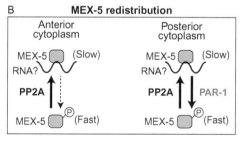
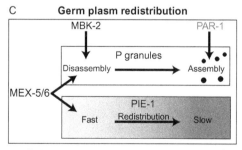

Fig. 2 Establishment of the anterior/posterior axis in the *C. elegans* zygote. In the polarized *C. elegans* zygote, the anterior PARs (*blue*) are enriched at the anterior cortex and PAR-1 and PAR-2 (*brown*) are enriched at the posterior cortex. Antagonistic interactions between the PAR proteins mediate their mutual exclusion. PAR-1 directs the redistribution of MEX-5 to the anterior cytoplasm and MEX-5 contributes to the redistribution of germ plasm proteins to the posterior cytoplasm. All of these factors are

the presence of aPKC. Cortical PAR-2 then acts to recruit its binding partner PAR-1 from the cytoplasm to the cortex. Once at the cortex, PAR-1 is able to phosphorylate PAR-3, thereby promoting the further growth of the posterior domain by locally excluding the anterior PARs from the posterior cortex (Motegi et al., 2011). Either one of these symmetry-breaking mechanisms is sufficient to break symmetry, although when only one mechanism is active polarization is either delayed or proceeds relatively slowly (Goehring, Trong, et al., 2011; Motegi et al., 2011). Therefore, the presence of two symmetry-breaking pathways appears to ensure that establishment of the cortical PAR domains is both robust and rapid (Motegi & Seydoux, 2013).

After symmetry breaking, the establishment of cortical PAR polarity progresses as the anterior PAR domain retracts and the posterior PAR domain expands until, roughly 5 min later, the boundary between the two domains reaches the midpoint of the anterior/posterior axis (Boyd et al., 1996; Etemad-Moghadam et al., 1995; Guo & Kemphues, 1995; Tabuse et al., 1998; Watts et al., 1996). The expansion of the PAR-1 and PAR-2 posterior domain depletes the levels of these proteins in the cytoplasm, which eventually limits their cortical recruitment and the growth of the posterior domain (Goehring, Trong, et al., 2011). Following their establishment, the two PAR domains are maintained through cytokinesis, which occurs roughly 10 min later.

symmetrically distributed before symmetry breaking. (A) Asymmetric PAR domains are maintained in part by the recruitment of cytoplasmic PAR-6 and aPKC to the anterior, but not to the posterior, cortex. PAR-1 and CHIN-1 restrict PAR-3 and active CDC42 from the posterior cortex, thereby preventing the recruitment of PAR-6 and aPKC from the cytoplasm to the posterior cortex. Other mechanisms that contribute to PAR domain maintenance are described in the text and not illustrated for simplicity. (B) The redistribution of MEX-5 to the anterior cytoplasm is controlled by PAR-1. PAR-1 phosphorylates MEX-5 and increases MEX-5 mobility in the posterior cytoplasm while PP2A reverses this effect. As a result, MEX-5 mobility is relatively fast in the posterior cytoplasm and MEX-5 redistributes to the anterior cytoplasm. MEX-5 association with RNA is likely to contribute to its slow diffusion. (C) As MEX-5 and MEX-6 accumulate in the anterior, germ plasm factors segregate to the posterior cytoplasm. P granules partition to the posterior cytoplasm because MEX-5/6 and MBK-2 promote their disassembly in the anterior cytoplasm and PAR-1 promotes their stability in the posterior cytoplasm. MEX-5/6 act to increase PIE-1 mobility in the anterior cytoplasm, thereby stimulating the redistribution of PIE-1 to the posterior cytoplasm.

4.3 Polarity Maintenance

The PAR domains are persistent in the polarized zygote even though individual PAR proteins undergo both lateral diffusion within the cortex and continual exchange between the cortex and the cytoplasm (Cheeks et al., 2004; Goehring, Hoege, Grill, & Hyman, 2011; Nakayama et al., 2009; Robin, McFadden, Yao, & Munro, 2014; Sailer, Anneken, Li, Lee, & Munro, 2015). The maintenance of stable PAR domains depends on a combination of mechanisms that control where they are recruited to the cell cortex and that counteract their lateral diffusion within the cortex. During polarity maintenance, the recruitment of PAR-6/aPKC from the cytoplasm to the cortex depends on interactions with both PAR-3 and the active form of the small GTPase, CDC42 (Aceto, Beers, & Kemphues, 2006; Gotta, Abraham, & Ahringer, 2001; Kay & Hunter, 2001). Both PAR-3 and active CDC42 are concentrated at the anterior cortex, resulting in PAR-6 cortical recruitment rates that are ∼9 times higher in the anterior than the posterior cortex (Sailer et al., 2015). This dramatic asymmetry in cortical PAR-6 recruitment depends on two mechanisms that prevent PAR-3 and active CDC42 from localizing to the posterior cortex (Fig. 2A). One mechanism relies on the phosphorylation of PAR-3 by PAR-1, which excludes PAR-3 from the posterior cortex (discussed earlier) (Sailer et al., 2015). The second mechanism depends on CHIN-1, a posteriorly enriched CDC42 GAP that inactivates CDC42 in the posterior (Beatty, Morton, & Kemphues, 2013; Kumfer et al., 2010; Sailer et al., 2015). The presence of either one of these mechanisms is sufficient to prevent PAR-6 recruitment to the posterior cortex. However, in *par-1;chin-1* double mutant embryos, PAR-6 is efficiently recruited to the posterior cortex, resulting in the uniform cortical distribution of the anterior PARs (Sailer et al., 2015). By independently inhibiting PAR-3 and active CDC42 in the posterior, the combined activities of PAR-1 and CHIN-1 provide a robust and reliable mechanism by which to stabilize the polarized anterior PAR domain. The cortical association rates of PAR-1 and PAR-2 have not been reported, and it will be interesting to learn whether analogous mechanisms act to prevent their recruitment from the cytoplasm to the anterior cortex. Lateral diffusion within the cortex leads to mixing of the anterior and posterior PAR proteins at the interface of their two domains. The PAR domain boundary is maintained by antagonistic interactions between the anterior and posterior PAR proteins (discussed earlier) that cause them to mutually exclude each other from the cortex (Goehring, Hoege, et al., 2011; Hoege & Hyman, 2013).

4.4 Establishment of Cytoplasmic Asymmetries

A primary output of the PAR system in the *C. elegans* zygote is the partitioning of maternally deposited cytoplasmic cell fate determinants along the anterior/posterior axis. In response to cues from the PAR proteins, the highly similar RNA-binding proteins MEX-5 and MEX-6 (MEX-5/6 hereafter) redistribute toward the anterior cytoplasm, leading to their preferential inheritance by the anterior daughter cell (Schubert, Lin, de Vries, Plasterk, & Priess, 2000). At the same time, a mixture of cytoplasmic RNAs and proteins that specify the germline lineage, collectively called the germ plasm, redistribute to the posterior cytoplasm and are therefore inherited preferentially by the posterior daughter cell. Germ plasm components include nonmembranous RNA/protein assemblages called P granules and more diffusely concentrated RNA-binding proteins such as PIE-1 (Mello et al., 1996; Strome & Wood, 1983; Tenenhaus, Schubert, & Seydoux, 1998; Updike & Strome, 2010). The asymmetric inheritance of these factors causes the anterior daughter cell to adopt a somatic fate and the posterior daughter cell to adopt a germline fate (Rose & Kemphues, 1998; Wang & Seydoux, 2013).

The partitioning of cytoplasmic somatic and germline determinants along the zygotic anterior/posterior axis is remarkable in several respects. These cytoplasmic determinants are symmetrically distributed in the zygote prior to the symmetry breaking, and their segregation is largely completed during the ~5 min it takes to establish the cortical PAR domains. Segregation does not depend on new protein synthesis, local protein degradation, or directed transport (Griffin, 2015; Hoege & Hyman, 2013). Rather, segregation results from mechanisms that locally increase protein diffusivity, generating gradients in protein mobility along the anterior/posterior axis that cause proteins to concentrate in the region of low mobility (Daniels, Perkins, Dobrowsky, Sun, & Wirtz, 2009; Griffin, Odde, & Seydoux, 2011; Lipkow & Odde, 2008; Tenlen, Molk, London, Page, & Priess, 2008). In *par-1* mutants, there is a complete failure to generate cytoplasmic asymmetries despite the fact the cortical anterior PAR domain is established, which indicates PAR-1 plays a central role in transducing cortical polarity cues to the cytoplasm (Cuenca et al., 2003; Guo & Kemphues, 1995). PAR-1 does so by controlling the segregation of MEX-5/6 to the anterior cytoplasm, which, in turn drive the segregation of germline factors to the posterior cytoplasm (Griffin et al., 2011; Schubert et al., 2000; Tenlen et al., 2008).

In the polarized zygote, MEX-5 forms a threefold, anterior-rich concentration gradient that spans the 50 μm anterior/posterior axis of the cell (Daniels, Dobrowsky, Perkins, Sun, & Wirtz, 2010; Griffin et al., 2011; Schubert et al., 2000; Tenlen et al., 2008). During gradient formation, MEX-5 mobility is increased in the posterior cytoplasm by PAR-1 phosphorylation, which likely prevents the formation of slow-diffusing MEX-5/RNA complexes (Griffin et al., 2011; Tenlen et al., 2008). The uniformly distributed cytoplasmic phosphatase PP2A counteracts PAR-1 by acting to decrease MEX-5 mobility, resulting in the formation of an anterior-slow, posterior-fast MEX-5 mobility gradient (Fig. 2B) (Griffin et al., 2011; Schlaitz et al., 2007). As a consequence of this mobility gradient, MEX-5 is preferentially retained in the anterior cytoplasm (Griffin et al., 2011). Epistasis analysis demonstrated that PAR-1 functions downstream of the anterior PARs to control MEX-5 mobility, indicating the primary role of the anterior PARs in MEX-5 segregation is to restrict PAR-1 to the posterior (Griffin et al., 2011; Tenlen et al., 2008). Interestingly, in addition to the enrichment of PAR-1 at the posterior cortex, there is a slight enrichment of PAR-1 in the posterior cytoplasm. Analysis of *par-2* mutants, in which PAR-1 does not load onto the cortex but still concentrates in the posterior cytoplasm, indicates that the asymmetric activity of PAR-1 in the cytoplasm is sufficient to drive MEX-5 segregation (Boyd et al., 1996; Griffin et al., 2011; Labbé, Pacquelet, Marty, & Gotta, 2006).

4.5 Germ Plasm Segregation

As MEX-5/6 accumulate in the anterior cytoplasm, they simultaneously stimulate the segregation of germ plasm components to the posterior cytoplasm (Fig. 2C) (Schubert et al., 2000). MEX-5/6 act downstream of the PAR proteins and through an unknown mechanism to increase the mobility of the RNA-binding protein PIE-1 in the anterior cytoplasm, resulting in the formation of an anterior-fast, posterior-slow PIE-1 mobility gradient (Wu, Zhang, & Griffin, 2015). As a result of its mobility gradient, PIE-1 is preferentially retained in the posterior cytoplasm (Daniels et al., 2009; Wu, Zhang, et al., 2015). The ability of MEX-5/6 to increase PIE-1 mobility is concentration dependent, suggesting there is a direct coupling between the formation of the MEX-5/6 gradients and the PIE-1 mobility gradient (Wu, Zhang, et al., 2015).

P granules are nonmembranous organelles composed of RNA and of dozens of RNA-binding proteins, many of which contain intrinsically disordered domains (Updike & Strome, 2010). P granules are highly dynamic and behave like phase-separated liquid droplets (Brangwynne et al., 2009; Weber & Brangwynne, 2012). During the asymmetric division of the zygote, P granules become dramatically enriched in the posterior cytoplasm as a result of mechanisms that promote their disassembly in the anterior cytoplasm and their stabilization and growth in the posterior cytoplasm (Brangwynne et al., 2009; Gallo, Wang, Motegi, & Seydoux, 2010). PAR-1 regulates P granule segregation indirectly by controlling the segregation of MEX-5/6, which promote P granule disassembly in the anterior cytoplasm (Brangwynne et al., 2009; Gallo et al., 2010; Schubert et al., 2000). In vitro studies with the P granule protein PGL-3 indicate that PGL-3 and MEX-5 compete for binding to mRNA, and that mRNA-binding promotes PGL-3 phase separation (Saha et al., 2016). These findings support a model in which the high concentrations of MEX-5/6 in the anterior cytoplasm deplete the pool of mRNA available to participate in P granule assembly, thereby locally shifting the balance between assembly and disassembly toward disassembly. PAR-1 also acts independently of MEX-5/6 to stabilize P granules in the posterior cytoplasm, raising the possibility that PAR-1 may contribute more directly P granule segregation (Cheeks et al., 2004; Gallo et al., 2010). Apart from regulation by MEX-5/6 and PAR-1, P granule dynamics are also controlled by MBK-2 kinase, which stimulates P granule disassembly through phosphorylation of the intrinsically disordered P granule proteins MEG-3 and MEG-4 (Wang et al., 2014).

In summary, PAR-1 functions at a critical juncture in the mechanisms that control the asymmetric division of the *C. elegans* zygote. PAR-1 functions downstream of the anterior PAR proteins to control the segregation of MEX-5/6 to the anterior cytoplasm. In turn, MEX-5/6 stimulate the segregation of the germ plasm to the posterior cytoplasm. These asymmetries are generated through reactions that control the redistribution of diffusive proteins, which enables the efficient partitioning of factors during the rapid cell divisions of the early embryo.

5. ESTABLISHMENT OF THE ANTERIOR/POSTERIOR AXIS DURING *DROSOPHILA* OOGENESIS

5.1 Overview of *Drosophila* Oogenesis

The development of the *Drosophila* oocyte begins with a single cell, called a cystoblast, that undergoes four rounds of incomplete cell division, giving rise to a 16 cell cyst. One of these 16 cystoblasts will differentiate into the oocyte while the other 15 become nurse cells (reviewed in Roth & Lynch, 2009). The nurse cells remain connected to the oocyte through cytoplasmic bridges called ring canals through which they transport cytoplasmic contents into the oocyte, fueling the extensive growth of the oocyte and providing factors that are essential for the patterning of the oocyte and the initial development of the fertilized embryo. The maturing oocyte and its nurse cells are encased within a single layer of somatic epithelial cells called the follicle cells. During mid-oogenesis, signaling between the oocyte and the follicle cells initiates the polarization of the oocyte microtubule cytoskeleton, which provides the basis for the transport of *bicoid* and *oskar* mRNA to the anterior and posterior poles, respectively (Januschke et al., 2002; Zimyanin et al., 2008). Oskar is subsequently translated and concentrated at the posterior pole where it organizes the formation of the germ plasm (reviewed in Huynh & St Johnston, 2004; Lehmann, 2016). Bicoid is translated at the anterior pole following fertilization where it functions as a morphogen to specify anterior embryonic tissues (Driever & Nusslein-Volhard, 1988; Frohnhofer & Nusslein-Volhard, 1986). In the following sections, we focus on the key roles played by dPar-1 at several stages of oocyte development, including the initial stage of oocyte specification, the polarization of the microtubule cytoskeleton, and the organization of the germ plasm in the posterior. In addition to the functions discussed later, dPar-1 also regulates *bicoid* mRNA transport (Riechmann & Ephrussi, 2004), apical/basal polarity of the follicle cells (Cox, Lu, et al., 2001; Doerflinger, Benton, Shulman, & St Johnston, 2003), asymmetric stem cell divisions in the male germline (Inaba, Venkei, & Yamashita, 2015; Yuan, Chiang, Cheng, Salzmann, & Yamashita, 2012), and the migration of a subset of the follicle cells (the border cells) through the nurse cells (Mcdonald, Khodyakova, Aranjuez, Dudley, & Montell, 2008).

5.2 Oocyte Specification

During the initial cystoblast divisions, dPar-1 concentrates on an ER-like cytoplasmic organelle called the fusome, which traverses the cytoplasmic bridges that connect the cystoblast cells (Cox, Lu, et al., 2001; Shulman et al., 2000; Tomancak et al., 2000). Specification of the future oocyte is marked by the concentration of a number of factors, including dPar-1, Par-3, and the oocyte determinants Orb, BicD, and Egl, in one of the 16 cystoblast cells. Shortly thereafter, the oocyte is polarized in a process that depends on dPar-1 and the anterior PARs such that Orb, BicD, and Egl concentrate in the oocyte posterior (Benton et al., 2002; Cox, Lu, et al., 2001; Cox, Seyfried, Jan, & Jan, 2001; Huynh, Petronczki, Knoblich, & St Johnston, 2001; Huynh, Shulman, Benton, & St Johnston, 2001; Vaccari & Ephrussi, 2002). In *par* mutants, Orb, BicD, and Egl become enriched in the future oocyte, but these proteins fail to concentrate in the posterior of the cell and the oocyte dedifferentiates into a nurse cell (Benton et al., 2002; Cox, Lu, et al., 2001; Huynh, Shulman, et al., 2001; Vaccari & Ephrussi, 2002). Therefore, the PAR proteins are not required for the initial specification of the oocyte, but rather for the subsequent maintenance of oocyte fate. Whether dedifferentiation is a consequence of the failure to asymmetrically localize Orb, BicD, and Egl within the oocyte or represents a separate function of the PAR proteins is not known.

5.3 Anterior/Posterior Axis Specification During Mid-oogenesis

Several hours after its initial specification, the oocyte is positioned at the posterior end of the egg chamber where it sits adjacent to the posterior follicle cells. The establishment of the oocyte anterior/posterior axis begins early in oogenesis with the secretion of the EGF-like protein Gurken from the posterior end of the stage 4–6 oocyte (Chang et al., 2008; Peri, Bokel, & Roth, 1999). Gurken signals to the overlying posterior follicle cells, which subsequently signal back to the oocyte through an unknown mechanism (Fig. 3A) (Gonzalez-Reyes, Elliott, & St Johnston, 1995; Roth & Lynch, 2009). These signaling events lead to the establishment of opposing cortical PAR domains: dPar-1 concentrates in a small cortical domain at the posterior end and the anterior PARs occupy the anterior and lateral regions of the oocyte cortex (Benton & St Johnston, 2003; Doerflinger et al., 2006; Shulman et al., 2000; Tomancak et al., 2000). Although the final configuration of the PAR domains is similar to that in the *C. elegans* zygote, the dynamics by which these domains form are strikingly different. Whereas the PAR domains are established in ∼5 min in *C. elegans*, it takes roughly 12 h for these domains to mature between stage 7 and 9 fly oocytes. Furthermore, the mechanisms underlying the recruitment of dPar-1 to the oocyte posterior cortex differ from those that underlie the formation of the PAR-1 cortical domain in the *C. elegans* zygote. Recruitment of dPar-1 to the oocyte posterior does not involve cortical actomyosin flows or microtubules, and there is no clear PAR-2 homolog in *Drosophila* (Doerflinger et al., 2006). Rather, signaling from the posterior follicle cells

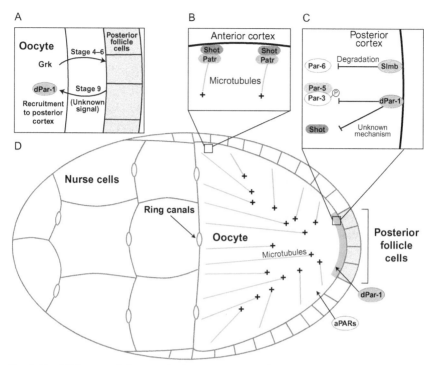

Fig. 3 Establishment of the anterior/posterior axis in the *Drosophila* oocyte. (A) Gurken (Grk) is secreted from the oocyte and signals to the neighboring follicle cells. The posterior follicle cells subsequently signal back to the oocyte, leading to the recruitment of dPar-1 to the posterior cortex. The signal from the follicle cells to the oocyte is not known. (B) Microtubules grow from the oocyte anterior cortex from foci containing Shot and Patronin, which binds microtubule minus ends. (C) At the oocyte posterior cortex, Par-6 is removed from the cortex by the E3 ubiquitin ligase, Slimb. Par-3 is removed from the posterior cortex by dPar-1 phosphorylation, which leads to interaction with Par-5 and sequestration of Par-3 in the cytoplasm. dPar-1 also acts through an unknown mechanism to restrict Shot from the posterior cortex. (D) The oocyte is positioned next to the posterior follicle cells and is connected to the nurse cells by ring canals. dPar-1 (*blue*) forms a posterior cortical domain that is reciprocal to the anterior/lateral anterior PAR domain (aPAR, *yellow*). Microtubules (*green*) are nucleated from the anterior and lateral cortex and not from the posterior cortex. As a result, microtubule density is higher in the anterior are microtubules tend to point toward the posterior.

initiates polarization of the oocyte by inducing the degradation of Par-6 and aPKC at the posterior cortex by the SCF E3 ubiquitin ligase Slimb (Fig. 3C) (Morais-de-Sa, Mukherjee, Lowe, & St Johnston, 2014). Slimb localizes to the posterior cortex and is required for the clearance of Par-6/aPKC, which allows for the recruitment of dPar-1 to the posterior cortex (Morais-de-Sa et al., 2014). Once dPar-1 is recruited to the posterior cortex, mutual antagonism between the PAR proteins stabilizes their reciprocal localization patterns. dPar-1 phosphorylates Par-3 (Bazooka), leading to its clearance from the posterior cortex and aPKC phosphorylates dPar-1, thereby restricting dPar-1 from spreading into the lateral and anterior oocyte cortex (Doerflinger et al., 2010).

The concentration of dPar-1 at the posterior cortex leads to a critical reorganization of the microtubule cytoskeleton along the anterior/posterior axis. During mid-oogenesis, noncentrosomal microtubules grow from the anterior and lateral cortex from foci containing the microtubule minus end-binding protein Patronin. Patronin foci also contain spectraplakin (Shot in *Drosophila*), which is required to recruit Patronin to the oocyte cortex (Nashchekin, Fernandes, & St Johnston, 2016). dPar-1 functions downstream of the anterior PAR proteins to suppress microtubule growth from the posterior cortex by acting to exclude Shot from localizing to the posterior cortex (Fig. 3B) (Nashchekin et al., 2016). It is not known whether Shot is a substrate of dPar-1 or if dPar-1 controls Shot localization more indirectly. As a result of the differences in cortical microtubule growth, there is both an anterior-high, posterior-low gradient in the density of microtubules and a weak bias in the orientation of the microtubules such that microtubule plus ends tend to be pointed toward the oocyte posterior (Khuc Trong, Doerflinger, Dunkel, St Johnston, & Goldstein, 2015; Parton et al., 2011; Zimyanin et al., 2008). These asymmetries provide the basis for differential transport of anterior and posterior determinants. *oskar* mRNA is transported by the plus end-directed motor Kinesin with a slight bias toward the posterior, which is sufficient to concentrate *oskar* mRNA at the posterior cortex (Khuc Trong et al., 2015; Zimyanin et al., 2008). Similarly, the minus end-directed motor dynein is responsible for the transport of *bicoid* mRNA to the anterior (St Johnston, 2005). In *dpar-1* mutant oocytes, microtubules are nucleated uniformly from the cortex with their plus ends oriented toward the center of the oocyte, resulting in the accumulation of *oskar* mRNA in a foci in the center of the oocyte (Fig. 3D) (Doerflinger et al., 2006; Parton et al., 2011; Shulman et al., 2000; Tomancak et al., 2000).

In addition to its role in directing the posterior transport of *oskar* mRNA through polarization of the microtubule cytoskeleton, dPar-1 also plays direct roles in the regulation of Oskar protein accumulation at the posterior pole. Oskar is translated in two forms, Long Oskar and Short Oskar. Long Oskar is required to anchor *oskar* mRNA at the posterior cortex during mid-oogenesis. Short Oskar does not accumulate to high levels until dPar-1 levels decrease during late oogenesis, at which point Short Oskar controls the organization of the germ plasm (reviewed in Lehmann, 2016). The delay in the accumulation of Short Oskar is controlled by its sequential phosphorylation by dPar-1 and Gsk-3, which target Short Oskar for degradation by the E3 ubiquitin ligase Slimb (Morais-De-Sa, Vega-Rioja, Trovisco, & St Johnston, 2013). dPar-1 has also been shown to stabilize Oskar through phosphorylation at a second site, which suggests dPar-1 may target different populations of Oskar for stabilization or degradation (Riechmann, Gutierrez, Filardo, Nebreda, & Ephrussi, 2002). Interestingly, overexpression of *oskar* mRNA results in the formation of an ectopic *oskar* mRNA dot in the cytoplasm that recruits dPar-1 (Zimyanin, Lowe, & St Johnston, 2007). This result suggests *oskar* mRNA and dPar-1 may participate in a positive feedback loop that promotes each other's localization to the posterior cortex, thereby stabilizing the formation of the posterior domain.

5.4 Comparison of Anterior/Posterior Polarization in Worms and Flies

Taken together, the studies described above have elucidated a collection of mechanisms by which the PAR proteins establish the anterior/posterior axis in flies and worms. In both cases, PAR-1 localizes and functions in opposition to the anterior PAR proteins and PAR-1 activity is intimately associated with organization and partitioning germ line factors to germ cells. At the mechanistic level, however, PAR-1 plays a different role in the two systems. In the *C. elegans* zygote, the PAR-1 drives cytoplasmic asymmetries through modulation of MEX-5/6 mobility, which then propagates these signals to drive the segregation of germline factors. These mechanisms are capable of rapidly generating asymmetries over relatively short length scales. In the *Drosophila* oocyte, dPar-1 controls asymmetries in large part by restricting the growth of noncentrosomal microtubules to the anterior and lateral cortex, thereby biasing the orientation of the microtubule cytoskeleton. Polarization of the microtubule cytoskeleton, in turn, provides a foundation for the asymmetric transport of factors toward the anterior or posterior cytoplasm. This mechanism is particularly well suited to the generation of long range asymmetries over relatively long time scales. Therefore, in different biological contexts, PAR-1 orchestrates polarities through mechanisms that are suited to the time scale and length scale of the cellular reorganization.

6. MARK KINASES AND NEUROGENESIS

The mammalian MARKs (microtubule affinity-regulating kinases) were initially purified from porcine brain extracts based on their ability to phosphorylate the microtubule-associated proteins TAU, MAP2, and MAP4 on conserved KXGS motifs present in their microtubule-binding domains (Drewes et al., 1997). Microtubule-associated proteins (MAPs) bind along the surface of microtubules and shift the equilibrium to microtubule polymerization by promoting microtubule nucleation, growth, and stability (Penazzi, Bakota, & Brandt, 2016). Phosphorylation by MARK kinases causes MAPs to dissociate from microtubules and thereby shifts the dynamics of microtubules toward depolymerization (Drewes et al., 1997; Schwalbe et al., 2013). Indeed, MARK overexpression in tissue culture cells results in destruction of the microtubule network (Drewes et al., 1997). Live imaging of microtubule dynamics demonstrated that MARK2 controls microtubule dynamics by decreasing the frequency of transitions from growth to depolymerization without altering the rate of plus end growth (Hayashi et al., 2011). In addition to regulating microtubule dynamics, MAPs can disrupt transport on microtubules by competing with motors for binding to microtubules (Penazzi et al., 2016). In this review, we will focus on the role of the MARK kinases in the initial polarization of neurons and in the morphogenesis of dendritic spines in mature neurons. Recent reviews have discussed the critical roles PAR-1/MARK kinases play in multiple steps of neurogenesis including the asymmetric division and migration of neural progenitors and the initial specification of axon/dendrite polarity (Knoblich, 2010; McDonald, 2014; Reiner & Sapir, 2014; Shelly & Poo, 2011).

6.1 Neuronal Polarization

The antagonistic relationship between MARK kinases and the anterior PARs is intimately involved in the initial polarization of neurons. During this process, the unpolarized neurons extend multiple neurites, one of which is selected as an axon while the remaining become dendrites. In cultured hippocampal neurons, MARK2 is active in nascent dendritic projections and inactive in axonal projections, which results in the accumulation of TAU specifically on axonal microtubules (Chen et al., 2006). Reduction of MARK2 expression results in the formation of multiple axons whereas the overexpression of MARK2 prevents axon formation, indicating that

MARK2 plays an important role in the initial control of axon/dendrite asymmetry (Chen et al., 2006). The anterior PAR proteins concentrate at the tip of the nascent axon and are required for its selection, at least in part because of their ability to inhibit MARK activity in the axon (reviewed in Insolera, Chen, & Shi, 2011). SAD kinases, which contain kinase domains similar to MARK kinases, also phosphorylate MAP protein at KXGS motifs, resulting in their dissociation from microtubules (Kishi, Pan, Crump, & Sanes, 2005). In vivo, SAD kinases are essential for neuronal polarization, likely due to their role in controlling the polarized accumulation of TAU and MAP2 axons and dendrites, respectively (Kishi et al., 2005).

In mature neurons, MARK1 and MARK2 concentrate in and are required for the morphogenesis of dendritic spines, which are postsynaptic protrusions through which dendrites receive synaptic inputs (Wu, DiBona, Bernard, & Zhang, 2012). MARK2 promotes the growth of microtubules and the associated transport of cargo into dendritic spines (Hayashi et al., 2011). MARK2 also phosphorylates PSD-95, a postsynaptic protein that scaffolds the assembly of proteins at the postsynaptic density and is required for spine morphogenesis (Wu et al., 2012). The role of MARK phosphorylation in PSD-95 function is not known, but may involve control of PSD-95 dynamics as dPar-1 phosphorylation of the *Drosophila* PSD-95 homologue Discs large (Dlg) stabilizes its postsynaptic localization at the neuromuscular junction (Zhang et al., 2007). Interestingly, the anterior PAR proteins are also essential for dendritic spine morphogenesis through their control of the actin cytoskeleton, suggesting that the PAR polarity network is deployed within each dendritic spine to orchestrate their organization (Zhang & Macara, 2006, 2008). It will be interesting to learn to what extent the principles and mechanisms underlying polarization at the cellular level are recapitulated in these small, subcellular compartments.

7. PAR-1 AND DISEASE

7.1 Alzheimer's Disease

During the progression of Alzheimer's disease, TAU becomes hyperphosphorylated, dissociates from microtubules, accumulates to abnormally high levels in the somatodendritic compartment, and forms paired helical filaments and neurofibrillary tangles (NFTs) (Iqbal, Liu, & Gong, 2016). Phosphorylation by MARK is thought to contribute to disease progression by causing TAU to dissociate from microtubules, thereby increasing the cytoplasmic pool of TAU that is available to aggregate. During the progression of Alzheimer's, one of the earliest TAU phosphorylation sites to be upregulated is the MARK phosphorylation site Ser262, suggesting that MARK activity may play an early role in disease progression (Matenia & Mandelkow, 2009). In *Drosophila*, Tau phosphorylation by dPar-1 increases the rate of subsequent phosphorylation at other residues by Gsk-3 and Cdk5, consistent with the idea that PAR-1 phosphorylation primes TAU for hyperphosphorylation (Nishimura, Yang, & Lu, 2004). Additionally, TAU phosphorylation also disrupts the sorting of TAU to the axonal compartment, leading to an increase in TAU levels in the somatodendritic compartments (Li et al., 2011). A later, more direct role for MARK in the formation of NFTs is suggested by the observation that MARK kinase is associated with NFTs (Chin et al., 2000). However, because TAU is subject to many different posttranslational modifications, including phosphorylation by a large number of kinases, it has been challenging to parse the contribution of individual enzymes to disease progression (Iqbal et al., 2016).

7.2 Cancer

The connections between cell polarity mechanisms and tumorigenesis are of great interest. Loss of cell polarity is a hallmark of many tumors and may contribute to oncogenesis through a number of mechanisms including loss of epithelial polarity and junctions, increased metastasis, and disrupted asymmetric progenitor cell divisions (reviewed in Halaoui & McCaffrey, 2015; Morrison & Kimble, 2006; Muthuswamy & Xue, 2012). Dysregulation of MARK kinases is associated with a number of tumors. MARK4 is amplified in glioblastoma, upregulated downstream of Wnt signaling in hepatocarcinoma and derepressed in breast and lung cancer cells (Beghini et al., 2003; Kato et al., 2001; Pardo et al., 2016). MARK2 is frequently upregulated in nonsmall cell lung carcinoma and its expression is correlated with malignant phenotypes (Hubaux et al., 2015). In addition, inherited mutations in LKB1 (PAR-4) kinase cause Peutz–Jeghers cancer syndrome and LKB1 is commonly inactivated in nonsmall cell lung carcinomas (reviewed in Sanchez-Cespedes, 2007). As discussed earlier, LKB1 activates AMPK family kinases including AMPK and MARK kinases and it is likely that dysregulation of several of these kinases contributes to tumor progression (Shorning & Clarke, 2016). A recent study identified a pathway that connects LKB1 inactivation and the consequent loss of MARK1 and MARK4 activity to activation of the epithelial-to-mesenchymal transition. MARK1 and MARK4 phosphorylate the scaffolding protein DIXDC1 to drive its localization to focal adhesions. Inactivation of MARK1 and MARK4 delocalizes DIXDC1 from focal adhesions, which activates a downstream signaling cascade that results in the transcriptional upregulation of Snail, a potent activator of the epithelial-to-mesenchymal transition (Goodwin et al., 2014).

More than half the human population is chronically infected with *Helicobacter pylori*, which colonizes the stomach mucosa and can cause a variety of gastric diseases, including gastric cancer (Stein, Ruggiero, Rappuoli, & Bagnoli, 2013). *H. pylori* strains that express the cytotoxin CagA are associated with a higher risk of gastric cancer (Stein et al., 2013). CagA is injected by *H. pylori* into epithelial cells lining the stomach where it interacts with a several cellular proteins and causes disruption of epithelial contacts, loss of epithelial polarity, and dramatic cell elongation (the "hummingbird" phenotype) that are thought to contribute to oncogenesis. One of the targets of CagA is MARK2, which is localized at the basolateral membrane in epithelial cells and is required to maintain apical/basal polarity (Saadat et al., 2007; Zeaiter et al., 2008). CagA inhibits MARK2 kinase activity via a short peptide that mimics MARK2 substrates and binds stably within the catalytic cleft (Nesić et al., 2010; Nishikawa, Hayashi, Arisaka, Senda, & Hatakeyama, 2016). Inhibition of MARK2 causes defects in cell polarity and in the assembly of tight junctions that disrupts the architecture of the gastric epithelia and is thought to contribute to an epithelial-to-mesenchymal transition (Saadat et al., 2007; Zeaiter et al., 2008). CagA also targets the host oncoprotein SHP-2 through a different mechanism, and the combined inhibition of MARK2 and SHP-2 is thought to underlie the ability of CagA to trigger carcinogenesis (Stein et al., 2013).

8. CONCLUDING REMARKS

In this review, we have discussed how the localized activity of PAR-1 kinase and the anterior PAR proteins underlies the establishment of cellular asymmetries in a few well-characterized systems. While a conserved module of antagonistic interactions between the anterior PAR proteins and PAR-1 provides a core means by which the activity of these proteins is localized, the mechanisms that trigger polarization and the speed and spatial scale at which the PAR domains are established are unique to different polarized cells. In addition, we have discussed how PAR-1 can drive the formation of downstream asymmetries through a variety of different mechanisms. As future studies of PAR-1 continue to extend beyond the well-characterized systems described in this review, it will be critical to identify the relevant PAR-1 substrates, to characterize how phosphorylation modulates the function of these substrates and ultimately contributes to the organization of the cell. In addition, future studies are likely to further elucidate how PAR-1 kinase activity is controlled and in particular how kinase activity may be coordinated with PAR-1 localization. These studies will provide a foundation for understanding how cells establish and maintain polarity and for understanding how disruption of PAR-1 activity contributes to pathogenesis in a variety of contexts, including Alzheimer's disease progression and tumorigenesis.

REFERENCES

Aceto, D., Beers, M., & Kemphues, K. J. (2006). Interaction of PAR-6 with CDC-42 is required for maintenance but not establishment of PAR asymmetry in C. elegans. *Developmental Biology, 299*, 386–397.

Beatty, A., Morton, D. G., & Kemphues, K. (2013). PAR-2, LGL-1 and the CDC-42 GAP CHIN-1 act in distinct pathways to maintain polarity in the C. elegans embryo. *Development, 140*, 2005–2014.

Beghini, A., Magnani, I., Roversi, G., Piepoli, T., Di Terlizzi, S., Moroni, R. F., et al. (2003). The neural progenitor-restricted isoform of the MARK4 gene in 19q13.2 is upregulated in human gliomas and overexpressed in a subset of glioblastoma cell lines. *Oncogene, 22*, 2581–2591.

Benton, R., Palacios, I. M., & Johnston, D. S. (2002). Drosophila 14-3-3/PAR-5 is an essential mediator of PAR-1 function in axis formation. *Developmental Cell, 3*, 659–671.

Benton, R., & St Johnston, D. (2003). Drosophila PAR-1 and 14-3-3 inhibit Bazooka/PAR-3 to establish complementary cortical domains in polarized cells. *Cell, 115*, 691–704.

Bessone, S., Vidal, F., Le Bouc, Y., Epelbaum, J., Bluet-Pajot, M. T., & Darmon, M. (1999). EMK protein kinase-null mice: Dwarfism and hypofertility associated with alterations in the somatotrope and prolactin pathways. *Developmental Biology, 214*, 87–101.

Boyd, L., Guo, S., Levitan, D., Stinchcomb, D. T., & Kemphues, K. J. (1996). PAR-2 is asymmetrically distributed and promotes association of P granules and PAR-1 with the cortex in C. elegans embryos. *Development, 122*, 3075–3084.

Brangwynne, C. P., Eckmann, C. R., Courson, D. S., Rybarska, A., Hoege, C., Gharakhani, J., et al. (2009). Germline P granules are liquid droplets that localize by controlled dissolution/condensation. *Science, 324*, 1729–1732.

Chang, W. L., Liou, W., Pen, H. C., Chou, H. Y., Chang, Y. W., Li, W. H., et al. (2008). The gradient of Gurken, a long-range morphogen, is directly regulated by Cbl-mediated endocytosis. *Development, 135*, 1923–1933.

Cheeks, R. J., Canman, J. C., Gabriel, W. N., Meyer, N., Strome, S., & Goldstein, B. (2004). C. elegans PAR proteins function by mobilizing and stabilizing asymmetrically localized protein complexes. *Current Biology, 14*, 851–862.

Chen, L., Jiao, Z. H., Zheng, L. S., Zhang, Y. Y., Xie, S. T., Wang, Z. X., et al. (2009). Structural insight into the autoinhibition mechanism of AMP-activated protein kinase. *Nature, 459*, 1146–1149.

Chen, Y. M., Wang, Q. J., Hu, H. S., Yu, P. C., Zhu, J., Drewes, G., et al. (2006). Microtubule affinity-regulating kinase 2 functions downstream of the PAR-3/PAR-6/atypical PKC complex in regulating hippocampal neuronal polarity. *Proceedings of the National Academy of Sciences of the United States of America, 103*, 8534–8539.

Chin, J. Y., Knowles, R. B., Schneider, A., Drewes, G., Mandelkow, E. M., & Hyman, B. T. (2000). Microtubule-affinity regulating kinase (MARK) is tightly associated with neurofibrillary tangles in Alzheimer brain: A fluorescence resonance energy transfer study. *Journal of Neuropathology and Experimental Neurology, 59*, 966–971.

Cowan, C. R., & Hyman, A. A. (2004). Centrosomes direct cell polarity independently of microtubule assembly in *C. elegans* embryos. *Nature, 431*, 92–96.

Cox, D. N., Lu, B., Sun, T. Q., Williams, L. T., & Jan, Y. N. (2001). Drosophila par-1 is required for oocyte differentiation and microtubule organization. *Current Biology, 11*, 75–87.

Cox, D. N., Seyfried, S. A., Jan, L. Y., & Jan, Y. N. (2001). Bazooka and atypical protein kinase C are required to regulate oocyte differentiation in the *Drosophila* ovary. *Proceedings of the National Academy of Sciences of the United States of America, 98*, 14475–14480.

Cuenca, A. A., Schetter, A., Aceto, D., Kemphues, K., & Seydoux, G. (2003). Polarization of the *C. elegans* zygote proceeds via distinct establishment and maintenance phases. *Development, 130*, 1255–1265.

Daniels, B. R., Dobrowsky, T. M., Perkins, E. M., Sun, S. X., & Wirtz, D. (2010). MEX-5 enrichment in the *C. elegans* early embryo mediated by differential diffusion. *Development, 137*, 2579–2585.

Daniels, B. R., Perkins, E. M., Dobrowsky, T. M., Sun, S. X., & Wirtz, D. (2009). Asymmetric enrichment of PIE-1 in the *Caenorhabditis elegans* zygote mediated by binary counter diffusion. *The Journal of Cell Biology, 184*, 473–479.

Doerflinger, H., Benton, R., Shulman, J. M., & St Johnston, D. (2003). The role of PAR-1 in regulating the polarised microtubule cytoskeleton in the *Drosophila* follicular epithelium. *Development, 130*, 3965–3975.

Doerflinger, H., Benton, R., Torres, I., Zwart, M., & St Johnston, D. (2006). *Drosophila* anterior-posterior polarity requires actin-dependent PAR-1 recruitment to the oocyte posterior. *Current Biology, 16*, 1090–1095.

Doerflinger, H., Vogt, N., Torres, I. L., Mirouse, V., Koch, I., Nüsslein-Volhard, C., et al. (2010). Bazooka is required for polarisation of the *Drosophila* anterior-posterior axis. *Development, 137*, 1765–1773.

Drewes, G., Ebneth, A., Preuss, U., Mandelkow, E. M., & Mandelkow, E. (1997). MARK, a novel family of protein kinases that phosphorylate microtubule-associated proteins and trigger microtubule disruption. *Cell, 89*, 297–308.

Drewes, G., Trinczek, B., Illenberger, S., Biernat, J., Schmitt-Ulms, G., Meyer, H. E., et al. (1995). Microtubule-associated protein/microtubule affinity-regulating kinase (p110mark). A novel protein kinase that regulates tau-microtubule interactions and dynamic instability by phosphorylation at the Alzheimer-specific site serine 262. *The Journal of Biological Chemistry, 270*, 7679–7688.

Driever, W., & Nusslein-Volhard, C. (1988). A gradient of bicoid protein in *Drosophila* embryos. *Cell, 54*, 83–93.

Dumont, N. A., Wang, Y. X., von Maltzahn, J., Pasut, A., Bentzinger, C. F., Brun, C. E., et al. (2015). Dystrophin expression in muscle stem cells regulates their polarity and asymmetric division. *Nature Medicine, 21*, 1455–1463.

Elbert, M., Rossi, G., & Brennwald, P. (2005). The yeast par-1 homologs kin1 and kin2 show genetic and physical interactions with components of the exocytic machinery. *Molecular Biology of the Cell, 16*, 532–549.

Etemad-Moghadam, B., Guo, S., & Kemphues, K. J. (1995). Asymmetrically distributed PAR-3 protein contributes to cell polarity and spindle alignment in early *C. elegans* embryos. *Cell, 83*, 743–752.

Feng, M., Tian, L., Gan, L., Liu, Z., & Sun, C. (2014). Mark4 promotes adipogenesis and triggers apoptosis in 3T3-L1 adipocytes by activating JNK1 and inhibiting p38MAPK pathways. *Biology of the Cell, 106*, 294–307.

Frohnhofer, H. G., & Nusslein-Volhard, C. (1986). Organization of anterior pattern in the Drosophila embryo by the maternal gene bicoid. *Nature, 324*, 120–125.

Gallo, C. M., Wang, J. T., Motegi, F., & Seydoux, G. (2010). Cytoplasmic partitioning of P granule components is not required to specify the germline in *C. elegans*. *Science, 330*, 1685–1689.

Goehring, N. W., Hoege, C., Grill, S. W., & Hyman, A. A. (2011). PAR proteins diffuse freely across the anterior-posterior boundary in polarized *C. elegans* embryos. *The Journal of Cell Biology, 193*, 583–594.

Goehring, N. W., Trong, P. K., Bois, J. S., Chowdhury, D., Nicola, E. M., Hyman, A. A., et al. (2011). Polarization of PAR proteins by advective triggering of a pattern-forming system. *Science, 334*, 1137–1141.

Goldstein, B., & Hird, S. N. (1996). Specification of the anteroposterior axis in *Caenorhabditis elegans*. *Development, 122*, 1467–1474.

Goldstein, B., & Macara, I. G. (2007). The PAR proteins: Fundamental players in animal cell polarization. *Developmental Cell, 13*, 609–622.

Gonzalez-Reyes, A., Elliott, H., & St Johnston, D. (1995). Polarization of both major body axes in *Drosophila* by gurken-torpedo signalling. *Nature, 375*, 654–658.

Goodwin, J. M., Svensson, R. U., Lou, H. J., Winslow, M. M., Turk, B. E., & Shaw, R. J. (2014). An AMPK-independent signaling pathway downstream of the LKB1 tumor suppressor controls Snail1 and metastatic potential. *Molecular Cell, 55*, 436–450.

Goransson, O., Deak, M., Wullschleger, S., Morrice, N. A., Prescott, A. R., & Alessi, D. R. (2006). Regulation of the polarity kinases PAR-1/MARK by 14-3-3 interaction and phosphorylation. *Journal of Cell Science, 119*, 4059–4070.

Gotta, M., Abraham, M. C., & Ahringer, J. (2001). CDC-42 controls early cell polarity and spindle orientation in *C. elegans*. *Current Biology, 11*, 482–488.

Griffin, E. E. (2015). Cytoplasmic localization and asymmetric division in the early embryo of *Caenorhabditis elegans*. *Wiley Interdisciplinary Reviews. Developmental Biology, 4*, 267–282.

Griffin, E. E., Odde, D. J., & Seydoux, G. (2011). Regulation of the MEX-5 gradient by a spatially segregated kinase/phosphatase cycle. *Cell, 146*, 955–968.

Guo, S., & Kemphues, K. J. (1995). par-1, a gene required for establishing polarity in *C. elegans* embryos, encodes a putative Ser/Thr kinase that is asymmetrically distributed. *Cell, 81*, 611–620.

Halaoui, R., & McCaffrey, L. (2015). Rewiring cell polarity signaling in cancer. *Oncogene, 34*, 939–950.

Hao, Y., Boyd, L., & Seydoux, G. (2006). Stabilization of cell polarity by the *C. elegans* RING protein PAR-2. *Developmental Cell, 10*, 199–208.

Hayashi, K., Suzuki, A., Hirai, S., Kurihara, Y., Hoogenraad, C. C., & Ohno, S. (2011). Maintenance of dendritic spine morphology by partitioning-defective 1b through regulation of microtubule growth. *The Journal of Neuroscience, 31*, 12094–12103.

Hoege, C., & Hyman, A. A. (2013). Principles of PAR polarity in *Caenorhabditis elegans* embryos. *Nature Reviews Molecular Cell Biology, 14*, 315–322.

Hubaux, R., Thu, K. L., Vucic, E. A., Pikor, L. A., Kung, S. H., Martinez, V. D., et al. (2015). Microtubule affinity-regulating kinase 2 is associated with DNA damage response and cisplatin resistance in non-small cell lung cancer. *International Journal of Cancer, 137*, 2072–2082.

Hurov, J. B., Stappenbeck, T. S., Zmasek, C. M., White, L. S., Ranganath, S. H., Russell, J. H., et al. (2001). Immune system dysfunction and autoimmune disease in mice lacking Emk (Par-1) protein kinase. *Molecular and Cellular Biology, 21*, 3206–3219.

Hurov, J. B., Watkins, J. L., & Piwnica-Worms, H. (2004). Atypical PKC phosphorylates PAR-1 kinases to regulate localization and activity. *Current Biology, 14*, 736–741.

Huynh, J. R., Petronczki, M., Knoblich, J. A., & St Johnston, D. (2001). Bazooka and PAR-6 are required with PAR-1 for the maintenance of oocyte fate in *Drosophila*. *Current Biology, 11*, 901–906.

Huynh, J. R., Shulman, J. M., Benton, R., & St Johnston, D. (2001). PAR-1 is required for the maintenance of oocyte fate in *Drosophila*. *Development, 128*, 1201–1209.

Huynh, J. R., & St Johnston, D. (2004). The origin of asymmetry: Early polarisation of the *Drosophila* germline cyst and oocyte. *Current Biology, 14*, R438–R449.

Inaba, M., Venkei, Z. G., & Yamashita, Y. M. (2015). The polarity protein Baz forms a platform for the centrosome orientation during asymmetric stem cell division in the *Drosophila* male germline. *eLife, 4*, e04960.

Insolera, R., Chen, S., & Shi, S.-H. (2011). Par proteins and neuronal polarity. *Developmental Neurobiology, 71*, 483–494.

Iqbal, K., Liu, F., & Gong, C. X. (2016). Tau and neurodegenerative disease: The story so far. *Nature Reviews Neurology, 12*, 15–27.

Jaleel, M., Villa, F., Deak, M., Toth, R., Prescott, A. R., Van Aalten, D. M., et al. (2006). The ubiquitin-associated domain of AMPK-related kinases regulates conformation and LKB1-mediated phosphorylation and activation. *The Biochemical Journal, 394*, 545–555.

Januschke, J., Gervais, L., Dass, S., Kaltschmidt, J. A., Lopez-Schier, H., St Johnston, D., et al. (2002). Polar transport in the *Drosophila* oocyte requires Dynein and Kinesin I cooperation. *Current Biology, 12*, 1971–1981.

Kato, T., Satoh, S., Okabe, H., Kitahara, O., Ono, K., Kihara, C., et al. (2001). Isolation of a novel human gene, MARKL1, homologous to MARK3 and its involvement in hepatocellular carcinogenesis. *Neoplasia, 3*, 4–9.

Kay, A. J., & Hunter, C. P. (2001). CDC-42 regulates PAR protein localization and function to control cellular and embryonic polarity in *C. elegans*. *Current Biology, 11*, 474–481.

Kemphues, K. J., Priess, J. R., Morton, D. G., & Cheng, N. S. (1988). Identification of genes required for cytoplasmic localization in early *C. elegans* embryos. *Cell, 52*, 311–320.

Khuc Trong, P., Doerflinger, H., Dunkel, J., St Johnston, D., & Goldstein, R. E. (2015). Cortical microtubule nucleation can organise the cytoskeleton of *Drosophila* oocytes to define the anteroposterior axis. *eLife, 4*, e06088.

Kishi, M., Pan, Y. A., Crump, J. G., & Sanes, J. R. (2005). Mammalian SAD kinases are required for neuronal polarization. *Science, 307*, 929–932.

Knoblich, J. A. (2010). Asymmetric cell division: Recent developments and their implications for tumour biology. *Nature Reviews Molecular Cell Biology, 11*, 849–860.

Kotak, S., & Gonczy, P. (2013). Mechanisms of spindle positioning: Cortical force generators in the limelight. *Current Opinion in Cell Biology, 25*, 741–748.

Kumfer, K. T., Cook, S. J., Squirrell, J. M., Eliceiri, K. W., Peel, N., O'Connell, K. F., et al. (2010). CGEF-1 and CHIN-1 regulate CDC-42 activity during asymmetric division in the *Caenorhabditis elegans* embryo. *Molecular Biology of the Cell, 21*, 266–277.

Labbé, J.-C., Pacquelet, A., Marty, T., & Gotta, M. (2006). A genomewide screen for suppressors of par-2 uncovers potential regulators of PAR protein-dependent cell polarity in *Caenorhabditis elegans*. *Genetics, 174*, 285–295.

Lehmann, R. (2016). Germ plasm biogenesis—An Oskar-centric perspective. *Current Topics in Developmental Biology, 116*, 679–707.

Lennerz, J. K., Hurov, J. B., White, L. S., Lewandowski, K. T., Prior, J. L., Planer, G. J., et al. (2010). Loss of Par-1a/MARK3/C-TAK1 kinase leads to reduced adiposity, resistance to hepatic steatosis, and defective gluconeogenesis. *Molecular and Cellular Biology, 30*, 5043–5056.

Leventis, P. A., & Grinstein, S. (2010). The distribution and function of phosphatidylserine in cellular membranes. *Annual Review of Biophysics, 39*, 407–427.

Li, X., Kumar, Y., Zempel, H., Mandelkow, E. M., Biernat, J., & Mandelkow, E. (2011). Novel diffusion barrier for axonal retention of Tau in neurons and its failure in neurodegeneration. *The EMBO Journal, 30*, 4825–4837.

Lipkow, K., & Odde, D. J. (2008). Model for protein concentration gradients in the cytoplasm. *Cellular and Molecular Bioengineering, 1*, 84–92.

Liu, Z., Gan, L., Chen, Y., Luo, D., Zhang, Z., Cao, W., et al. (2016). Mark4 promotes oxidative stress and inflammation via binding to PPARgamma and activating NF-kappaB pathway in mice adipocytes. *Scientific Reports, 6*, 21382.

Lizcano, J. M., Göransson, O., Toth, R., Deak, M., Morrice, N. A., Boudeau, J., et al. (2004). LKB1 is a master kinase that activates 13 kinases of the AMPK subfamily, including MARK/PAR-1. *The EMBO Journal, 23*, 833–843.

Lu, M. S., & Johnston, C. A. (2013). Molecular pathways regulating mitotic spindle orientation in animal cells. *Development, 140*, 1843–1856.

Manning, G., Whyte, D. B., Martinez, R., Hunter, T., & Sudarsanam, S. (2002). The protein kinase complement of the human genome. *Science, 298*, 1912–1934.

Marx, A., Nugoor, C., Müller, J., Panneerselvam, S., Timm, T., Bilang, M., et al. (2006). Structural variations in the catalytic and ubiquitin-associated domains of microtubule-associated protein/microtubule affinity regulating kinase (MARK) 1 and MARK2. *The Journal of Biological Chemistry, 281*, 27586–27599.

Marx, A., Nugoor, C., Panneerselvam, S., & Mandelkow, E. (2010). Structure and function of polarity-inducing kinase family MARK/Par-1 within the branch of AMPK/Snf1-related kinases. *The FASEB Journal, 24*, 1637–1648.

Matenia, D., Griesshaber, B., Li, X. Y., Thiessen, A., Johne, C., Jiao, J., et al. (2005). PAK5 kinase is an inhibitor of MARK/Par-1, which leads to stable microtubules and dynamic actin. *Molecular Biology of the Cell, 16*, 4410–4422.

Matenia, D., & Mandelkow, E.-M. (2009). The tau of MARK: A polarized view of the cytoskeleton. *Trends in Biochemical Sciences, 34*, 332–342.

McDonald, J. A. (2014). Canonical and noncanonical roles of Par-1/MARK kinases in cell migration. *International Review of Cell and Molecular Biology, 312*, 169–199.

Mcdonald, J. A., Khodyakova, A., Aranjuez, G., Dudley, C., & Montell, D. J. (2008). PAR-1 kinase regulates epithelial detachment and directional protrusion of migrating border cells. *Current Biology, 18*, 1659–1667.

Mello, C. C., Schubert, C., Draper, B., Zhang, W., Lobel, R., & Priess, J. R. (1996). The PIE-1 protein and germline specification in C. elegans embryos. *Nature, 382*, 710–712.

Morais-de-Sa, E., Mukherjee, A., Lowe, N., & St Johnston, D. (2014). Slmb antagonises the aPKC/Par-6 complex to control oocyte and epithelial polarity. *Development, 141*, 2984–2992.

Morais-De-Sa, E., Vega-Rioja, A., Trovisco, V., & St Johnston, D. (2013). Oskar is targeted for degradation by the sequential action of Par-1, GSK-3, and the SCF-slimb ubiquitin ligase. *Developmental Cell, 26*, 303–314.

Moravcevic, K., Mendrola, J. M., Schmitz, K. R., Wang, Y.-H., Slochower, D., Janmey, P. A., et al. (2010). Kinase associated-1 domains drive MARK/PAR1 kinases to membrane targets by binding acidic phospholipids. *Cell, 143*, 966–977.

Morrison, S. J., & Kimble, J. (2006). Asymmetric and symmetric stem-cell divisions in development and cancer. *Nature, 441*, 1068–1074.

Morton, D. G., Shakes, D. C., Nugent, S., Dichoso, D., Wang, W., Golden, A., et al. (2002). The *Caenorhabditis elegans* par-5 gene encodes a 14-3-3 protein required for cellular asymmetry in the early embryo. *Developmental Biology, 241*, 47–58.

Motegi, F., & Seydoux, G. (2014). The PAR network: Redundancy and robustness in a symmetry-breaking system. *Philosophical Transactions of the Royal Society of London. Series B, Biological Sciences, 368*, 20130010.

Motegi, F., & Sugimoto, A. (2006). Sequential functioning of the ECT-2 RhoGEF, RHO-1 and CDC-42 establishes cell polarity in *Caenorhabditis elegans* embryos. *Nature Cell Biology, 8*, 978–985.

Motegi, F., Zonies, S., Hao, Y., Cuenca, A. A., Griffin, E., & Seydoux, G. (2011). Microtubules induce self-organization of polarized PAR domains in *Caenorhabditis elegans* zygotes. *Nature Cell Biology, 13*, 1361–1367.

Munro, E., Nance, J., & Priess, J. R. (2004). Cortical flows powered by asymmetrical contraction transport PAR proteins to establish and maintain anterior-posterior polarity in the early *C. elegans* embryo. *Developmental Cell, 7*, 413–424.

Murphy, J. M., Korzhnev, D. M., Ceccarelli, D. F., Briant, D. J., Zarrine-Afsar, A., Sicheri, F., et al. (2007). Conformational instability of the MARK3 UBA domain compromises ubiquitin recognition and promotes interaction with the adjacent kinase domain. *Proceedings of the National Academy of Sciences of the United States of America, 104*, 14336–14341.

Muthuswamy, S. K., & Xue, B. (2012). Cell polarity as a regulator of cancer cell behavior plasticity. *Annual Review of Cell and Developmental Biology, 28*, 599–625.

Nakayama, Y., Shivas, J. M., Poole, D. S., Squirrell, J. M., Kulkoski, J. M., Schleede, J. B., et al. (2009). Dynamin participates in the maintenance of anterior polarity in the *Caenorhabditis elegans* embryo. *Developmental Cell, 16*, 889–900.

Nance, J., & Zallen, J. A. (2011). Elaborating polarity: PAR proteins and the cytoskeleton. *Development, 138*, 799–809.

Nashchekin, D., Fernandes, A. R., & St Johnston, D. (2016). Patronin/shot cortical foci assemble the noncentrosomal microtubule array that specifies the *Drosophila* anterior-posterior axis. *Developmental Cell, 38*, 61–72.

Nesić, D., Miller, M. C., Quinkert, Z. T., Stein, M., Chait, B. T., & Stebbins, C. E. (2010). Helicobacter pylori CagA inhibits PAR1-MARK family kinases by mimicking host substrates. *Nature Structural and Molecular Biology, 17*, 130–132.

Nishikawa, H., Hayashi, T., Arisaka, F., Senda, T., & Hatakeyama, M. (2016). Impact of structural polymorphism for the *Helicobacter pylori* CagA oncoprotein on binding to polarity-regulating kinase PAR1b. *Scientific Reports, 6*, 30031.

Nishimura, I., Yang, Y., & Lu, B. (2004). PAR-1 kinase plays an initiator role in a temporally ordered phosphorylation process that confers tau toxicity in *Drosophila*. *Cell, 116*, 671–682.

Panneerselvam, S., Marx, A., Mandelkow, E. M., & Mandelkow, E. (2006). Structure of the catalytic and ubiquitin-associated domains of the protein kinase MARK/Par-1. *Structure, 14*, 173–183.

Pardo, O. E., Castellano, L., Munro, C. E., Hu, Y., Mauri, F., Krell, J., et al. (2016). miR-515-5p controls cancer cell migration through MARK4 regulation. *EMBO Reports, 17*, 570–584.

Parton, R. M., Hamilton, R. S., Ball, G., Yang, L., Cullen, C. F., Lu, W., et al. (2011). A PAR-1-dependent orientation gradient of dynamic microtubules directs posterior cargo transport in the *Drosophila* oocyte. *The Journal of Cell Biology, 194*, 121–135.

Penazzi, L., Bakota, L., & Brandt, R. (2016). Microtubule dynamics in neuronal development, plasticity, and neurodegeneration. *International Review of Cell and Molecular Biology, 321*, 89–169.

Peri, F., Bokel, C., & Roth, S. (1999). Local Gurken signaling and dynamic MAPK activation during *Drosophila* oogenesis. *Mechanisms of Development, 81*, 75–88.

Reiner, O., & Sapir, T. (2014). Mark/Par-1 marking the polarity of migrating neurons. *Advances in Experimental Medicine and Biology, 800*, 97–111.

Riechmann, V., & Ephrussi, A. (2004). Par-1 regulates bicoid mRNA localisation by phosphorylating Exuperantia. *Development, 131*, 5897–5907.

Riechmann, V., Gutierrez, G. J., Filardo, P., Nebreda, A. R., & Ephrussi, A. (2002). Par-1 regulates stability of the posterior determinant Oskar by phosphorylation. *Nature Cell Biology, 4,* 337–342.

Robin, F. B., McFadden, W. M., Yao, B., & Munro, E. M. (2014). Single-molecule analysis of cell surface dynamics in *Caenorhabditis elegans* embryos. *Nature Methods, 11,* 677–682.

Rose, L. S., & Kemphues, K. J. (1998). Early patterning of the *C. elegans* embryo. *Annual Review of Genetics, 32,* 521–545.

Roth, S., & Lynch, J. A. (2009). Symmetry breaking during *Drosophila* oogenesis. *Cold Spring Harbor Perspectives in Biology, 1,* a001891.

Saadat, I., Higashi, H., Obuse, C., Umeda, M., Murata-Kamiya, N., Saito, Y., et al. (2007). *Helicobacter pylori* CagA targets PAR1/MARK kinase to disrupt epithelial cell polarity. *Nature, 447,* 330–333.

Sack, J. S., Gao, M., Kiefer, S. E., Myers, J. E., Jr., Newitt, J. A., Wu, S., et al. (2016). Crystal structure of microtubule affinity-regulating kinase 4 catalytic domain in complex with a pyrazolopyrimidine inhibitor. *Acta Crystallographica. Section F, Structural Biology Communications, 72,* 129–134.

Saha, S., Weber, C. A., Nousch, M., Adame-Arana, O., Hoege, C., Hein, M. Y., et al. (2016). Polar positioning of phase-separated liquid compartments in cells regulated by an mRNA competition mechanism. *Cell, 166,* 1572–1584.

Sailer, A., Anneken, A., Li, Y., Lee, S., & Munro, E. (2015). Dynamic opposition of clustered proteins stabilizes cortical polarity in the *C. elegans* zygote. *Developmental Cell, 35,* 131–142.

Sanchez-Cespedes, M. (2007). A role for LKB1 gene in human cancer beyond the Peutz-Jeghers syndrome. *Oncogene, 26,* 7825–7832.

Schlaitz, A.-L., Srayko, M., Dammermann, A., Quintin, S., Wielsch, N., MacLeod, I., et al. (2007). The *C. elegans* RSA complex localizes protein phosphatase 2A to centrosomes and regulates mitotic spindle assembly. *Cell, 128,* 115–127.

Schubert, C. M., Lin, R., de Vries, C. J., Plasterk, R. H., & Priess, J. R. (2000). MEX-5 and MEX-6 function to establish soma/germline asymmetry in early *C. elegans* embryos. *Molecular Cell, 5,* 671–682.

Schwalbe, M., Biernat, J., Bibow, S., Ozenne, V., Jensen, M. R., Kadavath, H., et al. (2013). Phosphorylation of human Tau protein by microtubule affinity-regulating kinase 2. *Biochemistry, 52,* 9068–9079.

Segu, L., Pascaud, A., Costet, P., Darmon, M., & Buhot, M. C. (2008). Impairment of spatial learning and memory in ELKL Motif Kinase1 (EMK1/MARK2) knockout mice. *Neurobiology of Aging, 29,* 231–240.

Shelly, M., & Poo, M.-M. (2011). Role of LKB1-SAD/MARK pathway in neuronal polarization. *Developmental Neurobiology, 71,* 508–527.

Shorning, B. Y., & Clarke, A. R. (2016). Energy sensing and cancer: LKB1 function and lessons learnt from Peutz–Jeghers syndrome. *Seminars in Cell and Developmental Biology, 52,* 21–29.

Shulman, J. M., Benton, R., & St Johnston, D. (2000). The *Drosophila* homolog of *C. elegans* PAR-1 organizes the oocyte cytoskeleton and directs oskar mRNA localization to the posterior pole. *Cell, 101,* 377–388.

St Johnston, D. (2005). Moving messages: The intracellular localization of mRNAs. *Nature Reviews Molecular Cell Biology, 6,* 363–375.

St Johnston, D., & Ahringer, J. (2010). Cell polarity in eggs and epithelia: Parallels and diversity. *Cell, 141,* 757–774.

Stein, M., Ruggiero, P., Rappuoli, R., & Bagnoli, F. (2013). *Helicobacter pylori* CagA: From pathogenic mechanisms to its use as an anti-cancer vaccine. *Frontiers in Immunology, 4,* 328.

Strome, S., & Wood, W. B. (1983). Generation of asymmetry and segregation of germ-line granules in early *C. elegans* embryos. *Cell, 35*, 15–25.
Sun, C., Tian, L., Nie, J., Zhang, H., Han, X., & Shi, Y. (2012). Inactivation of MARK4, an AMP-activated protein kinase (AMPK)-related kinase, leads to insulin hypersensitivity and resistance to diet-induced obesity. *The Journal of Biological Chemistry, 287*, 38305–38315.
Suzuki, A., Hirata, M., Kamimura, K., Maniwa, R., Yamanaka, T., Mizuno, K., et al. (2004). aPKC acts upstream of PAR-1b in both the establishment and maintenance of mammalian epithelial polarity. *Current Biology, 14*, 1425–1435.
Tabuse, Y., Izumi, Y., Piano, F., Kemphues, K. J., Miwa, J., & Ohno, S. (1998). Atypical protein kinase C cooperates with PAR-3 to establish embryonic polarity in *Caenorhabditis elegans*. *Development, 125*, 3607–3614.
Tenenhaus, C., Schubert, C., & Seydoux, G. (1998). Genetic requirements for PIE-1 localization and inhibition of gene expression in the embryonic germ lineage of *Caenorhabditis elegans*. *Developmental Biology, 200*, 212–224.
Tenlen, J. R., Molk, J. N., London, N., Page, B. D., & Priess, J. R. (2008). MEX-5 asymmetry in one-cell *C. elegans* embryos requires PAR-4- and PAR-1-dependent phosphorylation. *Development, 135*, 3665–3675.
Timm, T., Balusamy, K., Li, X., Biernat, J., Mandelkow, E., & Mandelkow, E. M. (2008). Glycogen synthase kinase (GSK) 3beta directly phosphorylates Serine 212 in the regulatory loop and inhibits microtubule affinity-regulating kinase (MARK) 2. *The Journal of Biological Chemistry, 283*, 18873–18882.
Timm, T., Li, X. Y., Biernat, J., Jiao, J., Mandelkow, E., Vandekerckhove, J., et al. (2003). MARKK, a Ste20-like kinase, activates the polarity-inducing kinase MARK/PAR-1. *The EMBO Journal, 22*, 5090–5101.
Timm, T., Marx, A., Panneerselvam, S., Mandelkow, E., & Mandelkow, E.-M. (2008). Structure and regulation of MARK, a kinase involved in abnormal phosphorylation of Tau protein. *BMC Neuroscience, 9*, S9.
Tomancak, P., Piano, F., Riechmann, V., Gunsalus, K. C., Kemphues, K. J., & Ephrussi, A. (2000). A *Drosophila melanogaster* homologue of *Caenorhabditis elegans* par-1 acts at an early step in embryonic-axis formation. *Nature Cell Biology, 2*, 458–460.
Updike, D., & Strome, S. (2010). P granule assembly and function in *Caenorhabditis elegans* germ cells. *Journal of Andrology, 31*, 53–60.
Vaccari, T., & Ephrussi, A. (2002). The fusome and microtubules enrich Par-1 in the oocyte, where it effects polarization in conjunction with Par-3, BicD, Egl, and dynein. *Current Biology, 12*, 1524–1528.
Wang, J. T., & Seydoux, G. (2013). Germ cell specification. *Advances in Experimental Medicine and Biology, 757*, 17–39.
Wang, J. T., Smith, J., Chen, B.-C., Schmidt, H., Rasoloson, D., Paix, A., et al. (2014). Regulation of RNA granule dynamics by phosphorylation of serine-rich, intrinsically disordered proteins in *C. elegans*. *eLife, 3*, e04591.
Watts, J. L., Etemad-Moghadam, B., Guo, S., Boyd, L., Draper, B. W., Mello, C. C., et al. (1996). par-6, a gene involved in the establishment of asymmetry in early *C. elegans* embryos, mediates the asymmetric localization of PAR-3. *Development, 122*, 3133–3140.
Watts, J. L., Morton, D. G., Bestman, J., & Kemphues, K. J. (2000). The *C. elegans* par-4 gene encodes a putative serine-threonine kinase required for establishing embryonic asymmetry. *Development, 127*, 1467–1475.
Weber, S. C., & Brangwynne, C. P. (2012). Getting RNA and protein in phase. *Cell, 149*, 1188–1191.
Wu, J. X., Cheng, Y. S., Wang, J., Chen, L., Ding, M., & Wu, J. W. (2015). Structural insight into the mechanism of synergistic autoinhibition of SAD kinases. *Nature Communications, 6*, 8953.

Wu, Q., DiBona, V. L., Bernard, L. P., & Zhang, H. (2012). The polarity protein partitioning-defective 1 (PAR-1) regulates dendritic spine morphogenesis through phosphorylating postsynaptic density protein 95 (PSD-95). *Journal of Biological Chemistry, 287*, 30781–30788.

Wu, P.-R., Tsai, P.-I., Chen, G.-C., Chou, H.-J., Huang, Y.-P., Chen, Y.-H., et al. (2011). DAPK activates MARK1/2 to regulate microtubule assembly, neuronal differentiation, and tau toxicity. *Cell Death and Differentiation, 18*, 1507–1520.

Wu, Y., Zhang, H., & Griffin, E. E. (2015). Coupling between cytoplasmic concentration gradients through local control of protein mobility in the *Caenorhabditis elegans* zygote. *Molecular Biology of the Cell, 26*, 2963–2970.

Yang, Z., Xue, B., Umitsu, M., Ikura, M., Muthuswamy, S. K., & Neel, B. G. (2012). The signaling adaptor GAB1 regulates cell polarity by acting as a PAR protein scaffold. *Molecular Cell, 47*, 469–483.

Yuan, H., Chiang, C. Y. A., Cheng, J., Salzmann, V., & Yamashita, Y. M. (2012). Regulation of cyclin A localization downstream of Par-1 function is critical for the centrosome orientation checkpoint in *Drosophila* male germline stem cells. *Developmental Biology, 361*, 57–67.

Zeaiter, Z., Cohen, D., Musch, A., Bagnoli, F., Covacci, A., & Stein, M. (2008). Analysis of detergent-resistant membranes of *Helicobacter pylori* infected gastric adenocarcinoma cells reveals a role for MARK2/Par1b in CagA-mediated disruption of cellular polarity. *Cellular Microbiology, 10*, 781–794.

Zhang, H., & Macara, I. G. (2006). The polarity protein PAR-3 and TIAM1 cooperate in dendritic spine morphogenesis. *Nature Cell Biology, 8*, 227–237.

Zhang, H., & Macara, I. (2008). The PAR-6 polarity protein regulates dendritic spine morphogenesis through p190 RhoGAP and the Rho GTPase. *Developmental Cell, 14*, 216–226.

Zhang, Y., Guo, H., Kwan, H., Wang, J., Kosek, J., & Lu, B. (2007). PAR-1 kinase phosphorylates Dlg and regulates its postsynaptic targeting at the drosophila neuromuscular junction. *Neuron, 53*, 201–215.

Zimyanin, V., Belaya, K., Pecreaux, J., Gilchrist, M., Clark, A., Davis, I., et al. (2008). In vivo imaging of oskar mRNA transport reveals the mechanism of posterior localization. *Cell, 134*, 843–853.

Zimyanin, V., Lowe, N., & St Johnston, D. (2007). An oskar-dependent positive feedback loop maintains the polarity of the *Drosophila* oocyte. *Current Biology, 17*, 353–359.

CHAPTER TWELVE

Receptor Tyrosine Kinases and Phosphatases in Neuronal Wiring: Insights From *Drosophila*

Carlos Oliva*,[1], Bassem A. Hassan[†],[1]
*Biomedical Neuroscience Institute, Faculty of Medicine, Universidad de Chile, Santiago, Chile
[†]Sorbonne Universités, UPMC Univ Paris 06, Inserm, CNRS, AP-HP, Institut du Cerveau et la Moelle (ICM)—Hôpital Pitié-Salpêtrière, Boulevard de l'Hôpital, Paris, France
[1]Corresponding authors: e-mail address: caoliva@uc.cl; bassem.hassan@icm-institute.org

Contents

1. Introduction 400
2. Model Circuits Used to Study the Genetic Control of Neuronal Wiring in *Drosophila* 401
   2.1 Studying Axon and Dendrite Growth and Guidance 401
   2.2 Studying Synaptogenesis and Physiology 404
3. Receptor Tyrosine Kinase 405
   3.1 Introduction 405
   3.2 Ephrin/Eph Signaling 406
   3.3 Ryk Receptors 408
   3.4 EGF Receptor 410
   3.5 Alk Receptor Signaling 411
   3.6 Off-Track Pseudokinase 412
   3.7 Insulin Receptor 413
   3.8 FGF Receptor 413
4. Receptor Protein Tyrosine Phosphatases 414
   4.1 Introduction 414
   4.2 Type IIa RPTPs in Brain Wiring 416
   4.3 Type IIa RPTPs in Vertebrate Brain Wiring 420
   4.4 Type III RPTPs in Brain Wiring 421
   4.5 Vertebrate Type III RPTPs in Brain Wiring 423
5. Future Directions 424
6. Conclusions 425
Acknowledgments 425
References 425

Current Topics in Developmental Biology, Volume 123
ISSN 0070-2153
http://dx.doi.org/10.1016/bs.ctdb.2016.10.003

Abstract

Tyrosine phosphorylation is at the crossroads of many signaling pathways. Brain wiring is not an exception, and several receptor tyrosine kinases (RTKs) and tyrosine receptor phosphates (RPTPs) have been involved in this process. Considerable work has been done on RTKs, and for many of them, detailed molecular mechanisms and functions in several systems have been characterized. In contrast, RPTPs have been studied considerably less and little is known about their ligands and substrates. In both families, we find redundancy between different members to accomplish particular wiring patterns. Strikingly, some RTKs and RPTPs have lost their catalytic activity during evolution, but not their importance in biological processes. In this regard, we have to keep in mind that these proteins have multiple domains and some of their functions are independent of tyrosine phosphorylation/dephosphorylation. Since RTKs and RPTPs are enzymes involved not only in early stages of axon and dendrite pathfinding but also in synapse formation and physiology, they have a potential as drug targets. *Drosophila* has been a key model organism in the search of a better understanding of brain wiring, and its sophisticated toolbox is very suitable for studying the function of genes with pleiotropic functions such as RTKs and RPTPs, from wiring to synaptic formation and function. In these review, we mainly cover findings from this model organism and complement them with discoveries in vertebrate systems.

1. INTRODUCTION

Protein phosphorylation is a key mechanism of control in a myriad of cellular processes. This biochemical modification is under the control of two classes of enzymes, kinases and phosphatases encoded by 518 and 200 genes, respectively, in mammals (Manning, Whyte, Martinez, Hunter, & Sudarsanam, 2002; Sacco, Perfetto, Castagnoli, & Cesareni, 2012). The most common modification occurs in serine and threonine, but the most important is the phosphorylation of tyrosines. Unlike Ser/Thr phosphorylation, Tyr phosphorylation is very short lived because of the high activity of tyrosine phosphatases (PTPs) that rapidly remove any residue that is not protected by phosphotyrosine (P-Tyr) binding motifs, because of this only less than 2% of the total phospho-proteome in a typical mammalian cell at a given time are P-Tyr residues (Hunter, 2014).

Interestingly, these modifications are also pivotal in signal transduction at the cell membrane, and numerous transmembrane protein bear a catalytic domain involved in kinase or phosphatase activities. In this regard, a special group of tyrosine kinases and phosphatases are receptor tyrosine kinases (RTKs) and receptor tyrosine phosphatases (RPTPs). These families are very important during signal transduction in response to extracellular stimuli.

A key step in nervous system (NS) development is the growth and guidance of axons and dendrites and the ensuing formation of synaptic connections. Over the last few years an increasing body of evidence supports the involvement of RTKs and PTPs in these processes. Much of the knowledge obtained comes from invertebrate models, especially from the fruit fly *Drosophila melanogaster*; for that reason in many sections, we will use this model to illustrate how these signaling regulators act in brain wiring.

We will begin with a brief introduction to *Drosophila* neuronal models, which are used to study wiring and synaptic development; then we will discuss RTKs and their involvement in different aspects of brain wiring; and finally, we will focus on RPTPs. The list of models and processes will be far from exhaustive and only serves to illustrate a few main points. We apologize to those authors whose work we were unable to cite as a result.

2. MODEL CIRCUITS USED TO STUDY THE GENETIC CONTROL OF NEURONAL WIRING IN *DROSOPHILA*

2.1 Studying Axon and Dendrite Growth and Guidance

2.1.1 CNS and Motor System of the Drosophila Embryo

The *Drosophila* embryo is a classical model to study axon guidance and growth. The embryonic nerve cord is divided into segments (Fig. 1A). Each segment bears two commissures—the anterior and posterior commissures (AC and PC)—formed by contralateral axons connecting the two sides of the organism. Axons of commissural neurons join their ipsilateral counterparts in one of three longitudinal pathways on each side of the midline. In both vertebrates and invertebrates, commissural axons cross the midline only once and ipsilateral axons never do (Dickson & Zou, 2010). This model has served as an excellent choice to study about the decision to cross, or not to cross, the midline (Araujo & Tear, 2003).

In the motor system of both vertebrates and invertebrates, motor axons leave the central nervous system (CNS) in common bundles that defasciculate in the periphery to make synaptic connections with their targets (Araujo & Tear, 2003) (Fig. 1B). In *Drosophila*, motor axons exit the CNS in the intersegmental nerve (ISN) and the segmental nerve (SN). When they reach the muscle field, they split into five pathways that innervate target muscles. SNa and SNc originate from the SN root, while ISN, ISNb, and ISNd do originate from the ISN root (Fig. 1B). A typical

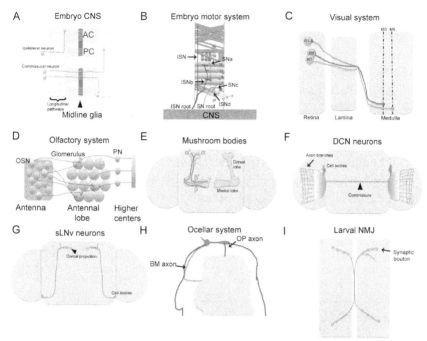

Fig. 1 *Drosophila* neurons used for nervous system wiring research. (A) Embryonic nervous system. *AC*, anterior commissure; *PC*, posterior commissure. (B) Embryonic motor system, only intersegmetal (ISN) axons are shown. Main target muscles are depicted in *orange* with their assigned numbers, and other muscles are in *gray*. (C) Visual system of the adult fly. Only the R-cells target neuropils of the optic lobe are depicted. M3 and M6 are medulla layers. (D) Olfactory system of the adult fly indicating the main elements. *OSN*, olfactory receptor neuron; *PN*, projection neuron. (E) Anterior view of the mushroom bodies. Lobes are indicated and branches of α/β, α'/β', and γ neurons are depicted. (F) Drawing depicting the dorsal cluster neurons (DCNs). Axon branches are indicated. (G) Drawing of the sLNv neurons. Dorsal projections are indicated. (H) Representation of the ocellar sensory system (OSS). The head capsule is represented in *dark gray* and the brain in *light gray*. *BM*, mechanosensory neurons; *OP*, ocellar pioneer. (I) Drawing depicting the larval neuromuscular junction (NMJ) system. Only one motoneuron connection is represented, and synaptic boutons are represented.

phenotype, that can be easily scored, is the failure of the ISNb axons to defasciculate at the correct point, which leads to axons bypassing their targets and following the main nerve, such as occurs for instance by overexpressing the adhesion molecule FasII in neurons (Lin & Goodman, 1994).

2.1.2 Adult Visual System

The *Drosophila* visual system is an excellent model system to study the molecular basis of axon targeting. The fly eye is composed of approximately

750 ommatidia. Each ommatidium contains eight subtypes of photoreceptor cells (R-cells, R1–R8) plus accessory cells (Hadjieconomou, Timofeev, & Salecker, 2011). R1–R6 photoreceptors are the so-called outer R-cells, which express rhodopsin-1 and are involved in motion detection. R7–R8 photoreceptors (inner R-cells) express green-, blue-, and ultraviolet-sensitive rhodopsins and are involved in color and polarized vision. R-cells axons make synapses in the optic lobe, which consists of four neuropils, the lamina, the medulla, the lobula, and the lobula plate. The R1–R6 cells project to the lamina where they form specific synaptic structures called cartridges and connect with Lamina neurons, amacrine cells, and centrifugal interneurons (Fig. 1C). The medulla neuropil is divided into 10 layers (M1–M10). R8 axons make synaptic connections with medulla neurons in the M3 layer, while R7 axons do it in the M6 layer (Fig. 1C) (Hadjieconomou et al., 2011). The targeting of R-cells is highly regulated and a myriad of signaling molecules required for this process have been identified.

2.1.3 Adult Olfactory System

In *Drosophila*, the sensory structures are localized in the antennae and maxillary palps. These structures are called sensilla and house sensory neurons (olfactory sensory neurons, OSNs) that harbor receptors for chemical compounds. Axons of OSNs connect with projection neurons (PNs) in the antennal lobe (the fly equivalent of the olfactory bulb). The organizational principle between OSNs and PNs is one to one, giving rise to a so-called discrete neural map. The antennal lobe is composed of 51 glomeruli, each of which harbors a specific class of PN dendrite and receives inputs from a single type of OSN (Fig. 1D; Laissue & Vosshall, 2008).

2.1.4 Mushroom Bodies

The mushroom bodies (MBs) are the centers of olfactory learning and memory in the fly brain. They are formed by two bilaterally symmetrical groups of approximately 2500 neurons called Kenyon cells. Kenyon cells are subdivided according to their axonal trajectories into $\alpha\beta$, $\alpha'\beta'$, and γ. $\alpha\beta$ and $\alpha'\beta'$ neurons have two major branches (α,β and α',β', respectively), while γ neurons have one. Axons are fasciculated into lobes, dorsal lobes bearing α and α' branches and the medial lobe containing β,β' branches and γ axons (Fig. 1E). However, there are further subdivisions within each main group (Keene & Waddell, 2007). MBs originate from four neuroblasts from each hemisphere of the embryonic brain. MB neuroblasts start dividing during embryonic development and continue until late pupal stages (Jefferis, Marin, Watts, & Luo, 2002). Axon guidance and branching, among other processes, have been studied using MBs extensively. For instance, it has been determined that the ubiquitin ligase Highwire is involved in the sorting of sister axons branched to different lobes. In *hiw* mutants, α and β branches can follow the dorsal lobe trajectory (Shin & DiAntonio, 2011).

2.1.5 Other Circuits

Several neuronal populations in the central brain are used to study distinct aspects of axon guidance. Dorsal cluster neurons (DCNs) are composed of two cluster of around 40 neurons located in the dorsal central brain adjacent to the optic lobe (Fig. 1F). These neurons form a commissure and innervate the lobula and medulla neuropils. Approximately two-thirds of the DCNs send axons to the lobula and 10–12 send projections to the medulla (Srahna et al., 2006), those DCNs which send axons to the medulla form a grid-like pattern composed by axonal branches. DCNs are a good model to study axon guidance, commissure formation, and branch formation.

Another group of neurons that can be used in neuronal wiring studies are the small lateral ventral neurons (sLNvs; Fig. 1G) whose main advantage is that it is a very discrete group of only four to five cells per hemisphere (Helfrich-Forster et al., 2007). sLNvs are involved in the regulation of the circadian cycle in the fly. The four to five neurons form a single fascicle that projects to the dorsal brain and generates terminal branches. These neurons are great to study axon guidance and, due to particular branch plasticity during the day, are also used to study structural plasticity related to circadian rhythm (Fernandez, Berni, & Ceriani, 2008).

The ocellar sensory system (OSS) (Garcia-Alonso, Fetter, & Goodman, 1996) is composed of the ocellar pioneer (OP) and bristle mechanosensory neurons (BMs) (Fig. 1H). These two groups of neurons differentiate side by side in the dorsal head of the pupae but project to distinct targets in the brain; the nature of this difference resides in the degree of interaction between each neuron and the epithelia of the head. OP axons grow in a groove along the epithelia and never get attached to it, while BMs grow attached to the head epithelia until they reach the dorsal antenna, where they leave the epithelia surface and start growing toward the brain.

2.2 Studying Synaptogenesis and Physiology

The larval neuromuscular junction (NMJ) is one of the most studied systems in the field of synaptogenesis and synaptic function. The typically studied muscles are located in the abdominal segments from A2 to A7 each bearing

30 muscles per hemisegment (Keshishian, Broadie, Chiba, & Bate, 1996). In the fly, motoneurons are mostly glutamatergic and reach their target muscles during embryogenesis (Fig. 1I).

3. RECEPTOR TYROSINE KINASE
3.1 Introduction

The human genome encodes 90 RTKs (Manning et al., 2002). They are involved in a huge number of cellular processes. Here, we cover the most representative exponents of this family involved in neuronal wiring using the fly as a model system. The general mechanism of RTK signaling consists of autophosphorylation, instead of phosphorylating target proteins directly. Upon this event, phosphorylated domains became docking sites for the binding of signaling partners that can themselves be kinases involved in phosphorylation cascades (Blume-Jensen & Hunter, 2001). Interestingly, some RTKs appear to have lost their kinase activity during evolution; here we also cover some examples of such members since they are also very important in axon wiring. The structure of some of these proteins is illustrated in Fig. 2. A summary of main RTK functions in *Drosophila* NS wiring can be found in Table 1.

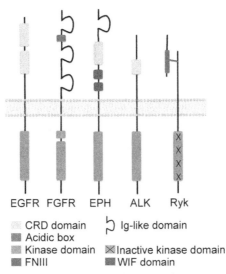

Fig. 2 Representation RTKs structural domains. Some characteristic RTKs are indicated. Note that FGFR bears two kinase domains and Ryk is processed in two fragments that are bound by a disulfide bridge (according to Blume-Jensen & Hunter, 2001).

Table 1 Summary of RTK Functions in Wiring

	Drosophila RTK	Ligand	Vertebrate Ortholog	Model System
Axon growth/ guidance	Eph	Ephrin	Eph	Visual system, MBs
	Ryk/Drl	Wnt	Ryk	Embryo CNS, MBs, ORN axons of the olfactory system
	EGFR	EGF	EGFR	Ocellar sensory system
	Alk	JEB	Alk	Visual system
	OTK	Wnt?	PTK7	Embryo CNS, visual system
	InR	Insulin	InR, IGF-1	Visual system
	FGFR	FGF	FGFR	Ocellar sensory system, DCNs
Axon branching	EGFR	EGF	EGFR	DCNs
Dendrite growth/ guidance	Ryk/Drl	Wnt	Ryk	PNS dendrites of the olfactory system, da neuron dendrites

3.2 Ephrin/Eph Signaling

Ephrin receptor activation (Eph) occurs upon clustering induced by their membrane bound ligands. Ephrin ligands come in two flavors, depending on the manner by which they attach to the membrane. Ephrins can be attached by a glycosyl-phosphatidylinositol linkage for type A Ephrins, or a transmembrane domain for type B Ephrins. Since both ligands are bound to membranes, Ephrin/Eph signaling generates a cell contact-mediated response (Wilkinson, 2001). Ephrin-Bs also bear a small cytoplasmatic domain that gets phosphorylated, upon binding to Eph, by cytoplasmic tyrosine kinases. This can then lead to a bidirectional signaling where both participating cells can transduce a signal (Wilkinson, 2001).

The vertebrate genome encodes 14 Eph receptors, 5 type A Ephrins, and 3 type B Ephrins. In contrast, the *Drosophila* genome encodes only one Ephrin and one Eph. A number of functions have been described for Eph–Ephrins in vertebrate NS wiring such as axon repulsion, axon branching, and formation of the retinotectal topographic maps (Wilkinson, 2001). Mutations in

Eph–Ephrin signaling components can lead to disease. For instance, mutations in the genes encoding EPHB2 and EPHA3 are found in several types of cancer and Ephrin-B1 is linked to craniofrontonasal syndrome, a disorder that leads to skeletal abnormalities (Pasquale, 2005). Interestingly, this disorder is also associated with agenesis of the corpus callosum (Cramer & Miko, 2016; Twigg et al., 2004). Furthermore, more recent genetic studies have linked EphA3 mutations with autism spectrum disorder, which is characterized by impaired cognitive abilities such as language and communication skills (Casey et al., 2012; Cramer & Miko, 2016).

In *Drosophila*, early studies using RNAi injections in embryos led to the conclusion that Eph is also involved in the patterning of the fly visual system (Dearborn, He, Kunes, & Dai, 2002). Eph is expressed in the developing eye, the growth cone of the arriving photoreceptor axons, and the optic ganglia. Interestingly an anteroposterior gradient is observed, which is consistent with a function in topographic mapping (Dearborn et al., 2002). Also ectopic expression of an Eph lacking most of the intracellular domain (that functions as a dominant negative) in photoreceptor cells and in optic lobe neurons using the Gal4/UAS system, a trademark of *Drosophila* genetics that allows to modify gene expression in time and space (Brand & Perrimon, 1993), produces fasciculation defects in photoreceptor axons and morphological defects in the medulla neuropil, indicating that the *Drosophila* visual system requires Eph for a correct axon pathfinding.

On the other hand, the unique *Drosophila* Ephrin has also been identified and is a transmembrane protein similar to vertebrate Ephrin-B. Initial experiments using RNAi showed that Ephrin was necessary to avoid the exit of interneuronal axons from the CNS in embryos (Bossing & Brand, 2002) and ectopic expression of Ephrin in embryonic midline cells leads to disruption of commissure formation. However, this evidence has been later disputed after an *eph* mutant was generated and showed no defects in the embryonic CNS (Boyle, Nighorn, & Thomas, 2006), indicating that some of the RNAi phenotypes are due to off-target effects or mutant phenotypes are masked by compensation during development. Instead, these *eph* mutants showed defects in the wiring of MBs. In *eph* mutants the major branches form but the dorsal branch fails to follow its normal path and instead collapses onto the medial branch.

One still uncompleted aspect of Eph signaling is the intracellular mechanisms implicated in its action. A more recent study used a modifier screen in the *Drosophila* eye as a readout to discover new interactors of the Eph cascade

(Dearborn et al., 2012). The study uncovered a putative nuclear protein, named regulator of Eph signaling (Reph). This study also confirmed the importance of Ephrin and Eph in the development of the visual system using new mutants that show similar phenotypes to the ones described previously using RNAi.

Thus, similar to vertebrates, Eph/ephrin signaling function in axon guidance in *Drosophila* and the possibility of performing highly sophisticated modifier screens can lead to the identification of new components of this pathway.

3.3 Ryk Receptors

Ryk receptors (for receptors related to tyrosine kinase (Patthy, 2000)) are highly conserved in worms, flies, zebrafish, and mammals. Flies, unlike the other model organisms where there is one member, have three paralogs encoded by the genes *derailed* (*drl*), *doughnut* (*dnt*), and *derailed 2* (*drl2*) (Fradkin, Dura, & Noordermeer, 2010). Ryk proteins have a conserved Wnt binding domain (WIF, named after Wnt inhibitory factor-1 protein (Patthy, 2000)) and a putative tetrabasic domain in their extracellular region which constitutes a cleavage site (Fradkin et al., 2010). In the intracellular portion, they bear a tyrosine kinase-related domain and a PDZ binding domain. However, due to mutations in specific key residues in their tyrosine kinase domain, they have lost most of that catalytic activity (Hovens et al., 1992).

The first evidence involving Ryk in axon guidance comes from studies in the *Drosophila* embryo (Bonkowsky, Yoshikawa, O'Keefe, Scully, & Thomas, 1999; Callahan, Muralidhar, Lundgren, Scully, & Thomas, 1995), showing that in *drl* mutants, axons of embryonic interneurons fail to fasciculate (Callahan et al., 1995). In the *Drosophila* embryo, Drl-expressing AC axons switch to the PC in the mutants. Conversely, missexpression of Drl in PC neurons reroutes their axons toward the AC, indicating that Drl is necessary and sufficient for pathway choice in axons (Bonkowsky et al., 1999). It was later shown that this is a Wnt-dependent phenomenon as Wnt5 from the PC leads to repulsion of the AC Drl-expressing axons (Yoshikawa, McKinnon, Kokel, & Thomas, 2003).

Drl is not only involved in the development of the embryo CNS but also in neuropils of the *Drosophila* brain. MB development also relies on Wnt5/Drl signaling although many other actors have been identified to be involved in Wnt5-dependent MB axon development. MB dorsal lobes are missing and medial lobes are fused in *drl* mutants (Grillenzoni, Flandre, Lasbleiz, & Dura, 2007) a phenotype that also depends on Wnt5. Interestingly, MB defects produced in *drl* mutants are only weakly rescued by expression of Drl in the MB. Instead, pan-neuronal expression restores the wild-type conditions. Consistently, Drl staining shows expression outside the MBs in axon tracks adjacent to them. On the other hand, Wnt5 defects can be rescued by expressing this protein in MBs. These data indicated that—unusually—the putative ligand plays a cell-autonomous role, while the receptor plays a noncell autonomous role. In a later study the mystery was revealed (Reynaud et al., 2015). Drl was shown to be expressed in dorsomedial lineages that are localized adjacent to MBs and give rise to neurons of the central complex. In order to guide dorsal MB projections, Drl is proposed to capture Wnt5 and present it to growing MB dorsal lobes, a process that requires Drl ectodomain shedding. In the MB axons, Drl2 (one of the other two paralogs) binds to the Drl/Wnt5 complex which triggers the repulsion of dorsal lobe axons from the WNT5 source.

The interplay between Drl and Wnt5 has also been observed during the development of the fly olfactory system (Yao et al., 2007). In *drl* mutants, the organization of the olfactory map is disturbed, ectopic midline glomeruli appear, and Wnt5 accumulates in the midline. During development of the olfactory map, Drl is expressed by a population of glial cells in the transient interhemispheric fibrous ring (TIFR) located in the midline and adjacent to developing glomeruli and in the dendritic fields of PNs, while Wnt5 is expressed by OSNs axons. The defects observed in the *drl* mutants could be rescued by Drl expression in glial cells and it was determined that the WIF domain is required, while the cytoplasmic region was largely dispensable (Yao et al., 2007). In this context, the function of Drl is to counteract Wnt5, since Wnt5 missexpression phenotypes are similar to *drl* loss of function and are more severe in *drl* mutant background. Thus, Wnt5 signal delivered by growing OSN axons is required for the correct patterning of the PN dendritic field but also requires an antagonistic regulation by Drl from TIFR glia.

Another example of the function of the Wnt5/Drl pair in dendritic patterning comes from the sensory neurons from the body wall. In the adult abdominal epidermis a structure called the sternite located on the most ventral part secretes Wnt5. Wnt5 in this system promotes dendrite termination in the periphery of the sternite to restrict their ventral field (Yasunaga et al., 2015). Mutants in *drl* and *drl2* show weak phenotypes of dendrite overgrowth; however the double mutants *drl/drl2* show strong defects similar to the *wnt5* mutant phenotype, indicating that these receptors act together to restrict dendrite ventral boundaries. Interestingly, *wnt5* and *drl/drl2* mutants have normal lateral boundaries, indicating that their function is specific. These receptors act autonomously in the neurons to transduce the Wnt5 signal, and consistently with this idea the intracellular domain is necessary for Drl in this context. Furthermore, the intracellular signaling molecules RacGEF Trio and Rho-1 are involved in transduction of the signal, as revealed by genetic interaction essays.

In summary, Ryk/Drl proteins play important roles in the patterning of axons and dendrites in many systems together with the Wnt5 ligand.

3.4 EGF Receptor

EGF receptors (EGFRs) have been known for a long time to be involved in multiple processes. Their roles in cell differentiation, proliferation, and cancer are well established (Jorissen et al., 2003). Similar to Eph receptors, upon binding of the EGF ligand, EGFR dimerizes and undergoes *trans*-phosphorylation of the intracellular region (Jorissen et al., 2003). After the induction of tyrosine phosphorylation, adaptor proteins recognize the phosphotyrosine-modified domains to trigger a signaling cascade (Jorissen et al., 2003). Only recently the participation of EGF signaling in NS wiring has been revealed. In *Drosophila* a single EGFR is the only receptor of this pathway and interacts with four ligands (Shilo, 2005) that act in different contexts: Spitz, Gurken, Keren, and Vein. Vein is produced as a secreted protein and does not need to process for its activation. On the other hand, the other three ligands are produced as inactive precursors bound to the membrane, and they are released upon proteolytic processing to become functional EGF ligands (Shilo, 2005).

One of the first pieces of evidence of EGF signaling in axon guidance comes from a study using the OSS from the adult head of *Drosophila*. In animals expressing a dominant negative of EGFR, OP axons are only partially affected, while BM axons stall in the epidermis and do not reach the brain. In this process, the cell adhesion molecule Neuroglian (Nrg) acts as a modulator of EGFR signaling.

EGFR also regulates axon branching. Work performed using the DCNs of the central brain has shown that optimal levels of EGFR signaling are required for the development of the wild-type branch pattern of DCN axons (Zschatzsch et al., 2014). During development, DCN axons form excessive branches that are later pruned. Strikingly, either gain or loss of EGFR signaling produces pruning defects. At the cellular level, defects in EGFR signaling lead to reduced filopodial dynamic in axon growth cones because of defects in actin polymerization. Interestingly, this function appears independent of the canonical MAPK pathway.

Thus, EGFR signaling controls different aspects of axon development and can involve a different set of downstream effectors.

3.5 Alk Receptor Signaling

Anaplastic lymphoma kinase (Alk) is a signaling pathway originally identified for its involvement in anaplastic large cell lymphoma (Pulford, Morris, & Turturro, 2004). It had been proposed to participate in NS development, based on expression pattern observations in vertebrates (Iwahara et al., 1997; Loren et al., 2001). Studies in the *Drosophila* visual system have shown that Alk and its ligand Jelly belly (Jeb), a secreted protein with LDL repeats (Weiss, Suyama, Lee, & Scott, 2001), are expressed in complementary populations in this system. During pupal development, Alk is expressed in developing lamina cartridge neurons, the targets of photoreceptor cells (R-cells), while Jeb is expressed in R-cells. Functional studies showed that Alk is required for the R-cell projection patterning but not in the R-cells themselves, instead in their targets. Loss of the ligand Jeb leads to defects in R1–R6 and R8 photoreceptors. Remarkably, loss of Jeb in photoreceptors affects the expression of three adhesion molecules in the medulla target region: Kirre, IrreC, and Flamingo, indicating that Jeb is an anterograde signal that regulates environmental factors in the photoreceptor targets (Bazigou et al., 2007). More recently a function of Jeb as a survival factor was revealed. Knockdown of Jeb in the retina produces the death of L3 lamina neurons during development, one of the target neurons of R1–R6 photoreceptors (Pecot et al., 2014). Similar results were found upon knockdown of Alk in the lamina and Alk MARCM clones in L3, indicating that the Alk receptor was required in the L3 neurons. Thus, the observed R1–R6- and R8-targeting defects before are a consequence of L3 loss and thus the loss of target space integrity. Furthermore, it is possible that the decrease in the expression in adhesion molecules observed by Bazigou et al. was due to cell death as opposite to lower expression. In this regard, Pecot et al. take the advantage of the previously known requirement of Netrin in L3 neurons for R8 targeting (Timofeev, Joly, Hadjieconomou, & Salecker, 2012) to conclude that lack of Netrin secondary to L3 loss is responsible for the observed R8 mistargeting in Alk loss-of-function conditions (Bazigou et al., 2007).

These data probably indicate that Alk acts as a neurotrophic factor rather than as a classic guidance factor during neuronal wiring.

3.6 Off-Track Pseudokinase

Off-track (OTK) is the ortholog of vertebrate protein tyrosine kinase 7 (PTK7). Like other members of the RTK family, OTK/PTK7 bears a dead kinase domain due to substitution of the conserved DFG triplet. In *Drosophila*, it was originally identified as the homolog of the TrkA receptor (Pulido, Campuzano, Koda, Modolell, & Barbacid, 1992); however, further research concluded that this was not the case and in *Drosophila* neurotrophin receptors are encode by Toll proteins (McIlroy et al., 2013).

Early studies in the embryonic CNS and motor axons showed that OTK cooperates with PlexinA during axon guidance (Winberg et al., 2001). RNAi knockdown of OTK in neurons, using the UAS/Gal4 tool, caused motor axons to fail to branch at proper locations and CNS axons to show stalling defects. These phenotypes are also observed in *plexinA* mutants. Strong genetic interactions are observed when *plexinA* and *otk* mutants are combined with transheterozygous animals, suggesting that they participate in a common process (Winberg et al., 2001). Otk also participates in targeting choice of photoreceptor cells. In *otk* mutants, R1–R6 cells, which express high levels of OTK under basal conditions, change their target choice from the lamina to the next neuropile, the medulla, indicating that OTK forms part of a receptor complex that is either required for target recognition (Cafferty, Yu, & Rao, 2004) or required for axonal stabilization, as recently shown for the cell adhesion molecule N-cadherin (Özel, Langen, Hassan, & Hiesinger, 2015). In either case, the study found that unlike in embryos, in this case OTK is not acting with Semaphorin-1A, the ligand of PlexinA. This suggests that OTK may act with different signaling molecules depending on the context.

More recently, it was shown in vertebrates and also in *Drosophila* that PTK7 forms part of the Wnt/PCP pathway (Hayes, Naito, Daulat, Angers, & Ciruna, 2013; Linnemannstons et al., 2014), where it acts as a Wnt coreceptor in several contexts. So far it is not clear if this complex plays a role in vertebrate NS wiring.

3.7 Insulin Receptor

The insulin receptor is well known for its function in cell metabolism (Saltiel & Kahn, 2001) and many of the components of insulin signaling are conserved in flies (Garofalo, 2002). The *Drosophila* genome encodes only one insulin receptor (dINR; Fernandez-Almonacid & Rosen, 1987; Petruzzelli et al., 1986). Interestingly, studies performed in *Drosophila* showed that it is also involved in the targeting of photoreceptor axons (Song, Wu, Chen, Kohanski, & Pick, 2003). General wiring defects are observed in R-cells when the eye primordium is mutant for dINR. Strikingly, this function is not through the *Drosophila* insulin receptor substrate (a protein called Chico), but it is mediated by the adaptor protein Dock, which has a well-known function in R-cell wiring (Hing, Xiao, Harden, Lim, & Zipursky, 1999; Newsome, Schmidt, et al., 2000). The *dinr/dock* transheterozygotes have phenotypes similar to *dock* homozygote mutants, indicating that dINR signals through Dock in R-cell axon guidance and a mutant form of dINR that lacks *chico* binding site is also able to rescue wiring defects, further proving a Chico-independent pathway of axon development (Li, Guo, & Pick, 2013).

Thus, outside its known functions in cell metabolism, insulin receptor also regulates brain wiring. In mammals, although so far no wiring function for the insulin receptor has been shown, a recent study has described the function of insulin-like growth factor 1 (IGF-1) and IGF-1 receptor in the wiring of mouse olfactory axons (Scolnick et al., 2008), indicating that this function in brain development may be conserved.

3.8 FGF Receptor

The fibroblast growth factor (FGF) has multiple functions in invertebrates and vertebrates. In *Drosophila*, there are two FGFRs and three FGFs in contrast to the 22 FGFs and 4 FGFRs found in vertebrates (Huang & Stern, 2005). Breathless (Btl) is one of the *Drosophila* FGFRs and is well known for its function in tracheal (the insect respiratory system) development (Lee, Hacohen, Krasnow, & Montell, 1996). The other *Drosophila* FGFR is Heartless (Htl) and it is essential for the patterning of the mesoderm (Huang & Stern, 2005). The *Drosophila* FGFs are Branchless (Btl ligand), Thisbe, and Pyramus (the last two are Htl ligands; Huang & Stern, 2005; Kadam, McMahon, Tzou, & Stathopoulos, 2009).

In contrast to what has been identified in vertebrate systems where FGFs play crucial roles in the neuronal wiring of diverse systems (Shirasaki,

Lewcock, Lettieri, & Pfaff, 2006; Webber, Chen, Hehr, Johnston, & McFarlane, 2005; Webber, Hyakutake, & McFarlane, 2003), only few studies report the function of FGF signaling in *Drosophila* neuronal wiring.

The first observation described a function for the Heartless FGFR in axon guidance. Htl collaborates with EGF and Nrg in the outgrowth of the ocellar neurons from the adult head mentioned earlier (Garcia-Alonso, Romani, & Jimenez, 2000). Further characterization of the involvement of Htl in axon outgrowth was obtained using cell culture of embryonic neurons. Neurons in culture respond to the NCAM *Drosophila* homologs Nrg and Fas2 to extend neurites. Interestingly, this outgrowth-promoting activity is ablated in the presence of an FGFR activity inhibitor or Htl mutant neurons (Forni, Romani, Doherty, & Tear, 2004), indicating that NCAM/FGFR signaling is involved in neurite outgrowth.

The other *Drosophila* FGFR also appears to be involved in brain wiring. In a study trying to identify signaling molecules involved in axon retraction during development, a dominant negative form of receptor Btl showed an increase in the medulla connecting axons of DCN neurons (Srahna et al., 2006). Further examination led to the conclusion that Btl promotes axon retraction during the DCN axonal development through the inhibition of JNK kinase. Interestingly, Btl signaling balances an axon extension signal provided by the Wnt/Frizzled pathway.

Thus, FGF signaling participates in axon extension and retraction during axon development probably depending on the context.

4. RECEPTOR PROTEIN TYROSINE PHOSPHATASES
4.1 Introduction

The human genome encodes 107 PTPs (Sacco et al., 2012). From this list, 21 are RPTPs in mammals (Stoker, 2015). RPTPs are classified into eight categories according to the homology of their extracellular domains (Ensslen-Craig & Brady-Kalnay, 2004). Five of these groups have known functions in neuronal wiring in mammals. In *Drosophila*, there are only six RPTPs and only two of the eight families are represented. These groups are type IIa and type III. Type IIa RPTPs are characterized by bearing two kinds of domains in their extracellular region, Ig domains and FNIII repeats which make them similar to the immunoglobulin superfamily of cell adhesion molecules (CAMs) in their motif constitution (Bixby, 2001) (Fig. 3). As with CAMs, it has been shown that some RPTPs can form homophilic interactions and in this way work as both ligand and receptor

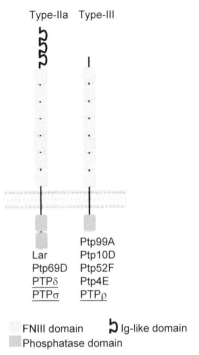

Fig. 3 Representation of RPTP structural domains. Type IIa and Type III families are represented (according to Ensslen-Craig & Brady-Kalnay, 2004). *Drosophila* and human (*underlined*) members belonging to each family are depicted.

(Bixby, 2001; Zondag et al., 1995). In *Drosophila*, Lar and Ptp69D belong to this category (Ensslen-Craig & Brady-Kalnay, 2004). The type III family is represented in *Drosophila* by Ptp99A, Ptp10D, and Ptp52F (Ensslen-Craig & Brady-Kalnay, 2004) (Fig. 3). These proteins were cloned in the 1990s when extensive efforts were undertaken to uncover RPTPs in *Drosophila*. First to be discovered were *Drosophila* Lar and *Drosophila* PTP (DPTP), using a consensus sequence obtained from the human genes LCA and LAR (Streuli, Krueger, Tsai, & Saito, 1989), to screen a library of embryonic cDNA. Further work using the same strategy and in situ hybridization on polytene chromosomes localized the locus of these genes and this resulted in the current nomenclature of all *Drosophila* RPTPs (4E, 10D, 52F, 69D, and 99A) except Lar (Hariharan, Chuang, & Rubin, 1991; Oon, Hong, Yang, & Chia, 1993; Tian, Tsoulfas, & Zinn, 1991). A summary of the findings presented below can be found in Table 2.

In this section, we will review functions of the *Drosophila* RPTPs making parallels with the mammalian homologs when possible. For more details

Table 2 Summary of RPTP Functions in Wiring

	Drosophila RPTP	Ligand	Substrates	Human Closest Homolog[b]	Model System
Axon growth/ guidance	LAR	Syndecan	Ena, Abl	LAR	Embryo ISN axons, visual system[a]
	PTP69D	nd	DSCAM	RPTPC	Embryo ISN axons, visual system, sLNv,[a] scutellar neurons
	PTP10D	Sas	nd	RPTPB	Embryo CNS
	PTP52F	nd	nd	nd	Embryo CNS, embryo SN axons
	PTP4E	nd	nd	RPTPB	Embryo CNS
Axon branching	PTP69d	nd	nd	RPTPC	Posterior scutellar neurons
Synaptic growth/ function	LAR	Sdc, Dlp	nd	LAR	Larva NMJ
	PTP69d	nd	nd	RPTPC	GFS
	PTP10D	nd	nd	RPTPB	MBs

[a]No catalytic activity is required.
[b]Flybase.

on mammalian RPTP functions in brain wiring, readers are referred to excellent reviews on the matter (Ensslen-Craig & Brady-Kalnay, 2004; Stoker, 2015).

4.2 Type IIa RPTPs in Brain Wiring
4.2.1 Lar-RPTP

The leukocyte common antigen-related (LAR) is one of the most studies RPTP families. DLAR belongs to the type IIA family which has three vertebrate members (LAR, PTPδ, and PTPσ). In *Drosophila*, Lar has been shown to regulate a myriad of steps in NS wiring. In *Drosophila* embryos that lack Lar expression a subset of axons of the ISN root, called ISNb and ISNd, cannot enter the region where their target muscles are located (Wills, Bateman, Korey, Comer, & Van Vactor, 1999) and instead bypass their ventral targets and continue growing dorsally toward inappropriate targets. This phenotype is suppressed by removing one copy of Abelson kinase (Abl), indicating that Lar and Abl have a functional relationship in vivo. Furthermore, Abl gain of function in neurons mimics the Lar loss of function phenotype. This was further reinforced by biochemical experiments, showing that Lar and Abl can associate (Wills et al., 1999). Enabled (Ena) was identified as another component of this complex that can also associate with Lar. *ena* mutants have similar defects to *lar* phenotypes although stronger, probably indicating that Ena is a common point for several inputs regulating ISNb neuron guidance. This study also showed that Abl and Ena are substrates of Lar.

Using the same model as above, Fox and Zinn (2005) discovered that Syndecan (Sdc), a heparan sulfate proteoglycan, is a ligand for Lar. They used a combination of stainings of a Lar-Alkaline-Phosphatase (AP) fusion in embryos together with a deficiency screen to determine which genomic deletions produced the loss of the signal. *Sdc* mutants also show genetic interactions with *Lar* mutants in the ISNb bypass phenotype, indicating that they act in vivo to regulate axon guidance. Sdc stainings show that its expression is confined to cells adjacent to motor nerves, which indicate that Lar and Sdc mediate cell–cell interactions in order to regulate axon guidance of ISNb neurons. Another DLAR-interacting protein, Caskin (Ckn), has also been shown to participate in motor axon guidance (Weng, Liu, DiAntonio, & Broihier, 2011). *ckn* mutants have the same ISNb bypass phenotype and interact genetically with Lar. The SAM domain of Ckn is required for its interaction with Lar as shown by a yeast two-hybrid assay. This interaction is Lar specific, since Ckn appears not to bind to the other *Drosophila* type IIa RPTP, Ptp69D, and also does not show genetic interactions with it (Weng et al., 2011).

Lar functions in axon targeting have been revealed in experiments using the *Drosophila* visual system (Clandinin et al., 2001; Hofmeyer & Treisman, 2009; Maurel-Zaffran, Suzuki, Gahmon, Treisman, & Dickson, 2001). In this context, DLAR is necessary for R1–R6 axons to choose their appropriate target neurons in the lamina. Furthermore, in *Lar* mutants, R7 axons initially choose the right layer of the medulla (M6) but then mistarget to the M3 layer where R8 axons normally connect (Fig. 1C) (Clandinin et al., 2001). Interestingly, expression of Lar in R8 photoreceptors can partially rescue the R7 phenotype, indicating that Lar can work in a cell nonautonomous manner (Maurel-Zaffran et al., 2001). The *Lar* hypomorphic mutant phenotype is enhanced by removing a copy of either *trio* or *ena*, and partially suppressed by overexpression of the same molecules, indicating that these two proteins could mediate Lar function. Strikingly, in this context the catalytic domain of Lar is not required (Maurel-Zaffran et al., 2001). The phenotype can be rescued even with the expression of a construct carrying its last three Fibronectin III domains (7–9FNIII) keeping the intracellular domain, indicating that Lar works through cell–cell interactions binding an unknown ligand that recognizes its FNIII domains (Hofmeyer & Treisman, 2009).

Synaptogenesis is also a process regulated by Lar. Using the *Drosophila* NMJ as a model system for synapse formation (Kaufmann, DeProto, Ranjan, Wan, & Van Vactor, 2002), it was observed that *Lar* mutants have abnormal synapse morphology and synaptic function. In this context, Lar does require its catalytic domain and acts together with its partner Liprin-alpha (LAR-interacting protein-related protein) (Serra-Pages, Medley, Tang, Hart, & Streuli, 1998; Stryker & Johnson, 2007). Furthermore, increasing levels of Lar increases the number of synaptic boutons per NMJ, indicating that Lar is involved in synaptic complexity.

Two *Drosophila* proteoglycans participate together with Lar in the regulation of synaptic growth and function (Johnson et al., 2006; Sdc and Dally-like (Dlp)). As Sdc, Dlp also binds Lar with high affinity. In *Sdc* mutants, although motoneuron axon branching and the distribution of synaptic markers are normal, a decrease in bouton number in NMJ 6/7 muscles is detected, indicating that Sdc is required for synaptic growth. Tissue-specific rescue experiments indicate that Sdc is mostly necessary in the presynaptic compartment. Concordantly, gain of function of Sdc either presynaptically or postsynaptically increases bouton number. On the other hand, *dlp* mutants have no clear defects in synapse morphology, but Dlp gain of function in *Sdc* mutant background further decreases bouton number. In *dlp*, unlike *Sdc* mutants, change in synaptic physiology is observed. *dlp* mutants have an increase in amplitude of excitatory junctional potentials. Genetic interaction experiments show that these phenotypes are modified by *Lar*. In *Sdc* hypomorphic alleles the bouton phenotype is increased by removing a copy of *Lar* and overexpression of Dlp can suppress Lar gain function phenotype, indicating that Sdc promotes Lar function, while Dlp is an antagonist.

4.2.2 Ptp69D

Ptp69D, like Lar, has been shown to participate in diverse aspects of neuronal development. Ptp69D is broadly expressed in the developing NS

(Kurusu & Zinn, 2008) and an extensive analysis of temperature-sensitive mutant alleles shows that Ptp69D is required for survival during the timing of axon growth (Desai & Purdy, 2003).

4.2.2.1 Axon Guidance

Analysis of the development of specific systems has revealed that Ptp69D has a function in CNS pathway choice of ipsilateral axons and muscle innervation of the embryo similar to what is observed in *Lar* mutants. In *Ptp69D* mutants, motoneurons show the typical ISNb bypass phenotype among others (Desai & Purdy, 2003). Lethality, CNS, and ISNb Ptp69D phenotypes are strongly suppressed in *Abl* mutants. In contrast, *Ptp69D* mutations do not modify *Abl* mutant phenotypes, indicating that Ptp69D may be an upstream regulator of Abl (Song, Giniger, & Desai, 2008).

Recently, we have shown that Ptp69D is also implicated in the axon growth of sLNv neurons (a group of neurons involved in circadian rhythms in *Drosophila*, Fig. 1F) in the protocerebrum. In the brain, unlike the embryonic VNC, the secreted ligand Slit (Dickson & Gilestro, 2006) is not expressed in midline glia, but rather in an extensive and complex neuronal structure, the MBs. sLNv axons terminate in the vicinity of the MB dendritic arbors and require the Robo family of Slit receptors so as not to overshoot their target area. Here Ptp69D is required to support Robo in limiting sLNv axonal growth. However, the phosphatase activity of Ptp69D appears not to be required in this context. Instead, data indicate that it acts as a chaperone increasing Robo levels at the cell membrane (Oliva et al., 2016).

Another circuit where the function of Ptp69D has been evaluated is the visual system. In *Ptp69D* mutants, R1–R6 axons do not stop in their correct layer the lamina, but instead continue growing toward the medulla. Although Ptp69D is required for R1–R6 axons to stop, it cannot retarget R7 or R8 axons to the lamina when overexpressed, indicating that it is either not sufficient for or not involved in targeting specificity (Garrity et al., 1999). In this context the catalytic activity of Ptp69D is essential. Improving on this work and introducing the eyFlp to make clones in the eye which allows to examine large patched of eye mutant tissue in otherwise wild-type animal and in this way overcome lethality issues of whole mutants, Newsome, Asling, and Dickson (2000) showed that not only R1–R6 have defects in *Ptp69D* mutants but also R7 axons. In this context, R7 axons terminate prematurely in the layer M3 (the normal R8 targeting layer).

4.2.2.2 Ptp69D and Axon Branching

In the formation of the adult mechanosensory system, neurons such as posterior scutellar neurons connect to multiple synaptic partners in the CNS through the formation of several axon collaterals, making them an excellent model to study axon branching. It recently has been shown that Dscam (a known guidance molecule) is necessary for axon branching in these neurons (Chen et al., 2006). Interestingly, Dscam is a substrate of Ptp69D, and upon dephosphorylation, Dscam is activated to promote axon branching (Dascenco et al., 2015).

4.2.2.3 Synapsis Formation and Function

Ptp69D also has a function in synaptic growth. This role has been tested in the giant fiber system (GFS) that controls the escape response of the fly, and is constituted by a small group of neurons. The sensory signal is received by the giant fiber and transmitted to the leg and wing muscles (Allen, Godenschwege, Tanouye, & Phelan, 2006). This system is composed by two interneurons in the brain that connect to tergotrochanteral motorneurons (TTMs) and peripheral synapsing interneurons (PSIs) in the fly thorax (Allen et al., 2006). The PSIs in turn connect to dorsolateral motoneurons regulating dorsolateral muscles. These neurons can be used to study axon growth and also synaptic function since they are suitable to electrophysiological studies. Mutations that affect the first Ig domain or the first catalytic domain ($Ptp69D^{10}$ and $Ptp69dD^{20}$) do not have defects in axon guidance or targeting but arrest of synaptic growth is observed (Lee & Godenschwege, 2015). Furthermore, these defects correlate with electrophysiological alterations. In control animals the GF-TTM synapse can respond to 10 pulses at 100 Hz with a 1:1 ratio. However, in *Ptp69D* mutants, this response is severely reduced. Structural and physiological defects are corrected when wild-type Ptp69D is expressed specifically in GFS neurons but not in TTM neurons, indicating a presynaptic defect.

Thus, Ptp69D appears to be involved in all steps of circuit wiring from axon growth and guidance, to branching to synapse formation and function.

4.3 Type IIa RPTPs in Vertebrate Brain Wiring

In mammals, there are three related members of the LAR family (LAR, RPTPσ, and RPTPδ). RPTPσ and RPTPδ are highly expressed in the developing NS, but no gross axon targeting defects are observed in single mutants, likely because they compensate for each other during CNS development (Uetani, Chagnon, Kennedy, Iwakura, & Tremblay, 2006). Careful examination shows that *rptpσ/rptpδ* double mutants have severe muscle dysgenesis and loss of motoneurons in the spinal cord, indicating that these proteins play essential roles during mammalian axonogenesis.

Interestingly, studies in cell culture of retinal axons exposed to ectodomain of RPTPσ show that this protein can indeed act as ligand to stimulate neurite outgrowth, supporting the view of RPTPs acting as both ligand and receptor at least in some contexts (Sajnani et al., 2005). This mechanism has not been studied in *Drosophila* so far.

These data together show that LAR and mammalian LAR-type RPTPs can exert different functions in distinct events of neuronal wiring and development.

4.4 Type III RPTPs in Brain Wiring
4.4.1 Ptp10D
4.4.1.1 Axon Guidance and Targeting

The first observation of a role for Ptp10D in axon guidance comes from studies in the embryonic CNS. *Ptp10D* mutants alone are viable, fertile, and do not have any obvious defects in embryos. Staining for FasII in the embryo shows also a normal pattern of longitudinal axons (Sun, Bahri, Schmid, Chia, & Zinn, 2000). Since it is known that other RPTPs have overlapping functions, *Ptp10D* mutants were evaluated together with other *RPtp* mutants. *Ptp10D/Ptp99A* double mutants do not exhibit embryonic defects. As explained earlier, *Lar* mutants have motor defects but not CNS defects in the embryo, but a combination of *Ptp10D* and *Lar* mutants does not modify the motor phenotype of *Lar* and has no defects in the CNS. *Ptp69D^1* mutants have defects in motor axon guidance and have only very weak defects in the CNS (1.4% of midline crossing defects). Interestingly, double mutant *Ptp69D/Ptp10D* shows strong defects in the CNS (Sun et al., 2000). In these animals, the longitudinal axon bundles are irregular and often fused to each other (Sun et al., 2000). These experiments indicate that Ptp10D and Ptp69D play redundant roles of CNS axon guidance in the embryo. Although only few ligands are known for RPTP receptors, a ligand of Ptp10D has been identified using a Ptp10D-AP probe (Lee, Cording, Vielmetter, & Zinn, Zinn, 2013). Using a collection of fly lines that allow ectopic expression of cell surface and secreted proteins by using the UAS/Gal4 system in embryos and later staining with the Ptp10D-AP probe, Lee and colleagues found that the stranded at second (Sas), a cell surface protein, binds to Ptp10D. Like Ptp10D, Sas is also expressed endogenously in axons. Sas is involved in the formation of the longitudinal tracks, since *sas* mutants show weak defects when these tracks are visualized with Fas2. *sas/Ptp10D* double mutants display stronger phenotypes than either mutant alone and tissue-specific rescue led to a model where Ptp10D in CNS axons interact with Sas on other axons or in a soluble form to control midline crossing.

4.4.1.2 Synaptic Function
Although there is no function described for Ptp10D in synaptic development, recent studies have shown that Ptp10D is involved in memory formation in the fly. In flies, memory has been divided into four phases (Margulies, Tully, & Dubnau, 2005), and in *Ptp10D* mutants, only the long-term memory (LTM) is affected, while short-term memory is normal. Ptp10D expression in the MBs is sufficient to recover the LTM phenotype in *Ptp10D* mutant background. Thus, Ptp10D is also involved in synaptic physiology.

4.4.2 Ptp99A
Together with Lar, Ptp69D, and Ptp10D, Ptp99A is also expressed in the embryonic CNS (Yang, Seow, Bahri, Oon, & Chia, 1991). Consistent with RPTP functions being highly redundant, so far *Ptp99A* mutants have no described phenotype. Analysis has been done in the embryo CNS where all the markers used show a normal architecture (Hamilton, Ho, & Zinn, 1995). More recently, it has been shown that together with other RPTPs, Ptp99A is expressed in young MB neurons (Kurusu & Zinn, 2008) during development. However, Ptp99A is not required for MB developments and mutant animals are normal. In agreement with overlapping functions of RPTPs however, *Ptp99A* mutations increase the severity of *Ptp69D* mutants in motor axons (Desai, Gindhart, Goldstein, & Zinn, 1996). In *Ptp99A/Ptp69D* double mutants, ISNb bypass and other phenotypes are strongly increased compared with *Ptp69D* alone. A striking observation was made when the ISNb axons were examined in *Lar/Ptp99A* double mutants, and *Lar* mutants have strong bypass phenotypes in ISNb axons; however, the removal of Ptp99A almost completely rescues this defect (Desai, Krueger, Saito, & Zinn, 1997). These data show that in *Drosophila*, RPTPs can cooperate but also play opposite roles depending on the context.

4.4.3 Ptp52F
4.4.3.1 Axon Guidance and Targeting
Ptp52F was the sixth identified *Drosophila* RPTP (Schindelholz, Knirr, Warrior, & Zinn, 2001). Similar to other RPTP, it is selectively expressed in the CNS, although only from late embryonic developmental stages

(Stages 13–17). *Ptp52F* mutant embryos display defects in the CNS longitudinal axons, and in the SNa (segmental neurons) in the motor system, the phenotypes are diverse, including lack of branching, stalling, and excessive branching depending on the hemisegment analyzed. Interestingly, single mutants of other RPTPs do not have defects in SNa axons, indicating that Ptp52F has very important roles in this subset of neurons. Furthermore, double mutants of *Ptp52F* with other RPTPs increase the penetrance of its phenotypes. On the other hand, although *Ptp52F* has only minor defects in ISNb axons, combinations with other *RPtp* mutants show synergistic interactions. Strikingly, removal of *Lar* produces a suppression of the *Ptp52F* phenotype in the CNS longitudinal axons, indicating a competitive action in this context; this suppression is specific, since in other cell populations (e.g., ISNb axons) it is not observed. This is another example of RPTPs having opposite roles in specific contexts.

4.4.4 Ptp4E

As mentioned earlier, Ptp4E is related to Ptp10D, since both are type III RPTPs and share high sequence homology. Sequence analyses indicate that Ptp4E and Ptp10D arose from a recent gene duplication event in the *Drosophila* genus (Jeon, Nguyen, Bahri, & Zinn, 2008). The catalytic domains of Ptp4E and Ptp10D shares 89% of identity, in contrast with 36–40% with others RPTPs.

Ptp4E mutants are viable, fertile, and have no obvious defects in the embryonic CNS. When combined with *Ptp10D* mutants, the resultant double mutant animals die at the beginning of the larval life (Jeon et al., 2008). This indicates that Ptp4E and Ptp10D can compensate for one another in vivo. The analysis of the NS phenotypes shows that *Ptp4E/Ptp10D* double mutants have defects in the patterning of longitudinal axons (visualized using FasII staining). These defects can be rescued by expressing Ptp4E specifically in neurons, indicating that its function is required in neurons.

However, Ptp4E is not equivalent to Ptp10D in all aspects: for instance, *Ptp4E/Ptp69D* double mutants have no CNS defects in contrast to *Ptp10D/Ptp69D* double mutants.

4.5 Vertebrate Type III RPTPs in Brain Wiring

In vertebrates, the RPTPO member of the type III family has been implicated in axonogenesis (Stepanek, Stoker, Stoeckli, & Bixby, 2005). In this work, targeting RPTPO in chick using RNAi produced growth defects of the anterior iliotibialis (AITIB) motor pool. Furthermore, combining

RNAis for RPTPO, RPTPδ, and RPTPσ (combination of RPTPδ and RPTPσ RNAis produced strong phenotypes in the AITIB nerve) leads to weaker defects than RPTPO alone or the RPTPδ/σ combination, suggesting that vertebrate type IIa and type III RPTPs can counteract in some contexts.

In sum, type III RPTPs, similar to Type IIa, have a variety of functions and can act in combination with Type IIa to coordinate the development of the NS.

5. FUTURE DIRECTIONS

Although there are several examples of RTKs and RPTPs acting in different stages of brain wiring in invertebrate and vertebrate models, still little is known about the molecular mechanisms involved in the signaling of some of these receptors. It is clear that the advances in the field of RTKs are far ahead of RPTPs. For instance, only a few RPTP ligands are known (see Table 2, Stoker, 2015, for RPTP ligands in vertebrates). Furthermore, only a few examples of RPTP substrates acting downstream as effectors of RPTP activation have been discovered. We expect that future approaches will close the gap to better understand how these molecules play their role in brain wiring. For instance the CRISPR revolution (Wright, Nunez, & Doudna, 2016; Xu et al., 2015) can greatly benefit the field by providing genomically tagged RTKs and RPTPs that can be used to purify signaling partners and substrates by creating for example genomic substrate trap mutations in RPTPs. Once identified, these partners can be later used in genetic experiments to sort out their functions in specific physiological contexts.

There are only few examples of known interactions between RTKs and RPTPs. In vertebrates RPTPO has been shown to dephosphorylate EphA and EphB receptors. Furthermore, this mechanism acts in vivo where RPTPO regulates the formation of retinotopic maps in the chick retina through Eph receptors (Shintani et al., 2006). More research in this vein will determine if interactions between these two classes of receptors are more general.

It is important to note that in some contexts, RPTPs have been shown to play a role independently of their catalytic activity. This adds an extra layer of complexity and has to be kept in mind when trying to tackle specific molecular mechanisms. In one case, we have found that Ptp69D can regulate Robo surface presentation (Oliva et al., 2016). In other cases, it can be mediated by interactions with extracellular ligands possibly participating in cell–cell adhesion as it has been proposed for Lar during R7 development (Hofmeyer & Treisman, 2009).

6. CONCLUSIONS

Tyrosine phosphorylation is an important control mechanism for CNS development. Here we describe that RTKs and RPTPs are involved in all the stages of brain wiring, from axon and dendrite growth to synaptic function. RTKs as well as RPTPs play partially redundant functions in several systems where they coexist. A deeper understanding of the molecular mechanisms involved in the functions of these enzymes, especially RPTPs, could contribute to drug design to treat for neurological diseases.

ACKNOWLEDGMENTS

Work in the Hassan lab is supported by ICM, IHU Pitié-Salpêtrière, VIB, KU Leuven, FWO, Belspo, and The Paul G. Allen Frontiers Group. C.O. is supported by Fondecyt 11150610 and ICM P09-015F grants. B.A.H. is an Allen Distinguished Investigator and an Einstein Visiting Fellow.

REFERENCES

Allen, M. J., Godenschwege, T. A., Tanouye, M. A., & Phelan, P. (2006). Making an escape: Development and function of the Drosophila giant fibre system. *Seminars in Cell & Developmental Biology*, 17(1), 31–41. http://dx.doi.org/10.1016/j.semcdb.2005.11.011.

Araujo, S. J., & Tear, G. (2003). Axon guidance mechanisms and molecules: Lessons from invertebrates. *Nature Reviews. Neuroscience*, 4(11), 910–922. http://dx.doi.org/10.1038/nrn1243.

Bazigou, E., Apitz, H., Johansson, J., Loren, C. E., Hirst, E. M., Chen, P. L., ... Salecker, I. (2007). Anterograde jelly belly and Alk receptor tyrosine kinase signaling mediates retinal axon targeting in drosophila. *Cell*, 128(5), 961–975. http://dx.doi.org/10.1016/j.cell.2007.02.024.

Bixby, J. L. (2001). Ligands and signaling through receptor-type tyrosine phosphatases. *IUBMB Life*, 51(3), 157–163. http://dx.doi.org/10.1080/152165401753544223.

Blume-Jensen, P., & Hunter, T. (2001). Oncogenic kinase signalling. *Nature*, 411(6835), 355–365. http://dx.doi.org/10.1038/35077225.

Bonkowsky, J. L., Yoshikawa, S., O'Keefe, D. D., Scully, A. L., & Thomas, J. B. (1999). Axon routing across the midline controlled by the Drosophila Derailed receptor. *Nature*, 402(6761), 540–544. http://dx.doi.org/10.1038/990122.

Bossing, T., & Brand, A. H. (2002). Dephrin, a transmembrane ephrin with a unique structure, prevents interneuronal axons from exiting the Drosophila embryonic CNS. *Development*, 129(18), 4205–4218.

Boyle, M., Nighorn, A., & Thomas, J. B. (2006). Drosophila Eph receptor guides specific axon branches of mushroom body neurons. *Development*, 133(9), 1845–1854. http://dx.doi.org/10.1242/dev.02353.

Brand, A. H., & Perrimon, N. (1993). Targeted gene expression as a means of altering cell fates and generating dominant phenotypes. *Development*, 118(2), 401–415.

Cafferty, P., Yu, L., & Rao, Y. (2004). The receptor tyrosine kinase off-track is required for layer-specific neuronal connectivity in drosophila. *Development, 131*(21), 5287–5295. http://dx.doi.org/10.1242/dev.01406.

Callahan, C. A., Muralidhar, M. G., Lundgren, S. E., Scully, A. L., & Thomas, J. B. (1995). Control of neuronal pathway selection by a Drosophila receptor protein-tyrosine kinase family member. *Nature, 376*(6536), 171–174. http://dx.doi.org/10.1038/376171a0.

Casey, J. P., Magalhaes, T., Conroy, J. M., Regan, R., Shah, N., Anney, R., ... Ennis, S. (2012). A novel approach of homozygous haplotype sharing identifies candidate genes in autism spectrum disorder. *Human Genetics, 131*(4), 565–579. http://dx.doi.org/10.1007/s00439-011-1094-6.

Chen, B. E., Kondo, M., Garnier, A., Watson, F. L., Puettmann-Holgado, R., Lamar, D. R., & Schmucker, D. (2006). The molecular diversity of Dscam is functionally required for neuronal wiring specificity in Drosophila. *Cell, 125*(3), 607–620. http://dx.doi.org/10.1016/j.cell.2006.03.034.

Clandinin, T. R., Lee, C. H., Herman, T., Lee, R. C., Yang, A. Y., Ovasapyan, S., & Zipursky, S. L. (2001). Drosophila LAR regulates R1-R6 and R7 target specificity in the visual system. *Neuron, 32*(2), 237–248.

Cramer, K. S., & Miko, I. J. (2016). Eph-ephrin signaling in nervous system development [version 1; referees: 2 approved]. *F1000Research*, (F1000 Faculty Rev) 5, 413. http://dx.doi.org/10.12688/f1000research.7417.1.

Dascenco, D., Erfurth, M. L., Izadifar, A., Song, M., Sachse, S., Bortnick, R., ... Schmucker, D. (2015). Slit and receptor tyrosine phosphatase 69D confer Spatial specificity to axon branching via Dscam1. *Cell, 162*(5), 1140–1154. http://dx.doi.org/10.1016/j.cell.2015.08.003.

Dearborn, R. E., Jr., Dai, Y., Reed, B., Karian, T., Gray, J., & Kunes, S. (2012). Reph, a regulator of Eph receptor expression in the Drosophila melanogaster optic lobe. *PLoS One, 7*(5). e37303. http://dx.doi.org/10.1371/journal.pone.0037303.

Dearborn, R., Jr., He, Q., Kunes, S., & Dai, Y. (2002). Eph receptor tyrosine kinase-mediated formation of a topographic map in the drosophila visual system. *The Journal of Neuroscience, 22*(4), 1338–1349.

Desai, C. J., Gindhart, J. G., Jr., Goldstein, L. S., & Zinn, K. (1996). Receptor tyrosine phosphatases are required for motor axon guidance in the Drosophila embryo. *Cell, 84*(4), 599–609.

Desai, C. J., Krueger, N. X., Saito, H., & Zinn, K. (1997). Competition and cooperation among receptor tyrosine phosphatases control motoneuron growth cone guidance in Drosophila. *Development, 124*(10), 1941–1952.

Desai, C., & Purdy, J. (2003). The neural receptor protein tyrosine phosphatase DPTP69D is required during periods of axon outgrowth in Drosophila. *Genetics, 164*(2), 575–588.

Dickson, B. J., & Gilestro, G. F. (2006). Regulation of commissural axon pathfinding by slit and its Robo receptors. *Annual Review of Cell and Developmental Biology, 22*, 651–675. http://dx.doi.org/10.1146/annurev.cellbio.21.090704.151234.

Dickson, B. J., & Zou, Y. (2010). Navigating intermediate targets: The nervous system midline. *Cold Spring Harbor Perspectives in Biology, 2*(8), a002055. http://dx.doi.org/10.1101/cshperspect.a002055.

Ensslen-Craig, S. E., & Brady-Kalnay, S. M. (2004). Receptor protein tyrosine phosphatases regulate neural development and axon guidance. *Developmental Biology, 275*(1), 12–22. http://dx.doi.org/10.1016/j.ydbio.2004.08.009.

Fernandez, M. P., Berni, J., & Ceriani, M. F. (2008). Circadian remodeling of neuronal circuits involved in rhythmic behavior. *PLoS Biology, 6*(3). e69. http://dx.doi.org/10.1371/journal.pbio.0060069.

Fernandez-Almonacid, R., & Rosen, O. M. (1987). Structure and ligand specificity of the Drosophila melanogaster insulin receptor. *Molecular and Cellular Biology, 7*(8), 2718–2727.

Forni, J. J., Romani, S., Doherty, P., & Tear, G. (2004). Neuroglian and FasciclinII can promote neurite outgrowth via the FGF receptor Heartless. *Molecular and Cellular Neurosciences, 26*(2), 282–291. http://dx.doi.org/10.1016/j.mcn.2004.02.003.

Fox, A. N., & Zinn, K. (2005). The heparan sulfate proteoglycan syndecan is an in vivo ligand for the Drosophila LAR receptor tyrosine phosphatase. *Current Biology, 15*(19), 1701–1711. http://dx.doi.org/10.1016/j.cub.2005.08.035.

Fradkin, L. G., Dura, J. M., & Noordermeer, J. N. (2010). Ryks: New partners for wnts in the developing and regenerating nervous system. *Trends in Neurosciences, 33*(2), 84–92. http://dx.doi.org/10.1016/j.tins.2009.11.005.

Garcia-Alonso, L., Fetter, R. D., & Goodman, C. S. (1996). Genetic analysis of laminin a in Drosophila: Extracellular matrix containing laminin a is required for ocellar axon pathfinding. *Development, 122*(9), 2611–2621.

Garcia-Alonso, L., Romani, S., & Jimenez, F. (2000). The EGF and FGF receptors mediate neuroglian function to control growth cone decisions during sensory axon guidance in Drosophila. *Neuron, 28*(3), 741–752.

Garofalo, R. S. (2002). Genetic analysis of insulin signaling in Drosophila. *Trends in Endocrinology and Metabolism, 13*(4), 156–162.

Garrity, P. A., Lee, C. H., Salecker, I., Robertson, H. C., Desai, C. J., Zinn, K., & Zipursky, S. L. (1999). Retinal axon target selection in Drosophila is regulated by a receptor protein tyrosine phosphatase. *Neuron, 22*(4), 707–717.

Grillenzoni, N., Flandre, A., Lasbleiz, C., & Dura, J. M. (2007). Respective roles of the DRL receptor and its ligand WNT5 in Drosophila mushroom body development. *Development, 134*(17), 3089–3097. http://dx.doi.org/10.1242/dev.02876.

Hadjieconomou, D., Timofeev, K., & Salecker, I. (2011). A step-by-step guide to visual circuit assembly in Drosophila. *Current Opinion in Neurobiology, 21*(1), 76–84. http://dx.doi.org/10.1016/j.conb.2010.07.012.

Hamilton, B. A., Ho, A., & Zinn, K. (1995). Targeted mutagenesis and genetic analysis of a Drosophila receptor-linked protein tyrosine phosphatase gene. *Roux's Archives of Developmental Biology, 204*, 187–192.

Hariharan, I. K., Chuang, P. T., & Rubin, G. M. (1991). Cloning and characterization of a receptor-class phosphotyrosine phosphatase gene expressed on central nervous system axons in Drosophila melanogaster. *Proceedings of the National Academy of Sciences of the United States of America, 88*(24), 11266–11270.

Hayes, M., Naito, M., Daulat, A., Angers, S., & Ciruna, B. (2013). Ptk7 promotes non-canonical Wnt/PCP-mediated morphogenesis and inhibits Wnt/beta-catenin-dependent cell fate decisions during vertebrate development. *Development, 140*(8), 1807–1818. http://dx.doi.org/10.1242/dev.090183.

Helfrich-Forster, C., Shafer, O. T., Wulbeck, C., Grieshaber, E., Rieger, D., & Taghert, P. (2007). Development and morphology of the clock-gene-expressing lateral neurons of Drosophila melanogaster. *The Journal of Comparative Neurology, 500*(1), 47–70. http://dx.doi.org/10.1002/cne.21146.

Hing, H., Xiao, J., Harden, N., Lim, L., & Zipursky, S. L. (1999). Pak functions downstream of dock to regulate photoreceptor axon guidance in drosophila. *Cell, 97*(7), 853–863.

Hofmeyer, K., & Treisman, J. E. (2009). The receptor protein tyrosine phosphatase LAR promotes R7 photoreceptor axon targeting by a phosphatase-independent signaling mechanism. *Proceedings of the National Academy of Sciences of the United States of America, 106*(46), 19399–19404. http://dx.doi.org/10.1073/pnas.0903961106.

Hovens, C. M., Stacker, S. A., Andres, A. C., Harpur, A. G., Ziemiecki, A., & Wilks, A. F. (1992). RYK, a receptor tyrosine kinase-related molecule with unusual kinase domain motifs. *Proceedings of the National Academy of Sciences of the United States of America, 89*(24), 11818–11822.

Huang, P., & Stern, M. J. (2005). FGF signaling in flies and worms: More and more relevant to vertebrate biology. *Cytokine & Growth Factor Reviews, 16*(2), 151–158. http://dx.doi.org/10.1016/j.cytogfr.2005.03.002.

Hunter, T. (2014). The genesis of tyrosine phosphorylation. *Cold Spring Harbor Perspectives in Biology, 6*(5), a020644. http://dx.doi.org/10.1101/cshperspect.a020644.

Iwahara, T., Fujimoto, J., Wen, D., Cupples, R., Bucay, N., Arakawa, T., ... Yamamoto, T. (1997). Molecular characterization of ALK, a receptor tyrosine kinase expressed specifically in the nervous system. *Oncogene, 14*(4), 439–449. http://dx.doi.org/10.1038/sj.onc.1200849.

Jefferis, G. S., Marin, E. C., Watts, R. J., & Luo, L. (2002). Development of neuronal connectivity in Drosophila antennal lobes and mushroom bodies. *Current Opinion in Neurobiology, 12*(1), 80–86.

Jeon, M., Nguyen, H., Bahri, S., & Zinn, K. (2008). Redundancy and compensation in axon guidance: Genetic analysis of the Drosophila Ptp10D/Ptp4E receptor tyrosine phosphatase subfamily. *Neural Development, 3,* 3. http://dx.doi.org/10.1186/1749-8104-3-3.

Johnson, K. G., Tenney, A. P., Ghose, A., Duckworth, A. M., Higashi, M. E., Parfitt, K., ... Van Vactor, D. (2006). The HSPGs syndecan and dallylike bind the receptor phosphatase LAR and exert distinct effects on synaptic development. *Neuron, 49*(4), 517–531. http://dx.doi.org/10.1016/j.neuron.2006.01.026.

Jorissen, R. N., Walker, F., Pouliot, N., Garrett, T. P., Ward, C. W., & Burgess, A. W. (2003). Epidermal growth factor receptor: Mechanisms of activation and signalling. *Experimental Cell Research, 284*(1), 31–53.

Kadam, S., McMahon, A., Tzou, P., & Stathopoulos, A. (2009). FGF ligands in Drosophila have distinct activities required to support cell migration and differentiation. *Development, 136*(5), 739–747. http://dx.doi.org/10.1242/dev.027904.

Kaufmann, N., DeProto, J., Ranjan, R., Wan, H., & Van Vactor, D. (2002). Drosophila liprin-alpha and the receptor phosphatase Dlar control synapse morphogenesis. *Neuron, 34*(1), 27–38.

Keene, A. C., & Waddell, S. (2007). Drosophila olfactory memory: Single genes to complex neural circuits. *Nature Reviews Neuroscience, 8*(5), 341–354. http://dx.doi.org/10.1038/nrn2098.

Keshishian, H., Broadie, K., Chiba, A., & Bate, M. (1996). The drosophila neuromuscular junction: A model system for studying synaptic development and function. *Annual Review of Neuroscience, 19,* 545–575. http://dx.doi.org/10.1146/annurev.ne.19.030196.002553.

Kurusu, M., & Zinn, K. (2008). Receptor tyrosine phosphatases regulate birth order-dependent axonal fasciculation and midline repulsion during development of the Drosophila mushroom body. *Molecular and Cellular Neurosciences, 38*(1), 53–65. http://dx.doi.org/10.1016/j.mcn.2008.01.015.

Laissue, P. P., & Vosshall, L. B. (2008). The olfactory sensory map in Drosophila. *Advances in Experimental Medicine and Biology, 628,* 102–114. http://dx.doi.org/10.1007/978-0-387-78261-4_7.

Lee, H. K., Cording, A., Vielmetter, J., & Zinn, K. (2013). Interactions between a receptor tyrosine phosphatase and a cell surface ligand regulate axon guidance and glial-neuronal communication. *Neuron, 78*(5), 813–826. http://dx.doi.org/10.1016/j.neuron.2013.04.001.

Lee, L. H., & Godenschwege, T. A. (2015). Structure-function analyses of tyrosine phosphatase PTP69D in giant fiber synapse formation of Drosophila. *Molecular and Cellular Neurosciences, 64,* 24–31. http://dx.doi.org/10.1016/j.mcn.2014.11.002.

Lee, T., Hacohen, N., Krasnow, M., & Montell, D. J. (1996). Regulated Breathless receptor tyrosine kinase activity required to pattern cell migration and branching in the Drosophila tracheal system. *Genes & Development, 10*(22), 2912–2921.

Li, C. R., Guo, D., & Pick, L. (2013). Independent signaling by Drosophila insulin receptor for axon guidance and growth. *Frontiers in Physiology, 4,* 385. http://dx.doi.org/10.3389/fphys.2013.00385.

Lin, D. M., & Goodman, C. S. (1994). Ectopic and increased expression of Fasciclin II alters motoneuron growth cone guidance. *Neuron, 13*(3), 507–523.

Linnemannstons, K., Ripp, C., Honemann-Capito, M., Brechtel-Curth, K., Hedderich, M., & Wodarz, A. (2014). The PTK7-related transmembrane proteins off-track and off-track 2 are co-receptors for Drosophila Wnt2 required for male fertility. *PLoS Genetics, 10*(7). e1004443. http://dx.doi.org/10.1371/journal.pgen.1004443.

Loren, C. E., Scully, A., Grabbe, C., Edeen, P. T., Thomas, J., McKeown, M., ... Palmer, R. H. (2001). Identification and characterization of DAlk: A novel Drosophila melanogaster RTK which drives ERK activation in vivo. *Genes to Cells, 6*(6), 531–544.

Manning, G., Whyte, D. B., Martinez, R., Hunter, T., & Sudarsanam, S. (2002). The protein kinase complement of the human genome. *Science, 298*(5600), 1912–1934. http://dx.doi.org/10.1126/science.1075762.

Margulies, C., Tully, T., & Dubnau, J. (2005). Deconstructing memory in Drosophila. *Current Biology, 15*(17), R700–R713. http://dx.doi.org/10.1016/j.cub.2005.08.024.

Maurel-Zaffran, C., Suzuki, T., Gahmon, G., Treisman, J. E., & Dickson, B. J. (2001). Cell-autonomous and -nonautonomous functions of LAR in R7 photoreceptor axon targeting. *Neuron, 32*(2), 225–235.

McIlroy, G., Foldi, I., Aurikko, J., Wentzell, J. S., Lim, M. A., Fenton, J. C., ... Hidalgo, A. (2013). Toll-6 and toll-7 function as neurotrophin receptors in the Drosophila melanogaster CNS. *Nature Neuroscience, 16*(9), 1248–1256. http://dx.doi.org/10.1038/nn.3474.

Newsome, T. P., Asling, B., & Dickson, B. J. (2000). Analysis of Drosophila photoreceptor axon guidance in eye-specific mosaics. *Development, 127*(4), 851–860.

Newsome, T. P., Schmidt, S., Dietzl, G., Keleman, K., Asling, B., Debant, A., & Dickson, B. J. (2000). Trio combines with dock to regulate Pak activity during photoreceptor axon pathfinding in Drosophila. *Cell, 101*(3), 283–294.

Oliva, C., Soldano, A., Mora, N., De Geest, N., Claeys, A., Erfurth, M. L., ... Hassan, B. A. (2016). Regulation of Drosophila brain wiring by neuropil interactions via a Slit-Robo-RPTP signaling complex. *Developmental Cell, 39*(2), 267–278. http://dx.doi.org/10.1016/j.devcel.2016.09.028. PubMed PMID: 27780041; PubMed Central PMCID: PMC5084709.

Oon, S. H., Hong, A., Yang, X., & Chia, W. (1993). Alternative splicing in a novel tyrosine phosphatase gene (DPTP4E) of Drosophila melanogaster generates two large receptor-like proteins which differ in their carboxyl termini. *The Journal of Biological Chemistry, 268*(32), 23964–23971.

Özel, M. N., Langen, M., Hassan, B. A., & Hiesinger, P. R. (2015). Filopodial dynamics and growth cone stabilization in Drosophila visual circuit development. *Elife, 4,* pii: e10721. http://dx.doi.org/10.7554/eLife.10721. PubMed PMID: 26512889; PubMed Central PMCID: PMC4728134.

Pasquale, E. B. (2005). Eph receptor signalling casts a wide net on cell behaviour. *Nature Reviews. Molecular Cell Biology, 6*(6), 462–475. http://dx.doi.org/10.1038/nrm1662.

Patthy, L. (2000). The WIF module. *Trends in Biochemical Sciences, 25*(1), 12–13.

Pecot, M. Y., Chen, Y., Akin, O., Chen, Z., Tsui, C. Y., & Zipursky, S. L. (2014). Sequential axon-derived signals couple target survival and layer specificity in the Drosophila visual system. *Neuron, 82*(2), 320–333. http://dx.doi.org/10.1016/j.neuron.2014.02.045.

Petruzzelli, L., Herrera, R., Arenas-Garcia, R., Fernandez, R., Birnbaum, M. J., & Rosen, O. M. (1986). Isolation of a Drosophila genomic sequence homologous to the kinase domain of the human insulin receptor and detection of the phosphorylated drosophila receptor with an anti-peptide antibody. *Proceedings of the National Academy of Sciences of the United States of America, 83*(13), 4710–4714.

Pulford, K., Morris, S. W., & Turturro, F. (2004). Anaplastic lymphoma kinase proteins in growth control and cancer. *Journal of Cellular Physiology, 199*(3), 330–358. http://dx.doi.org/10.1002/jcp.10472.

Pulido, D., Campuzano, S., Koda, T., Modolell, J., & Barbacid, M. (1992). Dtrk, a Drosophila gene related to the trk family of neurotrophin receptors, encodes a novel class of neural cell adhesion molecule. *The EMBO Journal, 11*(2), 391–404.

Reynaud, E., Lahaye, L. L., Boulanger, A., Petrova, I. M., Marquilly, C., Flandre, A., … Dura, J. M. (2015). Guidance of Drosophila mushroom body axons depends upon DRL-Wnt receptor cleavage in the brain dorsomedial lineage precursors. *Cell Reports, 11*(8), 1293–1304. http://dx.doi.org/10.1016/j.celrep.2015.04.035.

Sacco, F., Perfetto, L., Castagnoli, L., & Cesareni, G. (2012). The human phosphatase interactome: An intricate family portrait. *FEBS Letters, 586*(17), 2732–2739. http://dx.doi.org/10.1016/j.febslet.2012.05.008.

Sajnani, G., Aricescu, A. R., Jones, E. Y., Gallagher, J., Alete, D., & Stoker, A. (2005). PTPsigma promotes retinal neurite outgrowth non-cell-autonomously. *Journal of Neurobiology, 65*(1), 59–71. http://dx.doi.org/10.1002/neu.20175.

Saltiel, A. R., & Kahn, C. R. (2001). Insulin signalling and the regulation of glucose and lipid metabolism. *Nature, 414*(6865), 799–806. http://dx.doi.org/10.1038/414799a.

Schindelholz, B., Knirr, M., Warrior, R., & Zinn, K. (2001). Regulation of CNS and motor axon guidance in Drosophila by the receptor tyrosine phosphatase DPTP52F. *Development, 128*(21), 4371–4382.

Scolnick, J. A., Cui, K., Duggan, C. D., Xuan, S., Yuan, X. B., Efstratiadis, A., & Ngai, J. (2008). Role of IGF signaling in olfactory sensory map formation and axon guidance. *Neuron, 57*(6), 847–857. http://dx.doi.org/10.1016/j.neuron.2008.01.027.

Serra-Pages, C., Medley, Q. G., Tang, M., Hart, A., & Streuli, M. (1998). Liprins, a family of LAR transmembrane protein-tyrosine phosphatase-interacting proteins. *The Journal of Biological Chemistry, 273*(25), 15611–15620.

Shilo, B. Z. (2005). Regulating the dynamics of EGF receptor signaling in space and time. *Development, 132*(18), 4017–4027. http://dx.doi.org/10.1242/dev.02006.

Shin, J. E., & DiAntonio, A. (2011). Highwire regulates guidance of sister axons in the Drosophila mushroom body. *The Journal of Neuroscience, 31*(48), 17689–17700. http://dx.doi.org/10.1523/JNEUROSCI.3902-11.2011.

Shintani, T., Ihara, M., Sakuta, H., Takahashi, H., Watakabe, I., & Noda, M. (2006). Eph receptors are negatively controlled by protein tyrosine phosphatase receptor type O. *Nature Neuroscience, 9*(6), 761–769. http://dx.doi.org/10.1038/nn1697.

Shirasaki, R., Lewcock, J. W., Lettieri, K., & Pfaff, S. L. (2006). FGF as a target-derived chemoattractant for developing motor axons genetically programmed by the LIM code. *Neuron, 50*(6), 841–853. http://dx.doi.org/10.1016/j.neuron.2006.04.030.

Song, J. K., Giniger, E., & Desai, C. J. (2008). The receptor protein tyrosine phosphatase PTP69D antagonizes Abl tyrosine kinase to guide axons in Drosophila. *Mechanisms of Development, 125*(3-4), 247–256. http://dx.doi.org/10.1016/j.mod.2007.11.005.

Song, J., Wu, L., Chen, Z., Kohanski, R. A., & Pick, L. (2003). Axons guided by insulin receptor in Drosophila visual system. *Science, 300*(5618), 502–505. PubMed PMID: 12702880.

Srahna, M., Leyssen, M., Choi, C. M., Fradkin, L. G., Noordermeer, J. N., & Hassan, B. A. (2006). A signaling network for patterning of neuronal connectivity in the Drosophila brain. *PLoS Biology, 4*(11). e348. http://dx.doi.org/10.1371/journal.pbio.0040348.

Stepanek, L., Stoker, A. W., Stoeckli, E., & Bixby, J. L. (2005). Receptor tyrosine phosphatases guide vertebrate motor axons during development. *The Journal of Neuroscience, 25*(15), 3813–3823. http://dx.doi.org/10.1523/JNEUROSCI.4531-04.2005.

Stoker, A. W. (2015). RPTPs in axons, synapses and neurology. *Seminars in Cell & Developmental Biology, 37*, 90–97. http://dx.doi.org/10.1016/j.semcdb.2014.09.006.

Streuli, M., Krueger, N. X., Tsai, A. Y., & Saito, H. (1989). A family of receptor-linked protein tyrosine phosphatases in humans and Drosophila. *Proceedings of the National Academy of Sciences of the United States of America, 86*(22), 8698–8702.

Stryker, E., & Johnson, K. G. (2007). LAR, liprin alpha and the regulation of active zone morphogenesis. *Journal of Cell Science, 120*(Pt. 21), 3723–3728. http://dx.doi.org/10.1242/jcs.03491.

Sun, Q., Bahri, S., Schmid, A., Chia, W., & Zinn, K. (2000). Receptor tyrosine phosphatases regulate axon guidance across the midline of the Drosophila embryo. *Development, 127*(4), 801–812.

Tian, S. S., Tsoulfas, P., & Zinn, K. (1991). Three receptor-linked protein-tyrosine phosphatases are selectively expressed on central nervous system axons in the Drosophila embryo. *Cell, 67*(4), 675–685.

Timofeev, K., Joly, W., Hadjieconomou, D., & Salecker, I. (2012). Localized netrins act as positional cues to control layer-specific targeting of photoreceptor axons in Drosophila. *Neuron, 75*(1), 80–93. http://dx.doi.org/10.1016/j.neuron.2012.04.037.

Twigg, S. R., Kan, R., Babbs, C., Bochukova, E. G., Robertson, S. P., Wall, S. A., ... Wilkie, A. O. (2004). Mutations of ephrin-B1 (EFNB1), a marker of tissue boundary formation, cause craniofrontonasal syndrome. *Proceedings of the National Academy of Sciences of the United States of America, 101*(23), 8652–8657. http://dx.doi.org/10.1073/pnas.0402819101.

Uetani, N., Chagnon, M. J., Kennedy, T. E., Iwakura, Y., & Tremblay, M. L. (2006). Mammalian motoneuron axon targeting requires receptor protein tyrosine phosphatases sigma and delta. *The Journal of Neuroscience, 26*(22), 5872–5880. http://dx.doi.org/10.1523/JNEUROSCI.0386-06.2006.

Webber, C. A., Chen, Y. Y., Hehr, C. L., Johnston, J., & McFarlane, S. (2005). Multiple signaling pathways regulate FGF-2-induced retinal ganglion cell neurite extension and growth cone guidance. *Molecular and Cellular Neurosciences, 30*(1), 37–47. http://dx.doi.org/10.1016/j.mcn.2005.05.005.

Webber, C. A., Hyakutake, M. T., & McFarlane, S. (2003). Fibroblast growth factors redirect retinal axons in vitro and in vivo. *Developmental Biology, 263*(1), 24–34.

Weiss, J. B., Suyama, K. L., Lee, H. H., & Scott, M. P. (2001). Jelly belly: A Drosophila LDL receptor repeat-containing signal required for mesoderm migration and differentiation. *Cell, 107*(3), 387–398.

Weng, Y. L., Liu, N., DiAntonio, A., & Broihier, H. T. (2011). The cytoplasmic adaptor protein caskin mediates Lar signal transduction during Drosophila motor axon guidance. *The Journal of Neuroscience, 31*(12), 4421–4433. http://dx.doi.org/10.1523/JNEUROSCI.5230-10.2011.

Wilkinson, D. G. (2001). Multiple roles of EPH receptors and ephrins in neural development. *Nature Reviews. Neuroscience, 2*(3), 155–164. http://dx.doi.org/10.1038/35058515.

Wills, Z., Bateman, J., Korey, C. A., Comer, A., & Van Vactor, D. (1999). The tyrosine kinase Abl and its substrate enabled collaborate with the receptor phosphatase dlar to control motor axon guidance. *Neuron, 22*(2), 301–312.

Winberg, M. L., Tamagnone, L., Bai, J., Comoglio, P. M., Montell, D., & Goodman, C. S. (2001). The transmembrane protein Off-track associates with Plexins and functions downstream of Semaphorin signaling during axon guidance. *Neuron, 32*(1), 53–62.

Wright, A. V., Nunez, J. K., & Doudna, J. A. (2016). Biology and applications of CRISPR systems: Harnessing nature's toolbox for genome engineering. *Cell, 164*(1-2), 29–44. http://dx.doi.org/10.1016/j.cell.2015.12.035.

Xu, J., Ren, X., Sun, J., Wang, X., Qiao, H. H., Xu, B. W., ... Ni, J. Q. (2015). A toolkit of CRISPR-based genome editing systems in Drosophila. *Journal of Genetics and Genomics, 42*(4), 141–149. http://dx.doi.org/10.1016/j.jgg.2015.02.007.

Yang, X. H., Seow, K. T., Bahri, S. M., Oon, S. H., & Chia, W. (1991). Two Drosophila receptor-like tyrosine phosphatase genes are expressed in a subset of developing axons and pioneer neurons in the embryonic CNS. *Cell, 67*(4), 661–673.

Yao, Y., Wu, Y., Yin, C., Ozawa, R., Aigaki, T., Wouda, R. R., ... Hing, H. (2007). Antagonistic roles of Wnt5 and the Drl receptor in patterning the Drosophila antennal lobe. *Nature Neuroscience, 10*(11), 1423–1432. http://dx.doi.org/10.1038/nn1993.

Yasunaga, K., Tezuka, A., Ishikawa, N., Dairyo, Y., Togashi, K., Koizumi, H., & Emoto, K. (2015). Adult Drosophila sensory neurons specify dendritic territories independently of dendritic contacts through the Wnt5-Drl signaling pathway. *Genes & Development, 29*(16), 1763–1775. http://dx.doi.org/10.1101/gad.262592.115.

Yoshikawa, S., McKinnon, R. D., Kokel, M., & Thomas, J. B. (2003). Wnt-mediated axon guidance via the Drosophila Derailed receptor. *Nature, 422*(6932), 583–588. http://dx.doi.org/10.1038/nature01522.

Zondag, G. C., Koningstein, G. M., Jiang, Y. P., Sap, J., Moolenaar, W. H., & Gebbink, M. F. (1995). Homophilic interactions mediated by receptor tyrosine phosphatases mu and kappa. A critical role for the novel extracellular MAM domain. *The Journal of Biological Chemistry, 270*(24), 14247–14250.

Zschatzsch, M., Oliva, C., Langen, M., De Geest, N., Ozel, M. N., Williamson, W. R., ... Hassan, B. A. (2014). Regulation of branching dynamics by axon-intrinsic asymmetries in tyrosine kinase receptor signaling. *eLife, 3*. e01699. http://dx.doi.org/10.7554/eLife.01699.

CHAPTER THIRTEEN

VEGF Receptor Tyrosine Kinases: Key Regulators of Vascular Function

Alberto Álvarez-Aznar*, Lars Muhl[†,1,2], Konstantin Gaengel*[,1,2]

*Rudbeck Laboratory, Uppsala University, Uppsala, Sweden
[†]Karolinska Institutet, Stockholm, Sweden
[2]Corresponding authors: e-mail addresses: lars.muhl@ki.se; konstantin.gaengel@igp.uu.se

Contents

1. Introduction		434
2. Structure and Function of VEGFRs		437
2.1	VEGFR1 Ligand Binding Properties	437
2.2	VEGFR1 Expression	438
2.3	VEGFR1: A Decoy Receptor for VEGF-A	438
2.4	VEGFR1, a Kinase Impaired RTK?	438
2.5	A Soluble Variant of VEGFR1	440
2.6	VEGFR2 Expression and Vascular Function	441
2.7	VEGFR2 Ligand Binding Properties	441
2.8	VEGFR2 Signaling	442
2.9	A Soluble Variant of VEGFR2	445
2.10	VEGFR3 Structure and Ligand Binding Properties	446
2.11	VEGFR3 Function	446
2.12	VEGFR3 Expression Pattern	447
2.13	VEGFR3 in Development	447
2.14	VEGFR3 Signaling	448
2.15	Heterodimer Formation of VEGFR2 With VEGFR1 and VEGFR3	449
3. An Evolutionary Perspective on VEGFR Function		450
4. Neuropilins: Coreceptors Modulating VEGFR Signaling		453
5. VEGFR Ligands		455
5.1	Different Ligands for Different Receptors	455
5.2	Regulation of VEGF-A Expression	456
5.3	VEGF-A Splice Variants	457
5.4	The Antiangiogenic VEGF-A$_{xxxb}$ Isoforms	459
5.5	Proteolytic Processing as Regulatory Mechanism	461

---

[1] These authors contributed equally.

6. Perspective	463
Acknowledgments	464
References	464

Abstract

Vascular endothelial growth factor receptor (VEGFR) tyrosine kinases are key regulators of vascular development in vertebrates. Their activation is regulated through a family of secreted glycoproteins, the vascular endothelial growth factors (VEGFs). Expression, proteolytic processing, and diffusion range of VEGF proteins need to be tightly regulated, due to their crucial roles in development. While some VEGFs form concentration gradients across developing tissues and act as morphogens, others function as inhibitors of receptor activation and downstream signaling. Ligand-induced receptor dimerization leads to activation of the intrinsic tyrosine kinase activity, which results in autophosphorylation of the receptors and in turn triggers the recruitment of interacting proteins as well as the initiation of downstream signaling. Although many biochemical details of VEGFR signaling have been revealed, the in vivo relevance of certain signaling aspects still remains to be demonstrated. Here, we highlight basic principles of VEGFR signaling and discuss its crucial role during development of the vascular system in mammals.

1. INTRODUCTION

Within the receptor tyrosine kinase (RTK) superfamily, VEGFRs form the type-V subfamily (Shibuya, 2013a) and show closest resemblance to members of the type-III RTK subfamily, particularly the platelet-derived growth factor receptors (PDGFs) and the mast/stem cell growth factor receptor (SCFR, also known as KIT) with whom they share certain fundamental signaling mechanisms such as receptor dimerization, activation of intrinsic tyrosine kinase activity, autophosphorylation, and the formation of docking sites for downstream effectors (Lemmon & Schlessinger, 2010).

In mammals three different VEGFRs have been identified (see Fig. 1): VEGFR1 (also known as fms-related tyrosine kinase-1; *FLT1*; de Vries et al., 1992; Shibuya et al., 1990), VEGFR2 (also known as fetal liver kinase 1, *Flk1* in mice; Matthews et al., 1991; Quinn, Peters, de Vries, Ferrara, & Williams, 1993) or as kinase insert domain receptor (*KDR*) in humans (Terman et al., 1991, 1992), and VEGFR3 (fms-related tyrosine kinase-4; *FLT4*; Galland et al., 1993; Pajusola et al., 1992). Although an increasing body of literature indicates that VEGFRs play important roles outside the cardiovascular system, their core functions can, in a simplified manner, be described as follow: VEGFR1, the first VEGFR identified (Shibuya et al., 1990), mainly functions as a negative regulator of VEGFR2

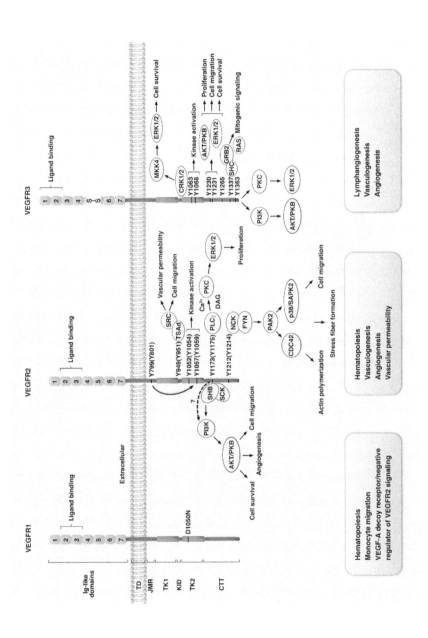

Fig. 1 See legend on next page.

(Fong, Rossant, Gertsenstein, & Breitman, 1995; Hiratsuka, Minowa, Kuno, Noda, & Shibuya, 1998) and also regulates monocyte and macrophage migration (Barleon et al., 1996). VEGFR2 is critical for multiple aspects of normal and pathological endothelial cell biology including hematopoiesis, vasculogenesis, angiogenesis, and vascular permeability. In contrast, VEGFR3 is mainly important for lymphatic EC development and function (Karkkainen et al., 2001; Mäkinen, Jussila, et al., 2001). In mammals, five ligands for VEGFRs have been identified (see Fig. 2) VEGF-A, -B,-C, -D, and placental growth factor (PlGF). VEGF-A, -B, and PlGF bind to VEGFR1 (Park, Chen, Winer, Houck, & Ferrara, 1994; Sawano, Takahashi, Yamaguchi, Aonuma, & Shibuya, 1996), VEGF-A binds VEGFR2 and VEGF-C and -D bind VEGFR3 (Achen et al., 1998; Joukov et al., 1996; Quinn et al., 1993; Terman et al., 1992). Proteolytic processing enables VEGF-C and -D to bind to VEGFR2, but with lower affinity compared to VEGFR3 (Achen et al., 1998; Joukov et al., 1996). In addition, VEGF-like proteins have been found in parapoxvirus (VEGF-E) and in certain snake venoms (VEGF-Fs). While all VEGF-E variants described so far exclusively bind to VEGFR2 (Takahashi & Shibuya, 2005), VEGF-Fs have been shown to bind both VEGFR1 and 2 (Yamazaki et al., 2009).

In their unprocessed form VEGFRs are characterized by an extracellular region that contains 7 immunoglobulin-like (IG) domains, a single transmembrane domain (TM), as well as a cytoplasmic tail that contains a split kinase domain (TK1, TK2), which is intersected by a kinase insert domain (KID) (see Fig. 1) (reviewed in Roskoski, 2008). A closer look at the

Fig. 1 Structure, signaling, and main functions of VEGFRs: VEGFRs are characterized by seven extracellular Ig-like domains, a single transmembrane domain (TD), a juxtamembrane region (JMR), and a tyrosine kinase domain that is split into two stretches (TK1, TK2) by a kinase insert domain (KID), followed by the carboxyl-terminal tail (CTT). VEGFR1 (*left*) has a unique amino acid substitution of aspartic acid to asparagine at position 1050 within the kinase domain activation loop (TK2), which is partially responsible for VEGFR1's low intrinsic kinase activity. VEGFR2 (*middle*) is represented with its most-studied phosphotyrosines (murine nomenclature [*left*] and human [*right in parentheses*]). Binding of downstream signaling effectors to individual phosphotyrosines and downstream signaling pathways are indicated (see text for details). In VEGFR3 (*right*), the fifth Ig-like domain undergoes proteolytic cleavage and the two resulting receptor parts are joined together through a disulfide bridge. The most-studied phosphotyrosines, corresponding adaptor proteins and downstream signaling pathways are indicated (see text for details). The main biological functions of each receptor are indicated in the respective *boxes*.

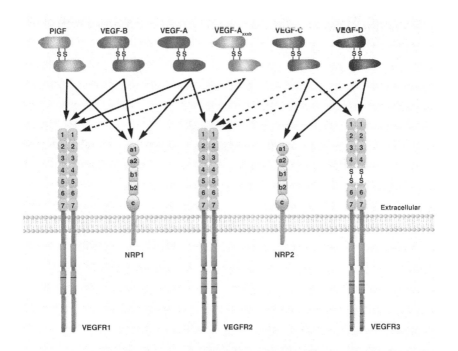

Fig. 2 VEGFR ligands and coreceptors: six different ligands (PlGF, VEGF-B, VEGF-A, VEGF-A$_{xxxb}$, VEGF-C, and VEGF-D) bind in partially overlapping manner to VEGFR1–3. The coreceptors NRP1 and NRP2 modulate signal transduction of VEGFRs. PlGF and VEGF-B specifically bind to VEGFR1 and convey their biological function through this receptor. VEGF-A binds to VEGFR1 and VEGFR2. The main biological functions of VEGF-A are transmitted by VEGFR2, while VEGFR1 is thought to serve as a decoy receptor for VEGF-A and thereby limits VEGF-A/VEGFR2 signaling. The antiangiogenic VEGF-A$_{xxxb}$ isoforms bind to VEGFR2 and presumably also VEGFR1. VEGF-C and VEGF-D bind to VEGFR3 that relays their main biological functions. In a proteolytically processed form, VEGF-C and VEGF-D also bind to VEGFR2, albeit with lower affinity.

structure of the VEGFRs helps understanding how the different domains contribute to receptor activation and subsequent downstream signaling.

2. STRUCTURE AND FUNCTION OF VEGFRs

VEGFR1

2.1 VEGFR1 Ligand Binding Properties

VEGFR1 is a 180-kDa high-affinity receptor for VEGF-A, VEGF-B, and PlGF (Olofsson et al., 1998; Park et al., 1994; Sawano et al., 1996) and in addition, has been shown to bind certain VEGF-F variants (Yamazaki et al., 2009). The crystal structure of parts of the extracellular domain of VEGFR1 in complex with VEGF-A or PlGF has revealed that the second and third IG domain are crucial for ligand binding (Christinger, Fuh, de Vos, & Wiesmann, 2004; Wiesmann et al., 1997). VEGF-A binds VEGFR1 with substantially higher affinity compared to VEGFR2 ($K_d = 10$–20 pM for VEGFR1 vs 75–125 pM for VEGFR2) (de Vries et al., 1992; Terman et al., 1992). However, ligand binding results only in a negligible increase of VEGFR1 kinase activity (Waltenberger, Claesson-Welsh, Siegbahn, Shibuya, & Heldin, 1994).

2.2 VEGFR1 Expression

VEGFR1 is expressed in vascular endothelial cells and various non-endothelial cells, including macrophages, monocytes (Sawano et al., 2001), and hematopoietic stem cells (Hattori et al., 2002). In addition, VEGFR1 was found to be transiently expressed in neurons of the cortex, striatum, and hippocampus within the first weeks of birth (Yang, Zhang, Huang, & Sun, 2003).

2.3 VEGFR1: A Decoy Receptor for VEGF-A

Global deletion of *Flt1 (the gene encoding* VEGFR1 in mice) results in vascular hyperplasia as well as failure to form a functional vessel network and leads to lethality at embryonic day (E) 8.5–9 (Fong et al., 1995; Fong, Zhang, Bryce, & Peng, 1999). Interestingly, however, expression of VEGFR1 lacking its kinase domain is sufficient to rescue embryonic lethality. This demonstrates that the kinase activity of VEGFR1 is not required for normal development (Hiratsuka et al., 1998). It is therefore believed that in the context of vascular development VEGFR1 does not function as a signaling receptor but rather as a trap for VEGF-A, restricting its accessibility for VEGFR2. It appears that the membrane-anchored ligand-binding domain is the essential part of VEGFR1 during development, as about half the mice with a deletion of both the transmembrane region and the tyrosine kinase domain, which results in a soluble receptor, are embryonic lethal (Hiratsuka et al., 2005).

2.4 VEGFR1, a Kinase Impaired RTK?

Although all the major motives required for kinase activity are present in VEGFR1 (Rahimi, 2006), the level of phosphorylation that occurs in response to VEGF-A is minor (Seetharam et al., 1995; Waltenberger et al., 1994). The catalytic loop that resides within the kinase domain is entirely conserved and it was long believed that there is no amino acid substitution in the kinase domain itself that could explain the poor kinase activity of VEGFR1 (Rahimi, 2006). However, Meyer and colleagues identified that an aspartic acid residue in the activation loop of the kinase domain, which is conserved in VEGFR2 and VEGFR3 as well as in the closely related type-III RTKs, is substituted to asparagine in VEGFR1. Exchange of this asparagine residue (Asn^{1050}) to aspartic acid (Asp^{1050}) promotes ligand-dependent tyrosine autophosphorylation and kinase activation as well as endothelial cell proliferation, but not tubulogenesis (Meyer, Mohammadi, & Rahimi, 2006). Further evidence of the importance of this asparagine residue was provided when it was introduced into VEGFR2, where it lowered autophosphorylation of the activation loop tyrosine residues Tyr^{1052} and Tyr^{1057} (Meyer et al., 2006).

However, the answer to VEGFR1's low intrinsic kinase activity might not only be found in the kinase domain itself. It has, for example, been suggested that the carboxyl-terminal tail of VEGFR1 (see Fig. 1) might function to inhibit its kinase activity (Meyer et al., 2004). In line with this hypothesis is that swapping the carboxyl-terminal tail of VEGFR1 with that of VEGFR2 promoted ligand-depended autophosphorylation of VEGFR1 and induction of endothelial cell proliferation (Meyer et al., 2004). Yet how the carboxyl-terminal tail of VEGFR2 is able to promote the kinase activity of VEGFR1 is not fully understood. Based on these experiments it was hypothesized that the carboxyl-terminal tail of VEGFR1 obstructs ligand-mediated autophosphorylation by interacting with its activation loop and keeping it in an inactive conformation. However, deletion of the entire carboxyl-terminal tail of VEGFR1 does not rescue VEGFR1's kinase activity (Meyer et al., 2004) and rather suggests that the carboxyl-terminal tail of VEGFR1 lacks elements required to promote its kinase activation.

In addition, the juxtamembrane region of VEGFR1 (see Fig. 1) has also been implicated in repressing some of its signaling properties, especially its ability to mediate cell migration. Exchange of the juxtamembrane region of VEGFR1 with that of VEGFR2 enables the VEGFR1/VEGFR2-juxtamembrane-chimera to mediate endothelial cell migration and activation of phosphatidylinositol 3-kinase in response to VEGF-A. Conversely, replacing the juxtamembrane region of VEGFR2 with that of VEGFR1 renders the VEGFR2/VEGFR1 juxtamembrane-chimera unable to mediate endothelial cell migration in response to VEGF-A. Further experiments revealed that a stretch of three nonconserved serine residues is responsible for the inhibitory function that resides in the VEGFR1 juxtamembrane region. Both, deletion of these three serine residues in VEGFR1, or their exchange with the corresponding amino acids in VEGFR2 (Ala-Asn-Gly-Gly), enabled the so-modified VEGFR1 to mediate endothelial cell migration in response to VEGF-A. However, deletion of the juxtamembrane region in VEGFR1 does not revert its inability to transmit effective proliferative signals, suggesting that the juxtamembrane region of VEGFR1 inhibits migration, but not proliferation (Gille et al., 2000). Although the tyrosine kinase activity of VEGFR1 is dispensable during development, it is required for numerous physiological and pathological processes later in life, for example, for fatty acid transport across the endothelium (Hagberg et al., 2010; Muhl et al., 2016), during monocyte recruitment (Hiratsuka et al., 1998), and during pathological angiogenesis (Hiratsuka et al., 2001). However, despite the fact that many potential interaction partners for the phosphorylated tyrosine residues have been identified through in vitro overexpression studies, the downstream signaling events, and their physiological relevance remain poorly understood (Koch & Claesson-Welsh, 2012).

2.5 A Soluble Variant of VEGFR1

Another aspect of VEGFR1 biology is that the receptor can undergo alternative splicing to form a truncated soluble version (referred to as sVEGFR1 or sFLT-1) that comprises only the first six extracellular IG domains and a short unique additional C-terminal sequence (Kendall & Thomas, 1993). sVEGFR1 either binds to and sequesters VEGF-A and thereby inhibits VEGFR2 activation or forms a heterodimer complex with VEGFR2 which is signaling-inactive (Kendall, Wang, & Thomas, 1996). sVEGFR1 has been suggested to facilitate the local guidance of angiogenic vessel sprouts. Its expression in endothelial cells directly adjacent to the emerging vessel sprout is thought to shape the tissue VEGF-A gradient, facilitating a straight outgrowth of the vessel sprout and limits the formation of new vessel sprouts in the direct vicinity (Chappell, Taylor, Ferrara, & Bautch, 2009). In addition, sVEGFR1 seems to limit VEGFR2 activation and thus angiogenesis in the endometrium during the early phase of post menstrual repair. sVEGFR1 downregulation in the late menstrual phase correlates with increased VEGFR2 phosphorylation and endothelial cell proliferation and contributes to the onset of angiogenesis and endothelial repair in the endometrium (Graubert, Asuncion Ortega, Kessel, Mortola, & Iruela-Arispe, 2001).

Furthermore, elevated plasma levels of sVEGFR1 have further been reported in patients with preeclampsia, ischemia, and cancer (Levine et al., 2004; Scheufler et al., 2003; Toi et al., 2002). Increased levels of sVEGFR1 in the circulation of preeclampsia patients are associated with decreased levels of VEGF-A and PlGF which contribute to endothelial dysfunction, hypertension, and proteinuria (Maynard et al., 2003).

VEGFR2

2.6 VEGFR2 Expression and Vascular Function

VEGFR2 is expressed in vascular and lymphatic endothelial cells as well as in endothelial precursor cells, megakaryocytes and in hematopoietic stem cells (Katoh, Tauchi, Kawaishi, Kimura, & Satow, 1995). Highest expression levels are observed during vasculogenesis and angiogenesis and during pathological neovascularization. In addition, expression of VEGFR2 has also been reported in the nervous system (Bellon et al., 2010). Global deletion of *Flk1* (the gene encoding VEGFR2 in mice) causes lethality between E8.5 and 9.5 due to defects in hematopoiesis and blood vessel formation (Shalaby et al., 1997). Conditional deletion at later developmental stages has revealed a crucial role of VEGFR2 in angiogenic sprouting, where it functions upstream of the delta-like ligand 4 (DLL4)/Notch signal transduction pathway (Zarkada, Heinolainen, Makinen, Kubota, & Alitalo, 2015). VEGFR2 is also considered the key receptor involved in pathological angiogenesis and vascular aspects of tumor biology and has been a target for both pro- and antiangiogenic therapy (Shibuya, 2013b).

2.7 VEGFR2 Ligand Binding Properties

The overall structure of VEGFR2 is very similar to that of VEGFR1 (see Fig. 1). VEGFR2 is initially synthesized as a 150 kDa protein which becomes glycosylated to a 200–230-kDa mature form that serves as the cell surface receptor (Takahashi & Shibuya, 1997). VEGFR2 is the main signaling receptor for VEGF-A (Shalaby et al., 1995) and mediates the vast majority of VEGF-A-dependent processes in the vasculature (reviewed in Koch & Claesson-Welsh, 2012; Olsson, Dimberg, Kreuger, & Claesson-Welsh, 2006; Takahashi & Shibuya, 2005). In addition, VEGFR2 has been shown to bind VEGF-E and certain snake venom VEGF-F variants (Yamazaki et al., 2009) and proteolytically cleaved variants of VEGF-C and D (Achen et al., 1998). Binding of proteolytically cleaved VEGF-C only requires the second IG domain (Jeltsch et al., 2006), while both the second and third IG domain are important for binding of VEGF-A (Fuh, Li, Crowley, Cunningham, & Wells, 1998) (see Fig. 2). This study further revealed that ligand binding to monomeric VEGFR2 is 100 times weaker than to VEGFR2 homodimers, suggesting that VEGF-A preferentially binds to and activates predimerized receptors. This hypothesis has recently been supported by quantitative FRET and biochemical experiments that demonstrated that VEGFR2 dimers can form in the absence of ligand, when expressed at physiological levels (Sarabipour, Ballmer-Hofer, & Hristova, 2016). Ligand binding leads to a conformational change in the TM domain, which results in increased phosphorylation of the kinase domain. Interreceptor contacts within the extracellular and TM domains are thought to be critical for the establishment of the unliganded dimer structure, and for the transition to the ligand-bound active conformation (Sarabipour et al., 2016). Experiments based on designed ankyrin repeat protein inhibitors (DARPins), targeting specific domains in VEGFR2, confirmed the critical roles of the second and third IG domain in ligand binding (see Fig. 1) and further revealed the importance of those domains in receptor dimerization and kinase activation. Inhibition of the fourth and seventh IG domains did not prevent ligand binding or receptor dimerization but effectively blocked receptor signaling, indicating that those domains allosterically regulate VEGFR2 activity (Hyde et al., 2012). This is further supported by electron microscopy data showing that the seventh and to a lesser extent the fourth IG domain of VEGFR2 form homotypic interactions with the corresponding domains of a neighboring receptor. This stabilizes the structure of the homodimeric complex and brings the intracellular domains of the dimerized receptors in close proximity (Ruch, Skiniotis, Steinmetz, Walz, & Ballmer-Hofer, 2007).

2.8 VEGFR2 Signaling

The kinase domain of VEGFR2 is similar to other kinase domains and folds into two lobes with a central cleft that allow ATP binding and phosphotransfer. Secondary structural elements include a catalytic loop, which contains an aspartic acid residue (Asp^{1028}) that is conserved among tyrosine kinases and crucial for kinase activity of VEGFR2. As in other tyrosine kinases, the ATP-binding site of VEGFR2 lies at the cleft between the N- and C-terminal lobes and contains a glycine-rich loop (referred to as the nucleotide-binding loop). In addition, the kinase domain includes an activation loop that contains an aspartic acid residue (Asp^{1056} in human, corresponding to Asp^{1054} in mouse VEGFR2), which is conserved among tyrosine kinases but substituted by asparagine (Asn^{1050} in both human and mouse VEGFR1; see Fig. 1). This aspartic acid residue is crucial for kinase activity of VEGFR2 and one of the reasons why VEGFR1 has low intrinsic kinase activity and is a poor signaling receptor (McTigue et al., 1999; Meyer et al., 2006). The activation loop also contains two tyrosine phosphorylation sites (Tyr^{1054} and Tyr^{1059} in human, corresponding to Tyr^{1052} and Tyr^{1057} in mouse VEGFR2). Further tyrosine phosphorylation sites exist in the KID (Tyr^{951} and Tyr^{996} in human, corresponding to Tyr^{949} and Tyr^{994} in mouse VEGFR2) and in the carboxyl-terminal tail (Tyr^{1175}, Tyr^{1214}, Tyr^{1223}, Tyr^{1305}, Tyr^{1309}, and Tyr^{1319} in human, corresponding to Tyr^{1173}, Tyr^{1212}, Tyr^{1121}, Tyr^{1303}, Tyr^{1307}, and Tyr^{1317} in mouse VEGFR2). Yet another phosphorylation site is found in the juxtamembrane region (Tyr^{801} in human, corresponding to Tyr^{799} in mouse VEGFR2; see Fig. 1). These tyrosine residues become autophosphorylated in trans upon ligand binding and dimerization of receptor monomers (Parast et al., 1998). Autophosphorylation of Tyr^{1054} and Tyr^{1059} in the activation loop stimulates the intrinsic kinase activity (Kendall et al., 1999), while other tyrosine residues become recruitment sites for adaptor proteins and downstream signaling effectors (reviewed in Koch & Claesson-Welsh, 2012; Roskoski, 2008; Takahashi, Yamaguchi, Chida, & Shibuya, 2001).

VEGF-A-mediated Tyr^{951} phosphorylation is observed in a subset of endothelial cells during development as well as in tumor vessels, but does not seem to occur in quiescent blood vessels (Matsumoto et al., 2005). Phosphorylated Tyr^{951} creates a binding site for T cell-specific adaptor molecule (TSAd, also known as VEGF receptor-associated protein, VRAP; Matsumoto et al., 2005; Wu et al., 2000) (see Fig. 1). Mutation of Tyr^{951} to phenylalanine, delivery of phosphorylated Y^{951} peptide, or TSAd siRNA into endothelial cells inhibits VEGF-A-induced actin stress fibers and cell migration, but not mitogenesis (Matsumoto et al., 2005). The in vivo relevance of Tyr^{951}-TSAd signaling was demonstrated in the context of pathological angiogenesis, where, in TSAd-deficient mice, both tumor vascularization and tumor growth were reduced (Wu et al., 2000). Tyr^{951} phosphorylation also plays a critical role in VEGF-A-induced vascular permeability. This involves TSAd-mediated activation of the cSRC tyrosine kinase, which in turn regulates permeability through modulation of endothelial cell junctions. TSAd silencing blocks VEGF-A-induced cSRC activation, but does not affect pathways involving phospholipase Cγ (PLCγ), extracellular signal-regulated kinase (ERK)1/2 (also known as mitogen-activated protein kinase (MAPK p42/p44)), and endothelial nitric oxide-synthase

(Sun et al., 2012). In a recent study Li and colleagues elegantly demonstrated the in vivo relevance of Tyr^{951} for vascular permeability. The authors created mice in which the corresponding tyrosine was substituted by phenylalanine ($Tyr^{949}Phe$) (Li et al., 2016). Homozygous $Tyr^{949}Phe$ mice are viable and the morphology of blood vessels, blood flow, and blood pressure appear normal. However, VEGF-A is no longer able to induce vascular permeability. Interestingly, the $Tyr^{949}Phe$ mutation does however not seem to affect the sensitivity of the endothelium to inflammatory cytokines (Li et al., 2016). The importance of this finding was further explored in the context of tumor biology. Tumor-bearing mice carrying the $Tyr^{949}Phe$ mutation displayed reduced vascular leakage, improved response to chemotherapy, and reduced metastatic spread (Li et al., 2016). In addition to its role in permeability, the VEGFR2-TSAd-cSRC signaling axis has been implicated to regulate sprouting angiogenesis in a subset of tissues (Gordon et al., 2016).

Tyr^{1173} is indispensable for normal development. Mice with a $Tyr^{1173}Phe$ point mutation die at E8.5–9 due to vascular defects that resemble those observed in VEGFR2 global knockout mice (Sakurai, Ohgimoto, Kataoka, Yoshida, & Shibuya, 2005). These results suggest that a single amino acid might control the downstream signaling that regulates crucial aspects of vascular development. Phosphorylation of Tyr^{1175} in human VEGFR2 (corresponding to Tyr^{1173} in mice) creates binding sites for SHC-like protein (SCK) (Warner, Lopez-Dee, Knight, Feramisco, & Prigent, 2000), SRC homology 2 protein B (SHB) (Holmqvist et al., 2004) and PLCγ (Takahashi et al., 2001) (see Fig. 1). The downstream signaling events have been the subject of intensive investigation (reviewed in Koch & Claesson-Welsh, 2012). Signaling through PLCγ seems most important to mediate the proliferative effects of VEGFR2 activation in endothelial cells. Experiments in cultured endothelial cells indicate that PLCγ does so in part through activation of ERK1/2 signaling (Takahashi et al., 2001) and through activation of protein kinase C (PKC) family members through generation of diacylglycerol and increases intracellular calcium levels (Wellner et al., 1999). While many aspects of PLCγ signaling in endothelial cells are yet to be elucidated, it is certain that PLCγ itself is of major importance for vascular development. *Plcg1* knockout mice (the gene encoding PLCγ1) die at E9.0 (Ji et al., 1997) with defects in erythropoiesis and vasculogenesis (Liao et al., 2002), a phenotype strikingly similar to that observed in VEGFR2 global knockout mice and in mice carrying the $Tyr^{1173}Phe$ point mutation that abrogates PLCγ binding. The physiological relevance of SHB and SCK binding to Tyr^{1175} is less clear. The phenotype of

SHB null mice is background dependent and does not result in severe vascular defects during embryogenesis. SHB does however seem to regulate certain aspects of vessel functionality in the adult and its deletion negatively impacted tumor angiogenesis and growth (Funa et al., 2009). SCK also seems dispensable for normal vascular development. SCK mutant mice are viable and do not develop obvious vascular defects. Instead SCK mutant mice are reported with neuronal defects—if this phenotype is related to the emerging role of VEGFR signaling in the nervous system remains to be elucidated (Sakai et al., 2000). Tyr^{1175} has further been implicated in phosphoinositide 3-kinase (PI3K) activation in vitro (Dayanir, Meyer, Lashkari, & Rahimi, 2001) and Graupera and colleagues have demonstrated that the p110alpha isoform is selectively required for angiogenesis and endothelial cell migration (Graupera et al., 2008). The pleiotropic roles of PI3K's in angiogenesis are reviewed in-depth in Soler, Angulo-Urarte, and Graupera (2015).

Tyr^{1214} has been implicated in the recruitment of a protein complex consisting of NCK, FYN, and PAK2, which conveys the signal to the CDC42/SAPK2/p38 MAPK module and leads to actin polymerization, stress fiber formation, and endothelial cell migration (Lamalice, Houle, & Huot, 2006) (see Fig. 1). However, the in vivo relevance of signaling downstream of Tyr^{1214} remains uncertain as mice in which the corresponding tyrosine is substituted with phenylalanine ($Tyr^{1212}Phe$) are viable and fertile (Sakurai et al., 2005).

2.9 A Soluble Variant of VEGFR2

More than a decade ago, a naturally occurring soluble truncated form of VEGFR2 (sVEGFR2) was detected in mouse and human plasma (Ebos et al., 2004). The role of this alternatively spliced VEGFR2 variant and its in vivo relevance has however remained somewhat elusive. Elevated levels of sVEGFR2 were detected in Malaria patients (Furuta, Kimura, & Watanabe, 2010), as well as in overweight and obese individuals (Silha, Krsek, Sucharda, & Murphy, 2005) but the relevance of these observations is not clear. Pavlakovic and colleagues suggested that sVEGFR2 might bind to and sequester VEGF-C and thus prevent VEGFR3-mediated lymphangiogenesis (Pavlakovic, Becker, Albuquerque, Wilting, & Ambati, 2010). These results indicate that possibly not all aspects of VEGFR2 biology require its kinase domain and activation of downstream signaling. However, further work is required to clarify the in vivo role of sVEGFR2.

VEGFR3

2.10 VEGFR3 Structure and Ligand Binding Properties

The overall structure of VEGFR3 is very similar to that of VEGFR1 and VEGFR2. The unprocessed precursor contains 7 immunoglobulin-like (IG) domains, a single transmembrane domain (TM), as well as a cytoplasmic tail that contains the split kinase domains (TK1, TK2), which are separated by a KID (see Fig. 1). However, in contrast to the other VEGFRs, the fifth IG domain of VEGFR3 undergoes proteolytic cleavage and the two resulting receptor parts are joined together through a disulfide bridge to create the 195-kDa active, membrane-bound receptor (Pajusola et al., 1994; Takahashi & Shibuya, 2005). VEGFR3 binds VEGF-C and VEGF-D. In their unprocessed forms both ligands have rather low affinity for VEGFR3, but proteolytic processing of the precursors increases their affinity substantially. The processed forms of VEGF-C and VEGF-D can also bind VEGFR2 but with lower affinity compared to VEGFR3 (Joukov et al., 1997; McColl et al., 2007; Stacker et al., 1999). In VEGFR3, the first and second IG domains are essential for ligand binding (of VEGF-C) (Jeltsch et al., 2006) (see Fig. 1). This is in contrast to VEGFR2, where ligand binding requires the second and third IG domains (Leppänen et al., 2010). Analysis of the crystal structure of VEGF-C in complex with the first and second IG domains of VEGFR3 identified a conserved ligand-binding interface in the second IG domain. The same study also described the crystal structure of a homodimer of the fourth and fifth IG domains of VEGFR3, which suggests that receptor dimerization requires homotypic interactions in the fifth IG domain (Leppänen et al., 2013). Accordingly, full activation of the receptor appears to require homotypic interactions in the fifth and, in addition, the seventh IG domains (Leppänen et al., 2013).

2.11 VEGFR3 Function

VEGFR3 is a key regulator of lymphatic development and function. This is demonstrated by the fact that naturally occurring mutations of VEGFR3 (*FLT4*) in humans cause lymphedema (Ghalamkarpour et al., 2006; Irrthum, Karkkainen, Devriendt, Alitalo, & Vikkula, 2000; Karkkainen et al., 2000, 2001). Patients with hereditary lymphedema type I (Milroy's disease), suffer from swellings of the extremities as defective cutaneous lymphatic vessels fail to drain lymphatic fluid. Hereditary lymphedema is an autosomal dominant trait, with incomplete penetrance, variable expression, and age of onset (Ferrell et al., 1998) that occurs in approximately 1:6000 births (Witte et al., 1998). In several families with hereditary lymphedema the disease has been linked to mutations in the kinase domain of VEGFR3 (Ghalamkarpour et al., 2006; Irrthum et al., 2000; Karkkainen et al., 2000), indicating that normal human lymphatic development is dependent on the kinase activity of VEGFR3. In addition, *Chy* mice, which were obtained through an ethylnitrosourea (ENU) mutagenesis approach carry a point mutation in the VEGFR3 kinase domain that renders the kinase inactive (Karkkainen et al., 2001). Like human patients with hereditary lymphedema, *Chy* mice have edema in the limbs due to hypoplastic cutaneous lymphatic vessels and have been proposed as an animal model for this disease (Karkkainen et al., 2001).

2.12 VEGFR3 Expression Pattern

During embryogenesis and early postnatal development, VEGFR3 is initially expressed in vascular endothelial cells, its expression declines however after their formation, and becomes restricted to lymphatic vessels (Kaipainen et al., 1995). VEGFR3 expression remains to some degree in certain quiescent endothelial cells, mainly in fenestrated capillaries (Partanen et al., 2000) and becomes upregulated during active angiogenesis, such as in tumor blood vessels (Valtola et al., 1999) and in angiogenic tip cells (Tammela et al., 2008). VEGFR3 is also expressed outside of the vasculature, in neuronal progenitors (Le Bras et al., 2006), osteoblasts (Orlandini et al., 2006), and macrophages (Schmeisser et al., 2006).

2.13 VEGFR3 in Development

Global deletion of VEGFR3 in mice revealed its essential role during development of the embryonic cardiovascular system, before the emergence of the lymphatic vessels. While vasculogenesis and angiogenesis occur normally in its absence, large vessels become abnormally organized with defective lumina and fluid accumulation in the pericardial cavity. These embryos die due to cardiovascular failure after E9.5 (Dumont et al., 1998). The critical role of VEGFR3 for lymphatic vessel development is demonstrated through a number of experiments. Overexpression of a soluble form of VEGFR3 in the skin inhibits embryonic lymphangiogenesis and induces the regression of already formed lymphatic vessels (Mäkinen, Jussila, et al., 2001). In addition, overexpression of VEGF-C Cys^{156}Ser in the skin (a VEGF-C variant that only binds to and activates VEGFR3), specifically induces the growth of lymphatic but not that of blood vessels suggesting that activation of VEGFR3 alone is sufficient to induce lymphangiogenesis

(Veikkola et al., 2001). In vitro work has further shown that VEGFR3 activation protects primary lymphatic endothelial cells from serum deprivation-induced apoptosis and promotes their growth and migration (Mäkinen, Veikkola, et al., 2001).

2.14 VEGFR3 Signaling

Signaling downstream of VEGFR3 is less well characterized than that downstream of VEGFR2. VEGF-C binding leads to phosphorylation of at least five tyrosine residues (Tyr^{1230}, Tyr^{1231}, Tyr^{1265}, Tyr^{1337}, and Tyr^{1363}) in the carboxyl-terminal tail of VEGFR3 in the case of VEGFR3 homodimer formation (Dixelius et al., 2003) (see Fig. 1). VEGF-C can also induce VEGFR2/VEGFR3 heterodimer formation, which results in phosphorylation of only three carboxyl-terminal tyrosine residues (Tyr^{1230}, Tyr^{1231}, and Tyr^{1265}). Thus, the two most carboxyl-terminal tyrosine residues in VEGFR3 are only substrate for VEGFR3 (Dixelius et al., 2003). It is further assumed, that at least two tyrosine residues in the kinase domain activation loop of VEGFR3 (Tyr^{1063} and Tyr^{1068}) are phosphorylated upon receptor activation (Dixelius et al., 2003; Salameh, Galvagni, Bardelli, Bussolino, & Oliviero, 2005). The position of those residues corresponds to Tyr^{1054} and Tyr^{1059} in VEGFR2, both of which are implicated in positive regulation of the receptors kinase activity. The precise role and in vivo relevance of most tyrosine residues in signal transduction downstream of VEGFR3 remains to be elucidated (see Fig. 1).

Tyr^{1337} is suggested to be a binding site for the SHC–GRB2 complex, which in turn may induce RAS activation, mitogenic signaling, and mediate transformation once the receptor is overexpressed in fibroblasts (Fournier, Dubreuil, Birnbaum, & Borg, 1995) (see Fig. 1). In accordance, the short VEGFR3 isoform which lacks part of the carboxyl-terminal tail (and thus Tyr^{1311}, Tyr^{1337}, and Tyr^{1363}) does not induce fibroblast transformation (Fournier et al., 1995) and mutation of Tyr^{1337} to phenylalanine ($Tyr^{1337}Phe$) reduces mitogenic signaling in human umbilical vein endothelial cells (HUVEC) (Salameh et al., 2005).

Tyr^{1230} and Tyr^{1231} have been suggested to contribute to proliferation, migration, and survival of HUVEC and to activate both protein kinase B (AKT) and ERK1/2 signaling. Tyr^{1063} was identified as the major phosphorylation site mediating survival and is proposed to associate with CRK I/II upstream of Jun N-terminal kinase (JNK) signaling (Salameh et al., 2005) (see Fig. 1).

VEGFR3 phosphorylation also leads to PI3K dependent activation of AKT as well as to PKC-dependent activation of ERK1/2 in vitro (Mäkinen, Veikkola, et al., 2001). Both signaling pathways are thought to mediate prosurvival effects for lymphatic endothelial cells. PI3K signaling downstream of VEGFR3 is supported by in vivo experiments. Inactivation of *Pik3r1*, the gene encoding three regulatory subunits (p58a, p55a, and p50a) of class 1A PI3Ks, results in accumulation of chylous ascites, indicating compromised lymphatic function (Fruman et al., 2000). Likewise, in vivo experiments have supported a requirement for AKT downstream of VEGFR3. However, the evidence is somewhat indirect. Single knockout mice lacking either *Akt1*, *Akt2*, or *Akt3* are viable and only *Akt1* knockout mice display a mild lymphatic phenotype with reduced lymphatic capillary diameter, decreased numbers of lymphatic endothelial cells, a partial loss of lymphatic valves and remodeling defects of collecting lymphatic vessels (Zhou et al., 2010). *Akt2* and *Akt3* double mutant mice are viable without a noticeable lymphatic phenotype (Zhou et al., 2010). In contrast, *Akt1* and *Akt3* double mutant mice are embryonic lethal before the emergence of lymphatic vessels, which precludes conclusions about their requirement during lymphangiogenesis (Yang et al., 2005). Taken together, these experiments suggest that the three AKT variants can partially compensate for the loss of each other but that AKT1 likely plays the primary role downstream of VEGFR3 in lymphatic development. In line with this hypothesis is the observation, that overexpression of VEGF-C in adult *Akt1* knockout mice induces lymphangiogenesis, likely through AKT2 and AKT3 (Zhou et al., 2010).

2.15 Heterodimer Formation of VEGFR2 With VEGFR1 and VEGFR3

An interesting aspect of VEGFR2 biology is the fact that its signaling properties can be modulated by the formation of heterodimers with either VEGFR1 or VEGFR3. Active VEGFR1/2 heterodimers have been identified in response to stimulation with VEGF-A (Huang, Andersson, Roomans, Ito, & Claesson-Welsh, 2001; Neagoe, Lemieux, & Sirois, 2005), and PlGF (Autiero et al., 2003). VEGF-A/PlGF heterodimerization (Cao et al., 1996; DiSalvo et al., 1995) might further lead to the formation of VEGFR1/2 heterodimers (Autiero et al., 2003). A computational study suggests that up to 50% of active VEGFR signaling complexes in endothelial cells are comprised of VEGFR1/2 heterodimers (Mac Gabhann & Popel, 2007). In some instances, VEGFR1/2 heterodimers do augment angiogenesis, while in some they may counteract angiogenic signaling (Eriksson et al., 2002), possibly due to altered downstream signaling (Cudmore et al., 2012).

VEGFR2/3 heterodimers are enriched in endothelial tip cells of growing angiogenic sprouts and can be activated by VEGF-C, and to a lesser extend by VEGF-A both of which promote angiogenesis, but not lymphangiogenesis in this context (Nilsson et al., 2010).

3. AN EVOLUTIONARY PERSPECTIVE ON VEGFR FUNCTION

VEGFRs are present in all Metazoan animals, from Cnidaria to humans, but are absent in fungi and plants (Kipryushina, Yakovlev, & Odintsova, 2015). Their role as key regulators of vascular development and function is highly conserved in vertebrates. However, far less is known about their diverse functions in invertebrates. Here we highlight some of these function and discuss a possible origin of VEGFRs in the animal kingdom. Beyond that, the interested reader is referred to reviews dedicated to that topic (Holmes & Zachary, 2005; Kipryushina et al., 2015).

Homologues of VEGFRs have been found in mammals, fish, birds, and amphibians (Stuttfeld & Ballmer-Hofer, 2009), but most of our knowledge of VEGFR function in vertebrates is based on experiments in mice. As described in detail earlier, VEGFR2 is crucial for most aspects of normal and pathological endothelial cell biology including hematopoiesis, blood vessel formation, and survival. VEGFR1 mainly functions as a decoy receptor for VEGF-A and thereby limits VEGFR2 activation, while VEGFR3 mainly regulates lymphatic development and function (Koch & Claesson-Welsh, 2012; Shibuya, 2013a). In zebrafish, four VEGFR-like genes have been identified: the VEGFR1 gene orthologue (*flt1*) (Bussmann, Bakkers, & Schulte-Merker, 2007; Rottbauer et al., 2005), the VEGFR3 gene orthologue (*flt4*) (Thompson et al., 1998), and two genes (*kdra* and *kdrb*) with highest similarity to the VEGFR2 gene (Bahary et al., 2007; Bussmann et al., 2007; Liao et al., 1997; Thompson et al., 1998). While it was initially believed that *kdra* and *kdrb* are the result of a teleost gene duplication, recent work suggests instead that *kdra* represents a forth class of vertebrate *VEGFRs* and it was therefore proposed to rename it *kdr-like*. Accordingly, it was suggested to rename the true VEGFR2 gene orthologue *kdrb* to *kdr* (Bussmann, Lawson, Zon, Schulte-Merker, & Zebrafish Nomenclature Committee, 2008). It is most likely that also the fourth VEGFR gene representative *kdr-like* (*kdra*), was present in the common ancestor of fish and mammals and that it was later lost specifically in placental mammals (eutherians), while it is still present in marsupials and platypus (monotremata) (Bussmann et al., 2008). However, although *kdr* (*kdrb*) and *kdr-like* (*kdra*) represent separate classes of VEGFR genes, they are both expressed in endothelial cells and are both activated by VEGF-A (Bahary et al., 2007). Both genes *also* function partially redundantly in blood vessel development as knock-down of *kdr* (*kdrb*) in a *flk1/kdr-like* (*kdra*) mutant background results in phenotypes similar to those observed when *vegf-A* or *phospholipase-cc1* (a common downstream signaling component) are knocked-down (Covassin, Villefranc, Kacergis, Weinstein, & Lawson, 2006). Knock down of the zebrafish VEGFR1 gene orthologue *flt1* results in increased tip cell numbers, increased angiogenesis, and hyperbranching of segmental arterial sprouts. Conversely, overexpression of its soluble form leads to a shortening of arterial sprouts and a reduction in filopodia number (Krueger et al., 2011). As in mammals, lymphatic development in zebrafish requires VEGF-C and VEGFR3 signaling (Küchler et al., 2006). Orthologues of VEGFR1–3 genes and a fourth *kdr-like* gene have also been identified in birds and amphibians (Bussmann et al., 2008). These results suggest that molecular mechanisms regulating hematopoiesis and vascular development have been conserved throughout vertebrate evolution.

The most primitive phylum in which components of the VEGF signaling system have been reported is Cnidaria. These animals are considered the simplest organisms at the tissue level of organization. Their body plan is characterized by radial symmetry and made up from only two germ layers, ectoderm and endoderm, which are separated by the mesoglea. Homologues of VEGF and VEGFR genes are reported in two classes of this phylum, Anthozoa (the jellyfish *Podocoryne carnea*) and Hydrozoa (the fresh water polyp *Hydra vulgaris*), where they are implicated in regulation of tube formation and development of nerve cells (Krishnapati & Ghaskadbi, 2013; Seipel et al., 2004). The evolutionary origin of the VEGF signaling pathway therefore appears to reside in the common ancestor of the Cnidaria and Bilateria (Seipel et al., 2004). Bilateria are subdivided into Protostomia and Deuterostomia. Within the Protostomia components of the VEGF signaling system have been reported in Annelida, Mollusca, Nematoda, and Arthropoda (Kipryushina et al., 2015). One of the best-studied representatives of the phylum Annelida is the leech *Hirudo medicinalis*, who has a closed circulatory system that consists of four blood vessels and capillary-like structures. Tettamanti and colleagues reported the existence of a VEGF-like molecule and homologues of VEGFR1 and VEGFR2 genes in this species and suggested a role for the VEGF/VEGFR system in neoangiogenesis

(Tettamanti, Grimaldi, Valvassori, Rinaldi, & de Eguileor, 2003). Studies that investigate the presence and function of VEGFRs in Mollusca are rare. It appears however, that Mollusca have a PDGFR/VEGFR-like molecule that is most similar to *Drosophila pvf1* (Setiamarga et al., 2013; Yoshida, Shigeno, Tsuneki, & Furuya, 2010). In *Caenorhabditis elegans (C. elegans)*, the best studied Nematode, four VEGFR-like genes (*ver1–4*) have been identified. *C. elegans* lacks a circulatory system and the expression of *ver1–3* in cells of neural origin suggests a function in neurogenesis (Popovici, Isnardon, Birnbaum, & Roubin, 2002). In addition, a single growth factor (PVF-1) with resemblance to PDGF/VEGF is described in *C. elegans*. PVF-1 is able to bind to human VEGFR1 and VEGFR2 and induces angiogenesis in the chicken chorioallantoic membrane assay (Tarsitano, De Falco, Colonna, McGhee, & Persico, 2006). Within Arthropoda, VEGF and VEGFR gene homologues have been identified in Crustacea and Hexapoda. In Crustacea the expression of VEGF- and VEGFR-like molecules suggests a neuronal function (Fusco et al., 2014) and a regulatory role in the immune response as well as the neuroendocrine system (Li et al., 2013). Within the subphylum Hexapoda VEGF signaling has been best characterized in *Drosophila melanogaster*, where the pathway controls cardiovascular-related and -unrelated functions. In *Drosophila*, three VEGF homologues (PVF-1–3) and one receptor (PVR), that appears equally related to PDGFR and VEGFR, have been described. *Drosophila* has an open circulatory system, and the hemolymph, a fluid analogous to the blood in vertebrates, is in direct contact with the tissue. Two key cardiovascular-related functions of the PVR/PVF system are to control the migration and survival of hemocytes during embryogenesis (Brückner et al., 2004; Cho et al., 2002; Heino et al., 2001). In addition, the PVF/PVR pathway is reported to control cardiac valve formation (Zeitouni et al., 2007). Outside of the cardiovascular system, PVF/PVR has been shown to control border cell migration (Duchek, Somogyi, Jékely, Beccari, & Rørth, 2001), nervous system development (Sears, Kennedy, & Garrity, 2003), and condensation of the nervous system (Olofsson & Page, 2005). However, the later two functions are a secondary consequence of defective hemocyte development (Olofsson & Page, 2005; Sears et al., 2003). VEGF and VEGFR gene homologues have also been found in Echinodermata, where they play a crucial role in gastrulation (reviewed in Adomako-Ankomah & Ettensohn, 2014; Kipryushina et al., 2015), in Tunicata, where they are implicated in vessel sprouting (Tiozzo, Voskoboynik, Brown, & De Tomaso, 2008) and in ascidians (Samarghandian & Shibuya, 2013).

In conclusion, the evolutionary origin of VEGFs and VEGFRs appears to reside in the common ancestor of Cnidaria and Bilateria. In vertebrates, the critical role of the VEGF/VEGFR signaling pathway in controlling vascular development and function is highly conserved. In invertebrates, VEGF/VEGFR signaling has been shown to regulate numerous developmental processes, some of which are related to cardiovascular development and function, while others are clearly not. Unfortunately, studies in invertebrates are still rare and the ancestral functions of the VEGF/VEGFR signaling system therefore remain only partially characterized. However, without doubt, the VEGF/VEGFR system also plays important roles outside the vasculature, for example, in the nervous system (reviewed in Mackenzie & Ruhrberg, 2012; Wittko-Schneider, Schneider, & Plate, 2013). One common theme of VEGF/VEGFR signaling that seems required for the majority of its reported functions in both vertebrates and invertebrates appears to be regulation of cell migration (Holmes & Zachary, 2005). Further studies in both invertebrates and vertebrates are necessary to obtain a more complete picture of the various nonvascular functions that are controlled by VEGF/VEGFR signaling.

4. NEUROPILINS: CORECEPTORS MODULATING VEGFR SIGNALING

Signaling properties of VEGFRs are modulated in part through interaction with their coreceptors: neuropilin (NRP)1 and NRP2 (see Fig. 2). During development, NRP1 is predominantly expressed in arteries, while NRP2 is found in veins and lymphatic endothelial cells (Eichmann, Makinen, & Alitalo, 2005; Herzog, Kalcheim, Kahane, Reshef, & Neufeld, 2001). NRPs do not have internal catalytic activity (Pellet-Many, Frankel, Jia, & Zachary, 2008) but instead modulate VEGFR internalization, intracellular trafficking, and downstream signaling (Prahst et al., 2008). NRP1 and NRP2 share 44% sequence identity. Both proteins contain a large extracellular part and a short cytoplasmic tail and share a similar domain structure (Chen, Chédotal, He, Goodman, & Tessier-Lavigne, 1997). The extracellular part of NRP1 is composed of five different subdomains (a1, a2, b1, b2, and c) (see Fig. 2). The two N-terminal domains, a1 and a2, are homologous to the complement system components C1r and C1s (CUB domain) followed by the b1 and b2 domain that resemble coagulation Factors V/VIII (FV/VII domains) and the C-terminal c domain that has homology with meprin, A5/NRP1, protein tyrosine-phosphatase μ (MAM domain). The CUB domains (a1-a2) mediate binding to class 3 semaphorins, the FV/VIII domains (b1-b2) mediate the binding to VEGFs, and the MAM domain (c), together with the transmembrane domain, confers oligomerization capability. The short cytoplasmic tail ends with the three amino acid C-terminus Ser-Glu-Ala, also called SEA motive that enables binding to PDZ (PSD-95/DLG/ZO-1) domain-containing proteins, such as GIPC1 (GAIP-interacting protein, C-terminus, and neuropilin-binding protein 1; also known as synectin) (Gu et al., 2002; Koch, Tugues, Li, Gualandi, & Claesson-Welsh, 2011).

Deletion of *Nrp1* in mice results, among other defects, in impaired vascular development and embryonic lethality around E13.5 (Kawasaki et al., 1999; Kitsukawa et al., 1997). Endothelial cell-specific deficiency of *Nrp1* is also embryonic lethal, due to defective vascularization and abnormal vascular morphology (Fantin et al., 2013; Gu et al., 2003), which demonstrates that endothelial cell expression of NRP1 is critical for developmental angiogenesis. In contrast, NRP2-deficient mice display no overt developmental phenotype of blood vasculature, but instead show abnormal development of distinct parts of the lymphatic vasculature (Yuan et al., 2002).

Specific VEGF-A isoforms, such as VEGF-A_{165}, can recruit NRP1 into the VEGF-A/VEGFR2 signaling complex. NRP1 association modulates VEGFR2 internalization, intracellular trafficking, and RTK-downstream signaling, which is mediated by binding of the NRP1-SEA motive to GIPC1 (Prahst et al., 2008). Guidance of VEGFR2 intracellular trafficking by NRP1 appears to be important for, e.g., arteriogenesis (Lanahan et al., 2010, 2013). Binding of VEGF-A to NRP1 has been studied in detail and two independent, single amino acid mutant NRP1 proteins have been engineered, aimed to inhibit VEGF-A binding to NRP1 without affecting the normal expression and binding of semaphorin to NRP1. Mice expressing the Tyr^{290}-Ala mutation survive past birth and have only minor vascular phenotypes, possibly in part due to *Nrp1*-hypomorphism (Fantin et al., 2014). The second mouse model expressing $Asp^{320}Lys$ mutation also survives after birth and displays no overt vascular phenotypes (Gelfand et al., 2014). Thus, it appears that direct binding of VEGF-A to NRP1 is dispensable for developmental angiogenesis. However, endothelial NRP1 itself is indispensible for vascular development, since mice with endothelial deficiency of *Nrp1* die prenatally. The precise mechanisms underlying these phenotypes remain to be completely deciphered. The interested reader is referred to excellent, recent reviews for more in-depth reading (Koch, 2012; Kofler & Simons, 2016; Zachary, 2014).

5. VEGFR LIGANDS
5.1 Different Ligands for Different Receptors

VEGFR signaling is controlled by the availability of its ligands and it is difficult to understand its complexity without having a closer look at the mechanisms that control ligand expression, tissue retention, diffusion range, and proteolytic modification. The mammalian family of VEGFs is comprised of five related glycoproteins, namely VEGF-A, VEGF-B, VEGF-C, VEGF-D, and PlGF, which are expressed from five different genes, respectively. These ligands bind in partially overlapping patterns to three different tyrosine kinase receptors; VEGFR1–3. VEGF-A binds to VEGFR1 and VEGFR2, while VEGF-B and PlGF bind specifically to VEGFR1 and VEGF-C, and VEGF-D bind to VEGFR3 and in a proteolytically processed form also VEGFR2 (reviewed by Koch et al., 2011) (see Fig. 2). Nonmammalian proteins with related structures and function to the mammalian VEGFs are VEGF-E (parapox virus *orf*) (Ogawa et al., 1998) and snake venom-derived VEGF-F (Suto, Yamazaki, Morita, & Mizuno, 2005).

5.1.1 VEGF-A

VEGF-A is the prototypical and most-studied VEGF-family member. Its key role during hematopoiesis, vasculogenesis, angiogenesis, and vascular permeability are mainly mediated by signaling through VEGFR2 (reviewed in Koch & Claesson-Welsh, 2012). Despite the fact that VEGF-A binds with higher affinity to VEGFR1, it only induces limited downstream signaling of the receptor. VEGFR1 is, for most aspects of vascular development, considered to be a decoy receptor for VEGF-A, which limits its availability for VEGFR2 and thereby restricts activation and signaling of the VEGF-A/VEGFR2 axis (the so called *decoy* or *sink*-hypothesis) (Hiratsuka et al., 2005; Shibuya, 2006). Due to its potency, VEGF-A levels are tightly regulated during embryonic development. This is illustrated by the fact that VEGF-A is haploinsufficient, as mice with deletion of only one *Vegfa* allele die prenatally due to severe vascular defects (Carmeliet et al., 1996; Ferrara et al., 1996). Similarly, mice deficient for VEGFR2, the main VEGF-A signaling receptor, are early embryonic lethal due to insufficient vascular development (Shalaby et al., 1995). In contrast, deletion of the decoy receptor VEGFR1 causes endothelial overgrowth, which results in vascular disorganization and embryonic lethality (Fong et al., 1995).

5.1.2 PlGF and VEGF-B

Signaling of both VEGFR1 specific ligands, PlGF (De Falco, 2012; Park et al., 1994; Sawano et al., 1996) and VEGF-B (Aase et al., 2001; Bellomo et al., 2000; Dijkstra et al., 2014), is dispensable during development. However, PlGF has been implicated in pathological situations, such as cancer or preeclampsia (Carmeliet et al., 2001; Dewerchin & Carmeliet, 2012). VEGF-B was shown to be neuroprotective (Falk et al., 2011; Poesen et al., 2008) and its overexpression, especially in the heart gives rise to varying phenotypes from ectopic lipid accumulation and hypertrophy (Kärpänen et al., 2008) to vascular growth and ischemia resistance (Kivelä et al., 2014). Furthermore, an intriguing function of VEGF-B in endothelial transcytosis of dietary fatty acids has been described (Hagberg et al., 2010) with possible implications for the management of metabolic diseases (Hagberg et al., 2012; Mehlem et al., 2016).

5.1.3 VEGF-C and -D

VEGF-C and -D bind to VEGFR3 and are primarily involved in lymphatic development and maintenance (Aspelund, Robciuc, Karaman, Makinen, & Alitalo, 2016; Nurmi et al., 2015). *Vegfc*-deficient mice lack lymphatic vessels and die in utero due to severe tissue edema (Karkkainen et al., 2004). Lack of VEGF-D, however, is dispensable for lymphatic development (Baldwin et al., 2005). Lately, VEGF-C/VEGFR3 (and VEGF-C/VEGFR2) (Jeltsch et al., 2014; Joukov et al., 1996; Nilsson et al., 2010) signaling has been implicated in sprouting angiogenesis, especially when the availability of VEGF-A/VEGFR2 signaling is limited (Benedito et al., 2012; Tammela et al., 2008), although the precise mechanism and the physiological relevance of this function of VEGF-C remain to be clarified (Villefranc et al., 2013). In a recent study, an unexpected, vital function of VEGF-C during initial fetal erythropoiesis was revealed. However, the corresponding receptor still remains to be defined (Fang et al., 2016).

5.2 Regulation of VEGF-A Expression

Most parenchymal cells express VEGF-A that subsequently acts in a paracrine manner to stimulate VEGFR2 on neighboring endothelial cells. The major regulator of *VEGFA* promoter activation is tissue oxygen tension. Low tissue oxygen concentration (hypoxia) stabilizes the transcription factor hypoxia-inducible factor (HIF)1α that then binds to hypoxia-response elements located within the *VEGFA* promoter region and activates gene transcription. HIF1α-dependent regulation of *VEGFA* expression governs physiological VEGF-A-mediated endothelial cell proliferation, migration, and differentiation during embryonic development; vascularization of growing tissues and wound healing; as well as pathologic angiogenesis, as seen in eye disease and cancer (Pugh & Ratcliffe, 2003). Other factors regulating *VEGFA* expression are members of the ETS (gene transduced by leukemia virus E26) family of transcription factors (Randi, Sperone, Dryden, & Birdsey, 2009), reactive oxygen species (Ushio-Fukai & Nakamura, 2008) and peroxisome proliferator activated receptor (PPAR)γ coactivator (PGC)-1α (Arany et al., 2008), separately or in concert with HIF1α.

5.3 VEGF-A Splice Variants

Different splice isoforms are described for VEGF-A, PlGF, VEGF-B, and mouse VEGF-D. Here we focus on VEGF-A, where the variety and functional difference among isoforms is the most intriguing. The human (and mouse) *VEGFA* gene contains 8 exons is located on chromosome 6 (murine *Vegfa* is located on chromosome 17) and is translated into various isoforms by differential splicing. The isoforms are denoted by the amino acid count of the secreted polypeptide, such as VEGF-A$_{111}$, VEGF-A$_{121}$, VEGF-A$_{145}$, VEGF-A$_{162}$, VEGF-A$_{165}$, VEGF-A$_{183}$, VEGF-A$_{189}$, and VEGF-A$_{206}$. Of those, VEGF-A$_{121}$, VEGF-A$_{165}$, and VEGF-A$_{189}$ represent the most abundantly expressed and studied isoforms. The murine *Vegfa* gene is spliced in a similar fashion to the human gene and most human isoforms are also found in mice with high homology. Notably, the sequence of murine VEGF-A is one amino acid shorter (due to an extra amino acid in exon 2 of the human VEGF-A gene), resulting in, e.g., VEGF-A$_{120}$, VEGF-A$_{164}$, and VEGF-A$_{188}$ (reviewed by Vempati, Popel, & Mac Gabhann, 2014).

Early studies showed that VEGF-A$_{165}$ is the most important VEGF-A isoform. Mice expressing only VEGF-A$_{120}$ (VEGF-A$_{121}$ in humans) have severe phenotypes, such as a 50% postnatal lethality due to defective cardiac vascularization (Carmeliet et al., 1999) and defective angiogenesis (Ruhrberg et al., 2002; Stalmans et al., 2002). Mice only expressing VEGF-A$_{188}$ (VEGF-A$_{189}$ in humans) have insufficient arterial outgrowth in the retina, dwarfism, and impaired bone growth (Maes et al., 2004). The sole expression of VEGF-A$_{164}$ (VEGF-A$_{165}$ in humans), in contrast, gives rise to mouse offspring that appears normal (Carmeliet, 2000; Stalmans et al., 2002). Tumor cells engineered to express isolated VEGF-A isoforms show different vascular recruitment and maintenance in vivo (Grunstein, Masbad, Hickey,

Giordano, & Johnson, 2000; Kanthou et al., 2014; Tozer et al., 2008). VEGF-A_{120} expression leads to capillary formation with increased diameter and fewer branch points, while expression of VEGF-A_{188} results in denser capillary networks with lower caliber and less hemorrhage. Those examples demonstrate the differential effect of the VEGF-A isoforms with respect to vascular morphogenesis (Grunstein et al., 2000; Ruhrberg et al., 2002).

All isoforms contain the sequence encoded by exons 1–5 that codes for the dimerization, receptor recognition, and plasmin cleavage site, also referred as the core growth factor domain. The only exception is the VEGF-A_{111} isoform found in cancer cells and induced by genotoxic agents, that is composed of exon 1–4 and 8 (Mineur et al., 2007). The cysteine-knot motif present in the core growth factor domain of VEGF-A conveys intra- and interchain disulfide bridges that link two chains to form antiparallel homo- (and potentially also hetero-) dimers that enable dimerization and autophosphorylation of VEGFR1 and VEGFR2 (Grünewald, Prota, Giese, & Ballmer-Hofer, 2010; Muller, Christinger, Keyt, & de Vos, 1997).

Incorporation of exon 6–7, coding for basic amino acid stretches that mediate binding to heparan sulfate proteoglycans (HSPG), extracellular matrix (ECM), and the coreceptor NRP1 differs to various degrees among the isoforms (Ruhrberg et al., 2002). VEGF-A_{121}, VEGF-A_{165}, and VEGF-A_{189} contain exon 8a that codes for a six amino acid C-terminus with the feature of a positive charged C-terminal arginine (Arg) exhibiting direct binding competence to the coreceptor NRP1 (Parker, Xu, Li, & Vander Kooi, 2012). Taken together the differential splicing of VEGF-A produces a multitude of VEGF-A isoforms that bear distinct characteristics implemented in the complex biological function of VEGF-A.

Especially the HSPG-binding domain encoded by exon 6–7 changes the properties of the respective isoform(s) for ECM or cell surface retention and affinity for the NRP1 coreceptor. VEGF-A_{189} contains both, exon 6a and 7, rendering it the strongest HSPG/ECM binding isoform, viewed as nondiffusible, while VEGF-A_{121} that lacks both exons 6 and 7 has the weakest binding to HSPG/ECM and is thus regarded as freely diffusible isoform. VEGF-A_{165} that contains only exon 7 exhibits intermediate retention in the ECM and is considered a dual, matrix bound, and diffusible isoform (Grunstein et al., 2000). These different matrix and cell surface retention characteristics were suggested to participate in the generation of VEGF-A gradients. During angiogenic processes these VEGF-A gradients are important for the correct patterning of the growing vasculature

(Gerhardt et al., 2003). The degree of HSPG/ECM retention of VEGF-A isoforms defines the slope of the VEGF-A gradients (Ruhrberg et al., 2002; Ruiz de Almodovar et al., 2010). Isoforms with strong heparin binding ability, such as VEGF-A$_{189}$, have a high concentration around the producing cell and a steep gradient, while diffusible isoforms, such as VEGF-A$_{121}$, can travel longer in the interstitial space and form shallower gradients compared to the HSPG-binding isoforms. Importantly, the shape and composition of the VEGF-A gradients are crucial determinants for vascular morphogenesis. However, more mechanisms than protein secretion and retention seem to be required to refine the VEGF-A gradients (Vempati, Popel, & Mac Gabhann, 2011). Proteolytic release and degradation are two additional processes representing powerful regulatory properties for the regulation of VEGF-A activity (Ferrara, 2010; Lee, Jilani, Nikolova, Carpizo, & Iruela-Arispe, 2005). Furthermore, matrix bound or freely diffusible VEGF-A-binding proteins, such as sVEGFR1 (see earlier), alpha-2-macroglobulin (Bhattacharjee, Asplin, Wu, Gawdi, & Pizzo, 2000), or thrombospondin 1 domain-containing proteins (Luque, Carpizo, & Iruela-Arispe, 2003) can modulate the activity of VEGF-A and assist to construct the gradient of VEGF-A (Gupta, Gupta, Wild, Ramakrishnan, & Hebbel, 1999; Luque et al., 2003).

5.4 The Antiangiogenic VEGF-A$_{xxxb}$ Isoforms

Almost one and a half decade ago yet another layer of complexity was added to VEGF-A biology when Bates and colleagues first described VEGF-A$_{xxxb}$ splice isoforms that arise through differential splicing of exon 8 (Bates et al., 2002). Interestingly, those splice isoforms were found to be antiangiogenic and have mainly been studied in the context of human cancer (Harper & Bates, 2008). Only, a few years ago the physiological relevance of this special subclass of VEGF-A isoforms was still in question, mainly because they had not been identified in mice (Dokun & Annex, 2011; Harris et al., 2012). However, recently, the existence and functionality of this special VEGF-A isoform(s) was also demonstrated in mice. Work by Kikuchi and colleagues link peripheral artery disease to the expression of VEGF-A$_{165b}$ and show that specific neutralization of the antiangiogenic VEGF-A$_{165b}$ can restore neovascularization in a mouse model of hind limb ischemia (Kikuchi et al., 2014).

Differential splicing of exon 8 of *VEGFA* by selection of the distal splice site (DSS) instead of the proximal splice site (PSS) leads to the expression of

VEGF-A$_{xxxb}$ isoforms (also called antiangiogenic isoforms) that contain exon 8b instead of 8a. This results in a changed amino acid sequence of the C-terminus (Cys-Asp-Lys-Pro-Arg-Arg or Ser-Lys-Thr-Arg-Lys-Asp in VEGF-A or VEGF-A$_{xxxb}$, respectively) (reviewed in Qiu, Hoareau-Aveilla, Oltean, Harper, & Bates, 2009). As a consequence VEGF-A$_{xxxb}$ isoforms lack the ability to bind to NRP1 and lose one cysteine for disulfide bridge generation, possibly leading to a miss-folded HSPG-binding domain in, e.g., VEGF-A$_{165b}$ and VEGF-A$_{189b}$. Despite similar binding affinity of VEGF-A$_{xxxb}$ isoforms to VEGFR2 (see Fig. 2), the activation of the intracellular kinase domain differs from the response to VEGF-A, e.g., Tyr^{1054}, which is important for stimulation of the internal kinase activity of VEGFR2 (see Fig. 1), is not phosphorylated in response to VEGF-A$_{xxxb}$ (Kawamura, Li, Harper, Bates, & Claesson-Welsh, 2008). As a consequence, VEGF-A$_{xxxb}$ binding to VEGFR2 results in a weaker and more transient activation of the VEGFR2 downstream effector kinases (Catena et al., 2010; Cébe Suarez et al., 2006). Using different in vivo models, it has been shown that overexpression of VEGF-A$_{165b}$ can counteract the effect of VEGF-A (Kikuchi et al., 2014); however, the physiological regulation of VEGF-A$_{xxxb}$ and its spatial-/temporal modulation of VEGF-A activity remains to become fully deciphered.

The underlying mechanisms for VEGF-A isoform splicing, both in general and at the 3′ end for pro- vs antiangiogenic isoforms is still elusive (Biselli-Chicote, Oliveira, Pavarino, & Goloni-Bertollo, 2012). However, recently several factors have been shown to be involved in the VEGF-A isoform splice selection, particularly for the selection of exon 8a and 8b. Serine/arginine protein kinase 1 (SRPK1) activates serine/arginine splicing factor 1 (SRSF1, also known as ASF/SF2) that favors the splicing of the PSS, thus augmenting expression of proangiogenic VEGF-A isoforms (Nowak et al., 2010, 2008; Oltean, Gammons, et al., 2012). SRSF2 and SRSF6 (also known as Srp30b and SRp55, respectively) promote the splicing of the DSS, increasing the expression of the antiangiogenic VEGF-A$_{xxxb}$ isoforms (Merdzhanova et al., 2010; Nowak et al., 2008).

In a high-throughput screen SRSF1 together with H3K9 methyltransferase (EHMT2) and the chromatin modulator HP1γ was identified to promote the inclusion of *VEGFA* exon 6a (Salton, Voss, & Misteli, 2014).

The zinc finger transcription factor WT1 (Wilms tumor suppressor 1) was found to bind and repress the *SRPK1* promoter, hence promoting the splicing of antiangiogenic VEGF-A$_{xxxb}$ isoforms (Amin et al., 2011).

Furthermore, different isoforms of WT1 have distinct effects on proangiogenic VEGF-A expression (McCarty, Awad, & Loeb, 2011) and isoform splicing (Katuri et al., 2014). Loss-of-function mutations in *WT1* that are related to acute myeloid leukemia result in a shift of VEGF-A isoform expression toward VEGF-A$_{121}$ in hematopoietic progenitor cells that due to the change in VEGF-A isoform balance undergo apoptosis (Cunningham, Palumbo, Grosso, Slater, & Miles, 2013).

Under physiological conditions, VEGF-A$_{xxxb}$ isoforms are found in most human tissues (Bevan et al., 2008; Perrin et al., 2005). One exception is the placenta, in which physiological angiogenesis is continuously ongoing and where VEGF-A$_{xxxb}$ protein levels are low (Bevan et al., 2008). Information about the role of the antiangiogenic VEGF-A$_{xxxb}$ isoforms during development is still limited. However, one can make predictions based on the reported studies. It seems likely that VEGF-A$_{xxxb}$ regulates and limits VEGF-A-induced angiogenic processes and promotes the maturation of vascular structures as well as other specialized tissue-cells, such as podocytes of the kidney glomeruli (Bevan et al., 2008; Oltean, Neal, et al., 2012) or epithelial cells in the remodeling mammary gland (Qiu et al., 2008). Based on this prediction, one would hypothesize that the splicing of VEGF-A$_{xxxb}$ isoforms is low during the initial phases of vasculogenesis and angiogenesis and gradually increases with the establishment of a functional vascular tree to promote maturation, stability, and quiescence. Dysfunctional regulation of the VEGF-A splicing events can be seen in many pathological situations that are linked with abnormal angiogenic processes, e.g., in cancer or eye disease, impaired fertility, and pregnancy, as well as systemic sclerosis (reviewed in Peiris-Pagès, 2012).

5.5 Proteolytic Processing as Regulatory Mechanism

In addition to differential splicing, which results in various VEGF-A isoforms, there is an additional layer of complexity, namely proteolytic processing of those isoforms. Various proteases have been described to cleave VEGF-A isoforms, which can result in either activation, degradation, or liberation from extracellular stores and thus affect VEGF-A signaling.

The best-described proteases processing VEGF-A are the serine proteases plasmin, elastase (neutrophil) (Kurtagic, Jedrychowski, & Nugent, 2009), and urokinase-type plasminogen activator (uPA) (Plouët et al., 1997). In addition, matrix metalloproteinases (MMPs), particularly MMP3, have been implicated in VEGF-A processing (Lee et al., 2005;

Roy, Zhang, & Moses, 2006). However, it is still debated whether MMPs can directly cleave VEGF-A, especially since mouse and human VEGF-A differ highly in their MMP cleavage site sequence (amino acids 110–114). The cleavage site for plasmin is located in exon 5 (R^{109}/T^{110} or R^{110}/A^{111} in mouse or human VEGF-A, respectively), implying that processing of VEGF-A by plasmin can occur in all isoforms with the exception of VEGF-A_{111} and remove potential HSPG/ECM-binding sites as well as the C-terminal NRP1-binding motif, resulting in VEGF-A_{109} in mice or VEGF-A_{110} in humans, respectively (Keyt et al., 1996). In contrast, uPA cleaves VEGF-A in the distal part of exon 6a encoded sequences and is therefore only processing HSPG domain-containing VEGF-A isoforms (VEGF-A_{145}, VEGF-A_{162}, VEGF-A_{183}, VEGF-A_{189}, and VEGF-A_{206}) (Plouët et al., 1997). The removal of the C-terminal part by cleavage of uPA is thought to partially liberate matrix-bound VEGF-A from the ECM, and in parallel removing the NRP1-binding motifs.

Sequences encoded by exon 6a mediate the strong binding of VEGF-A_{189} to the ECM and might additionally render the isoform a weaker mitogen as compared to VEGF-A_{165}, despite a 10-fold higher affinity to NRP1 (Lee, Folkman, & Javaherian, 2010; Vintonenko et al., 2011). Proteolytic removal of parts or the entire HSPG-binding domains (exon 6–7) releases VEGF-A_{189} from the ECM and removes the inhibitory sequences encoded by exon 6a, and thereby increases its activity (Plouët et al., 1997), while it simultaneously removes the binding sites for the coreceptor NRP1. This results in a weak VEGF-A isoform, compared to VEGF-A_{165}; however, more potent than the matrix bound, unprocessed VEGF-A_{189}. Strong evidence for the differential function and regulation of VEGF-A_{189} activity from in vivo studies not utilizing exogenous overexpression of the VEGF-A_{189} protein are still lacking. Nevertheless, this example nicely demonstrates one part of the complexity of VEGF-A regulation.

Furthermore, indirect processing by proteases can liberate VEGF-A proteins without changing their internal characteristics. Digestion of heparan sulfate chains or core proteins by heparanase or MMPs, respectively, frees bound VEGF-A proteins from their ECM stores, preserving their amino acid sequence and internal activity(s) (Hawinkels et al., 2008; Purushothaman et al., 2010; Rodriguez-Manzaneque et al., 2001).

Taken together the direct cleavage of VEGF-A or the liberation form extracellular stores participates in the generation and maintenance of VEGF-A gradients. Release from extracellular stores is widely accepted to increase the bioavailability and thus activity of VEGF-A. In addition, cleavage at additional, N-terminal sites by plasmin or elastase has been shown to result in inactivation or degradation of VEGF-A (reviewed in Vempati et al., 2014).

6. PERSPECTIVE

Here we have summarized some of the key aspects of VEGFR signaling during development of the vascular system with a focus on experiments in mice. The crucial role of this pathway in hematopoiesis, vasculogenesis, angiogenesis, vascular permeability, endothelial cell proliferation, and survival make it also of central importance in numerous human diseases. However, as excellent reviews dedicated to the role of VEGFR signaling in disease recently have been published (Aspelund et al., 2016; Dieterich & Detmar, 2016; Ferrara & Adamis, 2016; Kieran, Kalluri, & Cho, 2012; Shibuya, 2014) we do not attempt to cover this topic and consequently only want to highlight the role of the VEGF/VEGFR systems in tumor angiogenesis and in the tumor-associated lymphatic vasculature as examples of particular interest.

In 1971, Folkman proposed that tumors secrete factors that promote angiogenesis in their surroundings and so attract blood vessels that guarantee a steady source of nutrients for continuous tumor growth (Folkman, 1971; Folkman, Merler, Abernathy, & Williams, 1971). Later, the VEGF/VEGFR system was found to be overrepresented in different tumors (Millauer, Shawver, Plate, Risau, & Ullrich, 1994; Plate, Breier, Millauer, Ullrich, & Risau, 1993). Tumor angiogenesis has since been a target for anticancer therapy and the FDA approval of bevacizumab (Avastin®), a monoclonal antibody against VEGF-A, for the treatment of metastatic colon cancer in 2004 represented the first proof of principle for this approach and generated huge expectations in this research area. Since then, enormous efforts have been made to establish antiangiogenic treatment as a first line strategy in oncology. This includes protein-based VEGF-trap molecules that bind and sequester VEGF-A and potentially other growth factors. Apart from targeting VEGF-A, small molecule inhibitors against VEGFRs have been developed and tested in clinical trials. However, often occurring resistance to antiangiogenic therapy and severe side effects of systemic anti-VEGF-A/VEGFR2 treatment highlight the urge to develop more refined drugs for antiangiogenic cancer therapy (reviewed by Ferrara & Adamis, 2016).

In addition to the well-established role of VEGF-A in tumor angiogenesis, there is evidence that PlGF-dependent activation of VEGFR1 may promote tumor growth. The inhibition of PlGF/VEGFR1 signaling might therefore represent an additional option in cancer therapy (Dewerchin & Carmeliet, 2012).

Additionally, VEGF-C and VEGF-D signaling through VEGFR3 is emerging as a target for anticancer treatment. Not only does the proangiogenic role of VEGF-C/VEGFR3 (and potentially VEGF-C/VEGFR2) signaling represents the rational for these attempts, but also the crucial role of lymphatic recruitment, cooption, and lymphangiogenesis in tumor tissue, which are induced by VEGF-C- or VEGF-D-dependent activation of VEGFR3 (reviewed by Dieterich & Detmar, 2016). Furthermore, the lymphatic system is thought to be a main route for tumor metastasis and thus has emerged as a target for anticancer therapy (reviewed by Karaman & Detmar, 2014).

Our knowledge of the mechanisms through which VEGF/VEGFR signaling regulates vascular development and function has dramatically increased in recent years. However, there are still many aspects of VEGFR biology that remain incompletely understood. Further insight into the molecular mechanisms of angiogenesis, vascular stability, and vascular permeability seems a prerequisite in order to design meaningful novel treatments and to minimize adverse effects.

ACKNOWLEDGMENTS

We apologize to those authors whose work could not be mentioned or cited in this chapter due to space limitations. K.G. was supported by a young investigator grant from the Leducq Foundation (#14-CVD-02). L.M. was supported by the Swedish Society for Medical Research (SSMF).

REFERENCES

Aase, K., von Euler, G., Li, X., Pontén, A., Thorén, P., Cao, R., et al. (2001). Vascular endothelial growth factor-B-deficient mice display an atrial conduction defect. *Circulation*, *104*, 358–364. http://dx.doi.org/10.1161/01.CIR.104.3.358.

Achen, M. G., Jeltsch, M., Kukk, E., Mäkinen, T., Vitali, A., Wilks, A. F., et al. (1998). Vascular endothelial growth factor D (VEGF-D) is a ligand for the tyrosine kinases VEGF receptor 2 (Flk1) and VEGF receptor 3 (Flt4). *Proceedings of the National Academy of Sciences of the United States of America*, *95*, 548–553.

Adomako-Ankomah, A., & Ettensohn, C. A. (2014). Growth factors and early mesoderm morphogenesis: Insights from the sea urchin embryo. *Genesis*, *52*, 158–172. http://dx.doi.org/10.1002/dvg.22746.

Amin, E. M., Oltean, S., Hua, J., Gammons, M. V. R., Hamdollah-Zadeh, M., Welsh, G. I., et al. (2011). WT1 mutants reveal SRPK1 to be a downstream angiogenesis target by altering VEGF splicing. *Cancer Cell, 20*, 768–780. http://dx.doi.org/10.1016/j.ccr.2011.10.016.

Arany, Z., Foo, S.-Y., Ma, Y., Ruas, J. L., Bommi-Reddy, A., Girnun, G., et al. (2008). HIF-independent regulation of VEGF and angiogenesis by the transcriptional coactivator PGC-1alpha. *Nature, 451*, 1008–1012. http://dx.doi.org/10.1038/nature06613.

Aspelund, A., Robciuc, M. R., Karaman, S., Makinen, T., & Alitalo, K. (2016). Lymphatic system in cardiovascular medicine. *Circulation Research, 118*, 515–530. http://dx.doi.org/10.1161/CIRCRESAHA.115.306544.

Autiero, M., Waltenberger, J., Communi, D., Kranz, A., Moons, L., Lambrechts, D., et al. (2003). Role of PlGF in the intra- and intermolecular cross talk between the VEGF receptors Flt1 and Flk1. *Nature Medicine, 9*, 936–943. http://dx.doi.org/10.1038/nm884.

Bahary, N., Goishi, K., Stuckenholz, C., Weber, G., Leblanc, J., Schafer, C. A., et al. (2007). Duplicate VegfA genes and orthologues of the KDR receptor tyrosine kinase family mediate vascular development in the zebrafish. *Blood, 110*, 3627–3636. http://dx.doi.org/10.1182/blood-2006-04-016378.

Baldwin, M. E., Halford, M. M., Roufail, S., Williams, R. A., Hibbs, M. L., Grail, D., et al. (2005). Vascular endothelial growth factor D is dispensable for development of the lymphatic system. *Molecular and Cellular Biology, 25*, 2441–2449. http://dx.doi.org/10.1128/MCB.25.6.2441-2449.2005.

Barleon, B., Sozzani, S., Zhou, D., Weich, H. A., Mantovani, A., & Marmé, D. (1996). Migration of human monocytes in response to vascular endothelial growth factor (VEGF) is mediated via the VEGF receptor flt-1. *Blood, 87*, 3336–3343.

Bates, D. O., Cui, T.-G., Doughty, J. M., Winkler, M., Sugiono, M., Shields, J. D., et al. (2002). VEGF165b, an inhibitory splice variant of vascular endothelial growth factor, is down-regulated in renal cell carcinoma. *Cancer Research, 62*, 4123–4131.

Bellomo, D., Headrick, J. P., Silins, G. U., Paterson, C. A., Thomas, P. S., Gartside, M., et al. (2000). Mice lacking the vascular endothelial growth factor-B gene (Vegfb) have smaller hearts, dysfunctional coronary vasculature, and impaired recovery from cardiac ischemia. *Circulation Research, 86*, E29–E35.

Bellon, A., Luchino, J., Haigh, K., Rougon, G., Haigh, J., Chauvet, S., et al. (2010). VEGFR2 (KDR/Flk1) signaling mediates axon growth in response to semaphorin 3E in the developing brain. *Neuron, 66*, 205–219. http://dx.doi.org/10.1016/j.neuron.2010.04.006.

Benedito, R., Rocha, S. F., Woeste, M., Zamykal, M., Radtke, F., Casanovas, O., et al. (2012). Notch-dependent VEGFR3 upregulation allows angiogenesis without VEGF-VEGFR2 signalling. *Nature, 484*, 110–114. http://dx.doi.org/10.1038/nature10908.

Bevan, H. S., van den Akker, N. M. S., Qiu, Y., Polman, J. A. E., Foster, R. R., Yem, J., et al. (2008). The alternatively spliced anti-angiogenic family of VEGF isoforms VEGFxxxb in human kidney development. *Nephron. Physiology, 110*, 57–67. http://dx.doi.org/10.1159/000177614.

Bhattacharjee, G., Asplin, I. R., Wu, S. M., Gawdi, G., & Pizzo, S. V. (2000). The conformation-dependent interaction of alpha 2-macroglobulin with vascular endothelial growth factor. A novel mechanism of alpha 2-macroglobulin/growth factor binding. *The Journal of Biological Chemistry, 275*, 26806–26811. http://dx.doi.org/10.1074/jbc.M000156200.

Biselli-Chicote, P. M., Oliveira, A. R. C. P., Pavarino, E. C., & Goloni-Bertollo, E. M. (2012). VEGF gene alternative splicing: Pro- and anti-angiogenic isoforms in cancer. *Journal of Cancer Research and Clinical Oncology, 138*, 363–370. http://dx.doi.org/10.1007/s00432-011-1073-2.

Brückner, K., Kockel, L., Duchek, P., Luque, C. M., Rørth, P., & Perrimon, N. (2004). The PDGF/VEGF receptor controls blood cell survival in Drosophila. *Developmental Cell*, 7, 73–84. http://dx.doi.org/10.1016/j.devcel.2004.06.007.
Bussmann, J., Bakkers, J., & Schulte-Merker, S. (2007). Early endocardial morphogenesis requires Scl/Tal1. *PLoS Genetics*, 3, e140. http://dx.doi.org/10.1371/journal.pgen.0030140.
Bussmann, J., Lawson, N., Zon, L., Schulte-Merker, S., & Zebrafish Nomenclature Committee. (2008). Zebrafish VEGF receptors: A guideline to nomenclature. *PLoS Genetics*, 4, e1000064. http://dx.doi.org/10.1371/journal.pgen.1000064.
Cao, Y., Chen, H., Zhou, L., Chiang, M. K., Anand-Apte, B., Weatherbee, J. A., et al. (1996). Heterodimers of placenta growth factor/vascular endothelial growth factor. Endothelial activity, tumor cell expression, and high affinity binding to Flk-1/KDR. *The Journal of Biological Chemistry*, 271, 3154–3162.
Carmeliet, P. (2000). VEGF gene therapy: Stimulating angiogenesis or angioma-genesis? *Nature Medicine*, 6, 1102–1103. http://dx.doi.org/10.1038/80430.
Carmeliet, P., Ferreira, V., Breier, G., Pollefeyt, S., Kieckens, L., Gertsenstein, M., et al. (1996). Abnormal blood vessel development and lethality in embryos lacking a single VEGF allele. *Nature*, 380, 435–439. http://dx.doi.org/10.1038/380435a0.
Carmeliet, P., Moons, L., Luttun, A., Vincenti, V., Compernolle, V., De Mol, M., et al. (2001). Synergism between vascular endothelial growth factor and placental growth factor contributes to angiogenesis and plasma extravasation in pathological conditions. *Nature Medicine*, 7, 575–583. http://dx.doi.org/10.1038/87904.
Carmeliet, P., Ng, Y. S., Nuyens, D., Theilmeier, G., Brusselmans, K., Cornelissen, I., et al. (1999). Impaired myocardial angiogenesis and ischemic cardiomyopathy in mice lacking the vascular endothelial growth factor isoforms VEGF164 and VEGF188. *Nature Medicine*, 5, 495–502. http://dx.doi.org/10.1038/8379.
Catena, R., Larzabal, L., Larrayoz, M., Molina, E., Hermida, J., Agorreta, J., et al. (2010). $VEGF_{121}b$ and $VEGF_{165}b$ are weakly angiogenic isoforms of VEGF-A. *Molecular Cancer*, 9, 320. http://dx.doi.org/10.1186/1476-4598-9-320.
Cébe Suarez, S., Pieren, M., Cariolato, L., Arn, S., Hoffmann, U., Bogucki, A., et al. (2006). A VEGF-A splice variant defective for heparan sulfate and neuropilin-1 binding shows attenuated signaling through VEGFR-2. *Cellular and Molecular Life Sciences*, 63, 2067–2077. http://dx.doi.org/10.1007/s00018-006-6254-9.
Chappell, J. C., Taylor, S. M., Ferrara, N., & Bautch, V. L. (2009). Local guidance of emerging vessel sprouts requires soluble Flt-1. *Developmental Cell*, 17, 377–386. http://dx.doi.org/10.1016/j.devcel.2009.07.011.
Chen, H., Chédotal, A., He, Z., Goodman, C. S., & Tessier-Lavigne, M. (1997). Neuropilin-2, a novel member of the neuropilin family, is a high affinity receptor for the semaphorins Sema E and Sema IV but not Sema III. *Neuron*, 19, 547–559.
Cho, N. K., Keyes, L., Johnson, E., Heller, J., Ryner, L., Karim, F., et al. (2002). Developmental control of blood cell migration by the Drosophila VEGF pathway. *Cell*, 108, 865–876.
Christinger, H. W., Fuh, G., de Vos, A. M., & Wiesmann, C. (2004). The crystal structure of placental growth factor in complex with domain 2 of vascular endothelial growth factor receptor-1. *The Journal of Biological Chemistry*, 279, 10382–10388. http://dx.doi.org/10.1074/jbc.M313237200.
Covassin, L. D., Villefranc, J. A., Kacergis, M. C., Weinstein, B. M., & Lawson, N. D. (2006). Distinct genetic interactions between multiple Vegf receptors are required for development of different blood vessel types in zebrafish. *Proceedings of the National Academy of Sciences of the United States of America*, 103, 6554–6559. http://dx.doi.org/10.1073/pnas.0506886103.

Cudmore, M. J., Hewett, P. W., Ahmad, S., Wang, K.-Q., Cai, M., Al-Ani, B., et al. (2012). The role of heterodimerization between VEGFR-1 and VEGFR-2 in the regulation of endothelial cell homeostasis. *Nature Communications, 3*, 972. http://dx.doi.org/10.1038/ncomms1977.

Cunningham, T. J., Palumbo, I., Grosso, M., Slater, N., & Miles, C. G. (2013). WT1 regulates murine hematopoiesis via maintenance of VEGF isoform ratio. *Blood, 122*, 188–192. http://dx.doi.org/10.1182/blood-2012-11-466086.

Dayanir, V., Meyer, R. D., Lashkari, K., & Rahimi, N. (2001). Identification of tyrosine residues in vascular endothelial growth factor receptor-2/FLK-1 involved in activation of phosphatidylinositol 3-kinase and cell proliferation. *The Journal of Biological Chemistry, 276*, 17686–17692. http://dx.doi.org/10.1074/jbc.M009128200.

De Falco, S. (2012). The discovery of placenta growth factor and its biological activity. *Experimental & Molecular Medicine, 44*, 1–9. http://dx.doi.org/10.3858/emm.2012.44.1.025.

de Vries, C., Escobedo, J. A., Ueno, H., Houck, K., Ferrara, N., & Williams, L. T. (1992). The fms-like tyrosine kinase, a receptor for vascular endothelial growth factor. *Science, 255*, 989–991.

Dewerchin, M., & Carmeliet, P. (2012). PlGF: A multitasking cytokine with disease-restricted activity. *Cold Spring Harbor Perspectives in Medicine, 2*, a011056. http://dx.doi.org/10.1101/cshperspect.a011056.

Dieterich, L. C., & Detmar, M. (2016). Tumor lymphangiogenesis and new drug development. *Advanced Drug Delivery Reviews, 99*, 148–160. http://dx.doi.org/10.1016/j.addr.2015.12.011.

Dijkstra, M. H., Pirinen, E., Huusko, J., Kivelä, R., Schenkwein, D., Alitalo, K., et al. (2014). Lack of cardiac and high-fat diet induced metabolic phenotypes in two independent strains of Vegf-b knockout mice. *Scientific Reports, 4*, 6238. http://dx.doi.org/10.1038/srep06238.

DiSalvo, J., Bayne, M. L., Conn, G., Kwok, P. W., Trivedi, P. G., Soderman, D. D., et al. (1995). Purification and characterization of a naturally occurring vascular endothelial growth factor.placenta growth factor heterodimer. *The Journal of Biological Chemistry, 270*, 7717–7723.

Dixelius, J., Makinen, T., Wirzenius, M., Karkkainen, M. J., Wernstedt, C., Alitalo, K., et al. (2003). Ligand-induced vascular endothelial growth factor receptor-3 (VEGFR-3) heterodimerization with VEGFR-2 in primary lymphatic endothelial cells regulates tyrosine phosphorylation sites. *The Journal of Biological Chemistry, 278*, 40973–40979. http://dx.doi.org/10.1074/jbc.M304499200.

Dokun, A. O., & Annex, B. H. (2011). The VEGF165b "ICE-o-form" puts a chill on the VEGF story. *Circulation Research, 109*, 246–247. http://dx.doi.org/10.1161/CIRCRESAHA.111.249953.

Duchek, P., Somogyi, K., Jékely, G., Beccari, S., & Rørth, P. (2001). Guidance of cell migration by the Drosophila PDGF/VEGF receptor. *Cell, 107*, 17–26.

Dumont, D. J., Jussila, L., Taipale, J., Lymboussaki, A., Mustonen, T., Pajusola, K., et al. (1998). Cardiovascular failure in mouse embryos deficient in VEGF receptor-3. *Science, 282*, 946–949.

Ebos, J. M. L., Bocci, G., Man, S., Thorpe, P. E., Hicklin, D. J., Zhou, D., et al. (2004). A naturally occurring soluble form of vascular endothelial growth factor receptor 2 detected in mouse and human plasma. *Molecular Cancer Research, 2*, 315–326.

Eichmann, A., Makinen, T., & Alitalo, K. (2005). Neural guidance molecules regulate vascular remodeling and vessel navigation. *Genes & Development, 19*, 1013–1021. http://dx.doi.org/10.1101/gad.1305405.

Eriksson, A., Cao, R., Pawliuk, R., Berg, S. M., Tsang, M., Zhou, D., et al. (2002). Placenta growth factor-1 antagonizes VEGF-induced angiogenesis and tumor growth by the formation of functionally inactive PlGF-1/VEGF heterodimers. *Cancer Cell, 1*, 99–108.

Falk, T., Yue, X., Zhang, S., McCourt, A. D., Yee, B. J., Gonzalez, R. T., et al. (2011). Vascular endothelial growth factor-B is neuroprotective in an in vivo rat model of Parkinson's disease. *Neuroscience Letters*, *496*, 43–47. http://dx.doi.org/10.1016/j.neulet.2011.03.088.

Fang, S., Nurmi, H., Heinolainen, K., Chen, S., Salminen, E., Saharinen, P., et al. (2016). Critical requirement of VEGF-C in transition to fetal erythropoiesis. *Blood*, *128*, 710–720. http://dx.doi.org/10.1182/blood-2015-12-687970.

Fantin, A., Herzog, B., Mahmoud, M., Yamaji, M., Plein, A., Denti, L., et al. (2014). Neuropilin 1 (NRP1) hypomorphism combined with defective VEGF-A binding reveals novel roles for NRP1 in developmental and pathological angiogenesis. *Development*, *141*, 556–562. http://dx.doi.org/10.1242/dev.103028.

Fantin, A., Vieira, J. M., Plein, A., Denti, L., Fruttiger, M., Pollard, J. W., et al. (2013). NRP1 acts cell autonomously in endothelium to promote tip cell function during sprouting angiogenesis. *Blood*, *121*, 2352–2362. http://dx.doi.org/10.1182/blood-2012-05-424713.

Ferrara, N. (2010). Binding to the extracellular matrix and proteolytic processing: Two key mechanisms regulating vascular endothelial growth factor action. *Molecular Biology of the Cell*, *21*, 687–690. http://dx.doi.org/10.1091/mbc.E09-07-0590.

Ferrara, N., & Adamis, A. P. (2016). Ten years of anti-vascular endothelial growth factor therapy. *Nature Reviews. Drug Discovery*, *15*, 385–403. http://dx.doi.org/10.1038/nrd.2015.17.

Ferrara, N., Carver-Moore, K., Chen, H., Dowd, M., Lu, L., O'Shea, K. S., et al. (1996). Heterozygous embryonic lethality induced by targeted inactivation of the VEGF gene. *Nature*, *380*, 439–442. http://dx.doi.org/10.1038/380439a0.

Ferrell, R. E., Levinson, K. L., Esman, J. H., Kimak, M. A., Lawrence, E. C., Barmada, M. M., et al. (1998). Hereditary lymphedema: Evidence for linkage and genetic heterogeneity. *Human Molecular Genetics*, *7*, 2073–2078.

Folkman, J. (1971). Tumor angiogenesis: Therapeutic implications. *The New England Journal of Medicine*, *285*, 1182–1186. http://dx.doi.org/10.1056/NEJM197111182852108.

Folkman, J., Merler, E., Abernathy, C., & Williams, G. (1971). Isolation of a tumor factor responsible for angiogenesis. *The Journal of Experimental Medicine*, *133*, 275–288.

Fong, G. H., Rossant, J., Gertsenstein, M., & Breitman, M. L. (1995). Role of the Flt-1 receptor tyrosine kinase in regulating the assembly of vascular endothelium. *Nature*, *376*, 66–70. http://dx.doi.org/10.1038/376066a0.

Fong, G. H., Zhang, L., Bryce, D. M., & Peng, J. (1999). Increased hemangioblast commitment, not vascular disorganization, is the primary defect in flt-1 knock-out mice. *Development*, *126*, 3015–3025.

Fournier, E., Dubreuil, P., Birnbaum, D., & Borg, J. P. (1995). Mutation at tyrosine residue 1337 abrogates ligand-dependent transforming capacity of the FLT4 receptor. *Oncogene*, *11*, 921–931.

Fruman, D. A., Mauvais-Jarvis, F., Pollard, D. A., Yballe, C. M., Brazil, D., Bronson, R. T., et al. (2000). Hypoglycaemia, liver necrosis and perinatal death in mice lacking all isoforms of phosphoinositide 3-kinase p85 alpha. *Nature Genetics*, *26*, 379–382. http://dx.doi.org/10.1038/81715.

Fuh, G., Li, B., Crowley, C., Cunningham, B., & Wells, J. A. (1998). Requirements for binding and signaling of the kinase domain receptor for vascular endothelial growth factor. *The Journal of Biological Chemistry*, *273*, 11197–11204. http://dx.doi.org/10.1074/jbc.273.18.11197.

Funa, N. S., Kriz, V., Zang, G., Calounova, G., Akerblom, B., Mares, J., et al. (2009). Dysfunctional microvasculature as a consequence of shb gene inactivation causes impaired tumor growth. *Cancer Research*, *69*, 2141–2148. http://dx.doi.org/10.1158/0008-5472.CAN-08-3797.

Furuta, T., Kimura, M., & Watanabe, N. (2010). Elevated levels of vascular endothelial growth factor (VEGF) and soluble vascular endothelial growth factor receptor (VEGFR)-2 in human malaria. *The American Journal of Tropical Medicine and Hygiene*, *82*, 136–139. http://dx.doi.org/10.4269/ajtmh.2010.09-0203.

Fusco, M. A., Wajsenzon, I. J. R., de Carvalho, S. L., da Silva, R. T., Einicker-Lamas, M., Cavalcante, L. A., et al. (2014). Vascular endothelial growth factor-like and its receptor in a crustacean optic ganglia: A role in neuronal differentiation? *Biochemical and Biophysical Research Communications*, *447*, 299–303. http://dx.doi.org/10.1016/j.bbrc.2014.03.137.

Galland, F., Karamysheva, A., Pebusque, M. J., Borg, J. P., Rottapel, R., Dubreuil, P., et al. (1993). The FLT4 gene encodes a transmembrane tyrosine kinase related to the vascular endothelial growth factor receptor. *Oncogene*, *8*, 1233–1240.

Gelfand, M. V., Hagan, N., Tata, A., Oh, W.-J., Lacoste, B., Kang, K.-T., et al. (2014). Neuropilin-1 functions as a VEGFR2 co-receptor to guide developmental angiogenesis independent of ligand binding. *eLife*, *3*, e03720. http://dx.doi.org/10.7554/eLife.03720.

Gerhardt, H., Golding, M., Fruttiger, M., Ruhrberg, C., Lundkvist, A., Abramsson, A., et al. (2003). VEGF guides angiogenic sprouting utilizing endothelial tip cell filopodia. *The Journal of Cell Biology*, *161*, 1163–1177. http://dx.doi.org/10.1083/jcb.200302047.

Ghalamkarpour, A., Morlot, S., Raas-Rothschild, A., Utkus, A., Mulliken, J. B., Boon, L. M., et al. (2006). Hereditary lymphedema type I associated with VEGFR3 mutation: The first de novo case and atypical presentations. *Clinical Genetics*, *70*, 330–335. http://dx.doi.org/10.1111/j.1399-0004.2006.00687.x.

Gille, H., Kowalski, J., Yu, L., Chen, H., Pisabarro, M. T., Davis-Smyth, T., et al. (2000). A repressor sequence in the juxtamembrane domain of Flt-1 (VEGFR-1) constitutively inhibits vascular endothelial growth factor-dependent phosphatidylinositol 3′-kinase activation and endothelial cell migration. *The EMBO Journal*, *19*, 4064–4073. http://dx.doi.org/10.1093/emboj/19.15.4064.

Gordon, E. J., Fukuhara, D., Weström, S., Padhan, N., Sjöström, E. O., van Meeteren, L., et al. (2016). The endothelial adaptor molecule TSAd is required for VEGF-induced angiogenic sprouting through junctional c-Src activation. *Science Signaling*, *9*, ra72.

Graubert, M. D., Asuncion Ortega, M., Kessel, B., Mortola, J. F., & Iruela-Arispe, M. L. (2001). Vascular repair after menstruation involves regulation of vascular endothelial growth factor-receptor phosphorylation by sFLT-1. *The American Journal of Pathology*, *158*, 1399–1410. http://dx.doi.org/10.1016/S0002-9440(10)64091-6.

Graupera, M., Guillermet-Guibert, J., Foukas, L. C., Phng, L.-K., Cain, R. J., Salpekar, A., et al. (2008). Angiogenesis selectively requires the p110alpha isoform of PI3K to control endothelial cell migration. *Nature*, *453*, 662–666. http://dx.doi.org/10.1038/nature06892.

Grünewald, F. S., Prota, A. E., Giese, A., & Ballmer-Hofer, K. (2010). Structure-function analysis of VEGF receptor activation and the role of coreceptors in angiogenic signaling. *Biochimica et Biophysica Acta*, *1804*, 567–580. http://dx.doi.org/10.1016/j.bbapap.2009.09.002.

Grunstein, J., Masbad, J. J., Hickey, R., Giordano, F., & Johnson, R. S. (2000). Isoforms of vascular endothelial growth factor act in a coordinate fashion to recruit and expand tumor vasculature. *Molecular and Cellular Biology*, *20*, 7282–7291.

Gu, C., Limberg, B. J., Whitaker, G. B., Perman, B., Leahy, D. J., Rosenbaum, J. S., et al. (2002). Characterization of neuropilin-1 structural features that confer binding to semaphorin 3A and vascular endothelial growth factor 165. *The Journal of Biological Chemistry*, *277*, 18069–18076. http://dx.doi.org/10.1074/jbc.M201681200.

Gu, C., Rodriguez, E. R., Reimert, D. V., Shu, T., Fritzsch, B., Richards, L. J., et al. (2003). Neuropilin-1 conveys semaphorin and VEGF signaling during neural and cardiovascular development. *Developmental Cell*, *5*, 45–57.

Gupta, K., Gupta, P., Wild, R., Ramakrishnan, S., & Hebbel, R. P. (1999). Binding and displacement of vascular endothelial growth factor (VEGF) by thrombospondin: Effect on human microvascular endothelial cell proliferation and angiogenesis. *Angiogenesis, 3*, 147–158.

Hagberg, C. E., Falkevall, A., Wang, X., Larsson, E., Huusko, J., Nilsson, I., et al. (2010). Vascular endothelial growth factor B controls endothelial fatty acid uptake. *Nature, 464*, 917–921. http://dx.doi.org/10.1038/nature08945.

Hagberg, C. E., Mehlem, A., Falkevall, A., Muhl, L., Fam, B. C., Ortsater, H., et al. (2012). Targeting VEGF-B as a novel treatment for insulin resistance and type 2 diabetes. *Nature, 490*, 426–430. http://dx.doi.org/10.1038/nature11464.

Harper, S. J., & Bates, D. O. (2008). VEGF-A splicing: The key to anti-angiogenic therapeutics? *Nature Reviews. Cancer, 8*, 880–887. http://dx.doi.org/10.1038/nrc2505.

Harris, S., Craze, M., Newton, J., Fisher, M., Shima, D. T., Tozer, G. M., et al. (2012). Do anti-angiogenic VEGF (VEGFxxxb) isoforms exist? A cautionary tale. *PloS One, 7*, e35231. http://dx.doi.org/10.1371/journal.pone.0035231.

Hattori, K., Heissig, B., Wu, Y., Dias, S., Tejada, R., Ferris, B., et al. (2002). Placental growth factor reconstitutes hematopoiesis by recruiting VEGFR1(+) stem cells from bone-marrow microenvironment. *Nature Medicine, 8*, 841–849. http://dx.doi.org/10.1038/nm740.

Hawinkels, L. J. A. C., Zuidwijk, K., Verspaget, H. W., de Jonge-Muller, E. S. M., van Duijn, W., Ferreira, V., et al. (2008). VEGF release by MMP-9 mediated heparan sulphate cleavage induces colorectal cancer angiogenesis. *European Journal of Cancer, 44*, 1904–1913. http://dx.doi.org/10.1016/j.ejca.2008.06.031.

Heino, T. I., Kärpänen, T., Wahlström, G., Pulkkinen, M., Eriksson, U., Alitalo, K., et al. (2001). The Drosophila VEGF receptor homolog is expressed in hemocytes. *Mechanisms of Development, 109*, 69–77. http://dx.doi.org/10.1016/S0925-4773(01)00510-X.

Herzog, Y., Kalcheim, C., Kahane, N., Reshef, R., & Neufeld, G. (2001). Differential expression of neuropilin-1 and neuropilin-2 in arteries and veins. *Mechanisms of Development, 109*, 115–119.

Hiratsuka, S., Maru, Y., Okada, A., Seiki, M., Noda, T., & Shibuya, M. (2001). Involvement of Flt-1 tyrosine kinase (vascular endothelial growth factor receptor-1) in pathological angiogenesis. *Cancer Research, 61*, 1207–1213.

Hiratsuka, S., Minowa, O., Kuno, J., Noda, T., & Shibuya, M. (1998). Flt-1 lacking the tyrosine kinase domain is sufficient for normal development and angiogenesis in mice. *Proceedings of the National Academy of Sciences of the United States of America, 95*, 9349–9354.

Hiratsuka, S., Nakao, K., Nakamura, K., Katsuki, M., Maru, Y., & Shibuya, M. (2005). Membrane fixation of vascular endothelial growth factor receptor 1 ligand-binding domain is important for vasculogenesis and angiogenesis in mice. *Molecular and Cellular Biology, 25*, 346–354. http://dx.doi.org/10.1128/MCB.25.1.346-354.2005.

Holmes, D. I., & Zachary, I. (2005). The vascular endothelial growth factor (VEGF) family: Angiogenic factors in health and disease. *Genome Biology, 6*, 209. http://dx.doi.org/10.1186/gb-2005-6-2-209.

Holmqvist, K., Cross, M. J., Rolny, C., Hägerkvist, R., Rahimi, N., Matsumoto, T., et al. (2004). The adaptor protein Shb binds to tyrosine 1175 in vascular endothelial growth factor (VEGF) receptor-2 and regulates VEGF-dependent cellular migration. *The Journal of Biological Chemistry, 279*, 22267–22275. http://dx.doi.org/10.1074/jbc.M312729200.

Huang, K., Andersson, C., Roomans, G. M., Ito, N., & Claesson-Welsh, L. (2001). Signaling properties of VEGF receptor-1 and -2 homo- and heterodimers. *The International Journal of Biochemistry & Cell Biology, 33*, 315–324.

Hyde, C. A. C., Giese, A., Stuttfeld, E., Saliba, J. A., Villemagne, D., Schleier, T., et al. (2012). Targeting extracellular domains D4 and D7 of vascular endothelial growth factor receptor 2 reveals allosteric receptor regulatory sites. *Molecular and Cellular Biology, 32,* 3802–3813. http://dx.doi.org/10.1128/MCB.06787-11.

Irrthum, A., Karkkainen, M. J., Devriendt, K., Alitalo, K., & Vikkula, M. (2000). Congenital hereditary lymphedema caused by a mutation that inactivates VEGFR3 tyrosine kinase. *The American Journal of Human Genetics, 67,* 295–301. http://dx.doi.org/10.1086/303019.

Jeltsch, M., Jha, S. K., Tvorogov, D., Anisimov, A., Leppänen, V.-M., Holopainen, T., et al. (2014). CCBE1 enhances lymphangiogenesis via A disintegrin and metalloprotease with thrombospondin motifs-3-mediated vascular endothelial growth factor-C activation. *Circulation, 129,* 1962–1971. http://dx.doi.org/10.1161/CIRCULATIONAHA.113.002779.

Jeltsch, M., Karpanen, T., Strandin, T., Aho, K., Lankinen, H., & Alitalo, K. (2006). Vascular endothelial growth factor (VEGF)/VEGF-C mosaic molecules reveal specificity determinants and feature novel receptor binding patterns. *The Journal of Biological Chemistry, 281,* 12187–12195. http://dx.doi.org/10.1074/jbc.M511593200.

Ji, Q. S., Winnier, G. E., Niswender, K. D., Horstman, D., Wisdom, R., Magnuson, M. A., et al. (1997). Essential role of the tyrosine kinase substrate phospholipase C-gamma1 in mammalian growth and development. *Proceedings of the National Academy of Sciences of the United States of America, 94,* 2999–3003.

Joukov, V., Pajusola, K., Kaipainen, A., Chilov, D., Lahtinen, I., Kukk, E., et al. (1996). A novel vascular endothelial growth factor, VEGF-C, is a ligand for the Flt4 (VEGFR-3) and KDR (VEGFR-2) receptor tyrosine kinases. *The EMBO Journal, 15,* 290–298.

Joukov, V., Sorsa, T., Kumar, V., Jeltsch, M., Claesson-Welsh, L., Cao, Y., et al. (1997). Proteolytic processing regulates receptor specificity and activity of VEGF-C. *The EMBO Journal, 16,* 3898–3911. http://dx.doi.org/10.1093/emboj/16.13.3898.

Kaipainen, A., Korhonen, J., Mustonen, T., van Hinsbergh, V. W., Fang, G. H., Dumont, D., et al. (1995). Expression of the fms-like tyrosine kinase 4 gene becomes restricted to lymphatic endothelium during development. *Proceedings of the National Academy of Sciences of the United States of America, 92,* 3566–3570.

Kanthou, C., Dachs, G. U., Lefley, D. V., Steele, A. J., Coralli-Foxon, C., Harris, S., et al. (2014). Tumour cells expressing single VEGF isoforms display distinct growth, survival and migration characteristics. *PloS One, 9,* e104015. http://dx.doi.org/10.1371/journal.pone.0104015.

Karaman, S., & Detmar, M. (2014). Mechanisms of lymphatic metastasis. *The Journal of Clinical Investigation, 124,* 922–928. http://dx.doi.org/10.1172/JCI71606.

Karkkainen, M. J., Ferrell, R. E., Lawrence, E. C., Kimak, M. A., Levinson, K. L., McTigue, M. A., et al. (2000). Missense mutations interfere with VEGFR-3 signalling in primary lymphoedema. *Nature Genetics, 25,* 153–159. http://dx.doi.org/10.1038/75997.

Karkkainen, M. J., Haiko, P., Sainio, K., Partanen, J., Taipale, J., Petrova, T. V., et al. (2004). Vascular endothelial growth factor C is required for sprouting of the first lymphatic vessels from embryonic veins. *Nature Immunology, 5,* 74–80. http://dx.doi.org/10.1038/ni1013.

Karkkainen, M. J., Saaristo, A., Jussila, L., Karila, K. A., Lawrence, E. C., Pajusola, K., et al. (2001). A model for gene therapy of human hereditary lymphedema. *Proceedings of the National Academy of Sciences of the United States of America, 98,* 12677–12682. http://dx.doi.org/10.1073/pnas.221449198.

Kärpänen, T., Bry, M., Ollila, H. M., Seppänen-Laakso, T., Liimatta, E., Leskinen, H., et al. (2008). Overexpression of vascular endothelial growth factor-B in mouse heart alters cardiac lipid metabolism and induces myocardial hypertrophy. *Circulation Research, 103,* 1018–1026. http://dx.doi.org/10.1161/CIRCRESAHA.108.178459.

Katoh, O., Tauchi, H., Kawaishi, K., Kimura, A., & Satow, Y. (1995). Expression of the vascular endothelial growth factor (VEGF) receptor gene, KDR, in hematopoietic cells and inhibitory effect of VEGF on apoptotic cell death caused by ionizing radiation. *Cancer Research, 55,* 5687–5692.

Katuri, V., Gerber, S., Qiu, X., McCarty, G., Goldstein, S. D., Hammers, H., et al. (2014). WT1 regulates angiogenesis in Ewing Sarcoma. *Oncotarget, 5,* 2436–2449. http://dx.doi.org/10.18632/oncotarget.1610.

Kawamura, H., Li, X., Harper, S. J., Bates, D. O., & Claesson-Welsh, L. (2008). Vascular endothelial growth factor (VEGF)-A165b is a weak in vitro agonist for VEGF receptor-2 due to lack of coreceptor binding and deficient regulation of kinase activity. *Cancer Research, 68,* 4683–4692. http://dx.doi.org/10.1158/0008-5472.CAN-07-6577.

Kawasaki, T., Kitsukawa, T., Bekku, Y., Matsuda, Y., Sanbo, M., Yagi, T., et al. (1999). A requirement for neuropilin-1 in embryonic vessel formation. *Development, 126,* 4895–4902.

Kendall, R. L., Rutledge, R. Z., Mao, X., Tebben, A. J., Hungate, R. W., & Thomas, K. A. (1999). Vascular endothelial growth factor receptor KDR tyrosine kinase activity is increased by autophosphorylation of two activation loop tyrosine residues. *The Journal of Biological Chemistry, 274,* 6453–6460.

Kendall, R. L., & Thomas, K. A. (1993). Inhibition of vascular endothelial cell growth factor activity by an endogenously encoded soluble receptor. *Proceedings of the National Academy of Sciences of the United States of America, 90,* 10705–10709.

Kendall, R. L., Wang, G., & Thomas, K. A. (1996). Identification of a natural soluble form of the vascular endothelial growth factor receptor, FLT-1, and its heterodimerization with KDR. *Biochemical and Biophysical Research Communications, 226,* 324–328. http://dx.doi.org/10.1006/bbrc.1996.1355.

Keyt, B. A., Berleau, L. T., Nguyen, H. V., Chen, H., Heinsohn, H., Vandlen, R., et al. (1996). The carboxyl-terminal domain (111-165) of vascular endothelial growth factor is critical for its mitogenic potency. *The Journal of Biological Chemistry, 271,* 7788–7795.

Kieran, M. W., Kalluri, R., & Cho, Y.-J. (2012). The VEGF pathway in cancer and disease: Responses, resistance, and the path forward. *Cold Spring Harbor Perspectives in Medicine, 2,* a006593. http://dx.doi.org/10.1101/cshperspect.a006593.

Kikuchi, R., Nakamura, K., MacLauchlan, S., Ngo, D. T.-M., Shimizu, I., Fuster, J. J., et al. (2014). An antiangiogenic isoform of VEGF-A contributes to impaired vascularization in peripheral artery disease. *Nature Medicine, 20,* 1464–1471. http://dx.doi.org/10.1038/nm.3703.

Kipryushina, Y. O., Yakovlev, K. V., & Odintsova, N. A. (2015). Vascular endothelial growth factors: A comparison between invertebrates and vertebrates. *Cytokine & Growth Factor Reviews, 26,* 687–695. http://dx.doi.org/10.1016/j.cytogfr.2015.04.001.

Kitsukawa, T., Shimizu, M., Sanbo, M., Hirata, T., Taniguchi, M., Bekku, Y., et al. (1997). Neuropilin-semaphorin III/D-mediated chemorepulsive signals play a crucial role in peripheral nerve projection in mice. *Neuron, 19,* 995–1005.

Kivelä, R., Bry, M., Robciuc, M. R., Räsänen, M., Taavitsainen, M., Silvola, J. M., et al. (2014). VEGF-B-induced vascular growth leads to metabolic reprogramming and ischemia resistance in the heart. *EMBO Molecular Medicine, 6,* 307–321. http://dx.doi.org/10.1002/emmm.201303147.

Koch, S. (2012). Neuropilin signalling in angiogenesis. *Biochemical Society Transactions, 40,* 20–25. http://dx.doi.org/10.1042/BST20110689.

Koch, S., & Claesson-Welsh, L. (2012). Signal transduction by vascular endothelial growth factor receptors. *Cold Spring Harbor Perspectives in Medicine, 2,* a006502. http://dx.doi.org/10.1101/cshperspect.a006502.

Koch, S., Tugues, S., Li, X., Gualandi, L., & Claesson-Welsh, L. (2011). Signal transduction by vascular endothelial growth factor receptors. *The Biochemical Journal, 437,* 169–183. http://dx.doi.org/10.1042/BJ20110301.

Kofler, N., & Simons, M. (2016). The expanding role of neuropilin: Regulation of transforming growth factor-β and platelet-derived growth factor signaling in the vasculature.

Current Opinion in Hematology, 23, 260–267. http://dx.doi.org/10.1097/MOH. 0000000000000233.

Krishnapati, L.-S., & Ghaskadbi, S. (2013). Identification and characterization of VEGF and FGF from Hydra. *The International Journal of Developmental Biology*, 57, 897–906. http://dx.doi.org/10.1387/ijdb.130077sg.

Krueger, J., Liu, D., Scholz, K., Zimmer, A., Shi, Y., Klein, C., et al. (2011). Flt1 acts as a negative regulator of tip cell formation and branching morphogenesis in the zebrafish embryo. *Development*, 138, 2111–2120. http://dx.doi.org/10.1242/dev.063933.

Küchler, A. M., Gjini, E., Peterson-Maduro, J., Cancilla, B., Wolburg, H., & Schulte-Merker, S. (2006). Development of the zebrafish lymphatic system requires VEGFC signaling. *Current Biology*, 16, 1244–1248. http://dx.doi.org/10.1016/j.cub.2006.05.026.

Kurtagic, E., Jedrychowski, M. P., & Nugent, M. A. (2009). Neutrophil elastase cleaves VEGF to generate a VEGF fragment with altered activity. *American Journal of Physiology. Lung Cellular and Molecular Physiology*, 296, L534–L546. http://dx.doi.org/10.1152/ajplung.90505.2008.

Lamalice, L., Houle, F., & Huot, J. (2006). Phosphorylation of Tyr1214 within VEGFR-2 triggers the recruitment of Nck and activation of Fyn leading to SAPK2/p38 activation and endothelial cell migration in response to VEGF. *The Journal of Biological Chemistry*, 281, 34009–34020. http://dx.doi.org/10.1074/jbc.M603928200.

Lanahan, A. A., Hermans, K., Claes, F., Kerley-Hamilton, J. S., Zhuang, Z. W., Giordano, F. J., et al. (2010). VEGF receptor 2 endocytic trafficking regulates arterial morphogenesis. *Developmental Cell*, 18, 713–724. http://dx.doi.org/10.1016/j.devcel.2010.02.016.

Lanahan, A., Zhang, X., Fantin, A., Zhuang, Z., Rivera-Molina, F., Speichinger, K., et al. (2013). The neuropilin 1 cytoplasmic domain is required for VEGF-A-dependent arteriogenesis. *Developmental Cell*, 25, 156–168. http://dx.doi.org/10.1016/j.devcel.2013.03.019.

Le Bras, B., Barallobre, M.-J., Homman-Ludiye, J., Ny, A., Wyns, S., Tammela, T., et al. (2006). VEGF-C is a trophic factor for neural progenitors in the vertebrate embryonic brain. *Nature Neuroscience*, 9, 340–348. http://dx.doi.org/10.1038/nn1646.

Lee, T.-Y., Folkman, J., & Javaherian, K. (2010). HSPG-binding peptide corresponding to the exon 6a-encoded domain of VEGF inhibits tumor growth by blocking angiogenesis in murine model. *PloS One*, 5, e9945. http://dx.doi.org/10.1371/journal.pone.0009945.

Lee, S., Jilani, S. M., Nikolova, G. V., Carpizo, D., & Iruela-Arispe, M. L. (2005). Processing of VEGF-A by matrix metalloproteinases regulates bioavailability and vascular patterning in tumors. *The Journal of Cell Biology*, 169, 681–691. http://dx.doi.org/10.1083/jcb.200409115.

Lemmon, M. A., & Schlessinger, J. (2010). Cell signaling by receptor tyrosine kinases. *Cell*, 141, 1117–1134. http://dx.doi.org/10.1016/j.cell.2010.06.011.

Leppänen, V.-M., Prota, A. E., Jeltsch, M., Anisimov, A., Kalkkinen, N., Strandin, T., et al. (2010). Structural determinants of growth factor binding and specificity by VEGF receptor 2. *Proceedings of the National academy of Sciences of the United States of America*, 107, 2425–2430. http://dx.doi.org/10.1073/pnas.0914318107.

Leppänen, V.-M., Tvorogov, D., Kisko, K., Prota, A. E., Jeltsch, M., Anisimov, A., et al. (2013). Structural and mechanistic insights into VEGF receptor 3 ligand binding and activation. *Proceedings of the National academy of Sciences of the United States of America*, 110, 12960–12965. http://dx.doi.org/10.1073/pnas.1301415110.

Levine, R. J., Maynard, S. E., Qian, C., Lim, K.-H., England, L. J., Yu, K. F., et al. (2004). Circulating angiogenic factors and the risk of preeclampsia. *The New England Journal of Medicine*, 350, 672–683. http://dx.doi.org/10.1056/NEJMoa031884.

Li, X., Padhan, N., Sjöström, E. O., Roche, F. P., Testini, C., Honkura, N., et al. (2016). VEGFR2 pY949 signalling regulates adherens junction integrity and metastatic spread. *Nature Communications*, 7, 11017. http://dx.doi.org/10.1038/ncomms11017.

Li, F., Xu, L., Gai, X., Zhou, Z., Wang, L., Zhang, H., et al. (2013). The involvement of PDGF/VEGF related factor in regulation of immune and neuroendocrine in Chinese mitten crab Eriocheir sinensis. *Fish & Shellfish Immunology*, 35, 1240–1248. http://dx.doi.org/10.1016/j.fsi.2013.07.042.

Liao, W., Bisgrove, B. W., Sawyer, H., Hug, B., Bell, B., Peters, K., et al. (1997). The zebrafish gene cloche acts upstream of a flk-1 homologue to regulate endothelial cell differentiation. *Development*, 124, 381–389.

Liao, H.-J., Kume, T., McKay, C., Xu, M.-J., Ihle, J. N., & Carpenter, G. (2002). Absence of erythrogenesis and vasculogenesis in Plcγ1-deficient mice. *The Journal of Biological Chemistry*, 277, 9335–9341. http://dx.doi.org/10.1074/jbc.M109955200.

Luque, A., Carpizo, D. R., & Iruela-Arispe, M. L. (2003). ADAMTS1/METH1 inhibits endothelial cell proliferation by direct binding and sequestration of VEGF165. *The Journal of Biological Chemistry*, 278, 23656–23665. http://dx.doi.org/10.1074/jbc.M212964200.

Mac Gabhann, F., & Popel, A. S. (2007). Dimerization of VEGF receptors and implications for signal transduction: A computational study. *Biophysical Chemistry*, 128, 125–139. http://dx.doi.org/10.1016/j.bpc.2007.03.010.

Mackenzie, F., & Ruhrberg, C. (2012). Diverse roles for VEGF-A in the nervous system. *Development*, 139, 1371–1380. http://dx.doi.org/10.1242/dev.072348.

Maes, C., Stockmans, I., Moermans, K., Van Looveren, R., Smets, N., Carmeliet, P., et al. (2004). Soluble VEGF isoforms are essential for establishing epiphyseal vascularization and regulating chondrocyte development and survival. *The Journal of Clinical Investigation*, 113, 188–199. http://dx.doi.org/10.1172/JCI19383.

Mäkinen, T., Jussila, L., Veikkola, T., Karpanen, T., Kettunen, M. I., Pulkkanen, K. J., et al. (2001). Inhibition of lymphangiogenesis with resulting lymphedema in transgenic mice expressing soluble VEGF receptor-3. *Nature Medicine*, 7, 199–205. http://dx.doi.org/10.1038/84651.

Mäkinen, T., Veikkola, T., Mustjoki, S., Karpanen, T., Catimel, B., Nice, E. C., et al. (2001). Isolated lymphatic endothelial cells transduce growth, survival and migratory signals via the VEGF-C/D receptor VEGFR-3. *The EMBO Journal*, 20, 4762–4773. http://dx.doi.org/10.1093/emboj/20.17.4762.

Matsumoto, T., Bohman, S., Dixelius, J., Berge, T., Dimberg, A., Magnusson, P., et al. (2005). VEGF receptor-2 Y951 signaling and a role for the adapter molecule TSAd in tumor angiogenesis. *The EMBO Journal*, 24, 2342–2353. http://dx.doi.org/10.1038/sj.emboj.7600709.

Matthews, W., Jordan, C. T., Gavin, M., Jenkins, N. A., Copeland, N. G., & Lemischka, I. R. (1991). A receptor tyrosine kinase cDNA isolated from a population of enriched primitive hematopoietic cells and exhibiting close genetic linkage to c-kit. *Proceedings of the National academy of Sciences of the United States of America*, 88, 9026–9030.

Maynard, S. E., Min, J.-Y., Merchan, J., Lim, K.-H., Li, J., Mondal, S., et al. (2003). Excess placental soluble fms-like tyrosine kinase 1 (sFlt1) may contribute to endothelial dysfunction, hypertension, and proteinuria in preeclampsia. *The Journal of Clinical Investigation*, 111, 649–658. http://dx.doi.org/10.1172/JCI17189.

McCarty, G., Awad, O., & Loeb, D. M. (2011). WT1 protein directly regulates expression of vascular endothelial growth factor and is a mediator of tumor response to hypoxia. *The Journal of Biological Chemistry*, 286, 43634–43643. http://dx.doi.org/10.1074/jbc.M111.310128.

McColl, B. K., Paavonen, K., Karnezis, T., Harris, N. C., Davydova, N., Rothacker, J., et al. (2007). Proprotein convertases promote processing of VEGF-D, a critical step for binding the angiogenic receptor VEGFR-2. *The FASEB Journal, 21*, 1088–1098. http://dx.doi.org/10.1096/fj.06-7060com.

McTigue, M. A., Wickersham, J. A., Pinko, C., Showalter, R. E., Parast, C. V., Tempczyk-Russell, A., et al. (1999). Crystal structure of the kinase domain of human vascular endothelial growth factor receptor 2: A key enzyme in angiogenesis. *Structure, 7*, 319–330.

Mehlem, A., Palombo, I., Wang, X., Hagberg, C. E., Eriksson, U., & Falkevall, A. (2016). PGC-1α coordinates mitochondrial respiratory capacity and muscular fatty acid uptake via regulation of VEGF-B. *Diabetes, 65*, 861–873. http://dx.doi.org/10.2337/db15-1231.

Merdzhanova, G., Gout, S., Keramidas, M., Edmond, V., Coll, J.-L., Brambilla, C., et al. (2010). The transcription factor E2F1 and the SR protein SC35 control the ratio of pro-angiogenic versus antiangiogenic isoforms of vascular endothelial growth factor-A to inhibit neovascularization in vivo. *Oncogene, 29*, 5392–5403. http://dx.doi.org/10.1038/onc.2010.281.

Meyer, R. D., Mohammadi, M., & Rahimi, N. (2006). A single amino acid substitution in the activation loop defines the decoy characteristic of VEGFR-1/FLT-1. *The Journal of Biological Chemistry, 281*, 867–875. http://dx.doi.org/10.1074/jbc.M506454200.

Meyer, R. D., Singh, A., Majnoun, F., Latz, C., Lashkari, K., & Rahimi, N. (2004). Substitution of C-terminus of VEGFR-2 with VEGFR-1 promotes VEGFR-1 activation and endothelial cell proliferation. *Oncogene, 23*, 5523–5531. http://dx.doi.org/10.1038/sj.onc.1207712.

Millauer, B., Shawver, L. K., Plate, K. H., Risau, W., & Ullrich, A. (1994). Glioblastoma growth inhibited in vivo by a dominant-negative Flk-1 mutant. *Nature, 367*, 576–579. http://dx.doi.org/10.1038/367576a0.

Mineur, P., Colige, A. C., Deroanne, C. F., Dubail, J., Kesteloot, F., Habraken, Y., et al. (2007). Newly identified biologically active and proteolysis-resistant VEGF-A isoform VEGF111 is induced by genotoxic agents. *The Journal of Cell Biology, 179*, 1261–1273. http://dx.doi.org/10.1083/jcb.200703052.

Muhl, L., Moessinger, C., Adzemovic, M. Z., Dijkstra, M. H., Nilsson, I., Zeitelhofer, M., et al. (2016). Expression of vascular endothelial growth factor (VEGF)-B and its receptor (VEGFR1) in murine heart, lung and kidney. *Cell and Tissue Research, 365*, 51–63. http://dx.doi.org/10.1007/s00441-016-2377-y.

Muller, Y. A., Christinger, H. W., Keyt, B. A., & de Vos, A. M. (1997). The crystal structure of vascular endothelial growth factor (VEGF) refined to 1.93 A resolution: Multiple copy flexibility and receptor binding. *Structure, 5*, 1325–1338.

Neagoe, P.-E., Lemieux, C., & Sirois, M. G. (2005). Vascular endothelial growth factor (VEGF)-A165-induced prostacyclin synthesis requires the activation of VEGF receptor-1 and -2 heterodimer. *The Journal of Biological Chemistry, 280*, 9904–9912. http://dx.doi.org/10.1074/jbc.M412017200.

Nilsson, I., Bahram, F., Li, X., Gualandi, L., Koch, S., Jarvius, M., et al. (2010). VEGF receptor 2/-3 heterodimers detected in situ by proximity ligation on angiogenic sprouts. *The EMBO Journal, 29*, 1377–1388. http://dx.doi.org/10.1038/emboj.2010.30.

Nowak, D. G., Amin, E. M., Rennel, E. S., Hoareau-Aveilla, C., Gammons, M., Damodoran, G., et al. (2010). Regulation of vascular endothelial growth factor (VEGF) splicing from pro-angiogenic to anti-angiogenic isoforms: A novel therapeutic strategy for angiogenesis. *The Journal of Biological Chemistry, 285*, 5532–5540. http://dx.doi.org/10.1074/jbc.M109.074930.

Nowak, D. G., Woolard, J., Amin, E. M., Konopatskaya, O., Saleem, M. A., Churchill, A. J., et al. (2008). Expression of pro- and anti-angiogenic isoforms of VEGF is differentially regulated by splicing and growth factors. *Journal of Cell Science*, *121*, 3487–3495. http://dx.doi.org/10.1242/jcs.016410.

Nurmi, H., Saharinen, P., Zarkada, G., Zheng, W., Robciuc, M. R., & Alitalo, K. (2015). VEGF-C is required for intestinal lymphatic vessel maintenance and lipid absorption. *EMBO Molecular Medicine*, *7*, 1418–1425. http://dx.doi.org/10.15252/emmm.201505731.

Ogawa, S., Oku, A., Sawano, A., Yamaguchi, S., Yazaki, Y., & Shibuya, M. (1998). A novel type of vascular endothelial growth factor, VEGF-E (NZ-7 VEGF), preferentially utilizes KDR/Flk-1 receptor and carries a potent mitotic activity without heparin-binding domain. *The Journal of Biological Chemistry*, *273*, 31273–31282.

Olofsson, B., Korpelainen, E., Pepper, M. S., Mandriota, S. J., Aase, K., Kumar, V., et al. (1998). Vascular endothelial growth factor B (VEGF-B) binds to VEGF receptor-1 and regulates plasminogen activator activity in endothelial cells. *Proceedings of the National Academy of Sciences of the United States of America*, *95*, 11709–11714.

Olofsson, B., & Page, D. T. (2005). Condensation of the central nervous system in embryonic Drosophila is inhibited by blocking hemocyte migration or neural activity. *Developmental Biology*, *279*, 233–243. http://dx.doi.org/10.1016/j.ydbio.2004.12.020.

Olsson, A.-K., Dimberg, A., Kreuger, J., & Claesson-Welsh, L. (2006). VEGF receptor signalling—In control of vascular function. *Nature Reviews. Molecular Cell Biology*, *7*, 359–371. http://dx.doi.org/10.1038/nrm1911.

Oltean, S., Gammons, M., Hulse, R., Hamdollah-Zadeh, M., Mavrou, A., Donaldson, L., et al. (2012). SRPK1 inhibition in vivo: Modulation of VEGF splicing and potential treatment for multiple diseases. *Biochemical Society Transactions*, *40*, 831–835. http://dx.doi.org/10.1042/BST20120051.

Oltean, S., Neal, C. R., Mavrou, A., Patel, P., Ahad, T., Alsop, C., et al. (2012). VEGF165b overexpression restores normal glomerular water permeability in VEGF164-overexpressing adult mice. *American Journal of Physiology. Renal Physiology*, *303*, F1026–F1036. http://dx.doi.org/10.1152/ajprenal.00410.2011.

Orlandini, M., Spreafico, A., Bardelli, M., Rocchigiani, M., Salameh, A., Nucciotti, S., et al. (2006). Vascular endothelial growth factor-D activates VEGFR-3 expressed in osteoblasts inducing their differentiation. *The Journal of Biological Chemistry*, *281*, 17961–17967. http://dx.doi.org/10.1074/jbc.M600413200.

Pajusola, K., Aprelikova, O., Korhonen, J., Kaipainen, A., Pertovaara, L., Alitalo, R., et al. (1992). FLT4 receptor tyrosine kinase contains seven immunoglobulin-like loops and is expressed in multiple human tissues and cell lines. *Cancer Research*, *52*, 5738–5743.

Pajusola, K., Aprelikova, O., Pelicci, G., Weich, H., Claesson-Welsh, L., & Alitalo, K. (1994). Signalling properties of FLT4, a proteolytically processed receptor tyrosine kinase related to two VEGF receptors. *Oncogene*, *9*, 3545–3555.

Parast, C. V., Mroczkowski, B., Pinko, C., Misialek, S., Khambatta, G., & Appelt, K. (1998). Characterization and kinetic mechanism of catalytic domain of human vascular endothelial growth factor receptor-2 tyrosine kinase (VEGFR2 TK), a key enzyme in angiogenesis. *Biochemistry*, *37*, 16788–16801. http://dx.doi.org/10.1021/bi981291f.

Park, J. E., Chen, H. H., Winer, J., Houck, K. A., & Ferrara, N. (1994). Placenta growth factor. Potentiation of vascular endothelial growth factor bioactivity, in vitro and in vivo, and high affinity binding to Flt-1 but not to Flk-1/KDR. *The Journal of Biological Chemistry*, *269*, 25646–25654.

Parker, M. W., Xu, P., Li, X., & Vander Kooi, C. W. (2012). Structural basis for selective vascular endothelial growth factor-A (VEGF-A) binding to neuropilin-1. *The Journal of Biological Chemistry*, *287*, 11082–11089. http://dx.doi.org/10.1074/jbc.M111.331140.

Partanen, T. A., Arola, J., Saaristo, A., Jussila, L., Ora, A., Miettinen, M., et al. (2000). VEGF-C and VEGF-D expression in neuroendocrine cells and their receptor,

VEGFR-3, in fenestrated blood vessels in human tissues. *The FASEB Journal, 14*, 2087–2096. http://dx.doi.org/10.1096/fj.99-1049com.

Pavlakovic, H., Becker, J., Albuquerque, R., Wilting, J., & Ambati, J. (2010). Soluble VEGFR-2: An anti-lymphangiogenic variant of VEGF receptors. *The Annals of the New York Academy of Sciences, 1207*, E7–E15. http://dx.doi.org/10.1111/j.1749-6632.2010.05714.x.

Peiris-Pagès, M. (2012). The role of VEGF 165b in pathophysiology. *Cell Adhesion & Migration, 6*, 561–568. http://dx.doi.org/10.4161/cam.22439.

Pellet-Many, C., Frankel, P., Jia, H., & Zachary, I. (2008). Neuropilins: Structure, function and role in disease. *The Biochemical Journal, 411*, 211–226. http://dx.doi.org/10.1042/BJ20071639.

Perrin, R. M., Konopatskaya, O., Qiu, Y., Harper, S., Bates, D. O., & Churchill, A. J. (2005). Diabetic retinopathy is associated with a switch in splicing from anti- to pro-angiogenic isoforms of vascular endothelial growth factor. *Diabetologia, 48*, 2422–2427. http://dx.doi.org/10.1007/s00125-005-1951-8.

Plate, K. H., Breier, G., Millauer, B., Ullrich, A., & Risau, W. (1993). Up-regulation of vascular endothelial growth factor and its cognate receptors in a rat glioma model of tumor angiogenesis. *Cancer Research, 53*, 5822–5827.

Plouët, J., Moro, F., Bertagnolli, S., Coldeboeuf, N., Mazarguil, H., Clamens, S., et al. (1997). Extracellular cleavage of the vascular endothelial growth factor 189-amino acid form by urokinase is required for its mitogenic effect. *The Journal of Biological Chemistry, 272*, 13390–13396.

Poesen, K., Lambrechts, D., Van Damme, P., Dhondt, J., Bender, F., Frank, N., et al. (2008). Novel role for vascular endothelial growth factor (VEGF) receptor-1 and its ligand VEGF-B in motor neuron degeneration. *The Journal of Neuroscience, 28*, 10451–10459. http://dx.doi.org/10.1523/JNEUROSCI.1092-08.2008.

Popovici, C., Isnardon, D., Birnbaum, D., & Roubin, R. (2002). Caenorhabditis elegans receptors related to mammalian vascular endothelial growth factor receptors are expressed in neural cells. *Neuroscience Letters, 329*, 116–120.

Prahst, C., Héroult, M., Lanahan, A. A., Uziel, N., Kessler, O., Shraga-Heled, N., et al. (2008). Neuropilin-1-VEGFR-2 complexing requires the PDZ-binding domain of neuropilin-1. *The Journal of Biological Chemistry, 283*, 25110–25114. http://dx.doi.org/10.1074/jbc.C800137200.

Pugh, C. W., & Ratcliffe, P. J. (2003). Regulation of angiogenesis by hypoxia: Role of the HIF system. *Nature Medicine, 9*, 677–684. http://dx.doi.org/10.1038/nm0603-677.

Purushothaman, A., Uyama, T., Kobayashi, F., Yamada, S., Sugahara, K., Raprager, A. C., et al. (2010). Heparanase-enhanced shedding of syndecan-1 by myeloma cells promotes endothelial invasion and angiogenesis. *Blood, 115*, 2449–2457. http://dx.doi.org/10.1182/blood-2009-07-234757.

Qiu, Y., Bevan, H., Weeraperuma, S., Wratting, D., Murphy, D., Neal, C. R., et al. (2008). Mammary alveolar development during lactation is inhibited by the endogenous anti-angiogenic growth factor isoform, VEGF165b. *The FASEB Journal, 22*, 1104–1112. http://dx.doi.org/10.1096/fj.07-9718com.

Qiu, Y., Hoareau-Aveilla, C., Oltean, S., Harper, S. J., & Bates, D. O. (2009). The anti-angiogenic isoforms of VEGF in health and disease. *Biochemical Society Transactions, 37*, 1207–1213. http://dx.doi.org/10.1042/BST0371207.

Quinn, T. P., Peters, K. G., de Vries, C., Ferrara, N., & Williams, L. T. (1993). Fetal liver kinase 1 is a receptor for vascular endothelial growth factor and is selectively expressed in vascular endothelium. *Proceedings of the National academy of Sciences of the United States of America, 90*, 7533–7537.

Rahimi, N. (2006). VEGFR-1 and VEGFR-2: Two non-identical twins with a unique physiognomy. *Frontiers in Bioscience, 11*, 818–829.

Randi, A. M., Sperone, A., Dryden, N. H., & Birdsey, G. M. (2009). Regulation of angiogenesis by ETS transcription factors. *Biochemical Society Transactions*, *37*, 1248–1253. http://dx.doi.org/10.1042/BST0371248.

Rodriguez-Manzaneque, J. C., Lane, T. F., Ortega, M. A., Hynes, R. O., Lawler, J., & Iruela-Arispe, M. L. (2001). Thrombospondin-1 suppresses spontaneous tumor growth and inhibits activation of matrix metalloproteinase-9 and mobilization of vascular endothelial growth factor. *Proceedings of the National Academy of Sciences of the United States of America*, *98*, 12485–12490. http://dx.doi.org/10.1073/pnas.171460498.

Roskoski, R. (2008). VEGF receptor protein-tyrosine kinases: Structure and regulation. *Biochemical and Biophysical Research Communications*, *375*, 287–291. http://dx.doi.org/10.1016/j.bbrc.2008.07.121.

Rottbauer, W., Just, S., Wessels, G., Trano, N., Most, P., Katus, H. A., et al. (2005). VEGF-PLCgamma1 pathway controls cardiac contractility in the embryonic heart. *Genes & Development*, *19*, 1624–1634. http://dx.doi.org/10.1101/gad.1319405.

Roy, R., Zhang, B., & Moses, M. A. (2006). Making the cut: Protease-mediated regulation of angiogenesis. *Experimental Cell Research*, *312*, 608–622. http://dx.doi.org/10.1016/j.yexcr.2005.11.022.

Ruch, C., Skiniotis, G., Steinmetz, M. O., Walz, T., & Ballmer-Hofer, K. (2007). Structure of a VEGF–VEGF receptor complex determined by electron microscopy. *Nature Structural & Molecular Biology*, *14*, 249–250. http://dx.doi.org/10.1038/nsmb1202.

Ruhrberg, C., Gerhardt, H., Golding, M., Watson, R., Ioannidou, S., Fujisawa, H., et al. (2002). Spatially restricted patterning cues provided by heparin-binding VEGF-A control blood vessel branching morphogenesis. *Genes & Development*, *16*, 2684–2698. http://dx.doi.org/10.1101/gad.242002.

Ruiz de Almodovar, C., Coulon, C., Salin, P. A., Knevels, E., Chounlamountri, N., Poesen, K., et al. (2010). Matrix-binding vascular endothelial growth factor (VEGF) isoforms guide granule cell migration in the cerebellum via VEGF receptor Flk1. *The Journal of Neuroscience*, *30*, 15052–15066. http://dx.doi.org/10.1523/JNEUROSCI.0477-10.2010.

Sakai, R., Henderson, J. T., O'Bryan, J. P., Elia, A. J., Saxton, T. M., & Pawson, T. (2000). The mammalian ShcB and ShcC phosphotyrosine docking proteins function in the maturation of sensory and sympathetic neurons. *Neuron*, *28*, 819–833. http://dx.doi.org/10.1016/S0896-6273(00)00156-2.

Sakurai, Y., Ohgimoto, K., Kataoka, Y., Yoshida, N., & Shibuya, M. (2005). Essential role of Flk-1 (VEGF receptor 2) tyrosine residue 1173 in vasculogenesis in mice. *Proceedings of the National Academy of Sciences of the United States of America*, *102*, 1076–1081. http://dx.doi.org/10.1073/pnas.0404984102.

Salameh, A., Galvagni, F., Bardelli, M., Bussolino, F., & Oliviero, S. (2005). Direct recruitment of CRK and GRB2 to VEGFR-3 induces proliferation, migration, and survival of endothelial cells through the activation of ERK, AKT, and JNK pathways. *Blood*, *106*, 3423–3431. http://dx.doi.org/10.1182/blood-2005-04-1388.

Salton, M., Voss, T. C., & Misteli, T. (2014). Identification by high-throughput imaging of the histone methyltransferase EHMT2 as an epigenetic regulator of VEGFA alternative splicing. *Nucleic Acids Research*, *42*, 13662–13673. http://dx.doi.org/10.1093/nar/gku1226.

Samarghandian, S., & Shibuya, M. (2013). Vascular endothelial growth factor receptor family in ascidians, Halocynthia roretzi (Sea Squirt). Its high expression in circulatory system-containing tissues. *International Journal of Molecular Sciences*, *14*, 4841–4853. http://dx.doi.org/10.3390/ijms14034841.

Sarabipour, S., Ballmer-Hofer, K., & Hristova, K. (2016). VEGFR-2 conformational switch in response to ligand binding. *eLife*, *5*, e13876. http://dx.doi.org/10.7554/eLife.13876.

Sawano, A., Iwai, S., Sakurai, Y., Ito, M., Shitara, K., Nakahata, T., et al. (2001). Flt-1, vascular endothelial growth factor receptor 1, is a novel cell surface marker for the lineage of monocyte-macrophages in humans. *Blood, 97*, 785–791.
Sawano, A., Takahashi, T., Yamaguchi, S., Aonuma, M., & Shibuya, M. (1996). Flt-1 but not KDR/Flk-1 tyrosine kinase is a receptor for placenta growth factor, which is related to vascular endothelial growth factor. *Cell Growth & Differentiation, 7*, 213–221.
Scheufler, K.-M., Drevs, J., van Velthoven, V., Reusch, P., Klisch, J., Augustin, H. G., et al. (2003). Implications of vascular endothelial growth factor, sFlt-1, and sTie-2 in plasma, serum and cerebrospinal fluid during cerebral ischemia in man. *Journal of Cerebral Blood Flow and Metabolism, 23*, 99–110.
Schmeisser, A., Christoph, M., Augstein, A., Marquetant, R., Kasper, M., Braun-Dullaeus,-R. C., et al. (2006). Apoptosis of human macrophages by Flt-4 signaling: Implications for atherosclerotic plaque pathology. *Cardiovascular Research, 71*, 774–784. http://dx.doi.org/10.1016/j.cardiores.2006.06.012.
Sears, H. C., Kennedy, C. J., & Garrity, P. A. (2003). Macrophage-mediated corpse engulfment is required for normal Drosophila CNS morphogenesis. *Development, 130*, 3557–3565.
Seetharam, L., Gotoh, N., Maru, Y., Neufeld, G., Yamaguchi, S., & Shibuya, M. (1995). A unique signal transduction from FLT tyrosine kinase, a receptor for vascular endothelial growth factor VEGF. *Oncogene, 10*, 135–147.
Seipel, K., Eberhardt, M., Müller, P., Pescia, E., Yanze, N., & Schmid, V. (2004). Homologs of vascular endothelial growth factor and receptor, VEGF and VEGFR, in the jellyfish Podocoryne carnea. *Developmental Dynamics, 231*, 303–312. http://dx.doi.org/10.1002/dvdy.20139.
Setiamarga, D. H. E., Shimizu, K., Kuroda, J., Inamura, K., Sato, K., Isowa, Y., et al. (2013). An in-silico genomic survey to annotate genes coding for early development-relevant signaling molecules in the pearl oyster, Pinctada fucata. *Zoological Science, 30*, 877–888. http://dx.doi.org/10.2108/zsj.30.877.
Shalaby, F., Ho, J., Stanford, W. L., Fischer, K. D., Schuh, A. C., Schwartz, L., et al. (1997). A requirement for Flk1 in primitive and definitive hematopoiesis and vasculogenesis. *Cell, 89*, 981–990.
Shalaby, F., Rossant, J., Yamaguchi, T. P., Gertsenstein, M., Wu, X. F., Breitman, M. L., et al. (1995). Failure of blood-island formation and vasculogenesis in Flk-1-deficient mice. *Nature, 376*, 62–66. http://dx.doi.org/10.1038/376062a0.
Shibuya, M. (2006). Vascular endothelial growth factor receptor-1 (VEGFR-1/Flt-1): A dual regulator for angiogenesis. *Angiogenesis, 9*, 225–230. http://dx.doi.org/10.1007/s10456-006-9055-8. discussion 231.
Shibuya, M. (2013a). VEGFR and type-V RTK activation and signaling. *Cold Spring Harbor Perspectives in Biology, 5*, a009092. http://dx.doi.org/10.1101/cshperspect.a009092.
Shibuya, M. (2013b). Vascular endothelial growth factor and its receptor system: Physiological functions in angiogenesis and pathological roles in various diseases. *Journal of Biochemistry, 153*, 13–19. http://dx.doi.org/10.1093/jb/mvs136.
Shibuya, M. (2014). VEGF-VEGFR signals in health and disease. *Biomolecules & Therapeutics, 22*, 1–9. http://dx.doi.org/10.4062/biomolther.2013.113.
Shibuya, M., Yamaguchi, S., Yamane, A., Ikeda, T., Tojo, A., Matsushime, H., et al. (1990). Nucleotide sequence and expression of a novel human receptor-type tyrosine kinase gene (flt) closely related to the fms family. *Oncogene, 5*, 519–524.
Silha, J. V., Krsek, M., Sucharda, P., & Murphy, L. J. (2005). Angiogenic factors are elevated in overweight and obese individuals. *International Journal of Obesity, 29*, 1308–1314. http://dx.doi.org/10.1038/sj.ijo.0802987.
Soler, A., Angulo-Urarte, A., & Graupera, M. (2015). PI3K at the crossroads of tumor angiogenesis signaling pathways. *Molecular & Cellular Oncology, 2*, e975624. http://dx.doi.org/10.4161/23723556.2014.975624.

Stacker, S. A., Stenvers, K., Caesar, C., Vitali, A., Domagala, T., Nice, E., et al. (1999). Biosynthesis of vascular endothelial growth factor-D involves proteolytic processing which generates non-covalent homodimers. *The Journal of Biological Chemistry, 274*, 32127–32136.

Stalmans, I., Ng, Y.-S., Rohan, R., Fruttiger, M., Bouché, A., Yuce, A., et al. (2002). Arteriolar and venular patterning in retinas of mice selectively expressing VEGF isoforms. *The Journal of Clinical Investigation, 109*, 327–336. http://dx.doi.org/10.1172/JCI14362.

Stuttfeld, E., & Ballmer-Hofer, K. (2009). Structure and function of VEGF receptors. *IUBMB Life, 61*, 915–922. http://dx.doi.org/10.1002/iub.234.

Sun, Z., Li, X., Massena, S., Kutschera, S., Padhan, N., Gualandi, L., et al. (2012). VEGFR2 induces c-Src signaling and vascular permeability in vivo via the adaptor protein TSAd. *The Journal of Experimental Medicine, 209*, 1363–1377. http://dx.doi.org/10.1084/jem.20111343.

Suto, K., Yamazaki, Y., Morita, T., & Mizuno, H. (2005). Crystal structures of novel vascular endothelial growth factors (VEGF) from snake venoms: Insight into selective VEGF binding to kinase insert domain-containing receptor but not to fms-like tyrosine kinase-1. *The Journal of Biological Chemistry, 280*, 2126–2131. http://dx.doi.org/10.1074/jbc.M411395200.

Takahashi, T., & Shibuya, M. (1997). The 230 kDa mature form of KDR/Flk-1 (VEGF receptor-2) activates the PLC-gamma pathway and partially induces mitotic signals in NIH3T3 fibroblasts. *Oncogene, 14*, 2079–2089. http://dx.doi.org/10.1038/sj.onc.1201047.

Takahashi, H., & Shibuya, M. (2005). The vascular endothelial growth factor (VEGF)/VEGF receptor system and its role under physiological and pathological conditions. *Clinical Science, 109*, 227–241. http://dx.doi.org/10.1042/CS20040370.

Takahashi, T., Yamaguchi, S., Chida, K., & Shibuya, M. (2001). A single autophosphorylation site on KDR/Flk-1 is essential for VEGF-A-dependent activation of PLC-gamma and DNA synthesis in vascular endothelial cells. *The EMBO Journal, 20*, 2768–2778. http://dx.doi.org/10.1093/emboj/20.11.2768.

Tammela, T., Zarkada, G., Wallgard, E., Murtomäki, A., Suchting, S., Wirzenius, M., et al. (2008). Blocking VEGFR-3 suppresses angiogenic sprouting and vascular network formation. *Nature, 454*, 656–660. http://dx.doi.org/10.1038/nature07083.

Tarsitano, M., De Falco, S., Colonna, V., McGhee, J. D., & Persico, M. G. (2006). The C. elegans pvf-1 gene encodes a PDGF/VEGF-like factor able to bind mammalian VEGF receptors and to induce angiogenesis. *The FASEB Journal, 20*, 227–233. http://dx.doi.org/10.1096/fj.05-4147com.

Terman, B. I., Carrion, M. E., Kovacs, E., Rasmussen, B. A., Eddy, R. L., & Shows, T. B. (1991). Identification of a new endothelial cell growth factor receptor tyrosine kinase. *Oncogene, 6*, 1677–1683.

Terman, B. I., Dougher-Vermazen, M., Carrion, M. E., Dimitrov, D., Armellino, D. C., Gospodarowicz, D., et al. (1992). Identification of the KDR tyrosine kinase as a receptor for vascular endothelial cell growth factor. *Biochemical and Biophysical Research Communications, 187*, 1579–1586.

Tettamanti, G., Grimaldi, A., Valvassori, R., Rinaldi, L., & de Eguileor, M. (2003). Vascular endothelial growth factor is involved in neoangiogenesis in Hirudo medicinalis (Annelida, Hirudinea). *Cytokine, 22*, 168–179.

Thompson, M. A., Ransom, D. G., Pratt, S. J., MacLennan, H., Kieran, M. W., Detrich, H. W., et al. (1998). The cloche and spadetail genes differentially affect hematopoiesis and vasculogenesis. *Developmental Biology, 197*, 248–269. http://dx.doi.org/10.1006/dbio.1998.8887.

Tiozzo, S., Voskoboynik, A., Brown, F. D., & De Tomaso, A. W. (2008). A conserved role of the VEGF pathway in angiogenesis of an ectodermally-derived vasculature. *Developmental Biology, 315*, 243–255. http://dx.doi.org/10.1016/j.ydbio.2007.12.035.

Toi, M., Bando, H., Ogawa, T., Muta, M., Hornig, C., & Weich, H. A. (2002). Significance of vascular endothelial growth factor (VEGF)/soluble VEGF receptor-1 relationship in breast cancer. *International Journal of Cancer, 98*, 14–18.

Tozer, G. M., Akerman, S., Cross, N. A., Barber, P. R., Björndahl, M. A., Greco, O., et al. (2008). Blood vessel maturation and response to vascular-disrupting therapy in single vascular endothelial growth factor-A isoform-producing tumors. *Cancer Research, 68*, 2301–2311. http://dx.doi.org/10.1158/0008-5472.CAN-07-2011.

Ushio-Fukai, M., & Nakamura, Y. (2008). Reactive oxygen species and angiogenesis: NADPH oxidase as target for cancer therapy. *Cancer Letters, 266*, 37–52.

Valtola, R., Salven, P., Heikkilä, P., Taipale, J., Joensuu, H., Rehn, M., et al. (1999). VEGFR-3 and its ligand VEGF-C are associated with angiogenesis in breast cancer. *The American Journal of Pathology, 154*, 1381–1390. http://dx.doi.org/10.1016/S0002-9440(10)65392-8.

Veikkola, T., Jussila, L., Mäkinen, T., Karpanen, T., Jeltsch, M., Petrova, T. V., et al. (2001). Signalling via vascular endothelial growth factor receptor-3 is sufficient for lymphangiogenesis in transgenic mice. *The EMBO Journal, 20*, 1223–1231. http://dx.doi.org/10.1093/emboj/20.6.1223.

Vempati, P., Popel, A. S., & Mac Gabhann, F. (2011). Formation of VEGF isoform-specific spatial distributions governing angiogenesis: Computational analysis. *BMC Systems Biology, 5*, 59. http://dx.doi.org/10.1186/1752-0509-5-59.

Vempati, P., Popel, A. S., & Mac Gabhann, F. (2014). Extracellular regulation of VEGF: Isoforms, proteolysis, and vascular patterning. *Cytokine & Growth Factor Reviews, 25*, 1–19. http://dx.doi.org/10.1016/j.cytogfr.2013.11.002.

Villefranc, J. A., Nicoli, S., Bentley, K., Jeltsch, M., Zarkada, G., Moore, J. C., et al. (2013). A truncation allele in vascular endothelial growth factor c reveals distinct modes of signaling during lymphatic and vascular development. *Development, 140*, 1497–1506. http://dx.doi.org/10.1242/dev.084152.

Vintonenko, N., Pelaez-Garavito, I., Buteau-Lozano, H., Toullec, A., Lidereau, R., Perret, G. Y., et al. (2011). Overexpression of VEGF189 in breast cancer cells induces apoptosis via NRP1 under stress conditions. *Cell Adhesion & Migration, 5*, 332–343. http://dx.doi.org/10.4161/cam.5.4.17287.

Waltenberger, J., Claesson-Welsh, L., Siegbahn, A., Shibuya, M., & Heldin, C. H. (1994). Different signal transduction properties of KDR and Flt1, two receptors for vascular endothelial growth factor. *The Journal of Biological Chemistry, 269*, 26988–26995.

Warner, A. J., Lopez-Dee, J., Knight, E. L., Feramisco, J. R., & Prigent, S. A. (2000). The Shc-related adaptor protein, Sck, forms a complex with the vascular-endothelial-growth-factor receptor KDR in transfected cells. *The Biochemical Journal, 347*, 501–509.

Wellner, M., Maasch, C., Kupprion, C., Lindschau, C., Luft, F. C., & Haller, H. (1999). The proliferative effect of vascular endothelial growth factor requires protein kinase C-alpha and protein kinase C-zeta. *Arteriosclerosis, Thrombosis, and Vascular Biology, 19*, 178–185.

Wiesmann, C., Fuh, G., Christinger, H. W., Eigenbrot, C., Wells, J. A., & de Vos, A. M. (1997). Crystal structure at 1.7 A resolution of VEGF in complex with domain 2 of the Flt-1 receptor. *Cell, 91*, 695–704.

Witte, M. H., Erickson, R., Bernas, M., Andrade, M., Reiser, F., Conlon, W., et al. (1998). Phenotypic and genotypic heterogeneity in familial Milroy lymphedema. *Lymphology, 31*, 145–155.

Wittko-Schneider, I. M., Schneider, F. T., & Plate, K. H. (2013). Brain homeostasis: VEGF receptor 1 and 2-two unequal brothers in mind. *Cellular and Molecular Life Sciences, 70*, 1705–1725. http://dx.doi.org/10.1007/s00018-013-1279-3.

Wu, L. W., Mayo, L. D., Dunbar, J. D., Kessler, K. M., Ozes, O. N., Warren, R. S., et al. (2000). VRAP is an adaptor protein that binds KDR, a receptor for vascular endothelial cell growth factor. *The Journal of Biological Chemistry, 275*, 6059–6062.

Yamazaki, Y., Matsunaga, Y., Tokunaga, Y., Obayashi, S., Saito, M., & Morita, T. (2009). Snake venom Vascular Endothelial Growth Factors (VEGF-Fs) exclusively vary their structures and functions among species. *The Journal of Biological Chemistry*, *284*, 9885–9891. http://dx.doi.org/10.1074/jbc.M809071200.

Yang, Z.-Z., Tschopp, O., Di-Poï, N., Bruder, E., Baudry, A., Dümmler, B., et al. (2005). Dosage-dependent effects of Akt1/protein kinase Balpha (PKBalpha) and Akt3/PKBgamma on thymus, skin, and cardiovascular and nervous system development in mice. *Molecular and Cellular Biology*, *25*, 10407–10418. http://dx.doi.org/10.1128/MCB.25.23.10407-10418.2005.

Yang, S.-Z., Zhang, L.-M., Huang, Y.-L., & Sun, F.-Y. (2003). Distribution of Flk-1 and Flt-1 receptors in neonatal and adult rat brains. *The Anatomical Record Part A, Discoveries in Molecular, Cellular, and Evolutionary Biology*, *274*, 851–856. http://dx.doi.org/10.1002/ar.a.10103.

Yoshida, M.-A., Shigeno, S., Tsuneki, K., & Furuya, H. (2010). Squid vascular endothelial growth factor receptor: A shared molecular signature in the convergent evolution of closed circulatory systems. *Evolution & Development*, *12*, 25–33. http://dx.doi.org/10.1111/j.1525-142X.2009.00388.x.

Yuan, L., Moyon, D., Pardanaud, L., Bréant, C., Karkkainen, M. J., Alitalo, K., et al. (2002). Abnormal lymphatic vessel development in neuropilin 2 mutant mice. *Development*, *129*, 4797–4806.

Zachary, I. (2014). Neuropilins: Role in signalling, angiogenesis and disease. *Chemical Immunology and Allergy*, *99*, 37–70. http://dx.doi.org/10.1159/000354169.

Zarkada, G., Heinolainen, K., Makinen, T., Kubota, Y., & Alitalo, K. (2015). VEGFR3 does not sustain retinal angiogenesis without VEGFR2. *Proceedings of the National academy of Sciences of the United States of America*, *112*, 761–766. http://dx.doi.org/10.1073/pnas.1423278112.

Zeitouni, B., Sénatore, S., Séverac, D., Aknin, C., Sémériva, M., & Perrin, L. (2007). Signalling pathways involved in adult heart formation revealed by gene expression profiling in Drosophila. *PLoS Genetics*, *3*, e174. http://dx.doi.org/10.1371/journal.pgen.0030174.

Zhou, F., Chang, Z., Zhang, L., Hong, Y.-K., Shen, B., Wang, B., et al. (2010). Akt/Protein kinase B is required for lymphatic network formation, remodeling, and valve development. *The American Journal of Pathology*, *177*, 2124–2133. http://dx.doi.org/10.2353/ajpath.2010.091301.

CPSIA information can be obtained
at www.ICGtesting.com
Printed in the USA
BVHW01*1232100718
520466BV00013B/11/P